THE
INTERNATIONAL SERIES
OF
MONOGRAPHS ON PHYSICS

GENERAL EDITORS
W. MARSHALL   D. H. WILKINSON

H. S. W. MASSEY, E. H. S. BURHOP
AND H. B. GILBODY

ELECTRONIC
AND IONIC IMPACT
PHENOMENA

SECOND EDITION
IN FOUR VOLUMES

VOLUME II

Electron Collisions with Molecules
and Photo-ionization

BY H. S. W. MASSEY

OXFORD
AT THE CLARENDON PRESS
1969

Oxford University Press, Ely House, London W. 1

GLASGOW NEW YORK TORONTO MELBOURNE WELLINGTON
CAPE TOWN SALISBURY IBADAN NAIROBI LUSAKA ADDIS ABABA
BOMBAY CALCUTTA MADRAS KARACHI LAHORE DACCA
KUALA LUMPUR SINGAPORE HONG KONG TOKYO

PRINTED IN GREAT BRITAIN

PHYSICS

PREFACE TO THE SECOND EDITION

THE immense growth of the subject since the first edition was produced has raised many problems in connection with a new edition. Apart from the sheer bulk of new material, the interconnections between different parts of the subject have become very complex, while the sophistication of both theoretical and experimental techniques has greatly increased.

It became clear at the outset that it was no longer possible to attempt a nearly comprehensive treatment. As sources of data on cross-sections, reaction rates, etc., for use in various applications are now available and are becoming more comprehensive, it seemed that in the new edition the emphasis should be on describing and discussing experimental and theoretical techniques, and the interpretation of the results obtained by their use, rather than on compilation of data. Even so, a greater selectivity among the wide range of available material has been essential. Within these limitations the level of the treatment has been maintained roughly as in the first edition, although some allowance has been made for the general increase in the level of sophistication.

When all these considerations were taken into account it became clear that the new edition would be between four and five times larger than the first. To make practicable the completion of the task of writing so much against the rate of production of new results, it was decided to omit any discussion of phenomena occurring at surfaces (Chapters V and IX of the first edition) and to present the new edition in four volumes, the correspondence with the first edition being as follows:

Second edition	First edition
Volume I	Chapters I, II, III
Volume II	Chapters IV, VI
Volume III	Chapter VII
Volume IV	Chapters VIII, X

In this way Volumes I and II deal with electron impact phenomena: Volume I with electron–atom collisions and Volume II with electron–molecule collisions. Volume II also includes a detailed discussion of photo-ionization and photodetachment which did not appear in the first edition.

Volumes III and IV deal, in general, with collisions involving heavy particles. Thus, Volume III is concerned with thermal collisions involving neutral and ionized atoms and molecules, Volume IV with higher energy collisions of this kind. In addition, recombination is included in this volume as well as a description of collision processes involving slow positrons and muons, which did not appear in the first edition.

Because of the complicated mesh of cross-connections many difficult decisions had to be made as to the place at which a new technique should be described in detail. Usually it was decided to do this in relation to one of the major applications of the technique rather than attempting, in a wholly artificial way, to avoid forward references at all stages.

In covering such a wide field in physics, and indeed also in chemistry, acute difficulties of notation are bound to arise. The symbol k for wave number is now so universally used in collision theory that we have been so impious as to use κ instead of k for Boltzmann's constant. Unfortunately, k is also used very widely by physical chemists to denote a rate constant. In some places we have adhered to this but elsewhere, to avoid confusion with wave numbers, we have substituted a less familiar symbol. Another unfamiliar usage we have employed is that of f for oscillator strength to distinguish it from f for scattered amplitude. Again, we have been unfashionable in using F instead of E for electric field strength because of risk of confusion with E for energy.

No attempt has been made to adopt a set of symbols of universal application throughout the book, although we have stuck grimly to k for wave number and Q for cross-section as well as to e, h, κ and c.

We have tried not to be too pedantic in choice of units though admitting to a predilection for eV as against kcal/mole. When dealing with phenomena of strongly chemical interest we have at times used kcal/mole, but always with the value in eV in brackets.

The penetration of the work into chemistry, though perhaps occurring on a wider front, is no deeper than before. The deciding factor has always been the complexity of the molecules involved in the reactions under consideration.

In order to complete a volume it was necessary, at a certain stage, to close the books, as it were, and turn a blind eye to new results coming in after a certain date—unless, of course, they rendered incorrect anything already written. The closing date for Volumes I and II was roughly early 1967, for Volume III mid-1968, and for Volume IV about six months later. Notes on later advances over the whole field will be included at the end of Volume IV.

H. S. W. M.

London   E. H. S. B.
February 1969   H. B. G.

PREFACE TO THE FIRST EDITION

THERE are very many directions in which research in physics and related subjects depends on a knowledge of the rates of collision processes which occur between electrons, ions, and neutral atoms and molecules. This has become increasingly apparent in recent times in connection with developments involving electric discharges in gases, atmospheric physics, and astrophysics. Apart from this the subject is of great intrinsic interest, playing a leading part in the establishment of quantum theory and including many aspects of fundamental importance in the theory of atomic structure. It therefore seems appropriate to describe the present state of knowledge of the subject and this we have attempted to do in the present work.

We have set ourselves the task of describing the experimental techniques employed and the results obtained for the different kinds of collision phenomena which we have considered within the scope of the book. While no attempt has been made to provide at all times the detailed mathematical theory which may be appropriate for the interpretation of the phenomena, wherever possible the observations have been considered against the available theoretical background, results obtained by theory have been included, and a physical account of the different theories has been given. In some cases, not covered in *The theory of atomic collisions*, a more detailed description for a particular theory has been provided. At all times the aim has been to give a balanced view of the subject, from both the theoretical and experimental standpoints, bringing out as clearly as possible the well-established principles which emerge and the obscurities and uncertainties, many as they are, which still remain.

It was inevitable that some rigid principles of exclusion had to be practised in selecting from the great wealth of available material. It was first decided that phenomena involving the collisions of particles with high energies would not be considered, and that other phenomena associated with the properties of atomic nuclei such as the behaviour of slow neutrons would also be excluded. It was also natural to regard work on chemical kinetics as such, although clearly involving atomic collision phenomena, as outside the scope of the book, but certain of the more fundamental aspects are included. Phenomena involving neutral atoms or molecules only have otherwise been included on an equal footing with those involving ions or electrons. A further extensive class of phenomena have been excluded by avoiding any discussion of collision processes occurring within solids or liquids, confining the work to processes occurring in the gas phase or at a gas–solid interface. Among the latter phenomena electron diffraction at a solid surface has been rather arbitrarily excluded as it is a subject already adequately dealt with in other texts. Secondary electron emission and related effects are, however, included.

By limiting the scope of the book in this way it has just been possible to provide a fairly comprehensive account of the subjects involved. It is perhaps too much to hope that even within these limitations nothing of importance has been missed, but it is believed that the account given is fairly complete. Extensive tables of observed and theoretical data have been given throughout for reference purposes and the extent to which the data given are likely to be reliable has been indicated. Every effort has been made to provide a connected and systematic account but it is inevitable that there will be differences of opinion as to the relative weight given to the various parts of the subject and to the different contributions which have been made to it.

We are particularly indebted to Professor D. R. Bates for reading and criticizing much of the manuscript and for many valuable suggestions. Dr. R. A. Buckingham has also assisted us very much in this direction while Dr. Abdelnabi has checked some of the proofs. We also wish to express our appreciation of the remarkable way in which the Oxford University Press maintained the high standard of their work under the present difficult circumstances.

H. S. W. M.

E. H. S. B.

London
August 1951

ACKNOWLEDGEMENTS

VOLUMES I AND II

WE must express our indebtedness to a number of people whose help has been invaluable. First among these we would like to thank Mrs. J. Lawson for her assistance, so cheerfully and ably afforded, in preparing the diagrams and tables, obtaining and checking references, and in many other ways. In these tasks her husband, through the use of the computer which he so skilfully constructed, and otherwise, has provided valuable help. The task of transferring indecipherable manuscript to typescript has fallen to Mrs. M. Harding, who has remained unwaveringly in good humour throughout, no matter how bulky the material became. We are grateful for the speed and quality of the work she has done and for her assistance in many other details of checking, etc.

We have had the benefit of many discussions with colleagues, including particularly Professors M. J. Seaton, L. Castillejo, I. C. Percival, and J. B. Hasted, and Drs. A. Burgess, R. F. Stebbings, S. Zienau, D. W. O. Heddle, and R. G. W. Keesing. Dr. G. Peach and Professor L. Castillejo have read the proofs of some of the theoretical chapters and made a number of valuable suggestions. Dr. R. W. Lunt has similarly assisted with some of the chapters of Volume II. Professors J. B. Hasted, I. C. Percival, P. Burke, and Dr. A. Burgess have helped by providing us with advance information on the results of work in which they were engaged. We wish to express our thanks to all of these as well as to Professor L. O. Brockway, Drs. L. S. Bartell, R. P. Madden and P. Marmet for providing us with photographic illustrations from their original work.

Finally, it is a pleasure to express our appreciation of the work done by the Clarendon Press and by the printers in the speed of publication and the excellence of the format. Their assistance in checking was of great value in handling such a bulk of material.

CONTENTS

10

ELECTRON COLLISIONS WITH MOLECULES—TOTAL AND ELASTIC SCATTERING

THERE are a number of additional effects of importance that arise when electrons collide with molecules instead of single atoms. In elastic scattering, interference occurs between the electron waves scattered from the different atoms and is apparent in the angular distribution of the scattered electrons. Considerable use has been made of this for studying molecular structure (see § 1.5). Inelastic collisions include the possibility of molecular dissociation, of negative ion formation, and of the excitation of nuclear rotation and vibration. Molecular dissociation due to electron impact is usually a consequence of excitation of some electronic level of the molecule, but a great variety of possible modes of dissociation occur if the molecule is a complicated one. A detailed experimental study has been made for a number of molecules and much valuable information has been obtained.

1. Diffraction of fast electrons by molecules

If the separate atoms of a molecule are considered as separate scattering systems, interference will occur between the electron waves scattered by the individual atoms of a molecule. The nature of these effects will depend on the shape and size of the molecule. This raises the possibility of obtaining information about molecular structure by studying the scattering of electrons by the molecules concerned in the gas phase. Debye,[†] in 1915, developed the theory of the scattering of X-rays by gas molecules and initiated experimental work on the subject which had already begun to yield valuable results[‡] as early as 1927. Wierl[§] in 1931 was the first to use electrons for this purpose and since then a great deal of valuable work has been done using essentially his technique.[||] Electron diffraction is now one of the most powerful methods of investigating molecular structure.

[†] DEBYE, P., *Annln Phys.* **46** (1915) 809.
[‡] DEBYE, P., *Phys. Z.* **28** (1927) 135; **31** (1930) 142 and 419; DEBYE, P., BEWILOGUA, L., and EHRHARDT, F., ibid. **30** (1929) 84.
[§] WIERL, R., *Annln Phys.* **8** (1931) 521.   [||] See § 1.5.

1.1. *The independent scattering-centre approximation*

Under certain assumptions that are valid for collisions with fast electrons, the theory of the elastic scattering by a molecule can be derived very simply from that for the separate atoms. We assume:

(*a*) each atom scatters independently,

(*b*) any redistribution of atomic electrons due to the molecular binding is unimportant so that each atom scatters as if it were free,

(*c*) multiple scattering within the molecule is negligible.

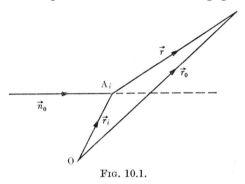

FIG. 10.1.

It is clearly necessary for (*a*) to be a good approximation that the electron wavelength be small compared with the atomic separations. This requires that the electrons be fast so that we would expect Born's approximation to be applicable to the determination of the atomic scattering. This will often be so but we shall not make this additional assumption at the beginning, as it seems likely that there are many cases in which the assumptions (*a*), (*b*), and (*c*) are still valid while Born's approximation is inadequate. We consider an electron beam of wavelength $\lambda = 2\pi/k$ incident on the molecule in the direction of the unit vector $\mathbf{n}_0$, which we shall take as the z-direction.

If we take the centre of the ith atom as origin, an incident wave of unit amplitude, with the corresponding scattered wave due to this atom, can be written

$$e^{ikz} + r^{-1}e^{ikr}f(\theta),\tag{1}$$

where r is the distance from the centre A_i of the atom (see Fig. 10.1) and θ is the angle of scattering. Now if we change the origin to a point O within the molecule, with respect to which the position vector of the centre of the ith atom is $\mathbf{r}_i$, (1) becomes

$$e^{ik(z_0-z_i)} + \frac{e^{ik|\mathbf{r}_0-\mathbf{r}_i|}}{|\mathbf{r}_0-\mathbf{r}_i|}f_i(\theta),$$

where coordinates referred to O are distinguished by the suffix 0. If we reckon phase with reference to the new origin this becomes

$$e^{ikz_0} + \frac{e^{ik|\mathbf{r}_0-\mathbf{r}_i|}}{|\mathbf{r}_0-\mathbf{r}_i|}\, e^{ikz_i}\, f_i(\theta).$$

Further, since $\mathbf{r}_0$, the position vector of the point of observation, is very large compared with $\mathbf{r}_i$ we may write, approximately

$$|\mathbf{r}_0-\mathbf{r}_i| = r_0 - \mathbf{n}.\mathbf{r}_i,$$

where $\mathbf{n}$ is a unit vector in the direction of observation, i.e. in the direction of scattering. This gives for the scattered wave due to the ith atom

$$r_0^{-1} e^{ikr_0} e^{ik(\mathbf{n}_0-\mathbf{n}).\mathbf{r}_i}\, f_i(\theta). \tag{2}$$

Since this expression allows for the differences of phase between different scattering centres, we obtain the amplitude of the scattered wave due to the molecule by summing (2) over all atoms i, giving for the differential cross-section

$$I(\theta) = \left| \sum_i e^{ik(\mathbf{n}_0-\mathbf{n}).\mathbf{r}_i}\, f_i(\theta) \right|^2$$

$$= \sum_i \sum_j f_i f_j^*\, e^{ik(\mathbf{n}_0-\mathbf{n}).\mathbf{r}_{ij}}, \tag{3}$$

where $\mathbf{r}_{ij} = \mathbf{r}_i - \mathbf{r}_j.$

This formula is still incomplete, for it assumes a fixed orientation of the molecule relative to the electron beam. In practice, molecules of a gas will be oriented at random and the observed differential cross-section will be obtained from the average of (3) over all molecular orientations. This average may be calculated by considering each term of (3) separately.

Choosing $\mathbf{n}_0-\mathbf{n}$ as polar axis, the average over-all orientations of the vector $\mathbf{r}_{ij}$ will be

$$\frac{1}{4\pi} \int_0^{2\pi} \mathrm{d}\phi_{ij} \int_0^{\pi} \sin\theta_{ij} \exp(isr_{ij}\cos\theta_{ij})\, \mathrm{d}\theta_{ij},$$

where θ_{ij}, ϕ_{ij} are the polar angles of $\mathbf{r}_{ij}$ relative to $\mathbf{n}_0-\mathbf{n}$ and

$$s = k|\mathbf{n}_0-\mathbf{n}| = 2k\sin\tfrac{1}{2}\theta.$$

Evaluating the integrals gives

$$\bar{I}(\theta) = \sum_i \sum_j f_i f_j^* \frac{\sin sr_{ij}}{sr_{ij}}. \tag{4}$$

Finally, we must allow for the fact that the molecule is vibrating so the distances r_{ij} are not fixed. If we denote by $P_{ij}(r)\, \mathrm{d}r$ the probability that the separation of atoms i and j should be between r and $r+dr$ we obtain

$$\bar{I}(\theta) = \sum_i \sum_j f_i f_j^* \int P_{ij}(r) \frac{\sin sr}{sr}\, dr. \tag{5}$$

Thus in the special case of a homonuclear diatomic molecule

$$\bar{I}(\theta) = 2|f|^2 \left\{ 1 + \int P(r) \frac{\sin sr}{sr}\, dr \right\}, \tag{6}$$

where $|f|^2\, d\omega = I_a\, d\omega$ is the differential cross-section for scattering by a free atom. If, further, molecular vibration is ignored so that $P(r)$ is replaced by a delta function $\delta(r-r_e)$, where r_e is the most probable nuclear separation,

$$\bar{I}(\theta) = 2I_a(\theta)\left\{ 1 + \frac{\sin x}{x} \right\}, \tag{7}$$

where

$$x = sr_e = 2kr_e \sin \tfrac{1}{2}\theta.$$

1.2. Use of Born's approximation

At this stage we make the further assumption that Born's approximation is valid to give (Chap. 7, § 4)

$$f_i(\theta) = \frac{2me^2}{\hbar^2} \frac{Z_i - F_i(\theta)}{s^2}, \tag{8}$$

and

$$\bar{I}(\theta) = \frac{4m^2e^4}{\hbar^4 s^4} \sum_i \sum_j (Z_i - F_i)(Z_j - F_j) \int P_{ij}(r) \frac{\sin sr}{sr}\, dr, \tag{9}$$

where $F_i(\theta)$ is the atom form factor defined in (28) of Chapter 7.

Before considering the practicability of using this result to explore molecular structure we must include the effect of inelastic collisions for, in any photographic technique, these will provide a background that will affect the contrast between diffraction maxima and minima. The total inelastic scattering can be regarded as incoherent and may therefore be built up as the simple sum of the contributions from different atoms. We have, then, for the complete differential cross-section

$$\bar{I}(\theta) = \frac{4m^2e^4}{\hbar^4 s^4} \left\{ \sum_i \sum_j (Z_i - F_i)(Z_j - F_j) \int P_{ij}(r) \frac{\sin sr}{sr}\, dr + \sum_i Z_i S_i \right\}, \tag{10}$$

where S_i is the function discussed and tabulated in Chapter 7, Table 7.7, which represents the total inelastic scattering from the atom.

To discuss the applicability of electron scattering to molecular structure determinations it is convenient to write the expression (10)

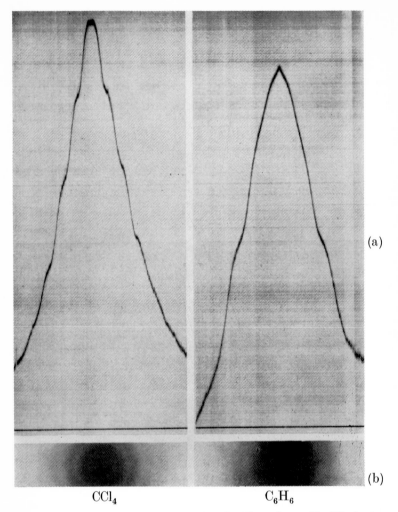

(a)

(b)

CCl₄   C₆H₆

FIG. 10.3. Diffraction of electrons by CCl_4 and C_6H_6 as observed by Wierl using electrons of energy 43 keV. (a) shows microphotometer records obtained under the same conditions as the photographs of (b).

in the form

$$\bar{I}(\theta) = \frac{4m^2e^4}{\hbar^4 s^4}\Bigg[\sum_i \{(Z_i-F_i)^2+4S_i\}\Bigg] +$$

$$+ \sum_i \sum_{j\neq i} (Z_i-F_i)(Z_j-F_j) \int P_{ij}(r)\frac{\sin sr}{sr}\, dr, \quad (11)$$

where the first summation exhibits no diffraction effects. The terms sensitive to molecular structure arise from the second sum in which i and j are always different. The fluctuations with angle due to these terms are superimposed upon a background that decreases very rapidly with increase of angle, due to the factor s^{-4}. This leads to difficulty in accurate structure determination that has been largely overcome in various ways.

1.3. *Application to carbon tetrachloride*

To obtain a clear idea of the nature of the various terms contributing to $\bar{I}(\theta)$ it is instructive to consider a special case, that of carbon tetrachloride, which has been studied extensively. Using the tables of F_i and S_i together with the known tetrahedral structure of CCl_4 with the carbon-chlorine equilibrium distance $1\cdot76$ Å and the chlorine-chlorine distance $2\cdot87$ Å, the curves of Fig. 10.2 may be derived, no allowance being made for molecular vibration. Curve I gives $\bar{I}(\theta)$ as a function of S, while curves II, III, and IV illustrate the respective contributions of the elastic atomic scattering $\sum_i (Z_i-F_i)^2/s^4$, the inelastic atomic scattering $4\sum S_i/s^4$ and the structure-sensitive scattering

$$\frac{1}{s^4}\sum_i \sum_{j\neq i} (Z_i-F_i)(Z_i-F_j)\frac{\sin sr_{ij}}{sr_{ij}}.$$

It will be seen that the decrease in the smooth background with increasing s is so rapid that the structure-sensitive terms are merely able to produce fluctuations and not true maxima and minima. This is clearly seen in the microphotometer records of diffraction patterns. Thus Fig. 10.3 (a) reproduces records taken by Wierl for CCl_4 and C_6H_6. The variation of intensity throughout the angular range observed is so great that, to obtain a measurable blackening at appreciable angles, such intensities must be used that at the small angles saturation of the plate or even reversal occurs. This subjectivity of the plate is exaggerated by visual observation of the diffraction photograph. The eye is very sensitive to rapid changes in intensity and enhances the contrast to give often the appearance of sharply defined rings of maximum and minimum intensity. This may be seen from Fig. 10.3 (b), which reproduces patterns obtained by Wierl† for CCl_4 and C_6H_6 under the same conditions as that from which the microphotometer records of Fig. 10.3 (a) were taken.

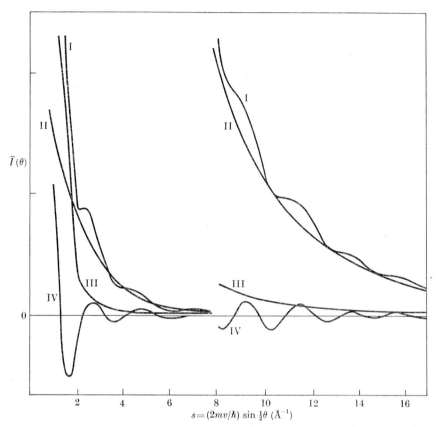

FIG. 10.2. Calculated angular distribution $\bar{I}(\theta)$ of electrons scattered in CCl_4 showing the diffraction effect due to the tetrahedral structure. The ordinate scale for $s > 8$ Å⁻¹ is increased 100 times. Curve I represents the total scattered intensity $\bar{I}(\theta)$. Curves II, III, and IV show respectively the contributions to $\bar{I}(\theta)$ from elastic atomic scattering, inelastic atomic scattering, and from the interference effect.

1.4. Analysis of observed data

In principle when (10) is applicable $P_{ij}(r)$ can be determined for each pair of atoms i,j. Given $P_{ij}(r)$ the most probable value of r_{ij} may be determined from the position of the maximum while the half-width is directly related to the amplitude of vibration. Furthermore, the existence of anharmonicity of the vibration will appear in a departure from symmetry of $P_{ij}(r)$ about its maximum.

† WIERL, R., loc. cit.

The main difficulty in carrying out a programme of this kind is the rapid decrease in the background scattering with s. It is convenient then to write (11) in the form

$$\bar{I}(\theta) = I_0(\theta)\{1 + M(s)\},$$

where

$$M(s) = [\sum \{(Z_i - F_i)^2 + Z_i S_i\}]^{-1} \sum_i \sum_{j \neq i} (Z_i - F_i)(Z_j - F_j) \int P_{ij} \frac{\sin sr}{sr} \, dr. \tag{12}$$

Thus if I is the total intensity of scattered electrons and B that part that depends only on the number and nature of the atoms in the molecule,

$$M = (I - B)/B. \tag{13}$$

B is then the background that falls monotonically as s increases. For large values of s, F_i, $F_j \to 0$ and $S_i \to 1$ so that

$$M(s) \sim \left\{ \sum_i (Z_i^2 + Z_i) \right\}^{-1} \sum_i \sum_{j \neq i} Z_i Z_j \int P_{ij} \frac{\sin sr}{sr} \, dr. \tag{14}$$

Under these conditions we have directly, by Fourier inversion,

$$\sum_i \sum_j Z_i Z_j r^{-1} P_{ij}(r) = (2/\pi) \sum_i (Z_i^2 + Z_i) \int s M(s) \sin sr \, ds. \tag{15}$$

In this way from the observed data at large s a first approximation to the P_{ij} can be found. This can then be substituted in the more elaborate expression (12) and a better approximation then found by trial and error.

1.5. *Experimental methods*

To obtain sharp diffraction patterns, a well-collimated electron beam, very homogeneous in velocity, is fired through a localized jet of gas, so that the scattering volume is very small, and the resulting scattering recorded in a short exposure on a photographic plate. Apparatus, therefore, has remained of the same general form as that used by Wierl,† the essential features of which are illustrated in Fig. 10.4. It consisted essentially of three parts, the electron tube, the diffraction chamber, and the camera. The cathode-ray beam passing through a hole A in a water-cooled anode was collimated by passage for 10 cm along a fine tube B of 0·1 mm bore. This tube also provided the only gas connection between the diffraction chamber C and the electron discharge tube, so that a considerable pressure could be maintained between them. The gas or vapour entered the diffraction chamber through a 0·1-mm hole in a conical jet D. This jet was so arranged as to intersect normally the

---

† WIERL, R., loc. cit.

electron beam issuing from B. The flow through D could be shut off by means of a stopcock E between the jet and the main container of the material whose vapour was being investigated. The vapour was condensed on a liquid air-cooled surface F immediately above the jet D. The electrons scattered from the jet were recorded photographically by a plate 15 cm from D, which could be placed in position and exposed when

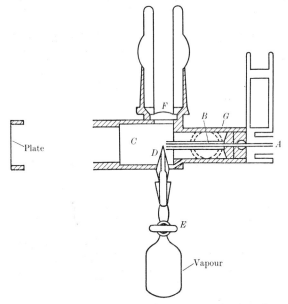

Fɪɢ. 10.4. Apparatus used by Wierl for investigating electron diffraction in gases and vapours.

required. A fluorescent screen for visual observation in effecting adjustments could be substituted for the plate. Pump connections G were provided to maintain as low a pressure as possible in the diffraction chamber. Some typical diffraction patterns obtained with this equipment are shown in Fig. 10.5.

The electron energy employed was, and still is, normally in the neighbourhood of 40 000 eV, corresponding to a wavelength of 6×10^{-2} Å. For precise determination of interatomic distances this wavelength must be known accurately. A convenient method of calibration of voltage measurements was to observe the diffraction of the electrons by a thin film of gold, the edge of the unit cell of which was known to be 4·700 Å. From the positions of the diffraction maxima the wavelength could then be determined with sufficient accuracy. To obtain the required homogeneity in velocity of the electron beam the high voltage

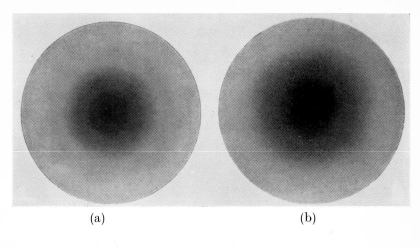

(a)   (b)

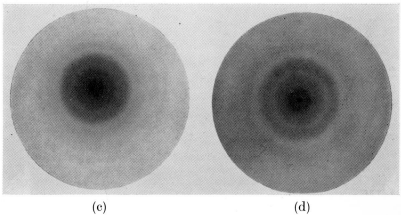

(c)   (d)

FIG. 10.5. Typical diffraction patterns obtained by Wierl for electron diffraction in (a) CS_2, (b) CO_2, (c) $SiCl_4$, (d) $GeCl_4$.

must be well regulated. A test of the homogeneity is the sharpness of definition of the rings obtained from a gold film. Alternatively, a magnetic field may be applied to the beam and its effect observed on the fluorescent screen.

Although there have been many minor improvements in the design for equipment for gaseous electron diffraction studies since the work of Wierl, such as ease and accuracy of adjustment, reduced extraneous

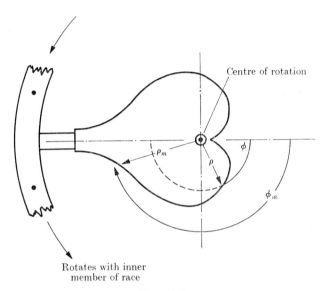

Fig. 10.6.

scattering, etc., the most important new feature that has proved very effective is the rotating sector, which automatically compensates to a large degree for the rapid fall off in the main background scattering B in (13). Otherwise very considerable difficulties are encountered in carrying out accurate photometry over a very great intensity range— much too great for a linear relation between photographic blackening and intensity to apply. Thus as s increases from 1·5 to 35 the background intensity falls by a factor of 10^4, whereas a linear blackening law holds only for an intensity range 100 to 1. To overcome this a sector† in the form of a heart-shaped mask is introduced that rotates about an axis lying along the incident electron beam in a plane parallel to and in front of the photographic plate. Fig. 10.6 illustrates the geometry of such a

† TRENDELENBURG, F., *Naturwissenschaften*, **21** (1933) 173; *Phys. Z.* **40** (1939) 727; DEBYE, P., ibid. **40** (1939) 404; FINBACK, C., HASSEL, O., and OTTAR, B., *Arch. Math. Naturv.* B**44** (1941) No. 13.

sector. In terms of plane polar coordinates ρ, ϕ with origin at the centre of rotation the equation of the sector boundary is $\phi = f(\rho)$, say. During an exposure with the sector rotating uniformly the ratio of the exposure times in the pattern at $\rho = \rho_1, \rho_2$ respectively is then given by $f(\rho_1)/f(\rho_2)$. As ρ is proportional to $\sin \theta$ this ratio may be expressed in terms of the corresponding scattering angles. Thus, for small θ, ρ is proportional to s, so that if the exposure time is to be proportional to s^n the equation of the sector is closely given by $\phi = a\rho^n$. The sector can be cut to give any desired dependence of exposure time on s and we shall refer to a sector in which the exposure time varies as s^n as an s^n sector. In practice s^2 and s^3 as well as s^4 sectors are often used.

As an example of a modern instrument incorporating a sector we show in Fig. 10.7 a sectional drawing of a unit designed by Brockway and Bartell.[†] The divergent beam issuing from the self-biased hot cathode gun is brought to a focus less than $20\,\mu m$ in diameter in the plane of the photographic plate by a magnetic lens. This is rendered possible by the electrostatic focusing within the gun, which provides a source in which less than one-half of the electrons diverge by $20\,\mu m$ from the axis of the beam when the plateau current is reached. Exposures are carried out with electron beam current of about $0\cdot5\,\mu\text{A}$. Larger beams charge up the surfaces and thereby enhance the continuous scattering.

Exposure times ranging from a fraction of a second to many seconds are regulated by an electrostatic shutter between gun and lens. Apertures of platinum are located at the shutter lens and gas-jet nozzle, one of the former two usually being the limiting aperture that is chosen so that the beam diameter is about $0\cdot1$ mm at the nozzle when this is at 10 cm from the plate.

The beam current is measured beneath the nozzle by a metal probe with a depression 8 mm deep to trap the electrons. During exposure this is always withdrawn to avoid casting a shadow on the pattern.

The lens and sector axis positions are aligned by bending the beam with electrostatic deflector plates and by moving the gun. Adjustment is facilitated by the presence of two fluorescent viewing screens, one just above the magnetic lens and the other close to the photographic plate holder. Alignment is achieved to an accuracy of about $0\cdot03$ mm.

It is essential to trap the main beam before it strikes the photographic emulsion. To do this effectively the beam must enter a deep depression in the trap and in the equipment of Brockway and Bartell the trapping tube is 20 mm deep mounted on the sector (see Fig. 10.9).

† BROCKWAY, L. O. and BARTELL, L. S., *Rev. scient. Instrum.* **25** (1954) 569.

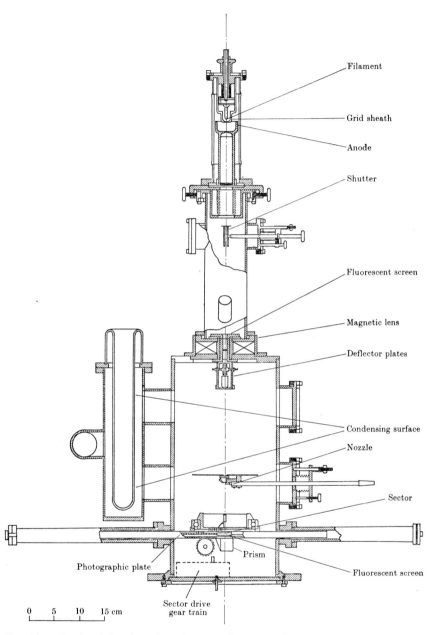

Filament

Grid sheath

Anode

Shutter

Fluorescent screen

Magnetic lens

Deflector plates

Condensing surface

Nozzle

Sector

Photographic plate

Prism

Fluorescent screen

Sector drive
gear train

0   5   10   15 cm

FIG. 10.7. Sectional drawing of an electron diffraction unit designed by Brockway and
Bartell.

The gas sample may be introduced through a fine nozzle of platinum 0·3 mm away from the centre of the beam. Fig. 10.8 is a sectional drawing of a typical nozzle. The pressure in the gas reservoir must be a compromise between too high a value that leads to multiple scattering or too low so that extraneous scattering is serious. It was found that for best results

$$p(\text{torr}) \sim 3 \times 10^4 \Big/ \sum_i Z_i^2,$$

where Z_i is the atomic number of the ith atom in a specimen molecule under investigation. The apparatus is evacuated to 10^{-5} torr before injection of the gas specimen.

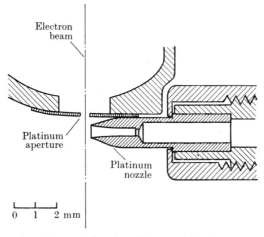

FIG. 10.8. Sectional drawing of the gas injection system in the electron diffraction unit designed by Brockway and Bartell.

Fig. 10.9 is a photograph showing a typical sector mounted on the inner ring of a ball bearing. For much of the work a sector of the shape $\phi = a\rho^3$ is used but for work at low angles of scattering $\phi = a\rho^2$ is more convenient. The sector rotates at 1200 rev/min. Care was taken to ensure that the details of the sector supports were arranged to avoid casting any shadow other than that of the sector. The sector may be located to within 5 mm of the photographic plate so as to minimize edge scattering and other undesirable effects.

The instrument is mounted on a non-magnetic table of welded aluminium to minimize disturbance of the electron beam by external magnetic fields.

In obtaining accurate microphotometer records the pattern must be spun about its centre while being scanned in order to average out

FIG. 10.9. Photograph of the sector used in the electron diffraction unit designed by Brockway and Bartell. Note the beam trap directly over the sector on the axis of rotation.

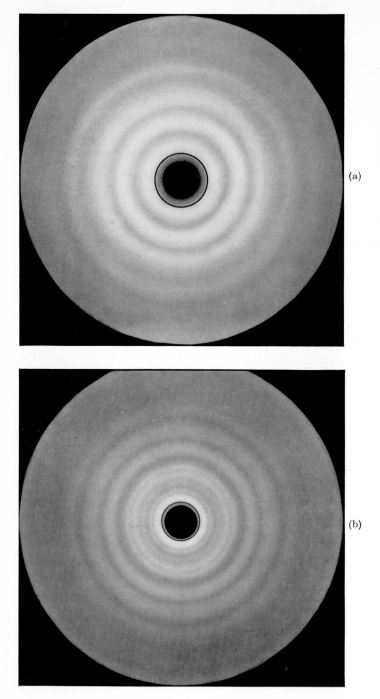

(a)

(b)

FIG. 10.10. Electron diffraction photographs taken with the electron
diffraction unit designed by Brockway and Bartell.

(a) AsF_5 (b) $Si[Si(CH_3)_3]_4$

irregularities in the emulsion. Fig. 10.10 and 10.11 illustrate some of the results obtained with this apparatus. The microphotometer intensity records obtained with the sector of shape function $\phi = a\rho^3$ should be compared with the calculated intensities shown in Fig. 10.2.

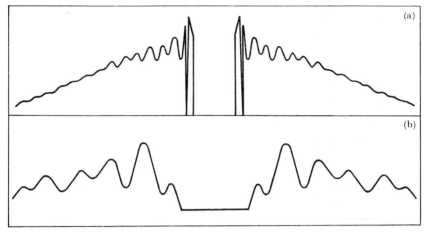

FIG. 10.11. Microphotometer traces of diffraction records obtained with CCl_4 by Brockway and Bartell using a sector of form $\phi = a\rho^3$. (a) With nozzle 25 cm from the plate. (b) With nozzle 10 cm from the plate.

1.6. *Attainable accuracy and typical results*

The internuclear separation between atoms in a diatomic molecule that do not differ too greatly in atomic number can be determined to 0·001 Å or, in some cases, even better. Thus we give in Table 10.1 a comparison between results† obtained by electron diffraction and by spectroscopic methods for the equilibrium nuclear separation r_e in O_2, NO, N_2, and Cl_2. A similar comparison is also given for the r.m.s. amplitude of vibration l_e.

The quantity that is obtained most directly from the diffraction pattern is the position r_g of the centre of gravity of the experimental radial distribution peak. If it is assumed that the nuclei vibrate under an interaction energy $V(r)$ of the Morse type

$$V(r) = D[1-\exp\{-a(r-r_e)\}]^2, \tag{16}$$

where a is the so-called asymmetry constant, then it may be shown‡ that the mean bond length $\bar{r}$ and equilibrium length r_e are given by

$$\bar{r} \simeq r_g + l_g^2/r_e, \tag{17 a}$$

$$r_e \simeq r_g + (l_g^2/r_e) - 3al_g^2/2, \tag{17 b}$$

† BARTELL, L. S. and KUCHITSU, K., *J. phys. Soc. Japan* **17** Supp. B II (1962) 20.
‡ BARTELL, L. S., *J. chem. Phys.* **23** (1955) 1219.

where l_g is the r.m.s. amplitude reckoned from the mean position. Thus

$$l_g^2 = \overline{(r-r_g)^2}, \qquad l_e^2 = \overline{(r-r_e)^2},$$

so
$$l_g^2 = l_e^2 - (r_e - r_g)^2. \tag{18}$$

In obtaining the results shown in Table 10.1 the asymmetry parameter was calculated from spectroscopic data, though in some cases it may be determined from analysis of the experimental radial distributions. It will be seen that the electron diffraction data for r_e are correct to 0·001 Å for all but Cl_2. For l_e the error is somewhat larger on the average but still not greater than 0·002 Å. Bracketed values for chlorine are given by a different investigation[†] in which the asymmetry parameter a was actually estimated from the data as 2 Å^{-1} in close agreement with the spectroscopic value.

TABLE 10.1

Structural parameters for certain diatomic molecules determined by electron diffraction and by spectroscopic methods

Parameter in Å		O_2	NO	N_2	Cl_2
r_g	(electron diff.)	1·212$_6$	1·155$_2$	1·102$_4$	1·991
r_e	(electron diff.)	1·207$_9$	1·150$_8$	1·099	1·983 (1·986$_5$)
	(spectroscop.)	1·2074	1·1507	1·0976	1·988
l_e	(electron diff.)	0·038$_9$	0·034$_7$	0·030$_9$	0·044$_9$ (0·050)
	(spectroscop.)	0·0370	0·0349	0·0323	0·0462

Although it is not in general possible to attain such high accuracies in the determination of the parameters for polyatomic molecules, in most cases an accuracy of 0·005 Å will certainly be attainable and may often be considerably better. For these molecules the interpretation of the basic data in terms of the molecular parameters is much simpler than that of spectroscopic data.

As an example of what may be achieved we give in Table 10.2 a comparison[‡] of the interatomic distances and r.m.s. amplitudes for CS_2, obtained by electron diffraction and by spectroscopic methods.

1.7. *Relaxation of assumptions*

1.7.1. *Failure of Born's approximation.* The accuracy attainable in the measurement of the intensity of scattering of fast electrons by molecules is now so high that it is necessary in some cases to improve

---

† SHIBATA, S., *J. phys. Soc. Japan* **17** Supp. B II (1962) 34.
‡ MORINO, Y. and IIJIMA, T., ibid. **17** Supp. B II (1962) 27.

the theory by relaxing the limitations imposed by the assumptions on which it is based.

Of these, we consider first the use of Born's approximation for the atomic scattering amplitudes. According to this approximation these amplitudes are all real. This will no longer be true in general so that we must write

$$f_j = |f_j| e^{i\delta_j}, \tag{19}$$

so that, using (4), we have, in place of (12),

$$M(s) = [\sum \{|f_i|^2 + g_i\}]^{-1} \sum_i \sum_{j \neq i} |f_i||f_j| \cos(\delta_i - \delta_j) \int P_{ij}(r) \frac{\sin sr}{sr} \, dr. \tag{20}$$

Here we have written g_i in place of $Z_i S_i$ because we are no longer assuming the applicability of Born's approximation.

TABLE 10.2

Structural parameters for CS_2 determined by electron diffraction and by spectroscopic method

	C–S	S–S
Most probable separation (r_g) (Å)		
(electron diffraction)	$1{\cdot}558_3$	$3{\cdot}114_4$
(spectroscopy.)	$1{\cdot}558$	$3{\cdot}109$
R.m.s. amplitude (l_e) (Å)		
(electron diffraction)	$0{\cdot}040_6$	$0{\cdot}039_1$
(spectroscopy)	$0{\cdot}044_7$	$0{\cdot}041_2$

If $\cos(\delta_i - \delta_j)$ differs appreciably from unity the procedure outlined in § 1.4 will not lead directly to the simple function $\sum_i \sum_{j \neq i} Z_i Z_j P_{ij}(r)$. Thus it was pointed out by Schomaker and Glauber† that when Z_i and Z_j are very different the function given from the right-hand side of (15) will show not one peak when r is nearly equal to the most probable separation of atoms i and j but two peaks. As Z_i and Z_j approach, the peaks move closer and eventually merge. The width of the single peak would at first be abnormally large so that if interpreted in terms of vibrational amplitude it would give erroneous results.

At the relatively high energies concerned, electron exchange and atom polarization effects (Chap. 8, § 2) will be very small so that a good approximation to $f(\theta)$ should be obtained by calculating the scattering by the Hartree–Fock field of the atom, using the method of partial cross-

† SCHOMAKER, V. and GLAUBER, R., *Nature, Lond.* **170** (1952) 290.

sections described in Chapter 6, § 3.2. According to this

$$|f(\theta)|^2 = f_I^2 + f_R^2, \tag{21}$$

$$\tan \delta_i = f_I/f_R, \tag{22}$$

where

$$f_I = \frac{1}{2k} \sum (2l+1)(1-\cos 2\eta_l) P_l(\cos \theta), \tag{23}$$

$$f_R = \frac{1}{2k} \sum (2l+1)\sin 2\eta_l \, P_l(\cos \theta). \tag{24}$$

Here k is the electron wave number and the real phase shifts η_l are such that the solution of the equation

$$\frac{d^2 G_l}{dr^2} + \left\{ k^2 - U(r) - \frac{l(l+1)}{r^2} \right\} G_l = 0, \tag{25}$$

which satisfies $G_l(0) = 0$, has the asymptotic form

$$G_l \sim \sin(kr - \tfrac{1}{2}l\pi + \eta_l). \tag{26}$$

$\hbar^2 U(r)/2m$ is the potential energy of interaction of the electron with the atom according to the Hartree–Fock approximation.

Ibers and Hoerni† used this method to calculate $|f(\theta)|$ and $\delta(\theta)$ for electrons of 40 keV and atoms with $Z = 9$, 18, 74, and 80, for which Hartree–Fock fields were available at the time. To obtain a representative set of values for all atoms they took for $\hbar^2 U(r)/2m$ in (25), the Fermi–Thomas statistical potential in an approximate form due to Rosental.‡ In each case they calculated η_l by the classical approximation discussed in Chapter 6, § 3.7 when $\eta_l \gg 1$, and by Born's approximation (Chap. 6, § 3.4 and Chap. 7, § 4) for $\eta_l \ll 1$. Some of these results are given in Table 10.3.

Bonham and Karle§ have also calculated $|f(\theta)|$ and $\delta(\theta)$ for uranium ($Z = 92$) and fluorine ($Z = 9$). For the former atom they took $U(r)$ to be given by the Fermi–Thomas statistical potential in an approximate form due to Molière,‖ while for fluorine they used an analytic representation of the Hartree–Fock field. Calculations were carried out using not only the non-relativistic formulation given above but relativistic formulations in terms of the Klein–Gordon and Dirac equations (see Chap. 6, § 6). In each case they calculated the small phases by Born's approximation and the others by accurate solution of the appropriate differential equations. Some of their results calculated non-relativistically are included in Table 10.3 for comparison with those of Ibers and

† IBERS, J. A. and HOERNI, J. A., *Acta crystallogr.* 7 (1954) 405.
‡ ROSENTAL, S., *Z. Phys.* 98 (1936) 742.
§ BONHAM, R. A. and KARLE, J., *J. phys. Soc. Japan* Supp. B II 16 (1962) 6.
‖ MOLIÈRE, G., *Z. Naturf.* 2a (1947) 133.

TABLE 10.3

Moduli $|f(\theta)|$ and arguments $\delta(\theta)$ of the amplitudes for scattering of 40-keV electrons by different atoms

θ (deg)	Uranium ($Z = 92$) $|f(\theta)|$ (units a_0)		$\delta(\theta)$ (rad)		Tungsten ($Z = 74$) $|f(\theta)|$		$\delta(\theta)$	
	(a)	(b)	(a)	(b)	(b)	(c)	(b)	(c)
0	15·20	16·8	0·469	0·29	16·2	15·2	0·26	0·23
1	11·14	10·73	0·604	0·43	9·94	8·7	0·39	0·38
2	6·53	5·44	0·898	0·73	4·96	4·33	0·68	0·62
4	2·81	2·15	1·450	1·35	1·97	1·93	1·24	1·13
6	1·51	1·18	1·960	1·89	1·08	1·07	1·72	1·61
8	0·938	0·751	2·332	2·38	0·696	0·688	2·14	2·03
10	0·643	0·532	2·715	2·82	0·486	0·483	2·52	2·41
12	0·466	0·403	3·058	3·20	0·363	0·363	2·85	2·74
16	0·268	0·253	3·681	3·86	0·228	0·228	3·43	3·30

θ	Germanium ($Z = 32$) $|f(\theta)|$	$\delta(\theta)$	Fluorine ($Z = 9$) $|f(\theta)|$			$\delta(\theta)$		
	(c)	(c)	(d)	(b)	(c)	(d)	(b)	(c)
0	13·3	0·14	2·31	9·1	2·1	0·115	0·05	0·09
1	6·81	0·26	1·90	3·12	1·8	0·134	0·11	0·10
2	3·22	0·45	1·24	1·36	1·22	0·182	0·19	0·13
4	1·29	0·77	0·513	0·500	0·504	0·309	0·31	0·24
6	0·688	1·05	0·258	0·243	0·251	0·419	0·41	0·34
8	0·425	1·29		0·150	0·148	0·566	0·50	0·43
10	0·292	1·50	0·100	0·100	0·098	0·617	0·57	0·50
12	0·211	1·68	0·071	0·070	0·070	0·698	0·63	0·56
16	0·128	1·98	0·041	0·041	0·042	0·862	0·72	0·65

(a) Calculated by Bonham and Karle† using the Fermi–Thomas field as represented analytically by Molière.‡

(b) Calculated by Ibers and Hoerni§ using the Fermi–Thomas field as represented by Rosental.§

(c) Calculated by Ibers and Hoerni§ using the Hartree–Fock field.

(d) Calculated by Bonham and Karle† using the Hartree–Fock field.

Hoerni.§ Relativistic effects were found to be quite small even for uranium at these energies.

It will be seen that $\delta(\theta)$ is considerably more sensitive to the assumed potential field than is $|f(\theta)|$. Moreover, it is quite large even for fluorine, particularly at large scattering angles.

In an earlier paper Hoerni and Ibers‖ calculated the differential cross-section for elastic scattering of electrons of 40- and 11-keV energy by

† loc. cit., p. 680.   ‡ loc. cit., p. 680.   § loc. cit., p. 680.
‖ HOERNI, J. A. and IBERS, J. A., *Phys. Rev.* **91** (1953) 1182.

UF_6 molecules. The amplitudes $f(\theta)$ for scattering by the respective atoms were calculated in the same way as described above, except that slightly less accurate analytical representations of the atomic fields were used and relativistic effects partly allowed by use of the Klein–Gordon rather than the Schrödinger equation for the electron motion. Quite good agreement was obtained with the observed molecular diffraction pattern.

Bonham and Ukaji† have developed a method for analysis of diffraction photographs based on representation of the argument $\delta(\theta)$ of the scattered amplitude in the form

$$\delta(\theta) = a + bs + cs^2, \tag{27}$$

where $s = 2\pi \sin \tfrac{1}{2}\theta/\lambda$ and a, b, c are constants depending on the atomic number. They found that the values given in the tables of Hoerni and Ibers could be fitted in this way with a maximum deviation less than 3 per cent for s ranging from 3 to 50 Å^{-1}. If it is assumed that the atomic distribution function has the Gaussian form

$$P_{ij}(r) = \{1/(2\pi)^{\frac{1}{2}}l_{ij}\}\exp\{-(r-r_{ij})^2/2l_{ij}^2\}, \tag{28}$$

(20) becomes for large s

$$\sum (Z_i^2 + Z_i)M(s) = \left\{\sum_{ij} Z_i Z_j \cos(\Delta a_{ij} + \Delta b_{ij}s + \Delta c_{ij}s^2)\exp(-l_{ij}^2 s^2/2)\right\} \times$$
$$\times \sin\{s(r_{ij} - l_{ij}^2/r_{ij})\}/sr_{ij}, \tag{29}$$

where

$$\Delta a_{ij} = a(Z_i) - a(Z_j), \text{ etc.}$$

If the Fourier transform of $sM(s)$ is taken as in (15) we find for the apparent radial distribution function

$$\{\sum (Z_i^2 + Z_i)\}P_{ij}(r) = 2^{-\frac{3}{2}}\pi^{-\frac{1}{2}} \sum_{ij} [Z_i Z_j\{\exp(-x_+^2/2\sigma_1)\cos(A - x_+^2/2\sigma_2) +$$
$$+ \exp(-x_-^2/2\sigma_1)\cos(A - x_-^2/2\sigma_2)\}/r'_{ij} l_{ij}^{\frac{1}{2}} \sigma_1^{\frac{1}{2}}], \tag{30}$$

where

$$r'_{ij} = r_{ij} - l_{ij}^2/r_{ij}, \qquad \sigma_1 = l_{ij}^2 + 4\Delta c_{ij}^2/l_{ij}^2,$$
$$\sigma_2 = 2\Delta c_{ij} + l_{ij}^4/2\Delta c_{ij}, \qquad x_\pm = r - r'_{ij} \pm \Delta b_{ij}, \tag{31}$$

and $A = \tfrac{1}{2}\arctan(2\Delta c_{ij}/l_{ij}^2) + \Delta a_{ij}(4\Delta c_{ij}^2 - l_{ij}^4)/l_{ij}^2 \sigma_1$. This formula explains why, when the usual method of analysis described in § 1.4 is used, double peaks are found in the distribution function when Z_i and Z_j are very different.

The calculation of the molecular scattering in terms of the coherent scattering from independent atomic centres of force, ignoring multiple scattering within the molecule and distortion of the atomic fields by valence forces, but allowing for the need to calculate the atomic scattering by the method of partial cross-sections instead of Born's approximation, has been extended with surprising success down to very much lower electron energies, of around 100 eV or even less. This is discussed in § 2.1.

1.7.2. *Multiple scattering within the molecule.*‡ Let A and B be two

† BONHAM, R. A. and UKAJI, T., *J. chem. Phys.* **36** (1962) 72.
‡ See MOTT, N. F. and MASSEY, H. S. W., *The theory of atomic collisions*, 3rd edn. p. 196 (Clarendon Press, Oxford, 1965).

similar atoms located at a distance R apart. O is an origin near the molecule with respect to which the position vectors of the atomic nuclei are $\mathbf{r}_a$, $\mathbf{r}_b$ and we write $\mathbf{R} = \mathbf{r}_a - \mathbf{r}_b$. An electron is incident in the direction $\mathbf{n}_0$ and is scattered into the direction $\mathbf{n}_1$. We discuss the contributions from single and double scattering to the total scattered amplitude.

Consider a plane wave of unit amplitude incident in the direction of the unit vector $\mathbf{q}$ on atom A. The analysis of § 1 shows that, if the atom were to scatter independently, the scattered wave in direction $\mathbf{n}$ is given at a large distance r_a from A by

$$\frac{e^{ik|\mathbf{r}_0 - \mathbf{r}_a|}}{|\mathbf{r}_0 - \mathbf{r}_a|}\, e^{ik\mathbf{q}\cdot\mathbf{r}_a} f(\mathbf{q}, \mathbf{n}). \tag{32}$$

To take account of double scattering we must consider that the wave incident on atom B is not just the incident plane wave $e^{ik\mathbf{n}_0\cdot\mathbf{r}_0}$ in the direction $\mathbf{n}_0$ but also includes a contribution due to the scattered wave arriving from A. This may be calculated from (32) if $k|\mathbf{r}_a - \mathbf{r}_b|\ (= kR)$ is large. In the neighbourhood of atom B where $\mathbf{r}_0 \simeq \mathbf{r}_b$

$$\frac{e^{ik|\mathbf{r}_0 - \mathbf{r}_a|}}{|\mathbf{r}_0 - \mathbf{r}_a|} \simeq \frac{e^{ik|\mathbf{r}_a - \mathbf{r}_b|}}{|\mathbf{r}_a - \mathbf{r}_b|}\, e^{-ik\hat{\mathbf{R}}\cdot(\mathbf{r}_0 - \mathbf{r}_b)}. \tag{33}$$

Hence the scattered wave from A is equivalent to a further plane wave

$$\alpha(A)e^{-ik\hat{\mathbf{R}}\cdot\mathbf{r}_0}, \tag{34}$$

incident in the direction $-\hat{\mathbf{R}}$ with $\alpha(A)$ given by

$$R^{-1}\, e^{ikR}\, e^{ik\mathbf{n}_0\cdot\mathbf{r}_a}\, e^{ik\hat{\mathbf{R}}\cdot\mathbf{r}_b}\, f(\mathbf{n}_0, -\hat{\mathbf{R}}). \tag{35}$$

The total wave in the direction $\mathbf{n}_1$ scattered from B at a large distance from the molecule can now be obtained from (32). It will be given by

$$\frac{e^{ik|\mathbf{r}_0 - \mathbf{r}_b|}}{|\mathbf{r}_0 - \mathbf{r}_b|} \{e^{ik\mathbf{n}_0\cdot\mathbf{r}_b}f(\mathbf{n}_0, \mathbf{n}_1) + e^{-ik\hat{\mathbf{R}}\cdot\mathbf{r}_b}\alpha(A)f(-\hat{\mathbf{R}}, \mathbf{n}_1)\}. \tag{36}$$

When $r_0 \gg r_b$ we have as in § 1

$$\frac{e^{ik|\mathbf{r}_0 - \mathbf{r}_b|}}{|\mathbf{r}_0 - \mathbf{r}_b|} \simeq \frac{e^{ikr_0}}{r_0}\, e^{-ik\mathbf{n}_1\cdot\mathbf{r}_b}, \tag{37}$$

so the total scattered wave from B at a large distance r_0 from O is given by

$$r_0^{-1}e^{ikr_0}\{e^{ik(\mathbf{n}_0-\mathbf{n}_1)\cdot\mathbf{r}_b}f(\mathbf{n}_0, \mathbf{n}_1) + R^{-1}e^{ikR}e^{ik(\mathbf{n}_0\cdot\mathbf{r}_a - \mathbf{n}_1\cdot\mathbf{r}_b)}f(\mathbf{n}_0, -\hat{\mathbf{R}})f(-\hat{\mathbf{R}}, \mathbf{n}_1)\}. \tag{38a}$$

In the same way we find for the total scattered wave from A

$$r_0^{-1}e^{ikr_0}\{e^{ik(\mathbf{n}_0-\mathbf{n}_1)\cdot\mathbf{r}_a}f(\mathbf{n}_0, \mathbf{n}_1) + R^{-1}e^{ikR}e^{ik(\mathbf{n}_0\cdot\mathbf{r}_b - \mathbf{n}_1\cdot\mathbf{r}_a)}f(\mathbf{n}_0, \hat{\mathbf{R}})f(\hat{\mathbf{R}}, \mathbf{n}_1)\}. \tag{38b}$$

Combining (38a) and (38b) we have for the differential cross-section

$$I(\theta)\, d\omega = |F|^2\, d\omega, \tag{39}$$

where

$$F = f(\mathbf{n}_0, \mathbf{n}_1)\{1 + e^{ik(\mathbf{n}_0-\mathbf{n}_1)\cdot\mathbf{R}}\} +$$
$$+ R^{-1}e^{ikR}\{e^{-ik\mathbf{n}_1\cdot\mathbf{R}}f(\mathbf{n}_0, \hat{\mathbf{R}})f(\hat{\mathbf{R}}, \mathbf{n}_1) + e^{ik\mathbf{n}_0\cdot\mathbf{R}}f(\mathbf{n}_0, -\hat{\mathbf{R}})f(-\hat{\mathbf{R}}, \mathbf{n}_1)\}. \tag{40}$$

We may write for $\bar{I}(\theta)$, the average over-all molecular orientations,

$$\bar{I}(\theta) = 2|f|^2\left(1 + \frac{\sin sR}{sR}\right) + \delta I, \tag{41}$$

where δI is the contribution from double scattering arising from the terms in (40) that include the factor $R^{-1}e^{ikR}$. To calculate δI, terms depending on $R^{-2}e^{2ikR}$ that occur in $|F|^2$ may be ignored, because the analysis is only really valid when δI is small compared with $|f|^2$.

Hoerni† has applied this analysis to the scattering of 40-keV electrons by a hypothetical molecule consisting of two uranium atoms 2 Å apart. The field of each atom was taken as

$$V(r) = Ze^2r^{-1}e^{-r/a}, \tag{42}$$

with $Z = 92$ and $a = 0 \cdot 468Z^{-\frac{1}{3}}$ Å, the effective range characteristic of the Fermi–Thomas statistical atomic field (Chap. 6, § 2.3). For the calculation of $|f|^2$ in (41) the method of partial cross-sections was used but in δI the amplitudes appearing in (40) were evaluated by Born's approximation. Table 10.4 gives the results obtained.

TABLE 10.4

The relative importance of single and double scattering terms for the elastic scattering of 40-keV electrons by a hypothetical U_2 molecule in which the atoms are 2 Å apart

Angle of scattering (°)	Direct atomic scattering $2|f|^2$	Molecular interference scattering divided by $2|f|^2/sR$	Double scattering correction δI
0	565		0·227
2	59·2	8·17	0·178
4	9·25	0·638	0·0953
6	2·79	0·128	0·0471
8	1·13	0·0390	0·0205
10	0·566	0·0157	0·00927
12	0·325	0·00749	0·00498
16	0·128	0·00222	0·00192
20	0·064	0·00089	0·00087

It will be seen that δI is small compared with $|f|^2$ at all angles of scattering. On the other hand, it is comparable with the molecular structure term throughout most of the angular range. This is not serious, however, because δI varies smoothly and gradually with angle so it contributes to the so-called background scattering which in any case is subtracted off before analysing the molecular structure.

Although double scattering within the molecule is not likely to be serious in the analysis of electron diffraction photographs because it introduces only an additional smooth background, it was pointed out

† HOERNI, J. A., *Phys. Rev.* **102** (1956) 1530.

by Bunyan† and Gjonnes‡ that triple scattering may be more serious in contributing fluctuating terms that could be comparable with the molecular interference scattering, particularly for polyatomic molecules with heavy atoms.

This possibility has been examined in some detail by Bonham§ who considered the case of a hypothetical molecule U_3 of equilateral triangular structure with atomic separations 2 Å. He found the maximum value of the triple scattering correction to be about 5 per cent of the molecular interference scattering for 40-keV electrons.

1.7.3. *Valence distortion effects—scattering by* H_2^+ *and* H_2. The chief effect of the distortion of the atomic charge distributions when they are bound in a molecule is likely to be the reduction in effective volume occupied by the electrons. This should lead to a smaller scattering cross-section than would be expected if the atoms scattered independently.

Detailed calculations using Born's approximation have been carried out only for H_2^+ and H_2 for which good molecular wave functions are available. The early work on $H_2\|$ has been corrected and extended by Bonham and Iijima†† who also carried out the first calculations for H_2^+.‡‡

Although it is not possible to investigate the case of H_2^+ experimentally it is nevertheless of interest because the exact electronic wave function for the ground state is known. Fig. 10.12 shows the ratio of the calculated total scattering $I(H_2^+)$ to the sum of that $I(H^+)$ from an isolated proton and an isolated hydrogen atom $I(H)$, it being assumed that (208) of Chapter 7 gives the inelastic scattering contribution. It will be seen that for $s < 5$ Å^{-1} the ratio falls not only substantially below unity, but also below that calculated by the model in which a proton and undistorted hydrogen atom at a nuclear separation equal to that in H_2^+ are assumed to scatter independently.

For H_2 no exact wave functions exist. Calculations were carried out by Bonham and Iijima†† using two approximate functions. If the protons are distinguished as A and B and the electrons by 1 and 2 these functions may be written in the form

$$\psi = N[u_A(1)u_B(2)+u_A(2)u_B(1)+\mu\{u_A(1)u_A(2)+u_B(1)u_B(2)\}], \quad (43)$$

† BUNYAN, P. J., *Proc. phys. Soc.* **82** (1963) 1051.
‡ GJONNES, J., *Acta crystallogr.* **17** (1964) 1095.
§ BONHAM, R. A., *J. chem. Phys.* **43** (1965) 1103.
‖ MASSEY, H. S. W. and MOHR, C. B. O., *Proc. R. Soc.* A**135** (1932) 258; ROSCOE, R., *Phil. Mag.* **26** (1938) 32.
†† BONHAM, R. A. and IIJIMA, T., *J. phys. Chem., Ithaca* **67** (1963) 2266.
‡‡ IIJIMA, T. and BONHAM, R. A., ibid. **67** (1963) 2769.

where $\qquad u_A(1) = e^{-Zr_{1A}}$, etc.,

r_{1A} being the distance of electron 1 from nucleus A. The following values were taken for Z and μ:

$$\text{Wang function}\dagger \quad Z = 1{\cdot}166/a_0, \qquad \mu = 0; \qquad (44\,\text{a})$$

$$\text{Weinbaum function}\ddagger \quad Z = 1{\cdot}193/a_0, \qquad \mu = 0{\cdot}256. \qquad (44\,\text{b})$$

The nuclear separation in the molecule is $0{\cdot}742$ Å.

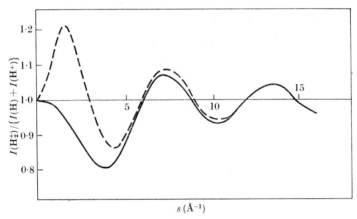

FIG. 10.12. Ratio of the total scattering from an H_2^+ ion to the sum of the scattering from an isolated proton and an isolated hydrogen atom, calculated by Born's approximation. ——— using the exact ground state wave function of H_2^+; – – – – assuming the H^+ and H in H_2^+ scatter as if independent and the H atom is undistorted by valence forces.

Fig. 10.13 shows the ratio $I(H_2)/2I(H)$ of the total scattered intensity for the molecule to that for two isolated hydrogen atoms with $I(H_2)$ calculated in three ways:

(a) as the total scattering from two undistorted H atoms at the separation equal to that in H_2 (independent atom model);

(b) using the Wang wave function for H_2;

(c) using the Weinbaum wave function for H_2.

The latter two allow for valence distortion and it will be seen from Fig. 10.13 that this reduces the scattering quite substantially below that for the independent atomic scattering model (a) for values of $s < 5$ Å^{-1}.

A similar result is found for the elastic scattering as shown in Fig. 10.14. Bonham and Iijima§ have obtained experimental evidence in support

† WANG, S. C., *Phys. Rev.* **31** (1928) 579.
‡ WEINBAUM, S., *J. chem. Phys.* **1** (1933) 593.
§ BONHAM, R. A. and IIJIMA, T., ibid. **42** (1965) 2612.

of their theoretical results for H_2. They obtained electron diffraction photographs from H_2 and He over the range 1 to 8 Å^{-1} of s, using electrons of 40-keV energy and a rotating s^3 sector. The observed intensity curves were corrected for extraneous scattering by taking a blank photograph

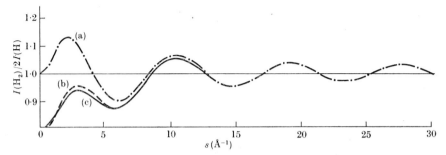

FIG. 10.13. Ratio of the total scattering from H_2 to the sum of the scattering from two isolated H atoms, calculated by Born's approximation. (a) Independent atom model. (b) Using the Wang wave function for H_2. (c) Using the Weinbaum wave function for H_2.

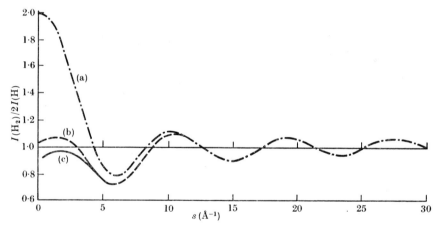

FIG. 10.14. Ratio of the elastic scattering from H_2 to the sum of the elastic scattering from two isolated H atoms, calculated by Born's approximation. (a) Independent atom model. (b) Using the Wang wave function for H_2. (c) Using the Weinbaum wave function for H_2.

with no injected gas present but otherwise under the same experimental conditions.

The observed intensities for helium were normalized to agree at $s = 6$ with the theoretical value obtained using a wave function for the helium ground state of the form

$$\psi = N\{\exp(-ar_1 - br_2) + \exp(-br_1 - ar_2)\}\{1 + \alpha\exp(-cr_{12})\}, \quad (45)$$

with the constants determined by the variational method. By comparison with the normalized intensities for He, results were obtained for H_2 that should be largely free of instrumental errors such as the shape of the sector.

Fig. 10.15 illustrates the comparison between the observed and calculated intensities. It will be seen that for helium the experimental intensities normalized at $s = 6$ Å^{-1} then agree quite well with the

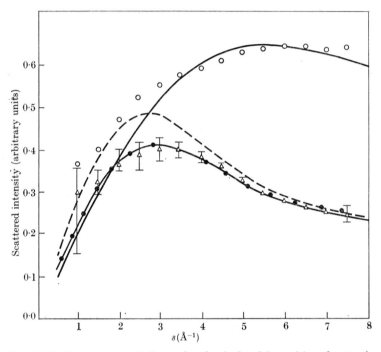

FIG. 10.15. Comparisons of observed and calculated intensities of scattering (both elastic and inelastic) of 40-keV electrons by helium and by hydrogen. The observed results were obtained using an s^3 sector. —— calculated for He using the wave function (45). ○ ○ ○ observed, for He normalized to agree with the calculated for $s = 6$. –●–●–● calculated for H_2 using the Weinbaum wave function. — — — calculated for H_2 ignoring valence distortion. △ observed for H_2.

theoretical intensities between $s = 3$ Å^{-1} and $s = 8$ Å^{-1} but begin to deviate for lower s. The observed relative intensity for H_2 is compared with two theoretical curves calculated using the Weinbaum wave function (44 b) and the independent atom model respectively. Good agreement is found with both for $s > 6$ Å^{-1} but for smaller s values the former is clearly favoured.

Further experimental evidence has been obtained by Geiger† for the

† GEIGER, J., Z. Phys. **181** (1964) 413.

elastic scattering of 25-keV electrons, using the technique described in Chapter 5, § 5.2.3 (see Chap. 7, § 4 for discussion of results for helium and for the heavier rare gases). Fig. 10.16 shows a comparison between the observed and calculated variation of elastically scattered intensity with scattering angle θ in the range $7 \times 10^{-4} \leqslant \theta \leqslant 4 \cdot 3 \times 10^{-2}$, corresponding to $0 \cdot 056 \text{ Å}^{-1} \leqslant s \leqslant 3 \cdot 44 \text{ Å}^{-1}$. The experimental and theoretical results are all normalized to unity at $\theta = 0$. It will be seen

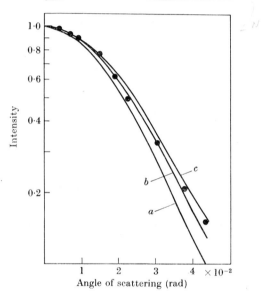

FIG. 10.16. Comparison of observed and calculated variation of the intensity of elastic scattering of 25-keV electrons by hydrogen molecules, with angle of scattering at small angles. ●●● experimental results. —— theoretical results: (a) undistorted hydrogen atom model, (b) molecular model with Wang wave function, (c) molecular model with Weinbaum wave function.

that the observed fall of intensity with scattering angle follows more closely that calculated using either the Wang or Weinbaum functions (44 a) and (44 b) respectively than that in which the molecule is represented by two undistorted hydrogen atoms at the equilibrium nuclear separation. It will be noted from Fig. 10.14 that, if the curves a, b, and c were normalized to the same value at $s = 0$ then, for $s < 2 \cdot 5 \text{ Å}^{-1}$, c would fall a little above b and both above a. Since a, b, and c correspond respectively to the undistorted atom model and the Wang and the Weinbaum molecular models respectively this is just the behaviour of the theoretical curves shown in Fig. 10.16.

2. Diffraction of electrons of medium speed by molecules

In the discussion of the various approximations made in deriving the formula (4) for the differential cross-section for scattering of electrons by molecules it emerged that, as the electron energy is reduced, the assumption of the validity of Born's approximation is likely to be the first to break down. It is of interest, therefore, to examine the range of validity of the approximation which, while treating the atoms in a molecule as independent undistorted scatterers and ignoring multiple scattering within the molecule, nevertheless calculates the scattered amplitude from each atom without resort to Born's approximation. It will be found that this gives good results at energies far below those normally used in electron diffraction experiments.

In this section we shall discuss the elastic scattering of electrons with these medium energies in terms of this approximation. Experimental and theoretical work in this direction has been concerned with the study of the angular distribution of electrons scattered by various molecules and of the fine structure of X-ray absorption edges in molecular gases. In the latter the electron ejected by absorption of the X-ray quantum is diffracted by the surrounding atoms.

2.1. *Angular distribution of elastically scattered electrons*

The first application of the formula (4), in which the amplitudes were calculated from the static field of the atom using the expression given by the method of partial cross-sections

$$f_i = \frac{1}{2ik} \sum_l (e^{2i\eta_l} - 1)(2l+1)P_l(\cos\theta), \qquad (46)$$

where the phases are as discussed in Chapter 6, § 3.2, was made by Bullard and Massey.† They calculated the angular distribution of electrons in the energy range 30–780 eV scattered elastically in nitrogen and found reasonable agreement with observation. Estimates made much later by Youssef‡ of the contribution from double scattering according to the formula (40) have shown that this at least is unimportant, even in this energy range, except at quite small angles of scattering.

Hill and Woodcock§ have carried out similar calculations on these lines for the much less favourable case of scattering of electrons with energies between 18 and 42 eV by the carbon tetrahalides. For all but

---

† BULLARD, E. C. and MASSEY, H. S. W., *Proc. Camb. phil. Soc. math. phys. Sci.* **29** (1933) 511.
‡ YOUSSEF, G. M. J., Thesis, London (1964).
§ HILL, S. and WOODCOCK, A. H., *Proc. R. Soc.* A**55** (1936) 231.

CF_4 the contribution from the carbon atom may be neglected so the scattered intensity becomes

$$I(\theta) = I_h(\theta)\left(1 + \frac{3\sin x}{x}\right),$$

where $I_h(\theta)$ is the scattered intensity for a single halogen atom and $x = 2kd\sin\frac{1}{2}\theta$, d being the distance between halogen atoms.

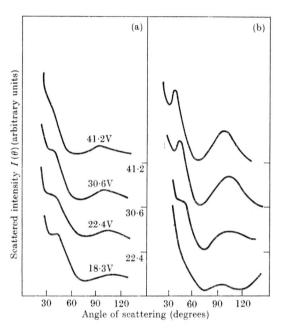

Fig. 10.17. Comparison of observed and calculated angular distributions of electrons of medium energy scattered in CCl_4 vapour. (a) observed; (b) calculated.

Fig. 10.17 illustrates $I(\theta)$ calculated from this formula for CCl_4 compared with observed distributions.† The agreement between this theory and observation is remarkably good, even for electrons with energies as low as 18 eV. Similar agreement was found for CF_4 but for CBr_4 only a rough qualitative correspondence was found with the observed curves. As the approximation depends on the assumption that the electron distribution in the molecular atoms is not disturbed by the binding and that scattering of an electron does not occur more than once within the molecule, it is not surprising that the agreement is not good for a molecule containing such large atoms as bromine, which

† HILL, S. and WOODCOCK, A. H., loc. cit., p. 690.

must overlap considerably. The surprising thing is rather that the agreement should be so close for CCl_4.

A discussion on these lines has also been given by Hughes and McMillen[†] of their results on the scattering of electrons by CH_4, C_2H_4, and C_2H_2. They found it to be useful in providing a qualitative description of the relative scattering in these cases.

An alternative procedure was introduced by Frye[‡] to analyse the results of electron scattering measurements in gaseous bromine.[§] Assuming the formula

$$I(\theta) = |f_a(\theta)|^2 \left(1 + \frac{\sin x}{x}\right),$$

where $f_a(\theta)$ is the amplitude scattered by a free bromine atom, he first obtained $|f_a(\theta)|$ and then derived the values of the phases in the formula (46) that gave the best agreement with the observed angular distributions in the electron energy range from 15 to 121 eV. These 'observed' phases are not very different from those estimated by Snyder and Shaw[||] from Holtsmark's calculated phases for krypton (see Chap. 6, § 4.2 and Fig. 6.8). Frye went further and derived an atomic field for bromine which would give the 'observed' phases.

2.2. *The fine structure of* X-*ray absorption edges in molecular gases*

X-radiation of quantum energy greater than that of the K electrons of the atoms of a gas may be absorbed by the gas in ejecting a K electron by the photoelectric effect. The absorption coefficient of a monatomic gas, due to this effect, falls off steadily as the frequency increases beyond the long wavelength limit. It was found, however, by Hanawalt[††] in 1931 that, if the gas is polyatomic, the variation of the absorption coefficient with the quantum energy may be more complicated. Within an energy range of a few hundred eV extending from the long wavelength limit, the absorption coefficient may exhibit a series of maxima and minima. This may be seen by reference to Fig. 10.19 (a), which illustrates the effect for germanium tetrachloride.[‡‡]

The theoretical explanation of the maxima and minima was first given by Kronig,[§§] who showed that they arose by diffraction of the

† HUGHES, A. L. and McMILLEN, J. H., *Phys. Rev.* **44** (1933) 876.
‡ FRYE, W. E., ibid. **60** (1941) 586.
§ ARNOT, F. L., *Proc. R. Soc.* A**144** (1934) 360.
|| SNYDER, T. M. and SHAW, C. H., *Phys. Rev.* **57** (1940) 881.
†† HANAWALT, J. D., ibid. **37** (1931) 715.
‡‡ GLASER, H., ibid. **82** (1951) 616.
§§ KRONIG, R. DE L., *Z. Phys.* **75** (1932) 468. See also PETERSEN, H., ibid. **80** (1933) 258 and *Diss. Groningen, Archs. néerl.* **14** (1933) 165.

photoelectron ejected from the central atom, by the surrounding atoms. This effect may be discussed in a very similar way to the treatment of electron scattering in the preceding section.

Let A be the central atom and B one of the surrounding atoms so that $\overrightarrow{AB} = r'\mathbf{n}'$. The absorption coefficient is proportional to the square of the magnitude of the mean dipole moment associated with the electronic transition, i.e. to $|\mathbf{M}|^2$, where

$$\mathbf{M} = e \int \psi_0(x, y, z)\mathbf{r}\psi_f^*(x, y, z)\, d\tau.$$

$\mathbf{M}$ is the dipole moment $e\mathbf{r}$ averaged over the wave functions ψ_0, ψ_f of the initial and final states. ψ_0 is the spherically symmetrical wave function for a K electron

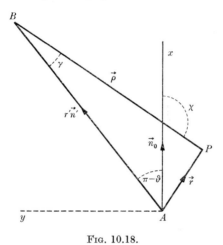

FIG. 10.18.

of atom A. If the electron is ejected in the direction of the unit vector $\mathbf{n}_0$ then, in the absence of the surrounding atoms, ψ_f has the form of a plane wave together with a scattered spherical wave. The latter arises from the distortion of the plane wave $e^{ik\mathbf{n}_0 \cdot \mathbf{r}}$ by the field of the atom A, k being as usual the wave number of the ejected electron.

When an atom such as B is present it also produces a distortion of the plane wave by introducing a spherical wave system centred round B. As viewed from A that part of this spherical wave system that travels in the direction $\overrightarrow{BA}$, i.e. of $-\mathbf{n}'$, will appear as a second incident plane wave $qe^{-ik\mathbf{n}' \cdot \mathbf{r}}$, q being an amplitude factor depending on the strength of the scattering from B. This plane wave will also be modified by the atomic field of A, so that we may write

$$\psi_f = \psi_f^a(x, y, z) + q\psi_f^b(x, y, z). \tag{47}$$

$\psi_f^a(x, y, z)$ includes the incident plane wave $e^{ik\mathbf{n}_0 \cdot \mathbf{r}}$ and the corresponding scattered wave. $q\psi_f^b(x, y, z)$ includes the apparent plane wave $qe^{-ik\mathbf{n}' \cdot \mathbf{r}}$, incident by scattering from B, and the corresponding scattered wave. $\psi_f^b(x, y, z)$ will differ from $\psi_f^a(x, y, z)$ only in that its axis of symmetry will be the vector $-\mathbf{n}'$ instead of $\mathbf{n}_0$.

We have now, dropping the superscript a,

$$\mathbf{M} = \mathbf{M}_1 + q^*\mathbf{M}_2,$$

where
$$\mathbf{M}_1 = e \int \psi_0(x,y,z)\mathbf{r}\psi_f^*(x,y,z)\,\mathrm{d}\tau,$$
$$\mathbf{M}_2 = e \int \psi_0(x,y,z)\mathbf{r}\psi_f^*(x_1,y_1,z_1)\,\mathrm{d}\tau.$$

To relate $\mathbf{M}_1$ and $\mathbf{M}_2$ choose the direction $\mathbf{n}_0$ as the x-direction and let $\mathbf{n}'$ lie in the plane of x and y. We then have (see Fig. 10.18)
$$M_{1y} = M_{1z} = 0,$$
$$x = x_1\cos\vartheta + y_1\sin\vartheta, \qquad y = x_1\sin\vartheta - y_1\cos\vartheta, \qquad z = z_1.$$

Changing the axes for the calculation of $\mathbf{M}_2$ to x_1, y_1, z_1 we have, since $\psi_0(x,y,z)$ (being spherically symmetrical) $= \psi_0(x_1,y_1,z_1)$,
$$M_{2x} = \cos\vartheta\, M_{1x}, \qquad M_{2y} = \sin\vartheta\, M_{1x}, \qquad M_{2z} = 0,$$
and
$$|\mathbf{M}|^2 = |\mathbf{M}_1|^2\{1 + (q+q^*)\cos\vartheta + |q|^2\}.$$

To complete the calculation it is necessary to sum over all surrounding atoms B and average over all orientations of the molecule relative to the direction of ejection. We then have, finally,

$$\frac{\chi_m}{\chi_a} = 1 + \tfrac{1}{2}\sum_B \int_0^\pi \{(q_B + q_B^*)\cos\vartheta + |q_B|^2\}\sin\vartheta_B\,\mathrm{d}\vartheta_B, \qquad (48)$$

where χ_m, χ_a are the respective absorption coefficients of molecule and atom.

It remains to calculate q_B. As in the preceding sections we assume that each atom scatters independently and make no allowance for multiple scattering back and forth between the different atoms. Referring to Fig. 10.18 we take B as origin so that the incident plane wave with the corresponding wave scattered by B is given, at a point $\boldsymbol{\rho}$, by
$$\psi = e^{ik\mathbf{n}_0\cdot\boldsymbol{\rho}} + \rho^{-1}e^{ik\rho}f(\chi),$$
assuming that $k\rho$ is large compared with unity. Writing now $\boldsymbol{\rho} = \mathbf{r} - \mathbf{r}'$ where $r \ll r'$ we have
$$\psi = e^{ik\mathbf{n}_0\cdot\mathbf{r}}e^{-ik\mathbf{n}_0\cdot\mathbf{r}'} + \frac{e^{ik|\mathbf{r}-\mathbf{r}'|}}{|\mathbf{r}-\mathbf{r}'|}f(\vartheta-\gamma).$$

Since $r \ll r'$ we may expand
$$|\mathbf{r} - \mathbf{r}'| \simeq r' - \mathbf{n}'\cdot\mathbf{r},$$
giving
$$\psi \simeq e^{ik\mathbf{n}_0\cdot\mathbf{r}}e^{-ik\mathbf{n}_0\cdot\mathbf{r}'} + r'^{-1}e^{ikr'}e^{-ik\mathbf{n}'\cdot\mathbf{r}}f(\vartheta)$$
$$= e^{-ik\mathbf{n}_0\cdot\mathbf{r}'}[e^{ik\mathbf{n}_0\cdot\mathbf{r}} + (r')^{-1}f(\vartheta)\exp\{ikr'(1+\cos\vartheta)\}e^{-ik\mathbf{n}'\cdot\mathbf{r}}].$$

Apart from the constant phase factor $e^{-ik\mathbf{n}_0\cdot\mathbf{r}'}$ this is of the form of two plane waves $e^{ik\mathbf{n}_0\cdot\mathbf{r}} + qe^{-ik\mathbf{n}'\cdot\mathbf{r}}$ where
$$q = (r')^{-1}\exp\{ikr'(1+\cos\vartheta)\}f(\vartheta). \qquad (49)$$

$f(\vartheta)$ may be calculated by the method of partial cross-sections (see (46)), if the field of the atom B is known.

Experimental study of the effects has been limited by the availability of volatile compounds of different elements. Further, if attention is concentrated on the K-level absorption and on X-ray wavelengths not in the vacuum region the choice is restricted virtually to Ga, Ge, As, Se, and Br as the central absorbing atom. Of the limited selection of volatile compounds available the halides are the most useful as the halogen atoms possess considerable scattering powers. Because germanium forms compounds with the most (4) halogen atoms and with the lowest boiling-points and is less dangerous to handle than arsenic, it has been the most intensively studied. However, observations have been made also for $AsCl_3$ and Br_2.

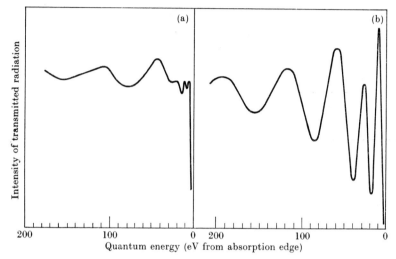

Fig. 10.19. The structure on the short wavelength side of the K absorption edge of $GeCl_4$. (a) Observed (Glaser). (b) Calculated (Hartree, Kronig, and Petersen).

For these reasons the most detailed calculations have been carried out by Hartree, Kronig, and Petersen[†] for $GeCl_4$. $f(\vartheta)$ was calculated for chlorine using the Hartree self-consistent field for that atom (see Chap. 6, § 2.2), the various phases appearing in the expression for $f(\vartheta)$ being evaluated by numerical integration of the appropriate differential equations (see Chap. 6, § 3.3). Assuming that the $GeCl_4$ molecule is tetrahedral in shape with the distance of each chlorine atom from the central atom 2·10 Å, the calculated variation of the X-ray absorption with energy excess above the K edge came out to be as shown in Fig. 10.19 (b).

The calculated structure agreed quite well with that found by the earlier observations[‡] but later work[§] at higher resolution using a double

† HARTREE, D. R., KRONIG, R. DE L., and PETERSEN, H., *Physica* 1 (1934) 895.

‡ COSTER, D. and KLAMER, G. H., ibid. 1 (1934) 889; DRYNSKI, T. and SMOLU-CHOWSKI, R., ibid. 6 (1939) 929; STEPHENSON, S. T., *Phys. Rev.* 71 (1947) 84.

§ GLASER, H., ibid. 82 (1951) 616.

crystal X-ray spectrometer showed a fine structure within 50 eV of the edge, which did not agree with that calculated. However, at shorter wavelengths the general agreement remained good as shown in Fig. 10.19. Comparison of the positions of absorption maxima and minima obtained in different observations and by calculation is given in Table 10.5. There is good agreement among all these results for ejected electron energies greater than 40 eV but at smaller energies this agreement

TABLE 10.5

Location of maxima and minima (in eV beyond the K edge) of the fine structure of the K X-ray absorption coefficient of GeCl$_4$

Structure†	Observed				Theoretical
	Glaser	Stephenson	Drynski and Smoluchowski	Coster and Klamer	Hartree, Kronig, and Petersen
A	3·85				
α	7·41				
B	9·48				2·8
β	11·8				10·5
C	14·4	16	18		19
γ	20·8				27
D	26·9				40
δ	43·8	46	48	50	50
E	77·3	74	79	86	85
ε	108	117	110	120	117
F	154	157	160	160	155
ζ			208	203	196
G			258	257	

† Absorption maxima are distinguished by *A*, *B*, *C*,..., minima by α, β, γ,... .

disappears. It is perhaps not surprising that the theory should prove inadequate at these energies as many of the neglected factors, such as multiple scattering within the molecule, are likely to be important. Also it would be surprising if the atomic scattering amplitudes for chlorine could be correctly evaluated at electron energies below 20 eV by simply calculating the scattering from the Hartree field neglecting exchange and polarization (see Chap. 8, § 7.1). A large part of the good agreement at higher energies may be due to the fact that the oscillations at these energies are largely determined by the size and shape of the molecule rather than details of atomic structure.

Snyder and Shaw† have used Kronig's theory to calculate the fine

† SNYDER, T. M. and SHAW, C. H., *Phys. Rev.* **57** (1940) 882.

structure of the X-ray absorption by gaseous bromine for which measurements are available. In this case a considerable sensitivity is found, particularly with regard to the behaviour of the η_1 phase at low electron energies. Phases derived by Frye† from Arnot's observations on electron scattering in bromine were used. Fig. 10.20 illustrates the comparison between the observed coefficient and those calculated using Frye's phases. General but not detailed agreement will be noted.

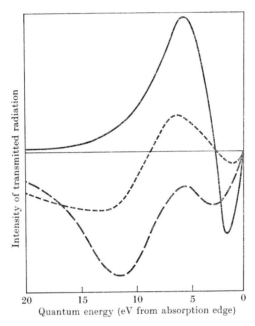

Fig. 10.20. The structure on the short wavelength side of the K absorption edge of Br_2. —— observed; — — — calculated from equations (48) and (49) using the phases derived by Frye for electron scattering by Br ; - - - - calculated using phases of Shaw and Snyder for krypton, adjusted appropriately.

3. Collisions of slow electrons with molecules

We now consider the scattering of electrons that have energies comparable with those of molecular binding and wavelengths of the same order or greater than interatomic separations. The scattering effects observed with these electrons should depend considerably on the modifications by the valence forces of the electron distributions in the various atoms within the molecule. The experimental results establish that this is so

† FRYE, W. E., loc. cit., p. 692.

in many cases, but it is very difficult to develop a satisfactory theory that would enable one to work back from the observed scattering to the charge distribution in the molecule. Some progress has been made in this direction but, as will be seen from § 3.2, a great deal more theoretical work is yet required.

3.1. *The broad features of observed total cross-sections*

Just as for atoms we distinguish between the broad features and the fine structure of the cross-sections, bearing in mind that the molecular vibration and rotation may contribute an additional source of fine structure in certain circumstances. In this section we discuss the broad features of observed total cross-sections.

Ramsauer's method (Chap. 1, § 4.1) has been applied to the determination of the total cross-sections of a number of molecules towards slow electrons.

In addition, a great deal of less precise information, essentially about momentum-loss cross-sections, is available from swarm experiments. The full analysis of data on electron transport properties obtained in this way is more complicated at low electron energies than for atoms because of the occurrence of inelastic collisions with molecules that lead to excitation of molecular vibration and rotation. Accordingly we shall defer consideration of the results of these analyses for particular molecules until Chapter 11 although reference will in some cases be made to them here.

Microwave and cyclotron resonance methods have been applied to measure momentum-loss cross-sections for molecular gases and some results obtained in this way, which refer to near thermal electrons, are discussed on p. 701.

The diatomic molecules that have been investigated by Ramsauer's method are hydrogen, oxygen, nitrogen, chlorine, carbon monoxide, nitric oxide, and hydrogen chloride.† The observed cross-section velocity

† H_2: RAMSAUER, C., *Annln Phys.* **64** (1921) 513; RUSCH, M., *Phys. Z.* **26** (1925) 748; BRODE, R. B., *Phys. Rev.* **25** (1925) 636; BRÜCHE, E., *Annln Phys.* **81** (1926) 537 and **82** (1927) 912; NORMAND, C. E., *Phys. Rev.* **35** (1930) 1217; RAMSAUER, C. and KOLLATH, R., *Annln Phys.* **4** (1930) 91; GOLDEN, D. E., BANDEL, H. W., and SALERNO, J. A., *Phys. Rev.* **146** (1966) 40.

O_2: BRÜCHE, E., *Annln Phys.* **83** (1927) 1065; RAMSAUER, C. and KOLLATH, R., ibid. **4** (1930) 91.

N_2: RAMSAUER, C., ibid. **64** (1921) 513; BRODE, R. B., *Phys. Rev.* **25** (1925) 636; BRÜCHE, E., *Annln Phys.* **81** (1926) 537 and **82** (1927) 912; NORMAND, C. E., *Phys. Rev.* **35** (1930) 1217; RAMSAUER, C. and KOLLATH, R., *Annln Phys.* **4** (1930) 91; FISK, J. B., *Phys. Rev.* **51** (1937) 25.

Cl_2: FISK, J. B., loc. cit.

CO: BRODE, R. B., ibid. **25** (1925) 636; BRÜCHE, E., *Annln Phys.* **83** (1927)

curves for these molecules are illustrated in Fig. 10.21.† (When curves have been given by more than one investigation, averaged results are given.)

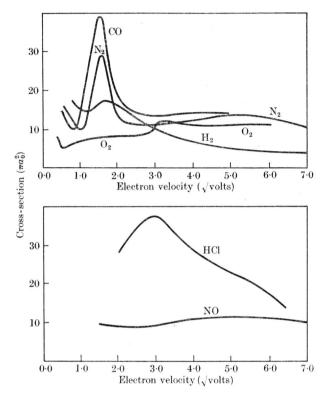

FIG. 10.21. Total collision cross-sections of diatomic molecules for slow electrons (a) H_2, N_2, O_2, CO, (b) HCl, NO.

The total cross-sections for a number of organic series of molecules have been measured in an attempt to observe correlations with the nature of the chemical binding. Thus Fig. 10.22 (a) exhibits the contrast between the behaviour of ethane, ethylene, and acetylene,‡ while Fig. 10.22 (b) shows the similarity of behaviour in the paraffin series from methane (CH_4) to butane (C_4H_{10}).§ It is also of interest to note

1065; NORMAND, C. E., *Phys. Rev.* **35** (1930) 1217; RAMSAUER, C. and KOLLATH, R., *Annln Phys.* **4** (1930) 91.
   NO: BRÜCHE, E., ibid. **83** (1927) 1065.
   HCl: BRÜCHE, E., ibid. **82** (1927) 25.
   † The curves for Cl_2 are given in Fig. 10.26 (d) on a different scale.
   ‡ BRÜCHE, E., ibid. **2** (1929) 909 and **4** (1930) 387.
   § BRÜCHE, E., ibid. **4** (1930) 387.

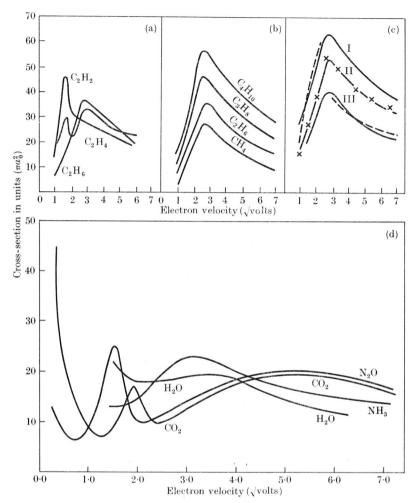

FIG. 10.22. Total collision cross-sections for slow electrons scattered by (a) C_2H_2, C_2H_4, C_2H_6; (b) molecules of the paraffin series CH_4, C_2H_6, C_3H_8, C_4H_{10}; (c) normal pentane (C_5H_{12}) curve I ———, iso-pentane (C_5H_{12}) curve I — — —, normal butane (C_4H_{10}) curve II ———, iso-butane (C_4H_{10}) curve II $\times\times\times$, methyl ether ($CH_3)_2O$ curve III ———, ethyl alcohol (C_2H_5OH) curve III — — —; (d) H_2O, NH_3, CO_2, N_2O.

the remarkable similarity in behaviour between methane† and argon. This may be seen by reference to Fig. 10.24. The detailed structure of a hydrocarbon does not seem to exert any important influence—isomers behave in very nearly the same way as may be seen in Fig. 10.22 (c),

† BRODE, R. B., *Phys. Rev.* **25** (1925) 636; BRÜCHE, E., *Annln Phys.* **83** (1927) 1065 and **4** (1930) 387; RAMSAUER, C. and KOLLATH, R., ibid. **4** (1930) 91.

where results are given for the normal and isomeric forms of butane and pentane.† Even isomers that differ very considerably in structure, such as ethyl alcohol (C_2H_5OH) and methyl ether ($(CH_3)_2O$,‡ do not seem to behave very differently towards slow electrons (see Fig. 10.22 (c)). Other groups of related organic molecules that have been investigated are the simply substituted methanes CH_3F, CH_3OH, CH_3NH compared with CH_4; $(CH_3)_3N$, $(CH_3)_3CH$,‡ and CH_3Cl, CH_2Cl_2, $CHCl_3$, CCl_4.§

Finally in Fig. 10.22 (d) results obtained for H_2O, NH_3, CO_2, and N_2O,‖ all of which, with the exception of CO_2, possess dipole moments, are illustrated. We also include in Fig. 10.23 results for the momentum-transfer cross-section Q_d at low electron energies obtained by Pack, Voshall, and Phelps†† from analysis of drift velocity measurements in these gases. These measurements were made at such low values of the ratio F/p of electric field to gas pressure that the electrons were nearly in thermal equilibrium with the gas molecules and so possessed a Maxwellian distribution of velocities about the gas temperature. Simple assumptions were then made about the variations of Q_d with electron velocity and the drift velocity u calculated in terms of the assumed Q_d using (35) of Chapter 2. This was carried out for a number of gas temperatures for which measurements of u were made and the constants in the assumed Q_d adjusted to give the best fit. It was not possible to determine Q_d unambiguously in this way because the temperature range covered by the observations was insufficient. Nevertheless the uncertainty is small. Mean momentum-transfer cross-sections derived using the cyclotron resonance technique‡‡ (Chap. 2, § 6) at 290 °K are indicated also in Fig. 10.23. They show general agreement with the drift velocity data.

It will be seen that, for all the molecules concerned, the cross-sections increase very rapidly at low electron energies. For NH_3 and N_2O this is almost certainly due to their possession of large permanent dipole moments. This cannot be the explanation for CO_2, which possesses no

† BRÜCHE, E., ibid. **5** (1930) 281.

‡ SCHMIEDER, F., Z. Elektrochem. **36** (1930) 700.

§ HOLST, W. and HOLTSMARK, J., K. norske Vidensk. Selsk. Mus. Oldsaksaml. Tilv. **4** (1931) 89.

‖ H_2O and NH_3: BRÜCHE, E., Annln Phys. **1** (1929) 93.
    CO_2: BRÜCHE, E., ibid. **83** (1927) 1065; RAMSAUER, C. and KOLLATH, R., ibid. **4** (1930) 91.
    N_2O: BRÜCHE, E., ibid. **83** (1927) 1065; RAMSAUER, C. and KOLLATH, R., ibid. **7** (1930) 176.

†† PACK, J. L., VOSHALL, R. E., and PHELPS, A. V., Phys. Rev. **127** (1962) 2084.

‡‡ BAYES, K. D., KIVELSON, D., and WONG, S. C., J. chem. Phys. **37** (1962) 1217.

such moment, but in this case it may be due to the large permanent quadrupole moment of the molecule.

In general the behaviour of the different molecules presents a range of variation in character and magnitude similar to that for atoms.

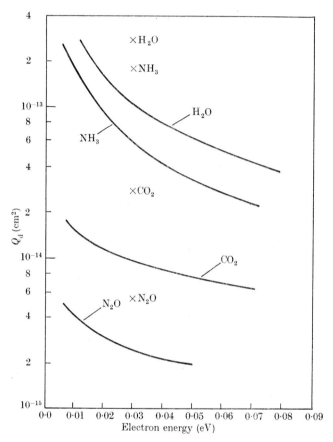

FIG. 10.23. Momentum-transfer cross-sections for slow electrons scattered in H_2O, NH_3, CO_2, N_2O derived from analysis of drift velocity data. × mean cross-sections measured by the cyclotron resonance method.

Before discussing the extent to which the details can be interpreted theoretically we shall describe the observations which have been carried out on the angular distribution of electrons scattered elastically by molecules.

The methods described in Chapter 5, § 5 for observing the angular distribution of slow electrons scattered elastically by atoms have also been applied to a number of molecular gases—the diatomic molecules

H_2, N_2, CO, Br_2, and I_2, the hydrocarbons CH_4, C_2H_6, C_2H_4, and C_2H_2, the inorganic hydrides PH_3 and H_2S, carbon dioxide, the carbon tetra-halides, CF_4, CCl_4, and CBr_4, and also CH_2Br_2.† Of these, the results for PH_3 and H_2S have already been described in Chapter 6, § 4.6 as being roughly representative of those expected for the P and S atoms respectively, and those for CF_4, CCl_4, and CBr_4 have been discussed in § 2.1 of this chapter.

The angular distribution curves for hydrogen resemble those for helium in exhibiting no marked maxima and minima. Those for nitrogen and carbon monoxide reveal very close similarity at all electron energies.

An interesting feature of the results for methane is noticed on com-parison with argon. For electrons with energies greater than 20 eV there is no resemblance, but for slower electrons there is quite a dis-tinct similarity between the two sets of angular-distribution curves that becomes more marked as the electron energy decreases (see Fig. 10.24).

3.2. *Theory of the broad features of the elastic scattering of slow electrons by molecules—semi-empirical treatment*

In order to obtain a theory of the elastic collisions of slow electrons with molecules as satisfactory as that for atoms it would be necessary to extend the method of partial cross-sections (Chap. 6, § 3) to scatter-ing by fields that no longer possess spherical symmetry. Even if this could be done the theory would be hampered by the greater ignorance of molecular as compared with atomic fields. Despite these difficulties

---

† H_2: BULLARD, E. C. and MASSEY, H. S. W., *Proc. R. Soc.* A**133** (1931) 637; ARNOT, F. L., ibid. 615; RAMSAUER, C. and KOLLATH, R., *Annln Phys.* **12** (1932) 837; MOHR, C. B. O. and NICOLL, F. H., *Proc. R. Soc.* A**138** (1932) 469; HUGHES, A. L. and McMILLEN J. H., *Phys. Rev.* **41** (1932) 39; WEBB, G. M., ibid. **47** (1935) 384.

N_2: BULLARD, E. C. and MASSEY, H. S. W., loc. cit.; ARNOT, F. L., loc. cit.; MOHR, C. B. O. and NICOLL, F. H., loc. cit.; EHRHARDT, H. and WILLMANN, K., *Z. Phys.* **204** (1967) 462.

CO: ARNOT, F. L., loc. cit.; RAMSAUER, C. and KOLLATH, R., *Annln Phys.* **12** (1932) 529 and *Phys. Z.* **32** (1931) 867.

Br_2, I_2, and CH_2Br_2: ARNOT, F. L., *Proc. R. Soc.* A**144** (1934) 360.

CH_4: BULLARD, E. C. and MASSEY, H. S. W., loc. cit.; ARNOT, F. L., *Proc. R. Soc.* A**133** (1931) 615; MOHR, C. B. O. and NICOLL, F. H., loc. cit.; HUGHES, A. L. and McMILLEN, J. H., *Phys. Rev.* **44** (1933) 876.

C_2H_6: CHILDS, E. C. and WOODCOCK, A. H., *Proc. R. Soc.* A**146** (1934) 199.

C_2H_4 and C_2H_2: CHILDS, E. C. and WOODCOCK, A. H., loc. cit.; HUGHES, A. L. and McMILLEN, J. H., *Phys. Rev.* **44** (1933) 876.

PH_3 and H_2S: MOHR, C. B. O. and NICOLL, F. H., loc. cit.

CO_2: ARNOT, F. L., *Proc. R. Soc.* A**133** (1931) 615; MOHR, C. B. O. and NICOLL, F. H., loc. cit.

CF_4, CCl_4, and CBr_4: see references, p. 690–1.

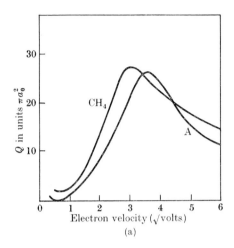

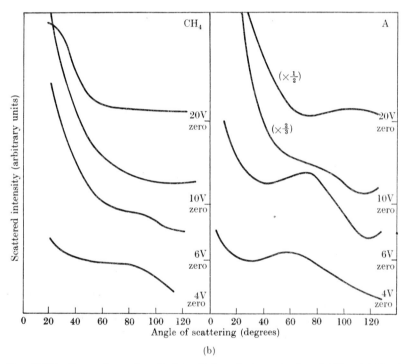

Fig. 10.24. Comparison of scattering of slow electrons in A and CH_4. (a) Total collision cross-section (Ramsauer). (b) Angular distribution.

it has, however, proved possible to increase the scope of the theory to a sufficient extent to obtain some interesting results for certain molecules.

Whereas the Schrödinger equation for the motion of electrons in a spherically symmetrical field of force can always be solved in principle, the corresponding problem for an axially symmetrical field is, in general, intractable even with present-day computers. However, by using spheroidal coordinates it is possible to obtain solutions for certain forms of axially symmetrical field that are likely to be rather similar to the actual fields of diatomic molecules. Stier† and, later, Fisk‡ have taken advantage of this to develop a theory of the elastic scattering of slow electrons by diatomic molecules that is a direct generalization of the method of partial cross-sections (see Chap. 6, § 3) which deals so successfully with the spherically symmetrical fields of atoms.

The total angular momentum of the incident electron about the centre of the molecule is no longer a constant of the motion. On the other hand, the component of the angular momentum in the direction of the nuclear axis is constant and is therefore quantized, the allowed values being $m\hbar$ ($m = 0, 1, 2,...$). The incident wave may therefore be resolved into partial waves for which $m = 0, 1, 2,...$. With the particular fields for which the Schrödinger equation is separable in spheroidal coordinates a further resolution may be made for a given value of m, as follows. When the two foci of the spheroidal coordinate system (the two nuclei) are allowed to come together (the united atom limit) the resolution in terms of the total angular momentum quantum number $l+m$ is again possible as the system is once more spherically symmetrical. In the spheroidal case a partial wave denoted by m, l is one in which the axial angular momentum is $m\hbar$ and the total angular momentum in the united atom limit is $\{(l+m)(l+m+1)\}^{\frac{1}{2}}\hbar$. For each such partial wave a phase-shift η_{lm} is introduced by the scattering field. The total elastic scattering cross-section Q_0, averaged over all orientations of the molecular axis, becomes

$$Q_0 = \sum_{m,l} q_{ml},$$

where

$$q_{ml} = \begin{cases} \dfrac{2\pi}{k^2} \sin^2\eta_{ml} & (m = 0), \\[2ex] \dfrac{4\pi}{k^2} \sin^2\eta_{ml} & (m \neq 0), \end{cases}$$

† STIER, H. C., Z. Phys. 76 (1932) 439.
‡ FISK, J. B., Phys. Rev. 49 (1936) 167.

and k is as usual equal to $mv/\hbar$, where v is the electron velocity. Associated with each partial wave is an angular distribution function. This depends, for a fixed orientation of the molecular axis, on the angles specifying the directions of the incident and scattered electrons with respect to the molecular axis. If θ is the angle between the direction of the scattered electron and the molecular axis and ϕ an azimuthal angle specifying the plane containing the molecular axis and the direction of scattering while ω, α are similar angles defining the direction of incidence, then under these conditions the scattered intensity $I(\vartheta)$ is found to be†

$$I(\vartheta) = k^{-2} \left| \sum_{m,l} (e^{2i\eta_{ml}} - 1) S_{ml}(kd, \cos\theta) S_{ml}(kd, \cos\omega) \cos m(\phi - \alpha) \right|^2, \quad (50)$$

with $\qquad\qquad \cos\vartheta = \cos\theta\cos\omega + \sin\theta\sin\omega\cos(\phi - \alpha).$

The first few functions $S_{ml}(c, \cos\theta)$ for $m, l = 0, 0; 1, 0;$ and $0, 1$ for a series of typical values of c are shown in Fig. 10.25.

It will be noticed that, in contrast to the scattering by a spherically symmetrical field, S_{ml}, and thence the angular distribution of the scattered electrons, depends on the energy of the electrons through kd, where d is the nuclear separation.

As pointed out above the expression (50) refers to a fixed molecular orientation and accordingly must be averaged over all possible orientations.

As $k \to 0$, i.e. for slow electrons, all the partial cross-sections tend to zero except q_{00}, which tends to a finite value. The corresponding angular function S_{00} also tends to a constant value as $k \to 0$, so the averaged angular distribution tends to become spherically symmetrical for sufficiently slow electrons just as for atoms.

In carrying out detailed calculations, involving determination of the phases η_{ml}, it is necessary to confine oneself to axially symmetrical scattering potentials that permit the separation of the spheroidal co-ordinates in the wave equation. It has, nevertheless, been possible to choose forms that provide a satisfactory representation of the observed scattering by many molecules. In the first detailed study on these lines Stier‡ used for the potential of the molecular field

$$V = -4Ze^2 d^{-1} \rho f(\rho)/(\rho^2 - \mu^2), \quad (51)$$

where $\qquad\qquad \rho = (r+p)/d, \qquad \mu = (r-p)/d. \quad (52)$

---

† MOTT, N. F. and MASSEY, H. S. W., *The theory of atomic collisions*, 3rd edn, p. 586 (Clarendon Press, Oxford, 1965).
  ‡ loc. cit.

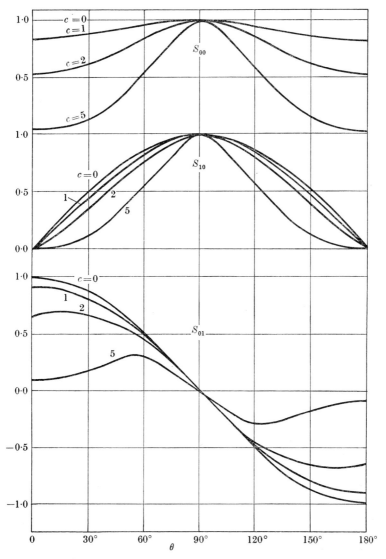

FIG. 10.25. The functions $S_{ml}(c, \cos\theta)$ for $m, l = 0, 0$; 1, 0; 0, 1; and $c = 0$, 1, 2, 5.

Z is an effective nuclear charge and r, p refer to the distances from the centres of the two atoms in the molecule. $f(\rho)$ was taken to be of the form

$$f(\rho) = \begin{cases} (\rho - \rho_0)^2/(\rho_0 - 1)^2 & (\rho < \rho_0), \\ 0 & (\rho > \rho_0). \end{cases}$$

Two parameters Z and ρ_0 are thus available for adjustment of theory to agree with observation. Choice of the values $\rho_0 = 3\cdot46$, $Z = 4\cdot08$ gave good agreement with observed cross-section curves for electrons of energy less than 10 eV.

The most extensive calculations have been carried out by Fisk,† who was guided by the method of Morse and Allis for dealing with scattering by atoms (see Chap. 6, § 4.5). He also took V in the form (51), but with

$$f(\rho) = 1 - \frac{\rho_0(\rho-1)^2}{\rho(\rho_0-1)^2},$$

and studied the variation of η_{ml}, and hence of the q_{ml}, with two parameters β and x exactly analogous to those introduced by Morse and Allis for atoms. Thus $\beta^2 = \frac{1}{4}Zd\rho_0$ and $x = \rho_0 kd$. Throughout the work ρ_0 was taken to be 2, as it was found that the results were rather insensitive to values of ρ_0 between 1·75 and 3·0. While x is proportional to the electron velocity, β depends only on the molecular field. For a fixed x the partial cross-sections exhibit a periodic behaviour with β.

The method was applied to hydrogen, nitrogen, oxygen, and chlorine. For hydrogen only the 0, 0 partial wave is affected if the electron energy is below 10 eV and good agreement is obtained with the observed total cross-section if β is taken to be 0·6 (see Fig. 10.26 (a)). The results for nitrogen and oxygen are of considerable interest in revealing the power and limitations of the method. Fig. 10.26 (b) and (c) illustrates the good agreement obtained and the contributions from the partial cross-sections. The small increase in β from 1·32 to 1·35 in going from nitrogen to oxygen has a very pronounced effect in removing the sharp maximum for 2·25 eV electrons in nitrogen which is due to the 1, 0 partial wave. The angular distribution at this maximum derived from averaging

$$\{S_{10}(kd, \cos\theta)S_{10}(kd, \cos\omega)\cos(\phi-\alpha)\}^2$$

is compared with the observations in Fig. 10.27. The form of the observed distributions perhaps approximates more closely to that derived from the function S_{01} (Fig. 10.25), and it is of interest to notice that Stier, with his representation of the nitrogen field, did in fact interpret the sharp maximum for 2·25-eV electrons as arising from the 0, 1 partial cross-section and not the 1, 0 as did Fisk. Stier's theoretical cross-section is compared with observation in Fig. 10.26, and the good agreement of his calculated angular distribution with the observed may be seen from Fig. 10.27. By reducing the value of β from 1·3 to about 1·0 in Fisk's nitrogen field a similar analysis to Stier's could be obtained, the 0, 1 partial cross-section becoming the most important

† loc. cit., p. 705.

one. The sensitivity in going from nitrogen to oxygen would presumably remain at this lower value of β.

It is of interest to compare the molecular fields of nitrogen used by Stier and by Fisk with that derived by Hund† by an approximate statistical method. Such a comparison is exhibited in Fig. 10.28.

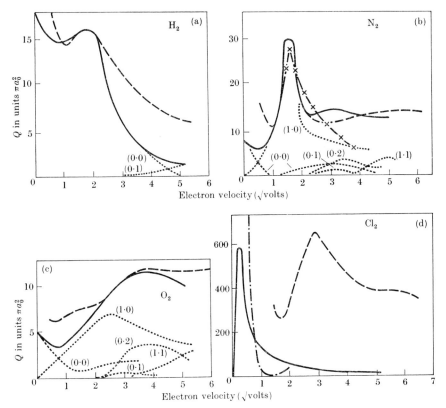

Fig. 10.26. Comparison with experiment of Fisk's calculations of the total collision cross-section of slow electrons in (a) H_2, (b) N_2, (c) O_2, (d) Cl_2. — — — experimental values. ——— Fisk's calculated curve. The dotted lines represent the contribution from the various partial cross-sections. —×—×— Stier's calculated curve for N_2. —·—·— momentum-transfer cross-section for Cl_2 derived from analysis of the swarm data of Bailey and Healey.

The agreement obtainable by Fisk's theory in the above cases does not appear when applied to chlorine—the observed total cross-section is very much greater than the theoretical over the range of electron energies investigated (see Fig. 10.26 (d)).

We also include for comparison the momentum-transfer cross-section

† HUND, F., *Z. Phys.* **77** (1932) 12.

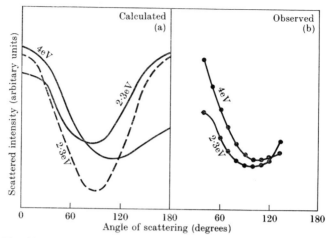

FIG. 10.27. Comparison of observed and calculated angular distributions of slow electrons scattered in N_2.
(a) calculated : —— Stier (2·3 and 4 eV); — — — Fisk (2·3 eV).
(b) observed : —·—·— (2·3 and 4 eV)

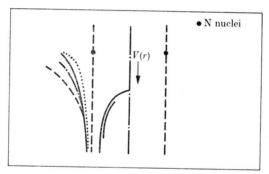

FIG. 10.28. Variation along the nuclear axis of the molecular
field of N_2 used by Stier and by Fisk compared with that calculated by Hund using a statistical method. Between the nuclei
the Stier and Fisk fields coincide with the Coulomb field.
—— Fisk ; ···· Stier ; —·—·— Hund ; — — — Coulomb.

derived from the swarm experiments of Bailey and Healey,† which is
inconsistent with both Fisk's semi-empirical and observed cross-
sections. It is hard to say which if any of the three results are correct
and more work is clearly needed.

3.3. *Calculations for hydrogen based on the actual molecular structure*

Molecular hydrogen is the only molecule of sufficient simplicity to
make possible calculations of cross-sections for elastic scattering of
slow electrons based on the theoretical electron structure of the molecule.

† BAILEY, V. A. and HEALEY, R. H., *Phil. Mag.* **19** (1935) 725.

Nagahara† calculated the mean interaction energy $V(\rho, \mu, \phi)$ of an electron with the molecule, in terms of spheroidal coordinates as defined by (52), by taking for the molecular wave function‡

$$\psi(\mathbf{r}_1, \mathbf{r}_2) = N e^{-\alpha(\rho_1+\rho_2)} \cosh\{\beta(\mu_1-\mu_2)\},$$

with $\alpha = 1\cdot 75$, $\beta = 1\cdot 375$, $R = 1\cdot 42$ Å. Exchange and molecular distortion effects were neglected so that the wave equation describing the motion of the electron in the molecular field was taken to be

$$\nabla^2 F + \{k^2 - 2mV(\rho, \mu, \phi)/\hbar^2\}F = 0.$$

Although this equation is not separable in spheroidal coordinates Nagahara calculated the scattering by expanding the solution in spheroidal harmonics. This leads to a series of coupled differential equations which he solved by successive approximation. The results he obtained are shown in Fig. 10.29.

Massey and Ridley§ used the variational method described in Chapter 6, § 3.8 and were able to take account of electron exchange. They chose as trial function, when exchange is included, a properly antisymmetrical linear combination of functions of the form

$$\psi(\mathbf{r}_1, \mathbf{r}_2)\, F(\mathbf{r}_3),$$

where the molecular wave function $\psi(\mathbf{r}_1, \mathbf{r}_2)$ was taken to be of the self-consistent form given by Coulson‖ and, in terms of spherical polar coordinates,

$$F(\mathbf{r}) = (1+a^2)^{-\frac{1}{2}}\{c(\rho-1)\}^{-1}[\sin\{c(\rho-1)\}+$$
$$+\{a+be^{-\gamma(\rho-1)}(1-e^{-\gamma(\rho-1)})\}\cos\{c(\rho-1)\}],$$

with $c = \frac{1}{2}kR$ where R is the nuclear separation. The parameters a and b were determined from the usual variational formula. The phase η_{00} in the notation of § 3.2 is then given by

$$\eta_{00} = (\arctan a) - c.$$

Their results, with and without inclusion of exchange, are also shown in Fig. 10.29.

Finally, Hara†† has attempted to include dipole distortion of the molecule during the impact. To do this the interaction of an electron with an undistorted molecule was calculated using the Wang wave function (44 a). This interaction was then averaged over all molecular

† NAGAHARA, S., *J. phys. Soc. Japan* **9** (1954) 52.
‡ IMRIE, T., *Proc. phys.-math. Soc. Japan* **20** (1938) 790.
§ MASSEY, H. S. W. and RIDLEY, R. O., *Proc. phys. Soc.* A**69** (1956) 659.
‖ COULSON, C. A., *Proc. Camb. phil. Soc. math. phys. Sci.* **34** (1938) 204.
†† HARA, S., *J. phys. Sec. Japan* **22** (1967) 710.

orientations. To this was added a similarly averaged dipole distortion potential

$$\frac{-\tfrac{1}{2}\alpha e^2}{(r^2+r_0^2)^2},$$

where α is the mean polarizability of the molecule and r_0 is an adjustable cut-off parameter. Electron exchange was taken into account by including a further central potential derived in a similar way to that suggested

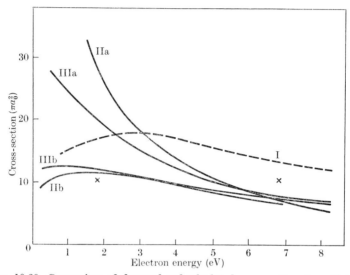

FIG. 10.29. Comparison of observed and calculated cross-sections for collisions of electrons with H_2 molecules. I observed Golden, Bandel, and Salerno; × calculated by Nagahara; IIa calculated by Massey and Ridley—exchange ignored; IIb calculated by Massey and Ridley—exchange included; IIIa calculated by Hara—exchange ignored; IIIb calculated by Hara—exchange included.

by Slater for atomic problems. In this way the problem was reduced to one of calculating the scattering by a centre of force. Results obtained by Hara with and without allowance for exchange are also shown in Fig. 10.29.

Reference to Fig. 10.29, which includes for comparison the observed results of Golden, Bandel, and Salerno,† shows that the inclusion of electron exchange is important. On the whole the agreement with the two calculations which include exchange is encouraging.

3.4. Scattering by highly symmetrical molecules

One further type of molecular scattering problem is amenable to approximate theoretical treatment. This arises when the molecule

† GOLDEN, D. E., BANDEL, H. W., and SALERNO, J. A., *Phys. Rev.* **146** (1966) 40.

possesses such a high degree of symmetry that its scattering field is nearly spherically symmetrical. An interesting illustration is provided by methane. The remarkable resemblance between the observed behaviour of methane and of argon towards slow electrons has already been stressed in § 3.1. Close similarity between the mean field outside the molecule and that of the atom at these distances must be responsible. This can only be represented by taking into account the highly symmetrical average field in which the outer electrons move in methane. Buckingham, Massey, and Tibbs[†] have calculated a spherically symmetrical average field for the methane molecule by first averaging the field due to the protons over a sphere and then applying the usual self-consistent field method of Hartree. They calculated the phases for scattering of electrons by this spherically symmetrical field, in the usual way (Chap. 6, § 3), and found that close similarity in behaviour to argon would indeed be expected for electrons with energies less than about 20 eV. For such electrons the calculated first- and second-order phases are about the same for both, while the zero-order phase for argon exceeds that for methane by π. As these are the only important phases the scattering effects should be much the same for both. The fact that the zero-order phase for methane is less by π than that for argon means that methane can be regarded as filling the place in the series xenon, krypton, and argon that one might have thought would be occupied by neon (see Chap. 6, §§ 4.2, 4.3). Thus the occurrence of the Ramsauer–Townsend effect in methane arises because the molecular field is just strong enough to introduce exactly two extra half-waves into the low-energy partial waves with zero angular momentum, whereas the argon field introduces exactly three, krypton four, and xenon five.

It is probable that much interesting information about the outer fields of molecules could be derived from a systematic study of the elastic scattering of slow electrons in conjunction with approximate theories, but a much wider range of molecular types would have to be investigated than have been examined up to the time of writing.

3.5. *Fine structure in total and elastic cross-sections of molecules*

The problem of distinguishing between broad features and fine structure in cross-sections for scattering of electrons by molecules is complicated by the possibility of vibrational and rotational excitation in which the energies involved are only a fraction of an eV. We shall,

[†] BUCKINGHAM, R. A., MASSEY, H. S. W., and TIBBS, S., *Proc. R. Soc.* **A178** (1941) 119.

in this section, confine ourselves to the description of experimental results in total and elastic cross-sections in which fine structure is observed, without attempting to interpret the results in any but a very general manner. More detailed theoretical analyses of these and other data are given in Chapter 12, § 6 and Chapter 13, §§ 1.6 and 3.7.

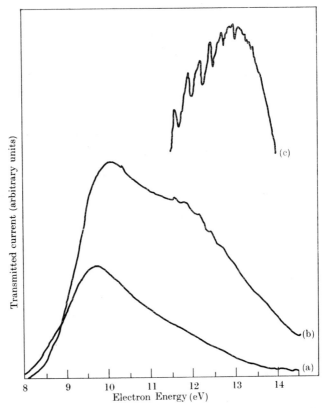

Fig. 10.30. Variation of transmitted current of electrons through hydrogen as a function of electron energy. Curve (a) for an evacuated chamber. Curve (b) with 0·04 torr cm of H_2 in the scattering chamber and current sensitivity increased 2000 times. Curve (c) as for curve (b) but with greatly increased sensitivity.

3.5.1. *Hydrogen.* Kuyatt, Mielczarek, and Simpson,† using the apparatus described in Chapter 1, § 5.1 have investigated the structure of the variations in the intensity of transmission of electrons through H_2 as a function of energy in the energy range 8–14·5 eV. Their results are illustrated in Fig. 10.30. It is to be remembered (see Chap. 1, § 5.1) that, in their experiments, attention was directed to the observation of

† Kuyatt, C. E., Mielczarek, S. R., and Simpson, J. A., *Phys. Rev. Lett.* **12** (1964) 293.

fine structure rather than the broad features so no correction has been made for the distortion due to the electron optical focusing effect† associated with electron voltage-scanning.

It will be seen from Fig. 10.30 (b) that fine structure appears in the region between 11·5 and 14 eV. A plot with a greatly increased current sensitivity shows the details in the form of a number of narrow peaks, the first appearing 4·28±0·1 eV below the appearance energy for positive ions. Table 10.6 gives the location of the eight clearly resolved

TABLE 10.6

Fine structure of the total cross-section for electron scattering by H_2

Energy (eV)	Spacing (eV)	Vibnl. spacing (eV) H_2 (C state)
11·62		
11·91	0·29	0·285
12·18	0·27	0·268
12·46	0·28	0·268
12·71	0·25	0·238
12·94	0·23	0·220
13·13	0·19	0·205
13·31	0·18	0·188

transmission maxima. The spacing between the peaks is quite close to that between the vibrational levels of the C state of H_2 (see Chap. 13, § 1.3.1), which may be related to the likelihood that the fine structure arises from resonant capture of electrons to produce different vibrational states of an H_2^- ion based on the C state of the neutral molecule. Further discussion of this point is given in Chapter 12, § 6.

3.5.2. *Nitrogen.* The fine structure in the transmission of electrons through nitrogen has been investigated at energies below 5 eV by Heideman, Kuyatt, and Chamberlain‡ who used the apparatus described in Chapter 1, § 5.1, by Golden and Nakano§ using the apparatus described in Chapter 1, § 4.1, and by Boness and Hasted.‖ The latter authors used a relatively large 127° electrostatic analyser of mean radius 3·18 cm with slits of dimensions 0·25×0·5 cm, to produce an incident electron beam with an energy spread of the order 0·05 eV. This beam was directed into a gas collision chamber so designed that only those electrons that were scattered through less than 1° emerged from the

† See footnote p. 35.
‡ HEIDEMAN, H. G. M., KUYATT, C. E., and CHAMBERLAIN, G. E., *J. chem. Phys.* **44** (1966) 355.
§ GOLDEN, D. E. and NAKANO, H., *Phys. Rev.* **144** (1966) 71.
‖ BONESS, M. J. W. and HASTED, J. B., *Phys. Lett.* **21** (1966) 526.

chamber to be collected. The gas pressure was adjusted to produce 20 per cent absorption.

Ehrhardt and Willmann† have also observed the transmission of electrons through nitrogen in the course of a much more extensive investigation primarily concerned with the observation of the dependence of resonance effects, both in elastic and inelastic scattering, on the

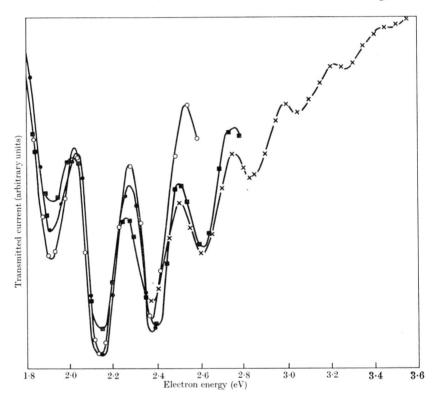

FIG. 10.31. Variation with electron energy of the transmission of electrons through N_2 corrected for electron optical effects, as observed by Boness and Hasted. The three curves with experimental points indicated by ○, ■, and × respectively were obtained on different occasions.

angle of scattering. An electron beam, rendered homogeneous in energy to 0·05 eV by passage through a 127° cylindrical analyser, was scattered from a molecular beam, and the scattered electrons velocity-analysed by a second 127° analyser that could be rotated so as to observe electrons in the range 0–110°. To minimize background effects due to electrons reflected from metal surfaces the energy of the incident electrons was modulated over a range of a few tenths of an eV.

† EHRHARDT, H. and WILLMANN, K., Z. Phys. 204 (1967) 462.

Boness and Hasted† made corrections for the electron-optical focusing effect in their experiment to obtain the true variation of the transmission with electron energy shown in Fig. 10.31. The broad variation corresponds to the peak in the total cross-section at about 2·5 eV shown in Fig. 10.21, while the separations of the peaks in the fine structure as listed in Table 10.7 are practically the same as between the first six vibrational states of the molecules NO and O_2^+ in their ground $^2\Pi$ states. As these molecules are iso-electronic with N_2^- there is support here for the interpretation of the structure as due to the formation, by resonant electron capture, of subsequently autodetaching N_2^- molecules in successive vibrational states. As, however, the first excited $(A\ ^3\Sigma_u^+)$ state of N_2 lies 6 eV above the ground state the situation is different from that with the low autodetaching states of negative atomic ions that lie much closer to the first excited atomic state. We defer further discussion of this important question until Chapter 12, § 6 and Chapter 13, § 3.7.

In earlier work Schulz and Koons‡ observed the variation of intensity of elastic scattering by nitrogen at 72°.

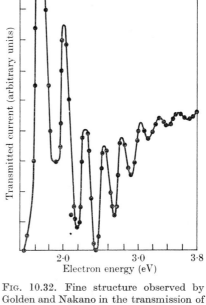

Fig. 10.32. Fine structure observed by Golden and Nakano in the transmission of electrons through N_2. The points are obtained from a number of plots of the transmitted current. Because of electron optical effects no significance attaches to the relative magnitudes of peaks and troughs.

Fig. 10.32 illustrates the results obtained by Golden and Nakano. It will be seen from Table 10.7, which lists the energies of the maxima and minima found by the different observers in transmission experiments, that there is good agreement as to the details. The structure observed by Schulz and Koons in the elastic scattering at 72° is also included and agrees well with the other data.

† loc. cit., p. 715.
‡ SCHULZ, G. J. and KOONS, H. C., *Westinghouse Research Report* 65–9E3–113–P2 (1965).

TABLE 10.7

Positions of maxima and minima in fine structure observed in electron collision cross-sections in nitrogen

Ehrhardt and Willmann		Boness and Hasted (corrected data)				Heideman et al.				Schulz and Koons				Golden and Nakano			
Peaks		Peaks		Troughs		Peaks		Troughs		Peaks		Troughs		Peaks		Troughs	
Posn. (eV)	Sepn. (eV)	Posn. (eV)	Sepn. (eV)	Posn. (eV)	Sepn. (eV)	Posn. (eV)	Sepn. (eV)	Posn. (eV)	Sepn. (eV)	Posn. (eV)	Sepn. (eV)	Posn. (eV)	Sepn. (eV)	Posn. (eV)	Sepn. (eV)	Posn. (eV)	Sepn. (eV)
1·89	0·26	1·60	0·44	1·92	0·23	1·66	0·46	1·98	0·25	1·92	0·33	1·78	0·30	1·71	0·32	1·90	0·27
2·15	0·25	2·03	0·24	2·14	0·24	2·12	0·25	2·23	0·23	2·25	0·23	2·08	0·27	2·03	0·27	2·17	0·26
2·40	0·25	2·27	0·25	2·38	0·22	2·37	0·22	2·46	0·23	2·48	0·27	2·35	0·27	2·30	0·26	2·43	0·26
2·65	0·24	2·52	0·25	2·61	0·22	2·59	0·23	2·69	0·23	2·75	0·23	2·62	0·27	2·56	0·24	2·69	0·22
2·89	0·24	2·77	0·24	2·83	0·21	2·82	0·24	2·92	0·23	2·98		2·89	0·20	2·80	0·26	2·91	0·24
3·13	0·23	3·01	0·22	3·04	0·22	3·05	0·23	3·16	0·24			3·09		3·06	0·24	3·15	0·25
3·36	0·22	3·23	0·22	3·26	0·21	3·28	0·23	3·40	0·24					3·30	0·24	3·40	
3·58		3·45		3·47		3·51	0·20	3·61	0·21					3·54			
						3·71											

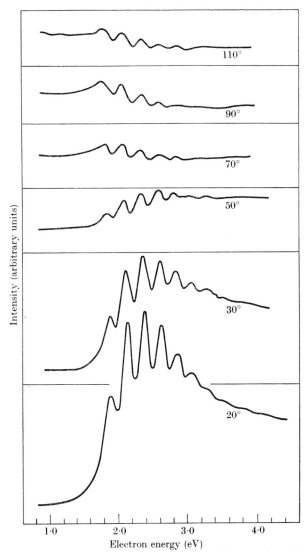

FIG. 10.33. Variation of the intensity of elastic scattering with incident electron energy in nitrogen at a number of fixed scattering angles, as observed by Ehrhardt and Willmann.

The results obtained by Ehrhardt and Willmann agree very well with others given in Table 10.7 except that they found no evidence of resonance structure below 1·8 eV either in the transmission experiments or in the measurements at a fixed scattering angle. Fig. 10.33 shows the structure that they observed for a number of scattering angles ranging from 20 to 110°.

Fig. 10.34 shows the angular distribution of the elastic scattering measured at a number of different energies over the resonance region. These distributions are readily explicable in terms of the semi-empirical

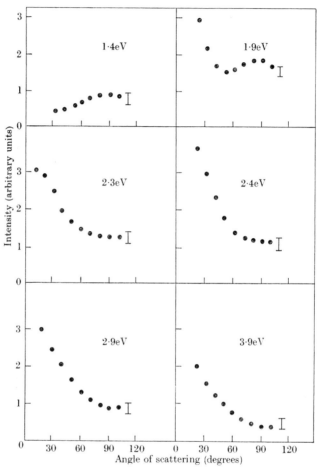

FIG. 10.34. Angular distribution of electrons of different incident electron energies scattered elastically in nitrogen as observed by Ehrhardt and Willmann.

two-centre potential scattering model discussed in § 3.2 (see Fig. 10.27) and support the idea that the major part of the elastic scattering does not arise from resonance effects.

A resonance effect much more similar to that found for certain atoms was observed by Heideman et al.† in the energy range 11–12·5 eV. Fig. 10.35 illustrates a typical transmission-voltage plot obtained. A large

† loc. cit., p. 715.

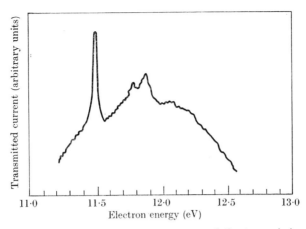

Fig. 10.35. Variation with electron energy of the transmission of electrons through N_2 as observed by Heideman, Kuyatt, and Chamberlain in the energy range between 11·2 and 12·6 eV.

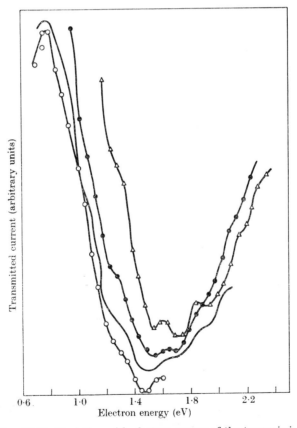

Fig. 10.36. Variation with electron energy of the transmission of electrons through CO corrected for electron optical effects, as observed by Boness and Hasted.

TABLE 10.8

Location in energy of peaks and troughs observed by Boness and Hasted in the transmission of electrons through CO, O_2, and NO

CO Peaks (eV)	1·26	1·33	1·62	1·85	2·12	2·31			
Troughs (eV)		1·54	1·74	1·95	2·18	2·37			
O_2 Peaks (eV)	0·39	0·51	0·62	0·73	0·83	0·93	1·04		
Troughs (eV)	0·32	0·43	0·55	0·66	0·76	0·88	0·99	1·08	
NO Peaks (eV)	0·60	0·70	0·83	0·99	1·15	1·34	1·51	1·67	1·83
Troughs (eV)	0·57	0·76	0·92	1·07	1·23	1·39	1·57	1·75	

and narrow resonance peak is clearly seen at 11·48 eV very similar to the principal resonance in helium at 19·3 eV (see Fig. 1.18). The energy scale was calibrated against this helium resonance and argon resonances. Because it is so narrow the nitrogen resonance is likely to be useful for

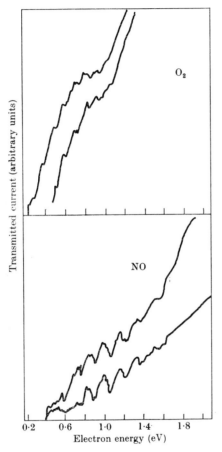

FIG. 10.37. Typical transmission–voltage plots for O_2 and NO as obtained by Boness and Hasted, no correction for electron-optical focusing being included.

calibration, particularly as its energy lies below the ionization energy of most gases.

The bump in the plot shown in Fig. 10.35 close to 11·9 eV is probably associated with excitation of the E state of N_2 for which the spectroscopic threshold is 11·87 eV. The sharp resonance, occurring at about 0·4 eV lower energy, would then be associated with an excited neutral molecular state in much the same way as in atoms. Further fine structure near

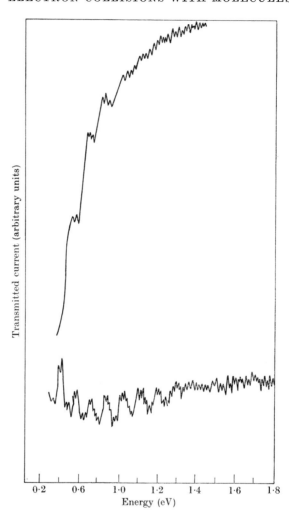

FIG. 10.38. Fine structure in the scattering of electrons by nitric oxide as observed by Ehrhardt and Willmann. Upper curve—intensity of transmitted electron current as a function of incident electron energy. Lower curve—intensity of elastic scattering at 20° as a function of incident electron energy.

11·75 eV would also appear to be a resonance effect involving a higher resonance state associated with the E state.

3.5.3. *Carbon monoxide.* Fig. 10.36 shows the transmission for carbon monoxide as a function of electron energy after correction for electron optical focusing, observed by Boness and Hasted.† The broad features are similar to those for nitrogen (see Fig. 10.32) and correspond

† loc. cit., p. 715.

to the peak in the total cross-section at about 2·5 eV shown in Fig. 10.21. On the other hand, the fine structure is much less clear than in nitrogen. This may well be due to the resonance level widths being greater than the vibrational level spacing so that overlap occurs between the resonances. The locations of the peaks and troughs are given in Table 10.8.

3.5.4. *Oxygen and nitric oxide.* Fig. 10.37 shows typical transmission–voltage plots for these two gases as obtained by Boness and Hasted,† no correction for electron-optical focusing being included. The observed locations of the resonances are as given in Table 10.8.

Ehrhardt and Willmann‡ have observed both the transmission through nitric oxide and the elastic scattering at 20° using the same equipment as in their observations for nitrogen (§ 3.5.2). Fig. 10.38 illustrates their results which show fine-structure spacing agreeing quite well with that observed by Boness and Hasted.

† loc. cit., p. 715.   ‡ loc. cit., p. 716.

COLLISIONS OF ELECTRONS WITH MOLECULES—EXCITATION OF VIBRATION AND ROTATION—ANALYSIS OF SWARM EXPERIMENTS AT LOW ENERGIES

In Chapter 10 we discussed the results of experiments on the elastic scattering of slow electrons by molecules that have been obtained by the use of electron beams of homogeneous energy. To derive information about elastic scattering and particularly about momentum-loss cross-sections from the analysis of experiments involving electron swarms possessing a distribution of energy, as in Chapter 2, § 7, we must take account of the fact that even very slow electrons passing through molecular gases may exchange energy with molecular vibration and rotation. The discussion of the cross-sections for collisions of this kind is of considerable interest in its own right so in this chapter we discuss this subject and the analysis of swarm experiments taking it into account. Throughout the chapter we shall restrict ourselves to collisions in which no electronic excitation is concerned. Collisions in which this is involved will form the subject of Chapters 12 and 13. Consideration of the effect of the internal degrees of freedom within a molecule on the nature of resonance effects in collisions of electrons with molecules will also be deferred to that chapter.

The energy losses involved in the excitation of molecular vibration and rotation are relatively small, at least if it can be assumed that the chance of exciting many vibrational quanta in a single collision is in-appreciable. Thus, the greatest separation between the ground and first excited vibrational states for any molecule, that in H_2, is only 0·54 eV. Rotational level separations are very much smaller (about $0·015J$ eV for H_2, where J is the rotational quantum number). Detailed investigation of this type of excitation requires advanced experimental and theoretical techniques which have only recently become available. The results obtained with the use of these techniques are of great interest. Before describing them we shall give a brief historical account of the development of the subject.

1. Excitation of molecular vibration and rotation—historical account

The first evidence that slow electrons suffer inelastic collisions with molecules came from swarm experiments in which electrons were allowed to diffuse in a gas under the influence of a uniform electric field F. Referring to the discussion in Chapter 2, § 1 we see from (23) and (24) of that chapter that, apart from factors of order unity, the mean fraction λ of its energy which an electron in the swarm loses in collisions with a gas molecule is given by

$$\lambda = 2u^2/c^2, \qquad (1)$$

where u is the drift velocity and c the mean random velocity. From measurements of c made by Townsend's method (Chap. 2, § 2) and of the drift velocity either by the method of Townsend and Bailey

TABLE 11.1

Average fractional energy loss λ due to collisions with gas molecules of electrons of different characteristic energy (ϵ_k) derived from early measurement of drift and mean random velocities

Gas $(2m/M) \times 10^4$	H$_2$ 5·45	N$_2$ 0·387	O$_2$ 0·341	CO 0·387	NO 0·363	HCl 0·298	N$_2$O 0·247	CO$_2$ 0·247	NH$_3$ 0·641
ϵ_k (eV)	$\lambda \times 10^4$								
0·1	25			45		250	600	360	285
0·2	29	7	65	70	110	600	1500	580	420
0·4	51	6	55	89	390	450	1300	650	370
0·8	26	6	24	54	380	290	1000	600	305
1·2	30	128	15	75	275		850	560	255
1·6	83	48	32	150	280		800	540	220
2·0	220	162	105	250			690	535	230
3·0	415	389	240	470			540		
4·0	580	500	355						
5·0	725	585							

The values given in this table are those quoted in HEALEY, R. H. and REED, J. W., *The behaviour of slow electrons in gases* (Amalg. Wireless (Aust.) Ltd., 1941).

(Chap. 2, § 4) or of Bradbury and Nielsen (Chap. 2, § 3.1) values of λ were derived as functions of the mean electron energy (or more properly the characteristic energy ϵ_k, see Chap. 2, § 2) for a number of molecular gases. In all cases the values found, some of which are given in Table 11.1, greatly exceeded those to be expected in purely elastic collisions ($2m/M$ where m, M are the respective masses of an electron and of a gas molecule).

These results showed clearly that inelastic collisions take place with molecules at such low energies of impact that electronic excitation is excluded. It was not possible to say in general how much of the excitation involved molecular rotation and how much vibration. However, the fact that λ for H_2 is still nearly five times the value expected from purely elastic losses when the mean energy is as low as 0·1 eV, suggests strongly that excitation of rotation is of importance. Another striking feature is the very rapid increase in λ for H_2 at mean electron energies above 1·6 eV. This points to onset of vibrational excitation, a conclusion strikingly confirmed in recent experiments as we shall see.

At an early stage also the Hertz diffusion method (see Chap. 5, § 2.2) was applied by Harries and Hertz[†] to nitrogen and by Ramien[‡] to hydrogen. Evidence of an energy loss due to excitation of a vibrational quantum was found in both gases. In H_2 the chance of excitation per collision was found to decrease from 0·03 to 0·02 as the electron energy increased from 3·5 to 7 eV. A similar but less probable effect was found in nitrogen, the chance per collision being $\frac{1}{79}$ for electrons of 5·2-eV energy.

Attempts to provide theoretical predictions of cross-sections for excitation of molecular vibration and rotation made during this period and even later met at first with only partial success.

A molecular state of a diatomic molecule is characterized (a) by a set n of quantum numbers that specify the state of electronic motion relative to the nuclei, (b) the vibrational quantum number v, and (c) the rotational quantum numbers J, M, it being supposed that the component of nuclear rotation about the nuclear axis is zero—there is no difficulty in generalizing the analysis to allow for cases in which this is not so. To a good approximation the wave function for the state can be written in the form

$$\Psi_{nvJM} = \psi_n(\mathbf{r}_i, R)\chi_{nv}(R)\rho_{nJM}(\Theta, \Phi), \tag{2}$$

where $\mathbf{r}_i$ denotes the coordinates of the electrons relative to the nuclei, R the nuclear separation, and Θ, Φ the polar angles of the nuclear axis relative to an axis fixed in space. The electronic wave function is calculated as a function of the nuclear separation assuming that, to a good approximation, the electronic motion adjusts equally to what are, from the electron's viewpoint, merely adiabatic changes due to nuclear motion. For each value of R there will be, corresponding to ψ_n, an electronic energy $E_n(R)$ (including also the energy of the Coulomb repulsion between the nuclei). The nuclei move under the influence of

† HARRIES, W. and HERTZ, G., Z. Phys. **46** (1927) 177.
‡ RAMIEN, H., ibid. **70** (1931) 353.

this energy, giving rise to the vibrational and rotational states (see Chap. 12, § 1).

According to Born's approximation (Chap. 7, § 1, (8), (16), and (17)) the differential cross-section for the excitation of a transition from the molecular state $nvJM$ to one $n'v'J'M'$ by impact of an electron with initial and final wave vectors $\mathbf{k}_0$, $\mathbf{k}_1$ respectively is given by

$$I_{nvJM}^{n'v'J'M'} \, d\omega = \frac{4\pi^2 m^2 e^4}{h^4} \frac{k_1}{k_0} \times$$

$$\times \left| \iiint \sum_i (|\mathbf{r}_i - \mathbf{r}|)^{-1} e^{i(\mathbf{k}_0 - \mathbf{k}_1) \cdot \mathbf{r}} \Psi_{nvJM}' \Psi_{n'v'J'M'}^* \, d\mathbf{R} d\mathbf{r} \Pi_i \, d\mathbf{r}_i \right|^2 d\omega. \quad (3)$$

Using (2) we see that

$$I_{nvJM}^{n'v'J'M'} = \left| \int \chi_{nv}(R) \chi_{nv'}^*(R) G(R,\Theta,\Phi) \rho_{JM}(\Theta,\Phi) \rho_{J'M'}^*(\Theta,\Phi) \, d\mathbf{R} \right|^2, \quad (4)$$

where

$$G(R,\Theta,\Phi) = \frac{2\pi m}{h^2} \left(\frac{k_1}{k_0}\right)^{\frac{1}{2}} \int V(\mathbf{r},\mathbf{R}) e^{i(\mathbf{k}_0 - \mathbf{k}_1)\cdot\mathbf{r}} \, d\mathbf{r}, \quad (5)$$

$$V(\mathbf{r},\mathbf{R}) = e^2 \int \sum_i (|\mathbf{r}_i - \mathbf{r}|)^{-1} |\psi_n(\mathbf{r}_i, R)|^2 \Pi_i \, d\mathbf{r}_i, \quad (6)$$

and is the mean interaction energy of the incident electron with the molecule in its ground state, for fixed nuclear separation $\mathbf{R}$.

It was first assumed that $V(\mathbf{r},\mathbf{R})$ can be written in the form

$$V(\mathbf{r},\mathbf{R}) = U(|\mathbf{r} + \tfrac{1}{2}\mathbf{R}|) + U(|\mathbf{r} - \tfrac{1}{2}\mathbf{R}|) \quad (7)$$

as the sum of two separate terms depending respectively on the distance of the electron from each nucleus.

With this approximation

$$G(R,\Theta,\Phi) = \frac{4\pi m e^2}{h^2} \left(\frac{k_1}{k_0}\right)^{\frac{1}{2}} \cos\{\tfrac{1}{2}(\mathbf{k}_0 - \mathbf{k}_1).\mathbf{R}\} \int U(s) e^{i(\mathbf{k}_0 - \mathbf{k}_1)\cdot\mathbf{s}} \, d\mathbf{s}, \quad (8)$$

and the differential cross-section for a particular $vJM \rightarrow v'J'M'$ transition is given by

$$I_{vJM}^{v'J'M'} \, d\omega = \frac{16\pi^2 m^2 e^4}{h^4} \frac{k_1}{k_0} \times$$

$$\times \left| \int \rho_{JM}(\Theta,\Phi) \chi_{nv}(R) \cos\{\tfrac{1}{2}(\mathbf{k}_0 - \mathbf{k}_1).\mathbf{R}\} \rho_{J'M'}^*(\Theta,\Phi) \chi_{nv'}(R) \, d\mathbf{R} \right|^2 |U_{k_0,k_1}|^2, \quad (9)$$

where

$$U_{k_0,k_1} = \int U(s) e^{i(\mathbf{k}_0 - \mathbf{k}_1)\cdot\mathbf{s}} \, d\mathbf{s}.$$

Calculations were carried out for hydrogen using this approximation by Massey,[†] Wu,[‡] and Morse§ and all yielded such small cross-sections

† MASSEY, H. S. W., *Trans. Faraday Soc.* **31** (1935) 556.
‡ WU, T. Y., *Phys. Rev.* **71** (1947) 111.   § MORSE, P. M., ibid. **34** (1929) 57.

that loss by vibrational excitation would not be much greater than that in purely elastic collisions.

This small cross-section for vibrational excitation arises from the fact that, for values of R close to the equilibrium nuclear separation R_0 the argument of the cosine in (9) is very small so the cosine is nearly unity. It follows that the integral over the nuclear coordinates will nearly vanish because of the orthogonality of the initial and final vibrational and rotational wave functions.

This means that the variation of U in (9) with R must be taken into account. Massey† therefore wrote

$$U = U(R_0) + \left(\frac{\partial U}{\partial R}\right)_{R=R_0} (R-R_0) \tag{10}$$

to give an additional term in $G(R,\Theta,\Phi)$

$$\frac{4\pi me^2}{h^2}\left(\frac{k_1}{k_0}\right)^{\frac{1}{2}} 2(R-R_0)\cos\{\tfrac{1}{2}(\mathbf{k}_0-\mathbf{k}_1).\mathbf{R}\} \int \left(\frac{\partial U}{\partial R}\right)_{R=R_0} e^{i(\mathbf{k}_0-\mathbf{k}_1).\mathbf{s}}\, \mathrm{d}s. \tag{11}$$

Detailed calculations were carried out for H_2 taking

$$U(s) = s^{-1}(1+Zsa_0)e^{-2Zsa_0}, \tag{12}$$

where Z is an effective charge on an atom in the molecule that depends on R through

$$Z = Z_0 + (R-R_0)\left(\frac{\partial Z}{\partial R}\right)_{R=R_0}. \tag{13}$$

For H_2, $R_0 = 1\cdot40\, a_0$ and variational calculations give $Z_0 = 1\cdot166$ and $(\partial Z/\partial R)_{R=R_0} = -0\cdot23/a_0$. Calculations carried out by Massey† and extended by Carson,‡ using these formulae, gave somewhat greater cross-sections but still far below those observed by Ramien.§ Thus Carson‡ found that the chance of exciting one vibrational quantum in a collision should never exceed $0\cdot1$ per cent, over one order of magnitude smaller than observed. Although no estimates were made for nitrogen the discrepancy could be expected to be larger for this molecule because of the greater mass of the nitrogen atom.

Carson's calculations also failed to provide an explanation of the large observed increase in λ above the elastic value at very low mean energies in hydrogen due to excitation of molecular rotation. He found very little increase due to such excitation for the following reasons.

To excite molecular rotation an incident electron must change its angular momentum. Thus an incident s-electron must emerge as a p- or d-electron. At low electron energies an electron with non-zero angular

† Massey, H. S. W., *Trans. Faraday Soc.* **31** (1935) 556.
‡ Carson, T. R., *Proc. phys. Soc.* A**67** (1954) 909.   § loc. cit., p. 728.

momentum is only to be found with appreciable probability at great distances from either atom in the molecule. The exponential interaction (12) will be very small at these distances and hence the transition probability will also be very small.

This does not apply if the molecule possesses a permanent dipole moment for the effective interaction with the electron then falls off quite slowly with distance, and Massey† showed that the cross-section for excitation of rotation in a dipolar molecule can be quite large. This probably is the explanation of the large values of λ found for heteronuclear molecules such as H_2O, N_2O, and NH_3 (see Table 11.1).

Although homonuclear molecules possess no permanent dipole moment they do possess a permanent quadrupole moment that is not included in the interaction U of (12). Gerjuoy and Stein‡ showed that, when allowance is made for this, the importance of rotational energy losses becomes sufficiently enhanced to lead to values of λ for very low energy electrons in H_2 and N_2 that are not far from those derived from experiment. Gerjuoy and Stein's theory has been refined and seems to provide the correct theoretical description of collisions of homonuclear diatomic molecules with very slow electrons that lead to rotational excitation. We shall return to a detailed discussion of the theory and of the analysis of experimental evidence in later sections.

The story turns back now to the investigation of vibrational excitation. In 1957 Haas§ repeated the Hertz-type diffusion experiment in nitrogen using more elaborate modern equipment. He was able to carry the observations down to energies as low as 2 eV and found that the chance of exciting a vibrational quantum increases quite rapidly as the electron energy falls (see Fig. 11.2), attaining a value as high as 0·15, corresponding to a cross-section $3·8 \times 10^{-16}$ cm², at the lowest energy investigated. This is so high as to defy explanation in terms of a theory such as that outlined above. Haas suggested that we must regard the collisions as taking place in two stages—the incident electron is first captured to form a negative ion N_2^- that is energetically unstable but has a lifetime greater than a vibrational period. It eventually breaks up, becoming a neutral molecule that may be in an excited vibrational state—in other words, the process is regarded as a resonance one of the same type as that found in elastic scattering of electrons by helium and other atoms and molecules (see Chap. 9).

† MASSEY, H. S. W., Proc. Camb. phil. Soc. math. phys. Sci. **28** (1932) 99.
‡ GERJUOY, E. and STEIN, S., Phys. Rev. **97** (1955) 1671.
§ HAAS, R., Z. Phys. **148** (1957) 177.

In 1959† Schulz applied his trapped-electron method (see Chap. 5, § 2.3 and § 3 of this chapter) to study inelastic collisions of slow electrons in N_2. Results obtained with a well depth of 0·2 V are illustrated in Fig. 11.1 (a) and only show a very small peak at 2·3 eV electron energy. This peak was greatly enhanced when the well depth was increased to 0·8 V as seen from Fig. 11.1 (b). The explanation of this marked change with well

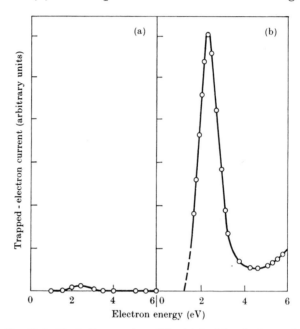

Fig. 11.1. Excitation spectra of N_2 obtained by the trapped-electron method. (a) Well depth 0·2 V. (b) Well depth 0·8 V.

depth provided definite evidence that several vibrational quanta were excited in some collisions. Thus if the applied accelerating voltage is V_A and the well depth is W the incident energy of the electrons (in eV) is $(V_A + W) = V_T$. Only those electrons with residual energy between 0 and W are collected. These electrons must therefore have lost an amount of energy between V_A and $V_A + W$, i.e. between $V_T - W$ and W_T. Thus in Fig. 11.1 (a) the energy loss is between 2·1 and 2·3 eV, whereas in Fig. 11.1 (b) it is between 1·5 and 2·3 eV. Since the vibrational level separation of the low-lying states in N_2 is about 0·3 eV this means that (a) several vibrational quanta may be excited, and (b) to understand the decreased prominence of the peak observed when the well depth is 0·2 V the probability must fall off with the number of quanta excited.

† Schulz, G. J., *Phys. Rev.* **116** (1959) 1141.

By varying W, keeping V_A at 0·15 V so that excitation of all vibrational states is included, Schulz† obtained the variation of the total cross-section for all vibrational excitation as a function of electron energy that is compared with Haas's results‡ in Fig. 11.2. The agreement is very good.

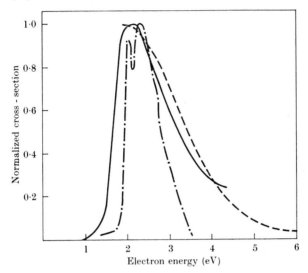

FIG. 11.2. Comparison of the energy variation of the cross-sections for vibrational excitation of N_2 as observed by the diffusion method (Haas) (— — —), the trapped-electron method (Schulz) (———), and from the sum of the differential cross-sections at 72° for all vibrational excitations (Schulz) (—·—·—).

Similar results were obtained for CO but no evidence of strong vibrational excitation was at first found in H_2 and O_2 (see § 3).

The establishment of the fact that more than one vibrational quantum could be excited with appreciable probability furnished strong evidence that the process proceeds through production of an intermediate negative ion. Final confirmation of the resonance nature of the process came from further experiments§ using a double electrostatic analyser that enabled Schulz to separate contributions from excitation of different vibrational states. We shall discuss the results of this work in detail in § 3.

An important check on observed and calculated cross-sections has been developed over the same period by Phelps and his collaborators (see § 7). Assuming values for the different cross-sections, the energy distribution function for the electrons diffusing in the gas under the

† loc. cit., p. 732.   ‡ loc. cit., p. 731.
§ SCHULZ, G. J., *Phys. Rev.* **125** (1962) 229.

action of a uniform electric field is calculated by a Boltzmann procedure. From these distributions various measurable quantities such as drift velocities, characteristic energies, etc., are calculated and compared with directly observed values. Complete consistency between observed cross-section data and observed average quantities for electron swarms, tested in this way, has not yet been achieved but a number of interesting conclusions have already been made.

2. Excitation of vibration by impact of fast electrons

At first sight there would appear to be little prospect of observing the excitation of molecular vibration, unaccompanied by electronic excitation, through impact of electrons with energy of the order of that used in electron diffraction studies (30–40 keV) (see Chap. 10, § 1.5). Apart from the technical difficulty of producing fast electron beams so homogeneous in energy that an energy loss of the order of 0·1 eV, less than 10^{-5} of the main beam energy, could be detected, a theory such as that of Massey or Carson discussed in § 1, which should be valid for excitation by high energy electrons, would predict extremely small cross-sections.

However, for molecules for which the vibrational transition is optically allowed the cross-sections for vibrational excitation are much larger. Boersch, Geiger, and Stickel† were able to develop an electron source with such a small energy spread (0·02 eV for 25 keV electrons) that it has been possible not only to observe energy losses due to vibrational excitation of such molecules as CO_2 and N_2O but also to determine the relative values of the generalized oscillator strengths (see Chap. 7, § 5.2.4) for excitation of different normal modes.

Turning to (43) of Chapter 7 we see that, according to Born's approximation, the differential cross-section $I_{0n}(K) \, d\omega$ for excitation of an optically allowed transition in which the incident electron suffers a momentum change $K\hbar$ is given by

$$I_{0n}(K) = \frac{4m^2}{\hbar^4} \frac{k_n}{k_0} \frac{e^4}{K^2} |z_{0n}|^2, \tag{14}$$

where k_0, k_n are the initial and final wave numbers of the colliding electrons so that, in terms of the angle of scattering θ,

$$k_0^2 + k_n^2 - 2k_0 k_n \cos \theta = K^2. \tag{15}$$

z_{0n} is the matrix element of the dipole moment in the z-direction associated with the transition. This formula is valid provided $K^2 \ll K_0^2$,

† BOERSCH, H., GEIGER, J., and STICKEL, W., *Z. Phys.* **180** (1964) 415.

where K_0 is of the order $1/a$, a being the dimension of the target atom or molecule. For impacts with high energy electrons in which the energy loss is a very small fraction of the initial energy $k_n \simeq k_0$ and the cross-section for a collision in which the electron is scattered into an angular range between 0 and Θ, where Θ is still sufficiently small for the condition $K^2 \ll K_0^2$ to hold, is given by

$$J(\Theta) = \frac{2\pi}{k^2} \int\limits_{k_0 - k_n}^{k\Theta} I_{0n}(K) K \, \mathrm{d}K = \frac{8\pi m^2}{\hbar^4} \frac{e^4}{k_0^2} |z_{0n}|^2 \ln\left(\frac{2E\Theta}{\Delta E}\right), \qquad (16)$$

where $E = k_0^2 \hbar^2/2m$ is the incident electron energy and ΔE the excitation energy. This formula is valid provided $2E\Theta/\Delta E \gg 1$, $k_0 a\Theta \ll 1$.

For application to the excitation of vibrational states of polyatomic molecules we use the relation (53) of Chapter 7, in the form

$$f_{0n} = \frac{8\pi^2 m \Delta E}{h^2} |z_{0n}|^2, \qquad (17)$$

where f_{0n} is the optical oscillator strength associated with the transition. This gives

$$J(\Theta) = \frac{16\pi^3 m e^4}{k_0^2 h^2} \frac{f_{0n}}{\Delta E} \ln\left(\frac{2E\Theta}{\Delta E}\right). \qquad (18)$$

f_{0n} may be obtained from measurement of the absorption coefficient of infra-red radiation emitted in a transition from the excited state n to the ground state.

It was explained in Chapter 5, § 4.1 how Boersch, Geiger, and Stickel[†] were able to improve the energy resolution of their apparatus for the observation of energy loss spectra of electrons with an incident energy of 25 keV so that the energy spread in the incident beam was as low as 0·02 eV. They were then able, by the method described in Chapter 5, § 4.1, to observe energy losses in CO_2,[‡] which is infra-red active, of 0·083 and 0·292 eV due to excitation of two of the normal vibrational modes associated with the ground electronic state.

The main difficulty in making quantitative observations of the relation probabilities of those energy losses arises from the intensity of the primary beam. Even though this was blocked off from the photographic detector by a screen it nevertheless, through scattering, contributed an intense background. Thus in Fig. 11.3 the observed variation of intensity with energy loss in CO_2 is shown. It will be seen that the 0·08 eV loss is apparent only as a 'dimple' on a background

† loc. cit., p. 311.
‡ BOERSCH, H., GEIGER, J., and STICKEL, W., Phys. Lett. 10 (1964) 285.

that falls sharply with the energy loss. To correct for this background the intensity variation was measured with no gas present. A typical example of such an observation over the energy range covering the first peak is shown in Fig. 11.3. Apart from the energy range (ΔE) covered by the dimple the variation is much the same, apart from a constant factor, as when gas is present. It is assumed that this relation will also

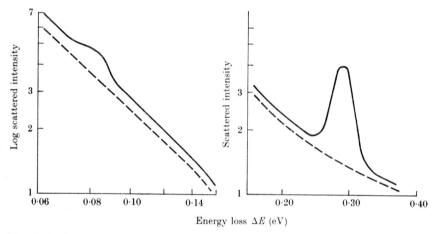

FIG. 11.3. Energy loss spectrum for electrons of incident energy 33 keV scattered through small angles in CO_2, as observed by Geiger and Wittmaack. ——— observed with CO_2 present. — — — observed with no gas present.

apply to the background in the energy range (ΔE), so that it may be subtracted off. The 0·292-eV loss shows up clearly as a peak but the same procedure could be adopted to eliminate the background.

This work was extended by Geiger and Wittmaack† to N_2O and C_2H_4. Fig. 11.4 illustrates the general form of the peaks observed due to vibrational excitation after correction for the background. The nature of the normal mode excited is indicated in each case.

In Table 11.2 the observed excitation energies are compared with spectroscopic values and it will be seen that the agreement is very close. The maximum observable angle of scattering Θ in the experiments was $1\cdot1\times10^{-4}$ rad. Values of $J(\Theta)$, calculated from (18) using oscillator strengths obtained from measurement of infra-red absorption, are also compared in Table 11.2 with values obtained directly from the relative intensity of the scattered as compared with the primary current at a particular gas pressure. Both the calculated and experimental values are subject to considerable uncertainties, the former because of the

† GEIGER, J. and WITTMAACK, K., *Z. Phys.* **187** (1965) 433.

range of measured infra-red absorption coefficients and the latter because of the difficulty of determining the ratio of scattered to primary intensity when this is very small. Allowing for this the agreement is quite good and is even better if relative instead of absolute intensities are compared.

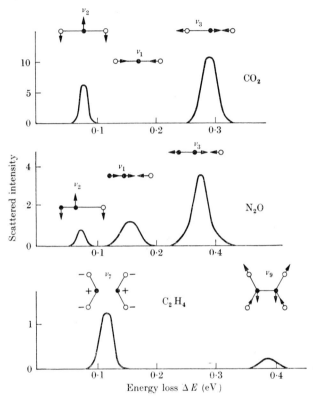

FIG. 11.4. Energy-loss spectra for electrons of incident energy 33 keV scattered through small angles in CO_2, N_2O, and C_2H_4 after correction for background. For each gas the relative heights of the peaks are as observed. The nature of the excited vibrational mode is indicated in each case.

3. Vibrational excitation by slow electron beams

In order to investigate cross-sections for excitation of separate vibrational states of diatomic molecules, Schulz employed the double electrostatic analyser technique that was described in Chapter 1, § 6.2.1 in connection with the observation of narrow resonances in the elastic scattering of electrons by atoms. With this equipment the energy resolution was as high as 0·06 eV, fully adequate to resolve separate vibrational excitation in the energy loss spectrum.

In the first apparatus of this kind† the scattered electron beam
was analysed in the forward direction. With its use, energy losses
corresponding to excitation of 2, 3,..., 7 vibrational quanta in nitrogen
were clearly resolved, and the energy variation of the excitation probability for each observed over the energy range 1·5–4 eV. Oscillatory
behaviour in all cases indicated resonance effects.

TABLE 11.2

*Comparison of energies and cross-sections for excitation of vibrational
states of CO_2, N_2O, and C_2H_4 as derived from infra-red absorption* (abs),
and from the energy loss spectra (el) *of 25-keV electrons*

Molecule	Symbol of normal mode excited	Excitation energy (meV)		Cross-section $(J(\Theta))$ 10^{-21} cm^2		Rel. cross-section	
		ΔE_{abs}	ΔE_{el}	$J_{abs}(\Theta)$	$J_{el}(\Theta)$	abs	el
CO_2	ν_2	83·75	82	1·44–2·17	4·0	0·36	0·4
	ν_3	291·31	291	4·67–5·29	10·0	1·00	1·00
N_2O	ν_2	73·01	73	0·23–0·42	0·35	0·10	0·10
	ν_1	159·34	158	0·98–1·54	1·2	0·37	0·35
	ν_3	275·14	276	3·03–3·70	3·4	1·00	1·00
C_2H_4	ν_7	117·70	117	2·07–2·70	1·2	1·00	1·00
	ν_{12}	178·99		0·14–0·23		0·09	0·05
	ν_{11}	370·70		0·08–0·12			
			385		0·2	0·12	0·17
	ν_0	385·08		0·14–0·24			

Analysis in the forward direction was subject to two important
limitations. It was not possible to distinguish elastically scattered
electrons, the resonance features of which are of special interest in
relation to those exhibited by the inelastically scattered. Furthermore,
the background due to the primary beam prevented observation of the
energy loss due to excitation of a single vibrational quantum. To remove
these limitations further observations‡ were carried out by analysing
at the scattering angle of 72° as in the atomic resonance investigations
(Chap. 1, § 6.2, Chap. 5, § 4.2.1, and Chap. 10, § 3.5).§

Ehrhardt and Willmann‖ have extended this work to cover a wide
range of scattering angles. They used the double electrostatic analyser
technique described in Chapter 10, § 3.5.2 in connection with the

† SCHULZ, G. J., *Phys. Rev.* **125** (1962) 229.
‡ SCHULZ, G. J., ibid. **135** (1964) A988.
§ HEIDEMAN, H. G. M., KUYATT, C. E., and CHAMBERLAIN, G. E., *J. chem. Phys.* **44**
(1966) 355.
‖ loc. cit., p. 716.

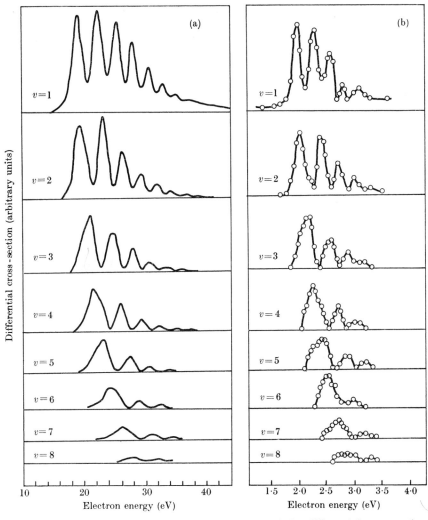

Fig. 11.5. Variation with incident electron energy of the differential cross-section for scattering of electrons after excitation of different vibrational states in N_2. (a) 20°. (b) 72°.

observation of fine-structure effects in elastic scattering by nitrogen. Fig. 11.5 shows results obtained in the form of relative differential cross-sections at a fixed angle of scattering for excitation of different vibrational states as functions of electron energy. Fig. 11.5 (b) gives results at 72° obtained by Schulz† and Fig. 11.5 (a) at 20° obtained by Ehrhardt and Willmann.‡ The behaviour is closely the same and is similar to

† loc. cit., p. 738.   ‡ loc. cit., p. 716.

that observed in the forward direction by Schulz with his earlier equipment.

Very similar results were also obtained by Heideman, Kuyatt, and Chamberlain† using the hemispherical electrostatic analyser technique of Simpson (see Chap. 1, § 5.1, Chap. 5, § 4.1, and Chap. 10, § 3.5). In particular they observed maxima at the same incident energies. The existence of fine structure in the elastic cross-section in the same energy range has already been described in Chapter 10, § 3.5.2.

An important feature of the results from the point of view of interpretation is that the resonance peaks in the cross-section–energy curves for the different vibrational excitations do not occur at the same energy in each case. This would be expected in terms of the simple concept of the processes proceeding via formation of a single relatively long-lived collision complex. On that model the positions of the peaks should occur at the same energies for all inelastic as well as elastic collisions. Further attention will be paid to this question in Chapter 12, § 6.

In the apparatus used by Ehrhardt and Willmann‡ the analysis of the scattered electrons could be carried out over a wide range of different scattering angles. This made it possible to observe the angular distributions of electrons of fixed incident energy scattered after excitation of a particular vibrational level. Fig. 11.6 shows such distributions for vibrational states with $v = 1$, 3, and 5 and certain incident electron energies in the resonance region. Reference to Fig. 10.34 shows that the angular distributions of the inelastically scattered electrons differ very considerably from those for elastic scattering. This is to be expected because, whereas almost all vibrational excitation proceeds only through formation of an N_2^- complex, only a small part of the elastic scattering arises in this way, most being due to simple potential scattering. The form of the observed inelastic angular distributions should prove of considerable importance in identifying the nature of the N_2^- state responsible.

It will be noted from Fig. 11.5 how gradually the differential cross-section at a fixed angle of scattering falls off with the number of vibrational quanta excited. Reference to Fig. 11.6 shows that the angular distributions for the different inelastic processes are quite similar and do not vary rapidly with electron energy throughout the resonance range. It is not surprising then that the differential cross-sections at 72°, summed over all vibrational states, should exhibit a variation with energy similar to that of the mean total inelastic

† loc. cit., p. 738.   ‡ loc. cit., p. 716.

cross-section as observed by Haas.† This may be seen by reference to Fig. 11.2. Assuming that the relative intensity of excitation of the different levels at 72° is characteristic of the separate cross-sections integrated over all angles, it is possible to normalize the latter from the absolute value 3.8×10^{-16} cm² of the total inelastic cross-section

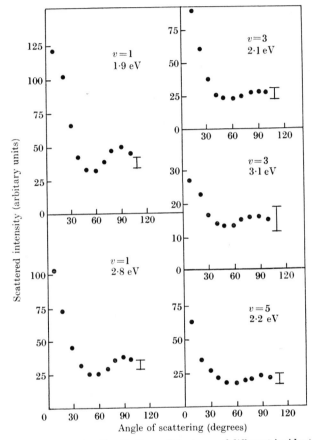

Fig. 11.6. Angular distributions of electrons of different incident electron energies scattered in nitrogen after excitation of particular vibrational states, as observed by Ehrhardt and Willmann.

observed by Haas. Further, more reliable, evidence about these cross-sections may be derived from the analysis of transport coefficients in N_2 as described in § 7.2.

It is of interest to note the existence of a 'tail' on the cross-section energy curve for the excitation of a single vibrational quantum, extending to low impact energies. This is probably due to direct excitation,

† loc. cit., p. 731.

in which no intermediate N_2^- ion is formed. The presence of this tail seems to be required also from the analysis of transport coefficients (see § 7.2).

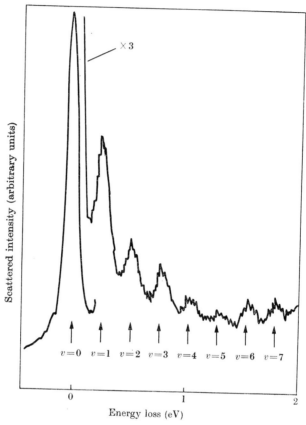

FIG. 11.7. Energy loss spectrum for electrons of incident energy 2·05 eV scattered through 72° in CO.

Results obtained for carbon monoxide† were essentially similar to those for nitrogen though rather less clear-cut. Fig. 11.7 shows an energy loss spectrum for electrons of incident energy 2·05 eV scattered through 72°. From the relative magnitudes of the elastic and inelastic peaks, assuming similar angular distributions for elastic and inelastic scattering (which is certainly a poor approximation), the total inelastic cross-section has a maximum of 8×10^{-16} cm² (about $\frac{1}{3}$ of the elastic, see Chap. 10, Fig. 10.21) at an incident energy of 1·75 eV as may be seen by reference to Fig. 11.8. The existence of a pronounced 'tail' at energies below 1 eV

† SCHULZ, G. J., *Phys. Rev.* **135** (1964) A988.

is noteworthy and is of importance theoretically and in the analysis of swarm data (§ 7.3).

Considerably different behaviour is exhibited by hydrogen. At most energies only one vibrational loss, corresponding to excitation of a single quantum, showed up in the energy loss spectrum of the scattered

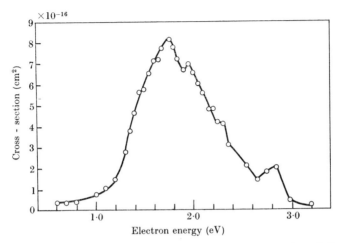

Fig. 11.8. Energy variation of the cross-section for excitation of all vibrational states of CO as a function of electron energy, obtained from differential cross-section data at 72° on the assumption of similar angular distributions of inelastically scattered electrons throughout the energy range.

electrons (see Fig. 11.9). Fig. 11.10 illustrates the observed variation with energy of the differential cross-section for this excitation.† Assuming similarity in the elastic and inelastic angular distributions the absolute value may be obtained from comparison with observed elastic scattering and use of measured total cross-sections (see Fig. 10.21). The peak inelastic cross-section is then found to be 6×10^{-17} cm^2, which is quite large. Quite good agreement is then found with the diffusion measurements of Ramien‡ but analysis of transport coefficients (see § 7.1) requires the considerably different cross-sections shown in Fig. 11.24 (b). The reason for this discrepancy, which will be further discussed in § 7.1, is still not clear.

In Fig. 11.10 two observations are included for excitation of the second vibrational level from which it can be seen that this is quite improbable.

Confirmation of the absence of fine structure in the total cross-section

---

† SCHULZ, G. J., loc. cit., p. 742.
‡ RAMIEN, H., *Z. Phys.* **70** (1931) 353.

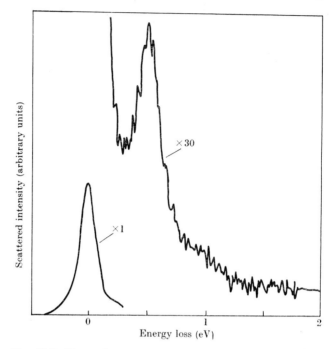

Fig. 11.9. Energy loss spectrum for electrons of incident energy 2·0 eV scattered through 72° in H_2.

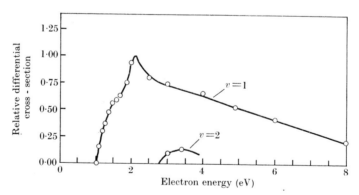

Fig. 11.10. Variation with incident electron energy of the differential cross-section for scattering of electrons through 72° after excitation of the first and second vibrational states of H_2. ○ observed by Schulz.

of H_2 in the energy range covered in these experiments comes from the experiments of Golden and Nakano† (see Chap. 1, p. 38, and Chap. 10, § 3.5.1).

† GOLDEN, D. E. and NAKANO, H., *Phys. Rev.* **144** (1966) 71.

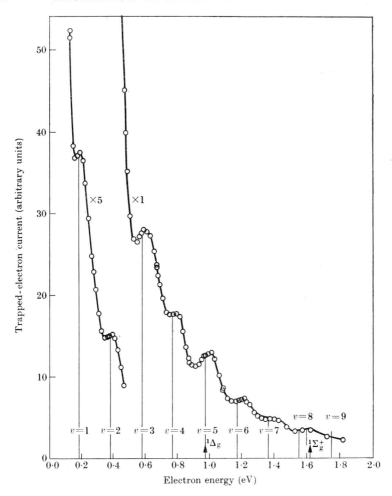

Fig. 11.11. Variation of trapped-electron current with incident electron energy in O_2, the well depth being 0·16 eV. The locations of vibrational levels of the ground electronic state are indicated as well as that of the $^1\Delta_g$ and $^1\Sigma_g^+$ excited electronic states.

For oxygen the cross-sections for vibrational excitation are very small, and energy losses due to such excitation have not been detected in experiments such as those carried out by Schulz, using the double electrostatic analyser that proved so effective for N_2, CO, and H_2. However, Schulz and Dowell† have been able to observe vibrational excitation in O_2 using the trapped-electron method (Chap. 5, § 2.3 and p. 732 of this chapter), which is much more sensitive. Fig. 11.11

† Schulz, G. J. and Dowell, J. T., ibid. 128 (1962) 174.

illustrates the variation with incident electron energy of the trapped-electron current, the well depth W being constant. As explained in Chapter 5, § 2.3, the trapped-electron current due to a discrete energy loss eV_{ex} would begin to be collected when the voltage V_A, accelerating the incident electrons, exceeds $V_{ex} - W$. If the cross-section for the excitation concerned rises linearly from the threshold, the peak of the trapped electron current will occur when $V_A = V_{ex}$, since, because of the potential well, the incident electron energy $eV_T = e(V + W)$ so that a peak will occur when $V_T = V_{ex} + W$. In the experiments that yielded the results shown in Fig. 11.11 W was 0·16 V, determined by the method described in Chapter 5, § 2.3. The locations of the vibrational energy levels of the ground state of O_2 are shown and there is clear evidence of peaks due to excitation of these states.

Analysis of the data is complicated by the fact that, at incident electron energies below 1·5 eV, a background of trapped-electron current is contributed by elastic scattering for reasons explained in Chapter 5, § 2.3. Also, there exist low-lying excited electronic states of O_2 at 0·98 and 1·63 eV respectively (see Fig. 13.69) so that interpretation is only unambiguous below 0·98 eV. To obtain estimates of the cross-sections for excitation of the different vibrational states at electron energies 0·16 eV above the respective thresholds, the 'elastic' background was assumed to vary smoothly with electron energy and could then be subtracted without too much error. The absolute cross-section was obtained by comparison of the trapped-electron current at a peak with the current of negative ions produced at an incident electron energy of 6·7 eV when the well depth was reduced to zero so that no trapped electron current was present. At this energy the cross-section for negative ion formation is $1·3 \times 10^{-18}$ cm^2 (see Chap. 13, § 4.4).

Fig. 11.12 illustrates the absolute cross-sections found in this way, at 0·16 eV above the respective thresholds. The cross-section, even for excitation of the lowest excited vibrational state is only $3·8 \times 10^{-19}$ cm^2, much smaller even than for H_2. Also, as in H_2, but not in N_2 and CO, the cross-section falls rapidly with increasing vibrational quantum number. However, it must be remembered that the results refer only to a particular energy close to the threshold in each case. Evidence from swarm experiments, discussed in § 7.4, suggest that at higher energies the cross-sections are much larger, though still considerably smaller than for H_2. It is also noteworthy that resonance structure was observed by Boness and Hasted (see Chap. 10, Table 10.8) in the elastic scattering of electrons with energies in the range 0·3–1·1 eV.

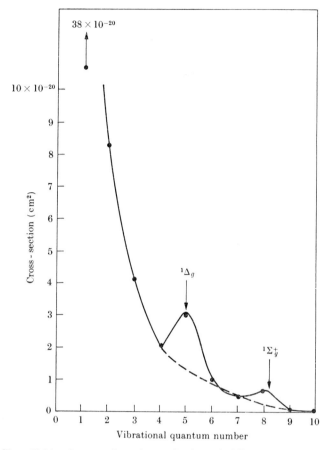

Fig. 11.12. Cross-sections for excitation of different vibrational states of O_2 by electrons with energies 0.16 eV above the respective excitation threshold.

4. Rotational excitation—application of Born's approximation

For reasons explained in § 1, rotational excitation by electron impact will arise primarily from long-range interactions. It is convenient, therefore, to write the interaction (6) between an electron and a molecule in the form

$$V(\mathbf{r}, \mathbf{R}) = V_s(\mathbf{r}, \mathbf{R}) - \tfrac{1}{2}\alpha(R)e^2 r^{-4} - \mu(R)er^{-2}P_1(\hat{\mathbf{r}}.\hat{\mathbf{R}}) -$$
$$- \{\tfrac{1}{2}\alpha'(R)e^2 r^{-4} + q(R)er^{-3}\}P_2(\hat{\mathbf{r}}.\hat{\mathbf{R}}) \qquad (19)$$
$$+ \text{higher order multipole terms.}$$

Here $\mathbf{R}$ denotes the molecular internal coordinates and $\mathbf{r}$ the coordinates of the electron relative to the centre of mass of the molecule. V_s is a short-range interaction falling off exponentially with r, μ and q are the permanent electric dipole and quadrupole moments respectively of the molecule, while α and α' are given in terms of the longitudinal and transverse electric polarizabilities $\alpha_{\parallel}$ and $\alpha_{\perp}$ of the molecule by

$$\alpha = \tfrac{1}{3}(\alpha_{\parallel}+2\alpha_{\perp}), \qquad \alpha' = \tfrac{2}{3}(\alpha_{\parallel}-\alpha_{\perp}). \tag{20}$$

We now suppose that coupling between molecular rotation and electron spin within the molecule can be ignored. If Born's approximation is assumed to be valid we see from (4) and (5) that the cross-section $Q_{JM}^{J'M'}$ for excitation of a rotational transition from a state with rotational quantum numbers J, M to one with J', M'

$$Q_{JM}^{J'M'} = \int |f_{JM}^{J'M'}(K)|^2 \, d\hat{\mathbf{K}}, \tag{21}$$

where

$$f_{JM}^{J'M} = \frac{2\pi m}{\hbar^2}\left(\frac{k'}{k}\right)^{\frac{1}{2}} \times$$

$$\times \iint V(\mathbf{r},\mathbf{R})e^{i\mathbf{K}\cdot\mathbf{r}}\rho_{JM}(\Theta,\Phi)\rho_{J'M'}^*(\Theta,\Phi)|\chi_{nv}(R)|^2 \, d\mathbf{r}d\mathbf{R}. \tag{22}$$

Here ρ_{JM}, $\rho_{J'M'}$ are the initial and final rotational wave functions that may be taken as normalized spherical harmonics Y_J^M, $Y_{J'}^{M'}$; $\chi_{nv}(R)$ is the vibrational wave function unchanged in the transition; $k\hbar$, $k'\hbar$ are the initial and final momenta of the electron and $\mathbf{K} = \mathbf{k}-\mathbf{k}' = K\hat{\mathbf{K}}$. To a close approximation the integration over the vibrational coordinates can be eliminated to give

$$f_{JM}^{J'M'} = \frac{2\pi m}{\hbar^2}\left(\frac{k'}{k}\right)^{\frac{1}{2}} \times$$

$$\times \iiint V(\mathbf{r},\mathbf{R_0})e^{i\mathbf{K}\cdot\mathbf{r}}\rho_{JM}(\Theta,\Phi)\rho_{J'M'}^*(\Theta,\Phi)\sin\Theta \, d\mathbf{r}d\Theta d\Phi, \tag{23}$$

where $V(\mathbf{r},\mathbf{R_0})$ is the value of V calculated for the equilibrium atomic separations in the molecule.

If the molecule possesses a permanent dipole moment, as does, for example, CO, this will be the most important term. Its angular dependence on $P_1(\hat{\mathbf{r}}\cdot\hat{\mathbf{R}})$ requires that $J' = J\pm1$ to give a finite cross-section. A simple calculation then gives, after averaging over all initial values of M and summing over final values of M',

$$Q_J^{J\pm1} = \frac{2\pi}{3k^2}\left(\frac{2m\mu_0 e}{\hbar^2}\right)^2 \frac{J+\tfrac{1}{2}\pm\tfrac{1}{2}}{2J+1} \ln\left(\frac{k+k'}{k\sim k'}\right), \tag{24}$$

where we have written $\mu(R_0) = \mu_0$.

Putting in numerical values for CO, for which $\mu_0 = 10^{-19}$ e.s.u., we find for excitation of $J = 1$ from $J = 0$, involving an energy transfer of $4 \cdot 8 \times 10^{-4}$ eV, the cross-section given in Fig. 11.13 as a function of electron energy.

The validity of Born's first approximation for cases in which a finite permanent dipole moment exists was discussed by Massey[†] who showed that it will be valid provided

$$8\pi^2\mu_0\,em/h^2 \ll 1. \qquad (25)$$

For CO this is well satisfied, the left-hand side being $0 \cdot 07$. On the other hand, for H_2O, $\mu_0 = 1 \cdot 85 \times 10^{-18}$ e.s.u. and the condition is far from satisfied.

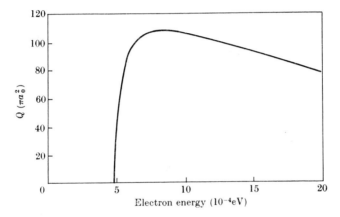

FIG. 11.13. Calculated cross-section for excitation of the rotational transition $J = 0$ to $J = 1$ by electron impact in carbon monoxide.

For homonuclear diatomic molecules μ_0 vanishes and the dominating term as far as rotational excitation is concerned arises from the term in (19) proportional to $P_2(\hat{\mathbf{r}}.\hat{\mathbf{R}})$. Because of this angular dependence the selection rule now becomes $J' = J \pm 2$. A similar calculation to that for the dipole case gives

$$Q_J^{J\pm 2} = \frac{8\pi}{15}\frac{k'}{k}\frac{(J+1\pm 1)(J\pm 1)}{(2J+1)(2J+1\pm 2)}\frac{q^2m^2e^2}{\hbar^4}\,\phi(k,\Delta k), \qquad (26)$$

where
$$\phi(k,\Delta k) = 1 + \frac{\alpha_0' e\pi}{4q_0 k}(k^2 - \tfrac{1}{4}\Delta k^2) + \frac{9(\alpha_0')^2 e^2\pi^2}{512 q_0}(k^2 - \tfrac{1}{2}\Delta k^2) \qquad (27)$$

with $\Delta k = k - k'$.

This formula was first derived by Gerjuoy and Stein[‡] who, however, omitted to include the term in α_0'. The need for allowing for this term

† MASSEY, H. S. W., *Proc. Camb. phil. Soc. math. phys. Sci.* **28** (1931) 99.

‡ GERJUOY, E. and STEIN, S., *Phys. Rev.* **97** (1955) 1671.

was first pointed out by Dalgarno and Moffatt† who derived the factor $\phi(k, \Delta k)$.

5. Evidence about rotational excitation cross-sections from swarm data—elementary analysis

The main way of testing the validity of the calculated cross-sections for rotational excitation is by analysing data on diffusion and drift of an electron swarm in a diatomic gas under the action of uniform electric and magnetic fields. Further information may be obtained by cross-modulation experiments in discharge afterglows.

We begin with a simplified description of the motion of an electron swarm that relates the energy loss due to rotational excitation to the quantity λ discussed in § 1, the mean fractional energy loss for collision. Applying this to diffusion in H_2 and N_2 shows semi-quantitative agreement between derived and observed values of λ, but reveals the need for a more complete analysis essentially on the same lines as that described in Chapter 5, § 3.3 for analysing drift data for monatomic gases when inelastic collisions are important. Before discussing in § 7 the method used and the results obtained in carrying out the analysis for H_2, D_2, N_2, O_2, CO, and CO_2 we shall describe in § 6 the cross-modulation method and results obtained by its use.

The rate at which an electron of velocity v and energy ϵ loses energy through rotational excitation in passing through a gas of homonuclear diatomic molecules is given by

$$\frac{d\epsilon}{dt} = v \sum_J N_J \{ Q_J^{J+2}(E_{J+2} - E_J) - Q_J^{J-2}(E_J - E_{J-2}) \}, \tag{28}$$

where N_J is the number/cm³ and E_J the rotational energy, of molecules with vibrational quantum number J. To a close approximation

$$E_J = BJ(J+1), \tag{29}$$

where $B = h^2/8\pi^2 I$, I being the moment of inertia of the molecule about the axis through the centre of mass perpendicular to the internuclear axis.

If $k'/k = 1$ we have from (26)

$$\frac{d\epsilon}{dt} = \frac{32\pi}{15} \frac{q_0^2 m^2 e^2}{\hbar^4} N B \phi(k) v, \tag{30}$$

where $\phi(k)$ is as (27) and N is the total number of molecules/cm³.

† DALGARNO, A. and MOFFATT, R. J., Indian Academy of Science Symposium on Collision Processes, 1962.

The mean fractional energy loss per collision is given by

$$\lambda = \overline{\frac{1}{\epsilon}\frac{d\epsilon}{dt}\frac{1}{NQv}} + \frac{2m}{M}, \tag{31}$$

where Q is the total collision cross-section. Hence, using (30),

$$\lambda = \frac{32\pi}{15}\frac{q_0^2\,m^2e^2}{\hbar^4}\,B\left(\overline{\frac{\phi(k)}{Q\epsilon}}\right) + \frac{2m}{M}, \tag{32}$$

where —— indicates an appropriate mean value.

In their original calculations for N_2 Gerjuoy and Stein used (32) with $\phi(k) = 1$ and Q taken as $4\cdot 8\pi a_0^2$, the value obtained from the earlier microwave measurements. For N_2 $|q_0|$ is nearly equal to ea_0^2 and $B = 0\cdot 25\times 10^{-3}$ eV. With these values λ as given by (32) was found to decrease from about 22×10^{-4} to 6×10^{-4} as the characteristic electron energy (Chap. 2, § 2) increased from $0\cdot 1$ to $0\cdot 6$ eV. The observed behaviour of λ shows it to be nearly constant and equal to 7×10^{-4} over this range. Although there was not very close agreement in detail the calculated magnitude for λ was clearly of the right order. The discrepancy was somewhat worse for hydrogen, particularly at higher characteristic energies. To analyse the situation thoroughly a more elaborate analysis is necessary. This we shall discuss in § 7 but first we shall examine the evidence available from cross-modulation experiments.

6. Cross-modulation experiments and rotational excitation

In Chapter 2, § 5 we discussed the use of measurements of the microwave conductivity of a discharge afterglow for the determination of electron collision frequencies under ordinary thermal conditions. An account was also given on p. 73 of that section of the technique introduced by Gould and Brown for heating the electrons at a suitable stage in the afterglow by means of a microwave pulse and so extending the measurements of collision frequency to higher electron temperatures. For the calculation of the temperature achieved with a given microwave pulse it is necessary to know the mean rate of energy loss per collision of an electron with a gas molecule. This offers no difficulty when dealing with monatomic gases for which there is no contribution to energy loss due to excitation of internal states of motion. In all other cases allowance must be made at least for excitation of rotation and, at higher temperatures, of vibration also. It is possible to reverse the situation so as to use microwave interaction methods to obtain information about the cross-sections for rotational excitation.

In discussing the energy relations for the motion of electrons of mean energy $\bar{\epsilon}$ in a gas of molecules of mean energy ϵ_0 we have introduced the quantity λ, which is such that the mean energy lost per collision of an electron with a gas atom is $\lambda\bar{\epsilon}$. This is convenient provided $\bar{\epsilon}$ is considerably greater than ϵ_0, but in the limit in which $\bar{\epsilon} \to \epsilon_0, \lambda \to 0$ so that, when $\bar{\epsilon} = \epsilon_0 + \Delta\epsilon$ where $\Delta\epsilon$ is small, it is more appropriate to introduce a new quantity G such that the mean loss per collision is $G\Delta\epsilon$. Thus

$$\lambda = G(1 - \epsilon_0/\bar{\epsilon}). \tag{33}$$

G may depend on $\bar{\epsilon}$ but must be such that $\lambda \to 0$ when $\bar{\epsilon} \to \epsilon_0$.

The rate of loss of energy per second will then be $\nu G \Delta\epsilon$, where ν is the collision frequency. We note also from the energy balance relation (1 a) of Chapter 2 that, for electrons drifting in a uniform electric field of strength F

$$G\nu = Feu/\Delta\epsilon, \tag{34}$$

where u is the drift velocity. $G\nu$ is therefore the ratio of the power input from the electric field per electron to the mean excess energy of an electron above the thermal value. It is sometimes referred to as the *energy exchange collision frequency* and, as it is determined essentially by the inelastic rather than the elastic cross-sections, it is a useful combination of transport coefficients to employ in the analysis of observed data from electron swarm experiments (see § 7, p. 769).

6.1. *Cross-modulation with radio waves—the Luxemburg effect*

It is of interest to note that a large-scale manifestation of wave interaction was observed[†] in 1933. This was the Luxemburg effect so-called because, in reception of radio waves from various medium-wave transmitters located in Western Europe, modulation of the long-wave transmitter at Luxemburg (252 kc/s) could also be traced. This interference was especially marked at night, indicating that the interaction occurred in the ionosphere, a result confirmed by later work. Since these initial observations extensive investigations of the effect have been carried out.

The theoretical explanation of the phenomenon was first given by Bailey and Martyn[‡] in 1934, and is now well established. According to this, absorption of the interfering waves increases the mean energy of the electrons in the ionosphere. This in turn leads to an increase in the collision frequency of the electrons. As the absorption of a wave

† TELLIGEN, B. D., *Nature, Lond.* 131 (1933) 840.
‡ BAILEY, V. A. and MARTYN, D. F., *Phil. Mag.* 18 (1934) 369; BAILEY, V. A., ibid. 23 (1937) 929 and 26 (1938) 425; HUXLEY, L. G. H., FOSTER, H. G., and NEWTON, C. C., *Proc. phys. Soc.* 61 (1948) 134; RATCLIFFE, J. A. and SHAW, I. J., *Proc. R. Soc.* A193 (1948) 311.

in a region of the ionosphere is determined by the collision frequency of the electrons in that region, the absorption of the 'wanted' wave is changed by the presence of the interfering wave. This modification takes the form of an impressed modulation of the wanted wave, characteristic of that in the interfering wave.

The detailed theory is complicated by the effect of the earth's magnetic field. We shall content ourselves here with merely outlining the relation of the cross-modulation to the electron collision frequency and the average energy loss experienced by an electron per collision with an atmospheric molecule, neglecting the earth's magnetic field.

Let W be the work per second done on an electron by the interfering wave. Part of the energy gained in this way by an electron will be lost in collisions with gas molecules, only the remainder leading to an increase in mean electron energy. If $\Delta\epsilon$ is the average excess energy at time t of an electron above that of the mean energy of the gas molecules, due to the interfering waves, then

$$W = \frac{d(\Delta\epsilon)}{dt} + \nu G \Delta\epsilon. \tag{35}$$

Also, as the mean velocity v of an electron is related to the collision frequency by the relation $v = \nu l$, where l is the mean free path, we have

$$\Delta\epsilon = ml^2\nu\Delta\nu,$$

where $\Delta\nu$ is the increase of ν due to the interfering wave. Hence

$$\frac{d}{dt}(\Delta\nu) + \nu G\Delta\nu = W/ml^2\nu. \tag{36}$$

We now consider the calculation of W. If the absorption coefficient of the interfering wave is κ_i, measured along the direction x of propagation, the magnitude of the electric vector may be written

$$E = E_0\,e^{-\kappa_i x}. \tag{37}$$

The energy absorbed per second in a slab of the ionosphere of thickness δx and unit cross-section will therefore be

$$-\frac{d}{dx}\left(\frac{KE^2}{4\pi}\right)\delta x,$$

where K is the dielectric constant of the region within the slab. In view of (37) this may be written

$$K\kappa_i E^2\,\delta x/2\pi. \tag{38}$$

As the slab contains $n_e\,\delta x$ electrons, n_e being the electron concentration,

$$W = \frac{K\kappa_i}{2\pi n_e}\,E^2.$$

Since the absorption coefficient κ_i is given by†

$$K\kappa_i = \frac{2\pi n_e e^2}{mc}\,\frac{\nu}{p^2+\nu^2}, \tag{39}$$

† BOOKER, H. G., Proc. R. Soc. A150 (1935) 267.

where p is the frequency of the interfering wave, we have

$$W = \frac{e^2}{mc}\, E^2\, \frac{\nu}{p^2+\nu^2}.$$

For sinusoidal modulation of the interfering wave with frequency $n/2\pi$,

$$E = E_1(1+M\sin nt),$$

so
$$W = \frac{e^2}{mc}\, E_1^2\, \frac{\nu}{p^2+\nu^2}\,\{1+2M\sin nt+\tfrac{1}{2}M^2(1-\cos 2nt)\}. \tag{40}$$

Hence, if we ignore the term in M^2, the modulation contributes an amount

$$Wn = ME_1^2\nu f(p,\nu)\sin nt. \tag{41}$$

This gives, on substitution in (36),

$$\Delta\nu = \Delta\nu_0\sin(nt-\phi), \tag{42}$$

where
$$\Delta\nu_0 = ME_1^2 f(p,\nu)/ml^2\nu G\{1+(n/G\nu)^2\}^{\frac{1}{2}}, \tag{43}$$

and
$$\tan\phi = n/G\nu. \tag{44}$$

It remains to determine the effect of this oscillation of $\Delta\nu$ on the wanted wave. According to (39), the absorption coefficient of the wanted wave is proportional to the collision frequency ν provided $p \gg \nu$, as it is for the cases of interest. The amplitude of the electric vector of the wanted wave after passage through the interacting region will therefore be given by

$$E_w = E_w^0\, e^{-\alpha\nu}, \tag{45}$$

where α is a constant independent of ν. Since, with the interfering wave present,

$$\nu = \nu_0+\Delta\nu_0\sin(nt-\phi),$$

we have
$$E_w = E_w^0\, e^{-\alpha\nu_0}\{1-M_i\sin(nt-\phi)\}, \tag{46}$$

where
$$M_i = \alpha\,\Delta\nu_0$$
$$= \frac{\alpha M(E_w^0)^2 f(p,\nu)}{ml^2\nu G\{1+(n/G\nu)^2\}^{\frac{1}{2}}}. \tag{47}$$

The wanted wave is thus modulated with the modulation frequency of the interfering wave, the coefficient of modulation M_i being as given in (47). It depends on the modulation frequency n of the interfering wave through the factor

$$\{1+(n/G\nu)^2\}^{\frac{1}{2}}.$$

Observations of M_i at different values of n thus enable $G\nu$ to be determined.

6.2. Laboratory investigation of wave interaction

Anderson and Goldstein† were the first to use a wave interaction technique to determine the energy-exchange collision frequency $G\nu$ for electrons in a gas. The method for determining collision frequencies from measurement of the real part of the microwave conductivity of a plasma filling a section of wave-guide has been described in Chapter 2, § 5. To determine $G\nu$ a perturbing microwave signal is passed through the plasma in the afterglow period when the electrons have reached

† ANDERSON, J. M. and GOLDSTEIN, L., *Phys. Rev.* **100** (1955) 1037.

thermal equilibrium with the neutral gas molecules at the ambient temperature T. This raises the electron temperature T_e above T. The perturbing signal is then cut off and the rate of decay of the electron temperature excess determined from observation of the collision frequency. The relaxation time so determined gives Gv provided certain conditions are satisfied. This may be shown as follows.

After the perturbing microwave field has been removed the energy balance equation for the electrons is

$$\frac{d\epsilon_e}{dt} = \sum_s \epsilon_s \nu_s - Gv(\epsilon_e - \epsilon_g). \tag{48}$$

Here ν_s is the frequency of superelastic collisions of electrons with excited species in the plasma that increase the electron energy by ϵ_s. ϵ_e and ϵ_g are the mean energies of an electron and gas molecule respectively.

Writing
$$\epsilon_e = \tfrac{3}{2}\kappa T_e, \qquad \epsilon_g = \tfrac{3}{2}\kappa T_g,$$

$$\frac{dT_e}{dt} = (2/3\kappa) \sum_s \epsilon_s \nu_s - Gv(T_e - T_g). \tag{49}$$

We now assume that the collision cross-sections, both for elastic collisions with neutral molecules and for superelastic collisions, are independent of electron energy so that the collision frequencies are proportional to $T_e^{\frac{1}{2}}$. (49) can then be written

$$\frac{dT_e}{dt} = - Gv_{T'}(T_e/T')^{\frac{1}{2}}(T_e - T'), \tag{50}$$

where
$$T' = T_g + \frac{2}{3\kappa} \frac{\sum\limits_s \epsilon_s \nu_{sT'}}{Gv_{T'}} \tag{51}$$

and $\nu_{T'}, \nu_{sT'}$ are collision frequencies when the electron temperature is T'.

If the relaxation time for the effect of the perturbation to disappear is short compared with the decay time of the afterglow, T' can be taken as constant and (50) solved to give

$$T_e^{\frac{1}{2}} = T'^{\frac{1}{2}}[1 + \exp\{-Gv_{T'}t - c\}][1 - \exp\{-Gv_{T'}t - c\}]^{-1}, \tag{52}$$

so
$$\nu(t) = \nu_0[1 + \exp\{-Gv_0 t - c\}][1 - \exp\{-Gv_0 t - c\}]^{-1}, \tag{53}$$

where ν_0 is the collision frequency in the absence of the perturbation. c is a constant determined by the initial conditions and it may be eliminated by introducing ν', the collision frequency at time $t = 0$.

We then have

$$\nu(t) = \nu_0 \left[1 + \frac{\nu' - \nu_0}{\nu' + \nu_0} e^{-G\nu_0 t} \right] \left[1 - \frac{\nu' - \nu}{\nu' + \nu_0} e^{-G\nu_0 t} \right]^{-1}$$

$$\rightarrow \nu_0 \left\{ 1 + \frac{\nu' - \nu_0}{\nu' + \nu_0} e^{-G\nu_0 t} \right\} \tag{54}$$

for large t.

The power absorption coefficient of the plasma for a probing signal of angular frequency ω is proportional to $\nu/(\nu^2 + \omega^2)$ (see Chap. 2 (78)) for a given electron concentration in the plasma. Hence if P is the power absorbed from the probing signal at time t after the perturbation ceases, P_0 that in the absence of the perturbation,

$$|P(t) - P_0| \simeq A e^{-G\nu_0 t}, \tag{55}$$

for large t, A being a constant provided $\nu(t)^2 + \omega^2 \simeq \nu_0^2 + \omega^2$.

Hence by observation of $P(t)$, $G\nu_0$ may be determined. It is important to remember that it has been assumed that the change of electron temperature due to the perturbation is small enough for the assumption of constant collision cross-sections to be valid and for $\nu(t)^2 + \omega^2$ to be replaceable by $\nu_0^2 + \omega^2$.

6.2.1. *Application to helium.* This method was first applied by Anderson and Goldstein† to helium at room temperature. In this case $G = 2m/M$, where m and M are the respective masses of an electron and a helium atom, so that a very good check of the technique can be made.

The plasma was produced in a thin-walled cylindrical Pyrex tube 152 cm long and 1·85 cm inside diameter, located coaxially in a square brass wave-guide with sides of inside dimensions 2·07 cm. Breakdown of the gas was produced by applying a high voltage d.c. pulse of 2 μs duration to a hot cathode and anode outside the wave-guide. The repetition frequency of these pulses was 25/s. Perturbation of the afterglow plasma was produced by a square pulse of microwave energy of frequency 8600 Mc/s, lasting from 1 to 100 μs, while the probing signal was a continuous wave of frequency 9600 Mc/s, its power level being 20 dB lower than that of the disturbing pulse.

Assuming $G = 2m/M$, the collision frequency ν was determined from the interaction relaxation time and compared with that derived from measurements of the real part of the conductivity of the plasma, for various gas pressures. The good agreement between them is illustrated in Fig. 11.14 in which both sets of values are plotted as functions of pressure. The linear form of these plots confirms that no important

† loc. cit., p. 754.

contribution to the collision frequency arises from collisions between electrons and positive ions.

An interesting extension of the technique has been made by Goldan and Goldstein† to study interaction effects in helium plasmas at very low temperatures (near 5 °K). The section of the wave guide containing the discharge tube was immersed in a bath of liquid helium at atmospheric pressure. Fig. 11.15 illustrates the general arrangement within a two-walled Dewar flask. The time sequence of operations is indicated in Fig. 11.16.

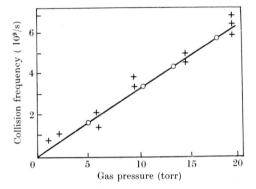

FIG. 11.14. Collision frequency of thermal electrons in helium as a function of gas pressure as observed by Anderson and Goldstein. ○, from measurements of the microwave conductivity of the plasma, +, from wave interaction studies.

In order to verify that the electrons in the afterglow came into thermal equilibrium with the gas at 4·2 °K within a millisecond or so, the radiation temperature of the electrons was measured, during the immediate afterglow phase, with a maser detector. Fig. 11.17 illustrates some observed results for the variation of electron temperature with time. The relaxation time is much longer than would be expected, due to collisions of the second kind with various excited species.

Even though the temperature relaxation time was longer than anticipated it was nevertheless small enough for wave interaction effects to be observed with electrons at a temperature of 4·2 °K, possessing a mean energy of only 0·004 eV. A useful check on the technique was made by determining the mean electron collision frequency when the electron temperature was still as high as 300 °K. This was done in the standard way by measuring the real component of the microwave

† GOLDAN, P. D. and GOLDSTEIN, L., *Phys. Rev.* **138A** (1965) 39.

conductivity of the plasma. From such observations at helium pressures of 0·51 and 1·3 torr in the discharge tube, mean momentum loss cross-sections of 5·7 and 5·3 × 10⁻¹⁶ cm² respectively were obtained. These

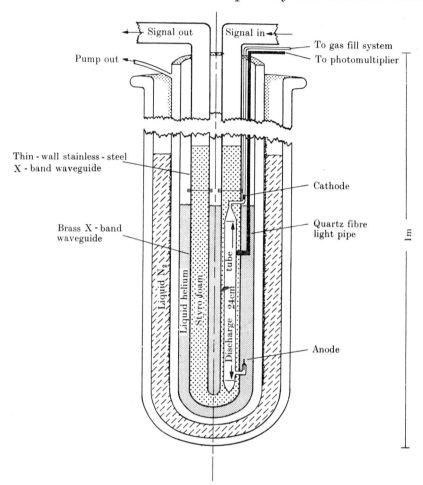

Fɪɢ. 11.15. General arrangement of apparatus used by Goldan and Goldstein for study-ing wave interaction effects in a helium afterglow at very low temperatures.

values agree quite well with those obtained in other experiments (see Chap. 2, § 7.1). On the other hand, when the electrons had cooled to 4·2 °K quite different behaviour was observed. Using the formulae (55, 57) of Chapter 2, the mean collision frequency was calculated as a function of gas pressure for different assumed constant collision cross-sections. The best fit at low pressures was obtained with a cross-section over three times that for electrons at 300 °K. Moreover, at higher

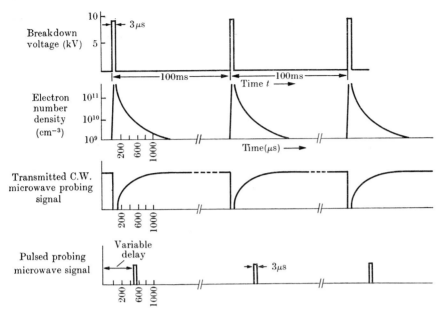

Fig. 11.16. Illustrating the time sequence of events in the wave interaction experiments of Goldan and Goldstein.

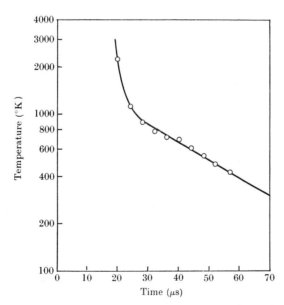

Fig. 11.17. Variation of electron temperature with time in a helium afterglow at 4·2 °K and 3·75 torr observed by Goldan and Goldstein.

pressures the measured collision frequency increased less rapidly than the calculated.

For the wave interaction studies the perturbing wave was of frequency 8800 Mc/s and power 50 mW, the probing signal of frequency 8532 Mc/s and 0·5 mW power. Fig.11.18 illustrates a typical record

(a)

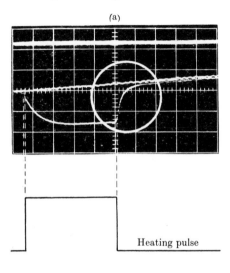

Heating pulse

(b)

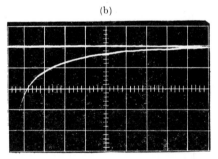

FIG. 11.18. Record taken by Goldan and Goldstein showing microwave cross modulation in a helium afterglow at 4·2 °K. (a) With a time scale 5 μs/cm. (b) With a time scale 1 μs/cm, displaying the portion in the circle in (a).

of the transmitted probing signal obtained at 1 ms duration of the afterglow, during and after application of the heating pulse. From such observations using formula (55) the relaxation time τ may be derived as a function of pressure, giving the results shown in Fig. 11.19. Assuming $G = 2m/M$ and using the measured variation of collision frequency with time shown we obtain an anticipated curve for τ also shown in Fig. 11.19. The observed values are about sixteen times larger, as if the scatterers possessed a mass equal to sixteen times that of a single helium atom.

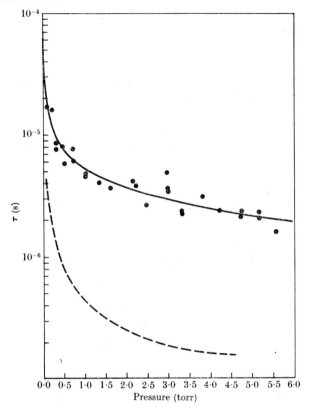

FIG. 11.19. Relaxation time for electron temperature as a function of pressure in a helium afterglow at 4·2 °K. ● observed points. ——— best fit of observed data. — — — expected from collision frequency data assuming binary collisions.

It is perhaps not surprising that a binary collision theory should not apply under the low-temperature conditions. Thus the wavelength of an electron with the mean temperature energy is $4·5 \times 10^{-6}$ cm. A sphere of this radius in helium at 0·1 torr pressure and 4·2 °K contains nearly twenty atoms. Under these conditions we would expect electron scattering to depend on simultaneous interactions with several atoms.

6.2.2. *Application to nitrogen and oxygen.* In their initial experiments Anderson and Goldstein† applied their wave interaction technique to nitrogen. They found $G = 9·8 \times 10^{-4}$ on the basis of their

† loc. cit., p. 754.

measurements of ν and $G\nu$. This was considerably smaller than the value, between 30 and 50×10^{-4}, derived by extrapolation of the observations of Crompton and Sutton[†] of the drift velocity and characteristic energies of electrons diffusing through nitrogen under the action of a uniform electric field. However, the values found by Anderson and Goldstein for ν were about four times larger than those that had been obtained by Phelps *et al.*, using the microwave technique described in Chapter 2, § 5. It therefore seemed likely that, while the measurement of $G\nu$ was correct that of ν was in error.

A few years later Mentzoni and Row[‡] approached the problem in a somewhat different manner, taking account of the theoretical values for the rotational excitation cross-sections given by Gerjuoy and Stein. They began with the expression (30) for the rate $d\epsilon/dt$ at which an electron of velocity v and energy ϵ loses energy through rotational excitation in passing through a gas of homonuclear molecules. The number n_J of molecules/cm³ with rotational quantum number J in a gas in thermal equilibrium at a temperature T is given by

$$N_J = N(2B/3\kappa T)(1+a)(2J+1)\exp\{-BJ(J+1)/\kappa T\}, \qquad (56)$$

where N is the total number of molecules/cm³, B is as in (29), and

$$a = \begin{cases} 1 & (J \text{ even}), \\ 0 & (J \text{ odd}). \end{cases}$$

The average rate of loss of energy per electron per second for electrons with a Maxwellian distribution of velocities at temperature T may now be calculated, using the expression (26) for the cross-sections $Q_J^{J\pm2}$ taking $\phi(k, \Delta k) = 1$. It is then found that, to a close approximation,

$$\frac{d\bar\epsilon}{dt} \simeq \frac{64\pi}{15}\left(\frac{2}{\pi}\right)^{\frac{1}{2}} \frac{q^2 m^2 e^2}{\hbar^4} Bm^{-\frac{1}{2}}\kappa N(\kappa T_e)^{-\frac{1}{2}}(T_e - T), \qquad (57)$$

where q is the quadrupole moment of the molecule. In terms of (31) and (33) this gives

$$(G - 2m/M)\nu = \frac{512}{45\pi}\frac{q^2 m^2 e^2}{\hbar^4} Bm^{-\frac{1}{2}}N(\kappa T_e)^{-\frac{1}{2}}, \qquad (58)$$

it being assumed that the frequency of inelastic collisions is small compared with that of the elastic so that ν is the momentum-transfer collision frequency.

Having obtained ν experimentally $G\nu$ may be calculated from (58) and hence the relaxation time τ for the return to thermal equilibrium after

† CROMPTON, R. W. and SUTTON, D. J., *Proc. R. Soc.* A**215** (1952) 467.
‡ MENTZONI, M. H. and Row, R. V., *Phys. Rev.* **130** (1963) 2312.

a heating pulse has been applied to the plasma. This calculated time is then compared with the experimentally observed values, thereby providing a check on the predictions of the theory of Gerjuoy and Stein.†

The experiments were carried out using a cylindrical discharge tube of quartz 80 cm long and of internal diameter 20 mm. The outer surface was gold-coated so it would operate as a wave guide. By insertion in an electric oven the whole system could be maintained at temperatures up to 900 °C with no deterioration of electrical properties. Special precautions were taken initially to outgas the tube and electrodes thoroughly so that the pressure was reduced to 5×10^{-9} torr. The probing wave was operated at 10 Gc/s and the heating pulse at 9·4 Gc/s. A pulse-sampling microwave radiometer was used to measure the electron temperature during the afterglow.

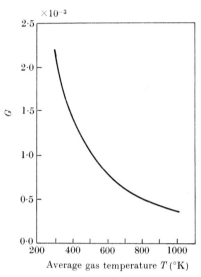

FIG. 11.20. The average excess fractional loss of energy G per collision for electrons in nitrogen as a function of gas temperature calculated from (58) using (59) and the quadrupole moment $0 \cdot 98 e a_0^2$.

The momentum-transfer collision frequency was measured and the results were in agreement with the formula

$$\nu = 1 \cdot 3 \times 10^{-7} N \bar{\epsilon} \, \text{s}^{-1}, \quad (59)$$

which was found by Frost and Phelps‡ from an analysis of drift velocity and characteristic energy measurements in nitrogen (see § 7.2). $\bar{\epsilon}$ is the mean electron energy in eV. Using this and taking the quadrupole moment§ $|q| = 0 \cdot 98 e a_0^2$ for nitrogen the values for G derived from (58) are as shown in Fig. 11.20. Fig. 11.21 illustrates a comparison between the observed and calculated relaxation time $\tau = Gv$ as a function of gas temperature at a fixed gas pressure. The agreement is seen to be quite close, showing that the formula of Gerjuoy and Stein† holds very well for N_2. In fact the agreement is less satisfactory if the correction factor $\phi(k, \Delta k)$ of (27) is included.

Fig. 11.22 shows a comparison between observed and calculated relaxation times τ for $T = 300$ °K as a function of gas pressure. Again the agreement is quite good except at the higher pressures.

† loc. cit., p. 749.
‡ FROST, L. S. and PHELPS, A. V., Phys. Rev. 127 (1962) 1621.   § See § 7.2.

Mentzoni and Rao† have carried out a similar investigation for oxygen based on an earlier measurement by Mentzoni‡ of the collision frequency ν in terms of the microwave conductivity of an oxygen afterglow.

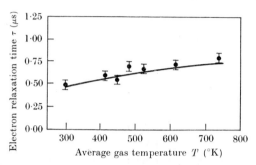

FIG. 11.21. Comparison of observed and calculated relaxation times τ for electrons in a nitrogen afterglow, at a pressure 6·0 torr, as a function of gas temperature. —— calculated. ● observed.

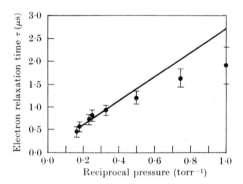

FIG. 11.22. Comparison of observed and calculated relaxation times τ for electrons in a nitrogen afterglow at a temperature of 300 °K as a function of pressure. —— calculated. ● observed.

He found that at an electron temperature T_e and gas pressure p torr the mean collision frequency is given by

$$\nu = 3{\cdot}0 \times 10^5 \, pT_e \text{ s}^{-1}. \tag{60}$$

This agrees very well with the results of microwave measurements made about the same time by Veatch, Verdeyen, and Cahn§ and is consistent with earlier measurements of van Lint, Wikner, and True-

† MENTZONI, M. H. and RAO, K. V. N., *Phys. Rev. Lett.* **14** (1965) 779.
‡ MENTZONI, M. H., *Radio Sci. J. R.* **69D** (1965) 213.
§ VEATCH, G. E., VERDEYEN, J. T., and CAHN, J. H., *Bull. Am. phys. Soc.* **11** (1966) 496.

blood† and of Fehsenfeld.‡ As will be seen by reference to § 7.4 it is considerably smaller than the values derived from the analysis of transport coefficients in oxygen. Further discussion of this matter is deferred till § 7.4.

Fig. 11.23 shows the observed values of the electron temperature relaxation time τ as a function of gas temperature. Earlier observations by Rao and Goldstein§ at 300 °K and by Gilardini‖ at 850 °K are also included. Using the collision frequency (60) it is found that the derived

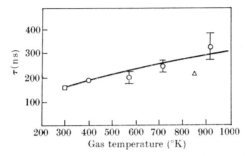

Fig. 11.23. Observed relaxation times τ for electrons in an oxygen afterglow, at a pressure of 3·0 torr, as a function of gas temperature T. ○ observed (Mentzoni and Rao); □ observed (Rao and Goldstein); △ observed (Gilardini). The solid curve follows the theoretical $T^{\frac{1}{2}}$ dependence and is normalized to agree with the observed result at 300 °K.

value of G agrees reasonably well with that given by (58) if q is taken as $2\cdot8ea_0^2$. This is much greater than the value $0\cdot33ea_0^2$ derived by Smith and Howard‡‡ for O_2 from microwave studies of line broadening. The evidence from the detailed analysis of transport coefficients for electrons in oxygen described in § 7.4 lends support to the rotational excitation cross-sections for O_2 being much larger than expected from direct application of the formula of Gerjuoy and Stein. Possible reasons for this discrepancy are discussed in § 8.

7. Detailed analysis of swarm data in terms of momentum loss and rotational and vibrational excitation cross-sections

Phelps and his collaborators have carried out detailed analysis of swarm data using the Boltzmann transport equation.

† VAN LINT, V. A. J., WIKNER, E. G., and TRUEBLOOD, D. L., ibid. **5** (1960) 122.
‡ FEHSENFELD, F. C., *J. chem. Phys.* **39** (1963) 1653.
§ RAO, K. V. N. and GOLDSTEIN, L., *Proc. 6th Int. Conf. Ioniz. Phenom. Gases*, ed. P. Herbert (Paris, 1963).
‖ GILARDINI, V., *Technical Rep. A.F.* 61 (052) -39 *Salenia S.p.A.* (Via Tibertina Kon 12, Rome, 1963).
†† SMITH, W. V. and HOWARD, R., *Phys. Rev.* **79** (1950) 132.

The velocity distribution function for electrons diffusing in a gas at temperature T under the influence of a uniform electric field F has been discussed in Chapter 2, § 1.2 for the case in which the electrons make only elastic collisions. Allowance for energy loss when inelastic collisions may also occur has been included in the discussion of Chapter 5, § 3.2 but only when the mean electron energy is small compared with the excitation threshold. This condition is certainly not satisfied for the diffusion of electrons in a molecular gas in which transfer of energy between the electrons and the rotation and vibration of the molecules may occur.

As in Chapter 2, § 1.2.1 and Chapter 5, § 3.2 the velocity distribution function $f(\mathbf{v})$ is taken as given by the first two terms in the harmonic expansion in terms of ξ/v, the ratio of the component of velocity in the direction of the field to the speed. Thus

$$f = f_0(v) + (\xi/v)f_1(v). \tag{61}$$

Differential equations for f_0 and f_1 are obtained by equating the numbers $c\,\mathrm{d}\gamma$ of representative points leaving the element $\mathrm{d}\gamma$ of velocity space per second due to the applied field F to $(b-a)\,\mathrm{d}\gamma$, where a is the number leaving per second due to inelastic collisions and b the number entering per second due to superelastic collisions. The calculation of c has already been carried out in Chapter 2, § 1.2.1 and gives (see Chap. 2 (13))

$$c = \frac{eF}{m}\left\{\cos\omega\,\frac{\mathrm{d}f_0}{\mathrm{d}v} + \frac{1}{3v^2}\frac{\mathrm{d}(v^2 f_1)}{\mathrm{d}v}\right\}, \tag{62}$$

where $\cos\omega = \xi/v$ and m is the electron mass. Also in the same section the contribution to $b-a$ due to elastic collisions has been calculated. Allowing for the thermal energy of the gas atoms this gives (see Chap. 2 (12) and (25))

$$(b-a)_{\mathrm{el}} = -NQ_{\mathrm{d}}\,vf_1\cos\omega + (mN/Mv^2)\frac{\mathrm{d}}{\mathrm{d}v}\left\{v^4 Q_{\mathrm{d}}\left(f_0 + \frac{\kappa T}{mv}\frac{\mathrm{d}f_0}{\mathrm{d}v}\right)\right\} \tag{63}$$

Here Q_{d} is the momentum loss cross-section, N the number of molecules/cm³, M the mass of a molecule. To this we must add the contributions $(b-a)_{\mathrm{in}}$, $(b-a)_{\mathrm{sup}}$ from inelastic and superelastic collisions respectively.

To calculate these consider the contribution from inelastic collisions that involve an energy loss $\epsilon_{\mathrm{s}} = \frac{1}{2}mv_{\mathrm{s}}^2$. Because of such collisions

$$a_{\mathrm{in}} = N_0\,vQ_{\mathrm{in}}^{\mathrm{s}}(v^2)f_0(v),$$

where $Q_{\mathrm{in}}^{\mathrm{s}}$ is the cross-section for the inelastic collision in which electrons of velocity v are concerned and N_0 the number/cm³ of ground state molecules. Similarly

$$b_{\mathrm{in}} = N_0\,v'Q_{\mathrm{in}}^{\mathrm{s}}(v'^2)f_0(v')\frac{\mathrm{d}\gamma'}{\mathrm{d}\gamma},$$

where

$$v'^2 = v^2 + v_{\mathrm{s}}^2 \tag{64}$$

and

$$\frac{\mathrm{d}\gamma'}{\mathrm{d}\gamma} = \frac{v'^2\,\mathrm{d}v'}{v^2\,\mathrm{d}v} = \frac{v'}{v}. \tag{65}$$

Hence

$$b_{\mathrm{in}} - a_{\mathrm{in}} = \frac{N_0}{v}\,[v'^2 Q_{\mathrm{in}}^{\mathrm{s}}(v'^2)f_0(v') - v^2 Q_{\mathrm{in}}^{\mathrm{s}}(v^2)f_0(v)]$$

$$= \frac{N_0}{v}\,[(v^2 + v_{\mathrm{s}}^2)Q_{\mathrm{in}}^{\mathrm{s}}(v^2 + v_{\mathrm{s}}^2)f_0\{(v^2 + v_{\mathrm{s}}^2)^{\frac{1}{2}}\} - v^2 Q_{\mathrm{in}}^{\mathrm{s}}(v^2)f_0(v)]. \tag{66}$$

For the corresponding contribution from superelastic collisions we have

$$b_{\text{sup}} - a_{\text{sup}} = \frac{N_s}{v}[(v^2-v_s^2)Q_{\text{sup}}^s(v^2-v_s^2)f_0\{(v^2-v_s^2)^{\frac{1}{2}}\} - v^2 Q_{\text{sup}}^s(v^2)f_0(v)]. \qquad (67)$$

Adding the contributions from all types of inelastic and superelastic collision to (63) equating separately the terms with and without the factor $\cos\omega$ of the corresponding terms in (62) we have

$$(eF/m)\,\mathrm{d}f_0/\mathrm{d}v = -NQ_\mathrm{d}\,vf_1, \qquad (68)$$

as in (14) of Chapter 2 and also†

$$\frac{eF}{3m}\frac{\mathrm{d}\{v^2 f_1(v)\}}{\mathrm{d}v} = \frac{mN}{M}\frac{\mathrm{d}}{\mathrm{d}v}\left[v^4 Q_\mathrm{d}\left\{f_0(v)+\frac{\kappa T}{mv}\frac{\mathrm{d}f_0}{\mathrm{d}v}\right\}\right] +$$

$$+ vN_0\sum_s[(v^2+v_s^2)Q_{\text{in}}^s(v^2+v_s^2)f_0\{(v^2+v_s^2)^{\frac{1}{2}}\} - v^2 Q_{\text{in}}^s(v^2)f_0(v)] +$$

$$+ v\sum_s N_s[(v^2-v_s^2)Q_{\text{sup}}^s(v^2-v_s^2)f_0\{(v^2-v_s^2)^{\frac{1}{2}}\} - v^2 Q_{\text{sup}}^s(v^2)f_0(v)]. \quad (69)$$

It is assumed that if the mean electron energy and hence also the thermal energy of the gas molecules is small compared with the excitation threshold then the terms arising from superelastic collisions are negligible as well as those depending on $f\{(v^2+v_s^2)^{\frac{1}{2}}\}$.

As the mean electron energy in the case we have considered will also need to be much greater than the gas temperature for inelastic collisions to make a significant contribution to the velocity distribution function, the term on the right-hand side of (69) that involves T can also be dropped. We then return to (28) of Chapter 5.

In our case none of these simplifications may be made in general, though there will be some circumstances in which superelastic collisions can be ignored. By eliminating f_1 between (68) and (69) we obtain a differential-difference equation for f_0. For present purposes it is convenient to express this equation in terms of the electron energy ϵ as variable instead of the electron velocity v. We then find

$$\tfrac{1}{3}F^2\frac{\mathrm{d}}{\mathrm{d}\epsilon}\left(\frac{\epsilon}{NQ_\mathrm{d}}\frac{\mathrm{d}f_0}{\mathrm{d}\epsilon}\right) + \frac{2m}{M}\frac{\mathrm{d}}{\mathrm{d}\epsilon}(\epsilon^2 NQ_\mathrm{d}f_0) + \frac{2m\kappa T}{Mc^2}\frac{\mathrm{d}}{\mathrm{d}\epsilon}\left(\epsilon^2 NQ_\mathrm{d}\frac{\mathrm{d}f_0}{\mathrm{d}\epsilon}\right) +$$

$$+ N\sum_s(\epsilon+\epsilon_s)f(\epsilon+\epsilon_s)Q_s'(\epsilon+\epsilon_s) - \epsilon f(\epsilon)N\sum_s Q_s'(\epsilon) +$$

$$+ N\sum_s(\epsilon-\epsilon_s)f(\epsilon-\epsilon_s)Q_{-s}'(\epsilon-\epsilon_s) - \epsilon f(\epsilon)N\sum_s Q_{-s}'(\epsilon) = 0. \quad (70)$$

Here

$$Q_{-s}' = \frac{N_s}{N}Q_{-s}, \qquad Q_s' = \frac{N_0}{N}Q_s,$$

where Q_{-s} is the cross-section for a collision in which the electron gains an energy ϵ_s, Q_s the corresponding cross-section for energy loss. Inclusion of the terms arising from collisions of the second kind is important in dealing with cases in which the electron characteristic energy ϵ_k and the significant transfer energies ϵ_s are less than about $10\kappa T$.

If we multiply all terms by $\epsilon\,\mathrm{d}\epsilon$ and integrate over all energies we obtain the

† Terms in $f_1(v)$ are omitted on the right-hand side of (69) because they are of higher order of smallness, consistent with the validity of the expansion (61).

energy balance equation

$$ueF = (8\pi N/m^2)\Big[(2m/M)\int_0^\infty \epsilon^2 Q_d\Big\{f_0(\epsilon)+\kappa T\,\frac{df_0}{d\epsilon}\Big\}\,d\epsilon +$$

$$+ \sum_s \epsilon_s \int_0^\infty \epsilon f_0(\epsilon)\{Q_s'(\epsilon)-Q_{-s}'(\epsilon)\}\Big]. \qquad (71)$$

$f_0(\epsilon)$ is taken to be normalized so

$$2\pi(2/m)^{\frac{3}{2}}\int_0^\infty \epsilon^{\frac{1}{2}} f_0(\epsilon)\,d\epsilon = 1, \qquad (72)$$

and
$$u = (-8\pi Fe/3m^2 N)\int_0^\infty \frac{\epsilon}{Q_d}\,\frac{df_0}{d\epsilon}\,d\epsilon \qquad (73)$$

is the drift velocity (see Chap. 2, § 1.3).

The left-hand side of (71) is the mean power input to an electron from the field. The first term on the right-hand side is the power loss due to elastic collisions, the second that due to excitation of rotation and vibration. Phelps and his collaborators have carried out a detailed analysis for H_2, D_2, N_2, CO, O_2, and CO_2 by solving the differential equation (70) for $f_0(\epsilon)$ directly by numerical methods after substitution of initially chosen forms for the cross-sections Q_d, Q_s, and Q_{-s}. Having obtained $f_0(\epsilon)$, the drift velocity and characteristic energy were calculated as functions of F/p or F/N and gas temperatures of 300 °K, using the formulae (18) and (21) of Chapter 2, and then compared with observation. Adjustments were then made in the cross-sections, $f_0(\epsilon)$ re-calculated, and new improved values obtained for u and ϵ_k. This procedure was continued until the final values of u and ϵ_k agreed well with the observed data. Further checks could be applied by using the cross-sections that give this good agreement to calculate the two co-efficients for drift in the gas at a lower temperature, usually 77 °K. In addition other transport coefficients could be calculated and compared with observation.

The transverse and longitudinal drift velocities respectively $u_\perp$, $u_\parallel$ (u_y, u_x in Chap. 2, § 4) in the case in which a magnetic field H is applied perpendicular to the electric field F have been discussed in Chapter 2, § 4 and are given by equations (45), (46), (47), and (48) of that chapter. If the collision frequency is constant then

$$\frac{u_\perp}{u_\parallel} = 1.06Hu/F,$$

where u is the drift velocity in the absence of the magnetic field. The departure of $u_\perp F/1.06Hu_\parallel$ from unity is a measure of the dependence of the collision frequency on the electron energy.

Further, it is possible to compare with observations made with an alternating electric field. It follows from (26) and (57) of Chapter 2, that, in this case, if ω is the angular frequency of the field, (71) remains valid if the constant electric field F is replaced by an energy dependent field $F(\epsilon)$ given by

$$\{F(\epsilon)\}^2 = \frac{Q_{\text{d}}^2}{Q_{\text{d}}^2 + (\omega/N)^2(m/2\epsilon)} F^2. \tag{74}$$

In carrying out the successive approximations to obtain cross-sections that give the best fit to the observations, Frost and Phelps† found it convenient to introduce two effective collision frequencies ν_m and ν_u determined directly from observed data on u and ϵ_k. The first, ν_m, is given by

$$\nu_m = \frac{e}{m}\frac{F}{Nu}. \tag{75}$$

If the collision frequency were accurately constant then it would be equal to ν_m. Otherwise ν_m, which we shall refer to as the *effective momentum-transfer collision frequency*, is directly obtained from the observations of u and is likely to be sensitive mainly to the form and magnitude of the momentum-loss cross-section. On the other hand, ν_u is most sensitive to the inelastic cross-sections. It is given by

$$\nu_u = Feu/(\epsilon_k - \kappa T)$$

$$= G\nu, \tag{76}$$

where ν is the mean collision frequency and G is as defined in (34) of § 6. As explained in that section ν_u, the energy-exchange collision frequency, can be expected to be sensitive to the inelastic rather than the elastic cross-section.

By working with ν_m and ν_u instead of u and ϵ_k in the first instance it is possible largely to separate the effects of changes in the elastic and inelastic cross-sections.

To use this procedure to check the validity of the rotational and vibrational excitation cross-sections it is convenient to proceed in stages. We first consider electrons with characteristic energies ϵ_k so small that energy loss through vibrational and electronic excitation is negligible. The range is then extended so that vibrational but not electronic excitation must be included. We now proceed to discuss the results obtained in this way. In Chapter 13, examples of extension

† loc. cit., p. 763.

of the analysis to still higher characteristic energies in which electronic excitation and ionization are important, will be described.

7.1. *Application to* H_2 *and* D_2

Analysis in the low energy region for H_2 was first carried out by Frost and Phelps[†] and then extended to higher energies and to D_2 by Engelhardt and Phelps[‡].

The vibrational excitation thresholds for H_2 and D_2 are at 0·516 and 0·360 eV and the first two stages of the analysis were carried out for the following ranges of characteristic energy ϵ_k.

Stage I. Elastic scattering and rotational excitation only:

$$\epsilon_k \leqslant 0 \cdot 08 \text{ eV.}$$

Stage II. Elastic scattering and both rotational and vibrational excitation: $0 \cdot 08 \leqslant \epsilon_k \leqslant 1 \cdot 0 \text{ eV.}$

The momentum-transfer cross-sections for H_2 and D_2 should be the same, and the first trial form assumed by Frost and Phelps was obtained by smoothing data obtained by Pack and Phelps[§] from drift velocity measurements using the method described in Chapter 2, § 3.1, by Bekefi and Brown[||] from microwave conductivity measurements in a hydrogen afterglow using the method described in Chapter 2, § 5, and by Brode[††] from measurements of the total cross-section (Chap. 1, § 4).

For the rotational excitation cross-sections it is necessary to know the quadrupole moment, which will be the same for H_2 and D_2 except for certain small corrections due to nuclear motion, the polarizabilities, and the rotational parameters B.

No difficulty arises about the latter, which have the values 0·00754 and 0·00377 eV for H_2 and D_2 respectively. The quadrupole moment q was taken as $0 \cdot 473 e a_0^2$ for H_2 on the basis of measurements by Harrick and Ramsey.[‡‡] It was found that a good fit could not be obtained if this value were used even when the polarizability correction factor $\phi(k, \Delta k)$ of (27) was included, so calculations were carried out on the following assumptions. In both H_2 and D_2, q was taken to have the value $0 \cdot 473 e a_0^2$ but the cross-sections were multiplied by an additional

---

† Frost, L. S. and Phelps, A. V., *Phys. Rev.* **127** (1962) 1621.
‡ Engelhardt, A. G. and Phelps, A. V., ibid. **131** (1963) 2115.
§ Pack, J. L. and Phelps, A. V., ibid. **121** (1961) 798.
|| Bekefi, G. and Brown, S. C., ibid. **112** (1958) 159.
†† Brode, R. B., *Rev. mod. Phys.* **5** (1933) 257.
‡‡ Harrick, N. J. and Ramsey, N. F., *Phys. Rev.* **88** (1952) 228.

empirical constant factor M_k as well as including the factor $\phi(k, \Delta k)$. The actual values used in the detailed calculations were given in Table 11.3.

It was assumed that only the first vibrational level is excited. The trial cross-section for vibrational excitation was taken to increase linearly from the threshold and then to pass through the values given by Ramien† (§ 3).

Fig. 11.24 illustrates a number of the cross-sections assumed, including those that give the best results for u and ϵ_k. Using these cross-sections the effective momentum transfer and energy-exchange collision frequencies ν_m and ν_u were calculated and are shown compared with the observed values in Fig. 11.25 for H_2 and Fig. 11.26 for D_2, at 77 °K.

TABLE 11.3

Assumptions made in calculating rotational excitation cross-sections for H_2 and D_2

	H_2			D_2	
	M_k	$\phi(k)$		M_k	$\phi(k)$
(i)	1·73	1·0	(i)	1·73	1·0
(ii)	1·54	> 1·0	(ii)	1·73	> 1·0
(iii)	1·0	> 1·0	(iii)	1·47	> 1·0

For ν_m the agreement is very good and as expected does not depend on assumptions made about the rotational cross-sections. On the other hand, for ν_u it will be seen that case (iii) for H_2 (see Table 11.3) is definitely not satisfactory in contrast to either (i) and (ii), which give indistinguishably satisfactory results. For D_2 we see that case (iii) gives the best, and quite good, results. There is definite evidence, for both gases, that the rotational excitation cross-section given by (26) with $\phi(k, \Delta k) = 1$ is about 50 per cent too small in magnitude and that a factor of the form of $\phi(k, \Delta k)$ exhibiting dependence on electron energy is also required.

To keep the discrepancies in proportion we have also included in Fig. 11.25 the energy exchange frequency ν_u if elastic collisions were alone important. There seems no doubt of the importance and order of magnitude of the contributions from rotational excitation.

Figs. 11.27 and 11.28 show a comparison of observed and calculated values of drift velocity and characteristic energy, using rotational cross-sections that give the best fit with the observed ν_u. It will be seen that the agreement is very good for H_2, being within 5 per cent for both coefficients. The situation is less satisfactory for D_2, arising apparently

† loc. cit., p. 743.

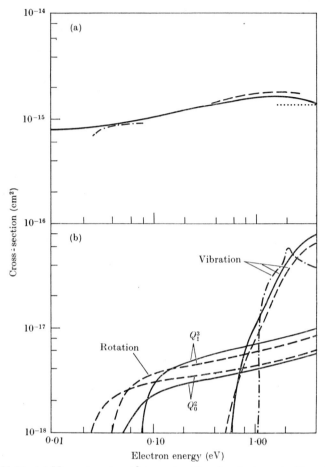

FIG. 11.24. (a) Momentum-transfer cross-section for electrons in H_2 and D_2. —— derived by Frost and Phelps from analysis of swarm data. —·—·— derived from drift velocity measurements of Pack and Phelps. ——— derived from microwave conductivity measurements of Bekefi and Brown. ···· from total cross-section measurements by Ramsauer technique. (b) Rotational and vibrational excitation cross-sections for electrons in H_2 and D_2 derived by Engelhardt and Phelps from analysis of swarm data. —— derived for H_2. ——— derived for D_2. $Q_J^{J'}$ denotes the cross-section for the rotational transition $J \to J'$. —·—·— vibrational excitation cross-sections observed by Schulz.

from an error as large as 20 per cent in the assumed shape of the theoretical rotational excitation cross-section. It seems that Born's approximation, though not too unsatisfactory, does not give results as accurate as required by the observations.

It is of interest to note the comparison between the observed and calculated values of the magnetic deflexion coefficient β (see Chap. 2 (50)). This is illustrated in Fig. 11.29 for H_2 and it will be seen that good agreement is obtained if the observations of Townsend and Bailey[†] are used but those of Hall[‡] are as much as 20 per cent lower.

To obtain the good agreement for H_2 at higher characteristic energies small adjustments only were found necessary in the assumed vibrational

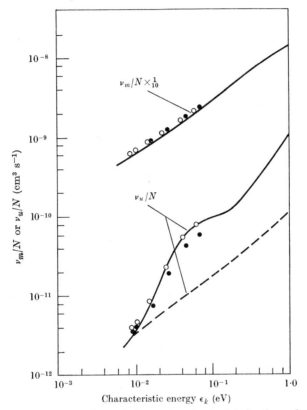

FIG. 11.25. Comparison of observed and calculated values of the collision frequencies ν_m and ν_u for electrons in H_2 at 77 °K. —— curve through points representing an average of best available experimental data. — — — calculated ignoring rotational transitions. ○ calculated using assumptions (i) and (ii) of Table 11.3. ● calculated using assumption (iii) of Table 11.3.

cross-section—the results are sensitive to the cross-section up to electron energies of about 4 eV. A slightly different form for the cross-section was found for D_2 (see Fig. 11.24 (b)).

The assumed vibrational cross-section does not agree with that observed by Schulz§ (see Fig. 11.24 (b)). The latter has a threshold at 1 eV in contrast to 0·52 eV, the maximum value is $5 \cdot 8 \times 10^{-17}$ cm^2 in

† TOWNSEND, J. S. and BAILEY, V. A., *Phil. Mag.* 42 (1921) 873.
‡ HALL, B. I. H., *Proc. phys. Soc.* B68 (1955) 334.   § loc. cit., p. 743.

contrast to $7 \cdot 8 \times 10^{-17}$ cm², and it is relatively much smaller at higher energies. The reason for the discrepancy is not known at the time of writing. Some of the evidence on which the cross-section at higher energies is based consists of the analysis of data at higher characteristic energies, which takes account of electronic excitation also and is described in Chapter 13, § 1.7.

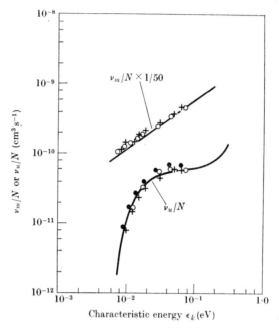

FIG. 11.26. Comparison of observed and calculated values of the collision frequencies ν_m and ν_u for electrons in D_2 at 77 °K. —— curve through points representing an average of best available experimental data. ○ calculated using assumption (i) of Table 11.3. ● calculated using assumption (ii) of Table 11.3. + calculated using assumption (iii) of Table 11.3.

Further useful results may be obtained from transport coefficients for electrons in mixtures of gases. Thus Engelhardt and Phelps† have pointed out how the peculiar form of the momentum-transfer cross-section Q_d for argon, exhibiting the Ramsauer minimum at an energy of about 0·4 eV and then increasing rapidly as the energy increases, increases the sensitivity of the transport coefficients to inelastic processes with thresholds between 0·5 and 10 eV. The rapid increase of Q_d above 0·4 eV reduces the high energy tail of the electron energy distribution

† loc. cit., p. 770.

so that the effective energy spread is comparatively small. This is further exaggerated in an H_2–Ar mixture because the threshold for the first vibrational excitation of H_2 occurs at 0·516 eV.

Using the cross-sections for H_2 shown in Fig. 11.24, which give satisfactory results for transport coefficients in that gas, and argon momentum-transfer and inelastic cross-sections discussed in Chapter 2, § 7.3,

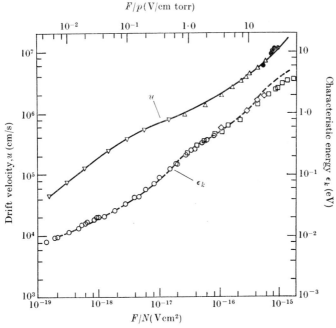

$F/p(\mathrm{V/cm\ torr})$

$F/N(\mathrm{V\,cm^2})$

FIG. 11.27. Comparison of observed and calculated values of drift velocity (u) and characteristic energy (ϵ_k) for electrons in hydrogen. —— calculated drift velocity u. △ u observed by Bradbury and Nielsen. ▽ u observed by Pack and Phelps. ● u observed by Frommhold. — — — calculated characteristic energy ϵ_k. □ ϵ_k observed by Townsend and Bailey. ◇ ϵ_k observed by Crompton and Sutton. ○ ϵ_k observed by Parker and Warren.

References: BRADBURY, N. E. and NIELSEN, R. A., *Phys. Rev.* **49** (1936) 388; PACK, J. L. and PHELPS, A. V., ibid. **121** (1961) 398; FROMMHOLD, L., *Z. Phys.* **160** (1960) 554; TOWNSEND, J. S. and BAILEY, V. A., *Phil. Mag.* **42** (1921) 873; CROMPTON, R. W. and SUTTON, D. J., *Proc. R. Soc.* A**215** (1952) 467; WARREN, R. W. and PARKER, J. H., *Phys. Rev.* **128** (1962) 2661.

Chapter 5, § 3.3.2, and Chapter 8, § 7.1, Engelhardt and Phelps calculated drift velocities, characteristic energies, and magnetic drift velocities for mixtures of 1, 1·5, 4, and 10 per cent of H_2 in Ar. In general quite good agreement was obtained but in certain cases such as for $\epsilon_k \simeq 1·0$ eV discrepancies of as much as 10 and 20 per cent were found between observed and calculated drift velocities. It seems that further accurate observations would be profitable.

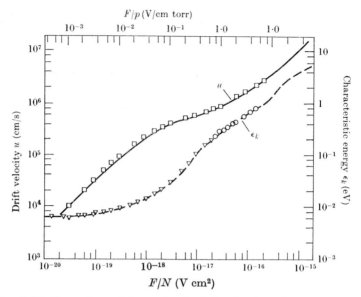

FIG. 11.28. Comparison of observed and calculated values of drift velocity (u) and characteristic energy (ϵ_k) for electrons in deuterium. —— calculated drift velocity u. ☐ u observed by Pack and Phelps. ——— calculated characteristic energy ϵ_k. ○ ϵ_k observed by Hall. ▽ ϵ_k observed by Parker and Warren.

References: PACK, J. L. and PHELPS, A. V., loc. cit. (Fig. 11.27); HALL, B. I. H., *Aust. J. Phys.* **8** (1955), 468; WARREN, R. W. and PARKER, J. H., loc. cit. (Fig. 11. 27).

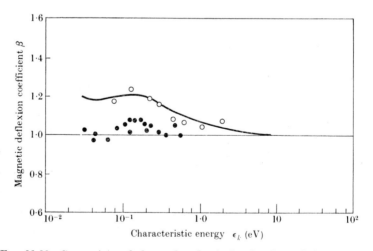

FIG. 11.29. Comparison of observed and calculated values of the magnetic deflexion coefficient β for electrons in H_2 at 300 °K. —— calculated. ○ observed by Townsend and Bailey (288 °K). ● observed by Hall (288 °K).

References: TOWNSEND, J. S. and BAILEY, V. A., loc. cit. (Fig. 11.27); HALL, B. I. H., loc. cit. (Fig. 11.28).

7.2. *Application to* N_2

For this gas a detailed analysis has been carried out by Frost and Phelps† and by Engelhardt, Phelps, and Risk.‡ We can ignore vibrational excitation for electrons of characteristic energy ϵ_k less than 0·1 eV and electronic excitation for $\epsilon_k < 1·4$ eV.

The momentum-transfer cross-section finally used is shown in Fig. 11.30. It does not differ seriously from that derived from the drift velocity observations of Pack and Phelps§ and of Crompton and Sutton,‖ the microwave conductivity measurements of Mentzoni and Row†† and the total cross-section measurements using the Ramsauer technique.‡‡

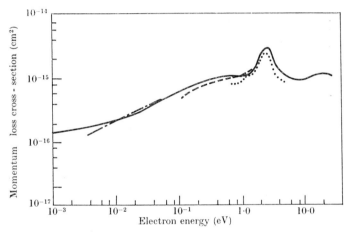

FIG. 11.30. Momentum-transfer cross-section for electrons in N_2. —— derived by Engelhardt, Phelps, and Risk from analysis of swarm data. —·—·— derived by Pack and Phelps from analysis of their drift velocity observations. — — — derived from drift velocity observations of Crompton and Sutton. ···· total cross-section measured by Ramsauer method.

The threshold for rotational excitation in N_2 is as low as $1·5 \times 10^{-3}$ eV, less than the mean thermal energy at 77 °K. Because it is necessary to take into account a large number of rotational levels, calculations were only carried out using the exact equation (70) for values of the characteristic energy ϵ_k less than $3\kappa T$. For $\epsilon_\kappa > 7\kappa T$ so many rotational levels are involved that the sum over their contributions may be approximated by an integral. At intermediate values of ϵ_k an approximate set of rotational excitation cross-sections was used with

† FROST, L. S. and PHELPS, A. V., *Phys. Rev.* **127** (1962) 1621.
‡ ENGELHARDT, A. G., PHELPS, A. V., and RISK, C. G., ibid. **135** (1964) A1566.
§ PACK, J. L. and PHELPS, A. V., ibid. **121** (1961) 798.
‖ CROMPTON, R. W. and SUTTON, D. J., *Proc. R. Soc.* A**215** (1952) 467.
†† MENTZONI, M. H. and ROW, R. V., *Phys. Rev.* **130** (1963) 2312.
‡‡ BRODE, R. B., *Rev. mod. Phys.* **5** (1935) 257.

thresholds doubled and magnitudes appropriately decreased. The reliability of this approximation was checked by comparison with exact results for $\epsilon_k = 3\kappa T$.

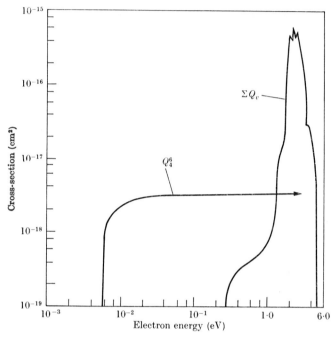

FIG. 11.31. Cross-sections for rotational and vibrational excitation of nitrogen. Q_4^6 is the cross-section for the rotational excitation $J = 4 \rightarrow J = 6$. $\Sigma_v Q_v$ is the sum of the cross-sections for vibrational excitation consistent with the swarm data.

It was found that the best fit with the observed data was obtained if the rotational cross-sections were calculated from (26) taking the quadrupole moment of N_2 as $1 \cdot 04 e a_0^2$ and the factor $\phi(k, \Delta k)$ as unity, i.e. making no allowance for the polarizability of the molecules. Fig. 11.31 illustrates one of the rotational cross-sections concerned.

The vibrational cross-sections assumed were based on those observed by Schulz† (see Fig. 11.2). For energies above $1 \cdot 7$ eV the relative magnitudes and shapes of the cross-sections for excitation of states with $v = 1$ to 8 were taken to be exactly as found by Schulz but it was necessary to add a low energy tail to the cross-section for $v = 1$. This is of interest because it may represent the contribution of the vibrational cross-section due to direct excitation instead of through intermediate

† loc. cit., p. 731.

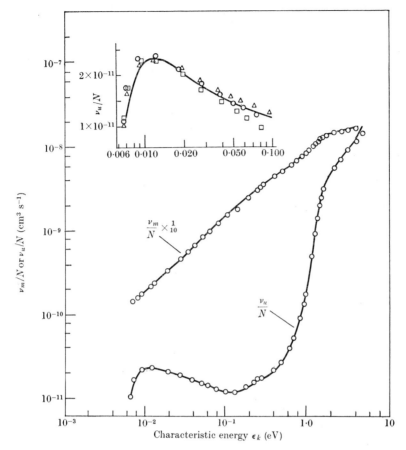

FIG. 11.32. Comparison of observed and calculated values of the collision frequencies v_m and v_u for nitrogen at 77 °K. —— curve through points representing an
average of best available experimental data. ○ calculated by Engelhardt, Phelps,
and Risk. *Inset.* Effect on v_u of different assumptions about the rotational cross-
sections. ○ calculated with $\phi(k) = 1$ and $q = 1·04ea_0^2$. △ calculated allowing for
polarization with $q = 0·974ea_0^2$. □ calculated allowing for polarization with
$$q = -1·10ea_0^2.$$

formation of an N_2^- ion. Without including this tail the calculated
value of the energy-exchange frequency was quite definitely too small
for $0·1 \leqslant \epsilon_k \leqslant 0·5$ eV. At higher energies the tail was unimportant
and it was found that the magnitude of the total vibrational cross-
section had to be normalized to $5·5 \times 10^{-16}$ cm² at 2·2 eV to give good
results for $0·5 \leqslant \epsilon_k \leqslant 1·4$ eV. This is considerably larger than the value
$3·8 \times 10^{-16}$ cm² observed by Haas† (see p. 741). Fig. 11.31 shows the
final form of the derived total vibrational cross-section.

† loc. cit., p. 731.

Comparison of observed and calculated effective momentum transfer and energy-exchange collision frequencies at 77 °K is shown in Fig. 11.32. The inset figure shows the effect of including the polarization correction

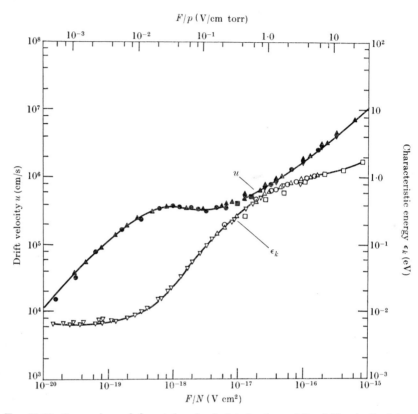

FIG. 11.33. Comparison of observed and calculated values of the drift velocity (u) and characteristic energy (ϵ_k) for electrons in nitrogen at 77 °K. —— calculated by Engelhardt, Risk, and Phelps. Drift velocity observations: ● Pack and Phelps (77 °K), ▲ Lowke (77·6 °K), ◆ Bradbury and Nielson (293 °K), ■ Errett (293 °K). Characteristic energy observations: ▽ Warren and Parker (77 °K), □ Townsend and Bailey (298 °K), △ Crompton and Elford (293 °K), ○ Cochran and Forrester (298 °K).

References: PACK, J. L. and PHELPS, A. V., loc. cit. (Fig. 11.27); LOWKE, J. J., *Aust. J. Phys.* **16** (1963) 115; BRADBURY, N. E. and NIELSEN, R. A., loc. cit. (Fig. 11.27); ERRETT, D., Ph.D. thesis, Purdue University (1951); WARREN, R. W. and PARKER, J. H., loc. cit. (Fig. 11.27); TOWNSEND, J. S. and BAILEY, V. A., loc. cit. (Fig. 11.27); CROMPTON, R. W. and ELFORD, M. T., *Proc. Sixth Int. Conf. Ioniz. Phenom. Gases* (1963); COCHRAN, L. W. and FORRESTER, D. W., *Phys. Rev.* **126** (1962) 1785.

factor $\phi(k)$ on the calculated ν_u for the two cases in which the quadrupole moment is taken to be $+0\cdot974ea_0^2$ and $-1\cdot10ea_0^2$ respectively. It will be seen that the best agreement is obtained if the polarization correction is ignored. Fig. 11.33 shows the good agreement obtained between observed and calculated values of the drift velocity and of the characteristic energy ϵ_k.

The behaviour expected of the magnetic deflexion coefficient β (Chap. 2, § 4) as a function of F/N is interesting, as shown in Fig. 11.34. Over the short range of values of F/N for which comparison with experiment is possible the agreement is good but it would clearly be of interest to extend the observations.

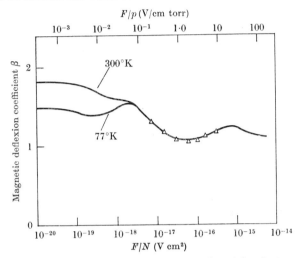

Fig. 11.34. The magnetic deflexion coefficient β for electrons in nitrogen. —— calculated. $\triangle$ observations of Townsend and Bailey (loc. cit. Fig. 11.27).

As an indication of the way in which the calculated energy distribution function differs from Maxwellian we show in Fig. 11.35 the calculated function for a characteristic energy of 0·067 eV as compared with a Maxwellian distribution for the same mean energy.

7.3. *Application to* CO

Analysis of low-energy transport data in carbon monoxide is of special interest because the molecule has a dipole moment sufficiently large to make a dominating contribution to rotational excitation cross-sections as well as to the direct excitation of vibration. Hake and Phelps† have carried out a detailed analysis for $\epsilon_k < 0.6$ eV. Above this range of ϵ_k no drift velocity data are available. However, the influence of the dipole moment can be demonstrated quite clearly in the low energy range.

The momentum-transfer cross-section derived from the analysis is shown in Fig. 11.36. Over the energy range covered by the drift velocity

† HAKE, R. D. and PHELPS, A. V., *Phys. Rev.* **158** (1967) 70.

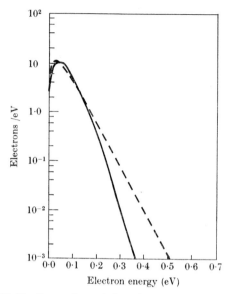

FIG. 11.35. Comparison of the calculated energy distribution function of electrons of mean energy 0·067 eV in nitrogen with a Maxwellian distribution about the same mean energy. ——— calculated distribution. — — — Maxwellian distribution.

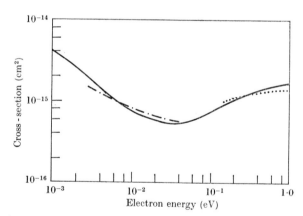

FIG. 11.36. Momentum-transfer cross-section for electrons in carbon monoxide. ——— derived by Hake and Phelps from analysis of swarm data. —·—·— derived by Pack, Voshall, and Phelps from analysis of their drift velocity observations. ···· total cross-section measured by Ramsauer method.

measurements of Pack, Voshall, and Phelps† there is good agreement with their derived values. There is only a small overlap with total

† PACK, J. L., VOSHALL, R. E., and PHELPS, A. V., *Phys. Rev.* **127** (1962) 2084.

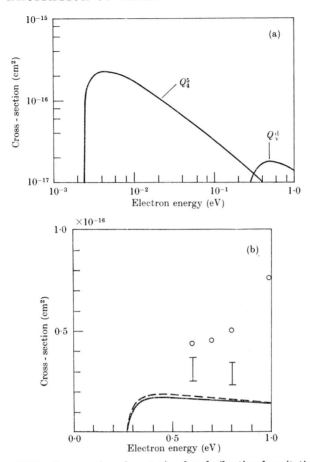

Fig. 11.37. Cross-sections for rotational and vibrational excitation of carbon monoxide. (a) Q_4^5 is the cross-section for the rotational excitation $J = 4 \to J = 5$. Q_v^d is the cross-section for direct vibrational excitation. (b) Vibrational cross-sections on an enlarged scale: —— in agreement with swarm data. — — — calculated by Takayanagi. I calculated by Breig and Lin. ◯ derived from the observations of Schulz at 72° angle of scattering (see p. 742).

cross-section measurements but the agreement is quite good over this limited region.

The rotational excitation cross-sections were calculated from (24) with the dipole moment adjustable to give the best agreement with observed transport coefficients. The best fit (see Fig. 11.39) was obtained with a dipole moment of $4 \cdot 6 \pm 0 \cdot 5 \times 10^{-2} e a_0$ compared with the Stark shift value $4 \cdot 4 \times 10^{-2} e a_0$. It is at first sight not surprising that the value derived from the transport coefficients should be somewhat large because no allowance was made for the contribution from the quadrupole moment. However, comparing the rotational cross-section for CO shown in Fig. 11.37 (a) with that for N_2 shown in Fig. 11.31 it will be seen that the former has a maximum value about 100 times larger so that the quadrupole correction for CO is not likely to be appreciable.

As for N_2 the rotational excitation threshold for CO is so small compared with κT that an exact solution of the Boltzmann equation (70) using the exact rotational cross-sections was only carried out for $\epsilon_k < 3\kappa T$ at 77 °K. For $\epsilon_k > 3\kappa T$ a continuous approximation was used in which the sum of contributions over the rotational states was represented by an integral, as in N_2. The validity of this approximation was checked by comparison with the exact solution at a low value of ϵ_k (see Fig. 11.38).

The cross-section for direct excitation of vibration† through the dipole moment is given according to Born's approximation by

$$Q_v^d = \int I_{0n}(K)\, d\omega, \tag{77}$$

where $I_{0n}(K)$ is as given in (14). Carrying out the integration

$$Q_v^d = \frac{8\pi m^2 e^2}{3\hbar^4 k^2}|\mu_{01}|^2 \ln\frac{k+k'}{k-k'}, \tag{78}$$

where k, k' are the wave numbers of the electron before and after impact and μ_{01} is the dipole moment associated with the vibrational transition from the ground to the first excited state.‡ As explained in § 2 this may be determined from infra-red absorption measurements in CO which give

$$|\mu_{01}|^2 = 1\cdot66 \times 10^{-3} e^2 a_0^2. \tag{79}$$

In the energy range of the analysis (for which the electron energy $< 1\cdot0$ eV) the most important contribution comes from this direct excitation. Fig. 11.37 (b) shows the cross-section calculated from (78) and (79) as well as that which gives the best fit with the transport data. The agreement is very close and provides a striking confirmation of the dipole effect. It is also of interest to note that the calculations of Breig and Lin§ that include induced polarization do not agree so well, while the values derived from the observations of Schulz‖ (see § 3, p. 743 and Fig. 11.7) on certain assumptions about the differential cross-sections for the elastically and inelastically scattered electrons, are far too large.

Fig. 11.38 shows the comparison between the calculated and observed values of the effective momentum-transfer and energy-exchange collision frequencies ν_m and ν_u. To make up for the paucity of time-of-flight

† Wu, T. Y., Phys. Rev. 71 (1947) 111; Takayanagi, K., J. phys. Soc. Japan 21 (1966) 507.   ‡ In the notation of (14) $e^2|z_{0n}|^2 = |\mu_{01}|^2/3$.
   § Breig, E. L. and Lin, C. C., J. Chem. Phys. 43 (1965) 3839.   ‖ loc. cit., p. 742.

drift velocity data at higher ϵ_k values for ν_m and ν_u are included in which the true drift velocity is replaced by the magnetic drift velocity (see Chap. 2, § 4). The agreement between calculation and observation is good and it is also clear that for ϵ_k as low as 0·1 eV the continuous representation of the rotational energy contribution gives results that agree well with the exact solution.

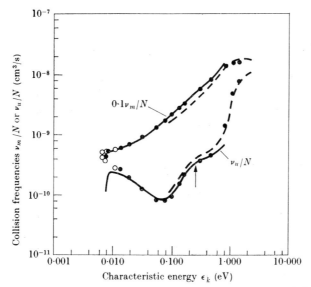

FIG. 11.38. Comparison of observed and calculated values of the collision frequencies ν_m and ν_u for carbon monoxide. —— curve through points representing an average of best available data. — — — curve derived using magnetic instead of actual drift velocity. ● calculated using the continuous approximation. ○ calculated by exact solution of Boltzmann equation.

The importance of the dipole moment in increasing the energy exchange collision frequency may be seen by comparison of ν_u in Fig. 11.38 with that for N_2 in Fig. 11.32.

Fig. 11.39 shows the degree of success achieved in reproducing the drift velocity and characteristic energy data. Results are also shown for the magnetic drift velocity.

7.4. Application to O_2

Despite its importance for application to atmospheric physics our knowledge of the cross-sections for slow electron collisions in oxygen is less reliable and complete than for the other gases we have already considered. This is partly due to experimental difficulties. Electron attachment to form negative ions occurs quite rapidly even for very low energy electrons (see Chap. 13, § 4.4) and this considerably increases the difficulty of making characteristic energy measurements (see Chap. 12, § 7.5.3 for an account of diffusion experiments when attachment is important). The result is that we have no measurements of ϵ_k below 0·15 eV. As the vibrational excitation threshold is at 0·195 eV this

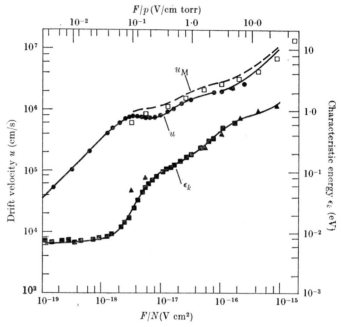

FIG. 11.39. Comparison of observed and calculated values of drift velocity (u), characteristic energy (ϵ_k), and magnetic drift velocity (u_M) of electrons in carbon monoxide. —— calculated for u and ϵ_k. — — — calculated for u_M. u, ● observations of Pack, Voshall, and Phelps (77 °K). ϵ_k, ▲ observations of Skinker and White (288 °K). ▉ observations of Warren and Parker (77 °K). u_M, □ observations of Skinker and White (288 °K).

References: PACK, J. L., VOSHALL, R. E., and PHELPS, A. V., Phys. Rev. 127 (1962) 2084; SKINKER, M. F. and WHITE, J. V., Phil. Mag. 46 (1923) 630; WARREN, R. W. and PARKER, J. H., loc. cit. (Fig. 11.27).

means that we have no chance of determining rotational excitation cross-sections from data at such low characteristic energies that vibrational energy loss is unimportant.

We also have little detailed information about the vibrational cross-sections as they appear to be too small to be studied by methods using energy-analysing devices. A further complication is introduced by the presence of low-lying excited electronic levels, the thresholds for excitation of which are 0·98 and 1·62 eV (see Fig. 11.11). Even for characteristic energies < 1 eV the excitation of these electronic states may be important.

It is quite clear from the observations which are available (see, for example, Table 11.1) that a large contribution to the rate of energy loss of slow electrons must come from the excitation of inner molecular motion. Thus the energy-exchange collision frequency (see Fig. 11.42) is much larger than for nitrogen.

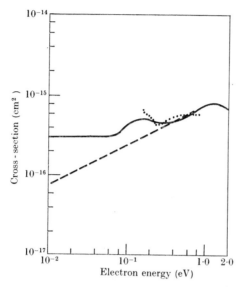

FIG. 11.40. Momentum-transfer cross-section for electrons in oxygen. —— derived by Hake and Phelps from analysis of swarm data. — — — derived from microwave conductivity measurements. . . . . total cross-section measured by Ramsauer method.

Notwithstanding the difficulties Hake and Phelps† have carried out an interesting analysis that brings out a number of special features. We consider here the details of this analysis for $\epsilon_k < 1$ eV.

The momentum-transfer cross-section, which gives the best fit to the observed data when taken with the rotational and vibrational excitation cross-sections discussed below, is shown in Fig. 11.40. While this cross-section agrees quite well with the total cross-section observed by the Ramsauer method, it is considerably larger than that obtained from microwave conductivity measurements by a number of authors‡ whose

† HAKE, R. D. and PHELPS, A. V., Phys. Rev. 158 (1967) 70.
‡ VAN LINT, V. A. J., WIKNER, E. G., and TRUEBLOOD, O. L., Bull. Am. phys. Soc. 5 (1960) 122; FEHSENFELD, F. C., J. chem. Phys. 39 (1963) 1653; MENTZONI, M. H.,

results agree quite well with each other. Electron drift velocities inferred from observations of the attachment–detachment equilibrium in O_2 (Chap. 19, § 4) are also consistent with the microwave data.

The initial choice of vibrational excitation cross-sections presents difficulty. The observations of Boness and Hasted† (Chap. 10, § 3.5.4), which show fine structure in the transmission of electrons through oxygen as a function of electron energy at quite low energies, suggest that vibrational excitation occurs by a resonance process as in N_2 though with

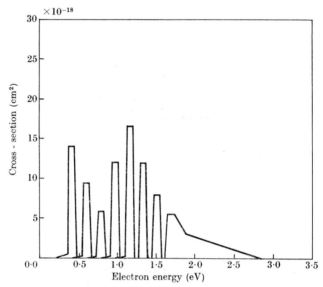

Fig. 11.41. Vibrational excitation cross-sections for electrons in oxygen derived from analysis of transport coefficients.

smaller maximum cross-sections. Hake and Phelps‡ assumed that the vibrational cross-sections are of this type, consisting of an energy plot of a number of narrow spikes commencing in each case at an energy beyond the corresponding threshold. This is necessary to maintain consistency with the observations of Schulz and Dowell§ made by the trapped-electron method (§ 3), which showed very small cross-sections at 0.15 eV above each threshold. Fig. 11.41 shows the cross-sections that give the best fit with the swarm data. No allowance was made for electronic excitation for $\epsilon_k < 1$ eV.

The rotational excitation cross-sections were taken to be of the form

Rad. Sci. J. R. **69D** (1965) 213; VEATCH, G. E., VERDEYEN, J. T., and CAHN, J. H., *Bull. Am. phys. Soc.* **11** (1966) 496.

† loc. cit., p. 715.   ‡ loc. cit., p. 787.   § loc. cit., p. 745.

(26) but with the magnitude of the quadrupole moment adjusted to give the best fit. It was found that $|q|$ needed to be as large as $1{\cdot}8ea_0^2$, which greatly exceeds the maximum value $0{\cdot}4ea_0^2$ derived from microwave studies of line broadening.† The large values required for the rotational excitation cross-section are in agreement with the evidence provided from microwave cross-modulation experiments (see § 6.2.2).

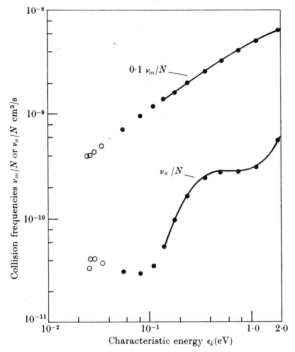

FIG. 11.42. Comparison of observed and calculated values of the collision frequencies ν_m and ν_u for oxygen. ——— curve through points representing an average of best available data. ● calculated by continuous approximation to rotational excitation. ○ calculated using a single-level approximation to the rotational excitation.

Fig. 11.42 shows the comparison between observed and calculated values of the momentum-transfer and electron-exchange collision frequencies ν_m and ν_u. The corresponding comparison for the drift velocities and characteristic energies is shown in Fig. 11.43.

A further check on the oxygen analysis is provided by analysis of the data for dry air discussed in § 7.6.

Extension of the analysis to higher characteristic energies is discussed in Chapter 13, § 4.4.5.

† SMITH, W. V. and HOWARD, R., *Phys. Rev.* **79** (1950) 132.

7.5. *Application to* CO_2

Both the effective momentum-transfer and energy-exchange collision frequencies for low energy electrons in CO_2 are exceptionally large as may be seen by reference to Fig. 11.46. No drift velocity measurements are available for $0.6 < \epsilon_k < 3$ eV so the analysis of Hake and Phelps,†

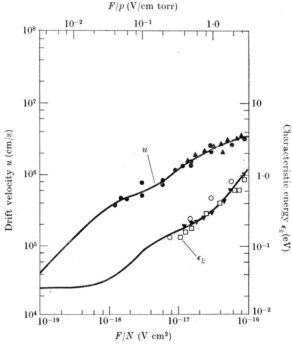

FIG. 11.43. Comparison of observed and calculated values of drift velocity (u) and characteristic energy (ϵ_k) for electrons in O_2. —— calculated. Observed for u: ● Pack and Phelps (300 °K); ▲ Nielsen and Bradbury (293 °K). Observed for ϵ_k: □ Brose (288 °K); ▼ Rees (293 °K); ○ Healey and Kirkpatrick (288 °K).

References: PACK, J. L. and PHELPS, A. V., *J. chem. Phys.* **44** (1966) 1870; BRADBURY, N. E. and NIELSEN, R. A.,*Phys. Rev.* **51** (1937) 69; BROSE, H. L., *Phil. Mag.* **50** (1925) 536; REES, J. A.,*Aust. J. Phys.* **18** (1965) 41; HEALEY, R. H. and KIRKPATRICK, C. B.,fromHEALEY, R. H. and REED, J. W., *The behaviour of slow electrons in gases*, p. 94 (Amalgamated Wireless, Sydney, 1941).

which we discuss here, is limited to $\epsilon_k < 0.6$ eV. Also, as the threshold for vibrational excitation is as low as 0.083 and the sublimation temperature T_s is so high that $\kappa T_s = 0.0168$ eV, it is not possible to obtain results that are determined by rotational but not vibrational energy loss. With the known quadrupole moment‡ of CO_2 ($3ea_0^2$) the rotational excitation

---

† loc. cit., p. 781.

‡ ORCUTT, R. H., *J. chem. Phys.* **39** (1963) 605; MARYOTT, A. A. and KRYDER, S. J., ibid. **41** (1964) 1580.

cross-sections are likely to contribute about as much to ν_u/N as for N_2. As this is about 2×10^{-11} cm^3 s^{-1} for $\epsilon_k = 0.04$ eV we see by reference to Fig. 11.46 that the rotational contribution is negligible for CO_2 in the range of the analysis.

Fig. 11.44 shows the momentum-transfer cross-section that is derived from the analysis. For low electron energies ϵ it increases as $\epsilon^{-\frac{1}{2}}$, consistent with the temperature independence of the drift velocity at

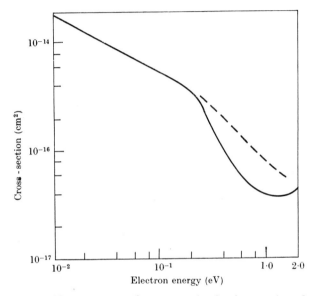

FIG. 11.44. Momentum-transfer cross-section for electrons in carbon dioxide. —— derived by Hake and Phelps from analysis of swarm data. — — — total cross-section measured by Ramsauer method.

thermal energies observed by Pack *et al.*† The marked minimum near 1 eV is not inconsistent with total cross-section data as the angular distribution of electrons of this energy scattered in CO_2 is known to show a pronounced concentration in the forward direction.

Observations made by Schulz‡ at a scattering angle of 72° have shown a resonance structure in the excitation cross-sections for the vibrational levels with thresholds at 0.3, 0.6, and 0.9 eV. They also reveal an extension of the cross-section for the 0.3-eV loss out to high energies that almost certainly arises through direct excitation. Geiger and Wittmaack§ (see § 2) observed two vibrational energy losses suffered by 33-keV electrons in CO_2, one at 0.3 eV and the other at 0.083 eV.

† loc. cit., p. 782.   ‡ Unpublished.   § loc. cit., p. 736.

Both of these excitations should persist to low energies as a direct excitation. The energy dependence of these 'direct' cross-sections was calculated from (78). For the 0·083-eV loss the magnitude of the cross-section was taken as that given by substituting for the effective dipole moment the value obtained from the infra-red absorption data as in § 2. It was found that for the 0·3-eV loss it was necessary to take a direct cross-section only 75 per cent of that derived in the same way.

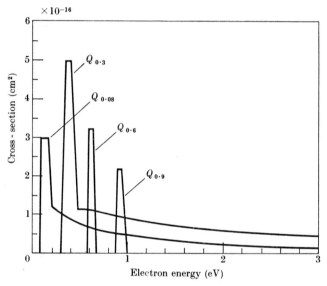

Fig. 11.45. Vibrational excitation cross-sections for electrons in carbon dioxide derived from analysis of transport coefficients. The suffixes refer to the threshold energies for each excitation.

Having determined the direct vibrational excitation cross-sections the magnitudes of the resonance portions for all four energy losses were adjusted to give the best fit to the swarm data. If the resonance portions are omitted and the direct cross-sections are both taken as given from (78) in conjunction with infra-red absorption data, good agreement is obtained with ν_u/N for $\epsilon_k \simeq 0·4$ eV but the values given are too low by a factor of 2 for $\epsilon_k \simeq 0·04$ eV.

Fig. 11.45 shows the set of vibrational cross-sections finally used. The important quantity in defining the resonance portions is the cross-section integrated over the resonance—the choice of width is dictated largely by convenience in the calculations.

The agreement between observed and calculated effective momentum transfer and energy-exchange collision frequencies is shown in Fig. 11.46

and the corresponding comparison between observed and calculated drift velocities and characteristic energies in Fig. 11.47.

Extension of the analysis to higher characteristic energies including energy loss through electron excitation is discussed in Chapter 13, § 8.

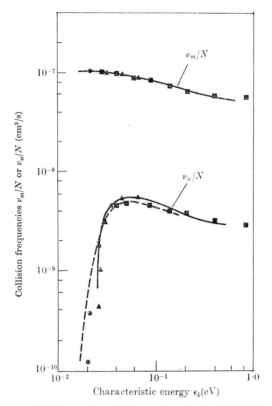

FIG. 11.46. Comparison of observed and calculated values of the collision frequencies ν_m and ν_u for carbon dioxide at 293 and 195 °K. Curves through points representing an average of best available data: —— 293 °K, — — — 195 °K. ▲ calculated (193 °K), ● calculated (195 °K). ■ calculated neglecting collisions of the second kind.

7.6. *Application to dry air*

Using the various cross-sections derived for N_2 and O_2 as described in §§ 7.2 and 7.4 the transport coefficients for dry air may be calculated. This has been done by Hake and Phelps with the results shown in Fig. 11.48, in which a comparison is made between calculated values of the effective momentum-transfer and energy-exchange collision frequencies and averaged experimental values. The agreement is not unsatisfactory.

The relative importance of the contribution from the two constituents to ν_u/N may be judged from the values of ν_u/N for the pure gases shown in Figs. 11.33 and 11.43.

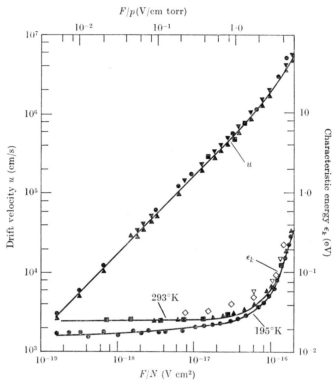

FIG. 11.47. Comparison of observed and calculated values of drift velocity (u) and characteristic energy (ϵ_k) for electrons in CO_2. —— calculated. u observed: ● Pack, Voshall, and Phelps (300 °K); ▲ Pack, Voshall, and Phelps (195 °K); ■ Errett (293 °K), ▼ Frommhold (293 °K); ◇ Riemann, ⊻ Elford (293 °K). ϵ_k observed: ● Warren and Parker (195 °K); ■ Warren and Parker (301 °K); ▲ Rees (293 °K); ◇ Skinker (288 °K); ▽ Rudd.

References: PACK, J. L., VOSHALL, R. E., and PHELPS, A. V., loc. cit. (Fig. 11.39); ERRETT, D., loc. cit. (Fig. 11.33); FROMMHOLD, L., loc. cit. (Fig. 11.27); RIEMANN, W., Z. Phys. 122 (1944) 216; ELFORD, M. T., Aust. J. Phys. 19 (1966) 629; WARREN, R. W. and PARKER, J. H., loc. cit. (Fig. 11.27); REES, J. A., Aust. J. Phys. 17 (1964) 462; SKINKER, M. F., Phil. Mag. 44 (1922) 994; RUDD, J. B., from HEALEY, R. H. and REED, J. W., The behaviour of slow electrons in gases (Amalgamated Wireless, Sydney, 1941).

8. Rotational excitation—improved theoretical description

It appears from the detailed analysis of swarm data and from the experiments on cross-modulation heating of discharge afterglows that, while the simple theory of Gerjuoy and Stein† (see (26)) gives results

† loc. cit., p. 731.

for the rotational excitation cross-sections that are of the correct order of magnitude, there are many features which it fails to predict.

Thus for nitrogen good agreement is obtained with experiment if the simple formula (26) is used with the observed value of the quadrupole

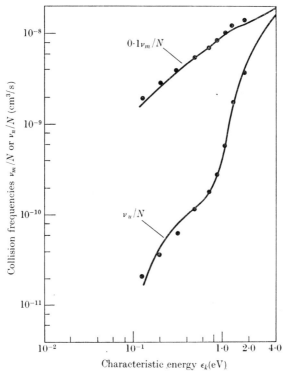

Fig. 11.48. Comparison of observed and calculated values of the collision frequencies ν_m and ν_u for dry air. —— smooth curve derived from drift velocity measurements by Nielsen and Bradbury and characteristic energy measurements of Townsend and Tizard; Crompton, Huxley, and Sutton; and Rees and Jory. ● calculated by Hake and Phelps.

References: NIELSEN, R. A. and BRADBURY, N. E., *Phys. Rev.* **51** (1937) 69; TOWNSEND, J. S. and TIZARD, H. T., *Proc. R. Soc.* A88 (1913) 336; CROMPTON, R. W., HUXLEY, L. G. H., and SUTTON, D. J., ibid. A218 (1953) 507; REES, J. A. and JORY, R. L., *Aust. J. Phys.* **17** (1964) 307.

moment and the factor $\phi(k, \Delta k)$, which includes allowance for polarization, put equal to unity. Because the quadrupole moment of nitrogen is negative the factor $\phi(k, \Delta k)$, according to the calculation of Dalgarno and Moffat,† should be less than unity (see Fig. 11.32).

For oxygen the situation is much less satisfactory. The experimental evidence is compatible with rotational excitation cross-sections of the

† loc. cit., p. 750.

simple form given by Gerjuoy and Stein but with a quadrupole moment over four times larger in magnitude than the best value observed in microwave line-broadening experiments.

Oxygen as well as nitrogen, has a negative quadrupole moment and Geltman and Takayanagi† have discussed the behaviour to be expected

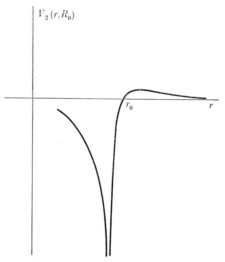

FIG. 11.49. General shape of the radial part of the non-central interaction $V_2(r, R_0)$ of electrons with molecules such as O_2 and N_2 which possess a negative quadrupole moment.

of the rotational excitation cross-section as a function of electron energy under these circumstances. The interaction between an electron and a diatomic molecule can be expanded in the form

$$V(\mathbf{r}, \mathbf{R}) = \sum V_l(r, R) P_l(\hat{\mathbf{r}} . \hat{\mathbf{R}}), \qquad (80)$$

where $\mathbf{R}$ is the vector separation of the nuclei and $\mathbf{r}$ the coordinate of the electron relative to the centre of mass of the molecule. The only term in this expansion that contributes to excitation of a transition in which the rotational quantum number changes by 2, as in the theory of Gerjuoy and Stein, is that involving $P_2(\hat{\mathbf{r}} . \hat{\mathbf{R}})$. Following (19) we may write

$$V_2(\mathbf{r}, \mathbf{R}_0) \simeq V_{s2}(r, R_0) - \{\tfrac{1}{2}\alpha'(R_0)e^2 r^{-4} + q(R_0)e r^{-3}\}, \qquad (81)$$

where R_0 is the equilibrium separation of the nuclei, α' is as defined in (20), and q the quadrupole moment. V_{s2} is the contribution from the short-range interaction. For N_2 and O_2

$$q < 0, \quad \alpha' > 0, \quad V_{s2}(r, R) < 0.$$

† GELTMAN, S. and TAKAYANAGI, K., Phys. Rev. 143 (1966) 25.

Under these circumstances the general shape of $V_2(r, R_0)$ as a function of r is as shown in Fig. 11.49. At large r it is repulsive due to the negative quadrupole moment but it vanishes at $r = r_0$, where

$$r_0 = \tfrac{1}{2}\alpha' e/|q|.$$

Inserting numerical values we find $r_0 \simeq 2 \cdot 0 a_0$ for N_2 and $8 \cdot 5 a_0$ for O_2. For $r < r_0$ the attractive potential rises quite rapidly as the short-range attraction becomes important.

To trace the effect of these features of V_2 on the variation of the rotational excitation cross-section with electron energy we note that, at low electron energies, the important contribution to the cross-section arises from incident p-electrons. It is true that s-electron waves overlap the molecular field strongly but they can only excite the rotational transitions concerned if they emerge as d-waves which overlap the molecular field only very slightly. A p-electron wave, on the other hand, can produce the transition while remaining a p-wave.

A plane p-wave of wave number k has its first maximum (see Fig. 6.2) when

$$kr \simeq 2,$$

so that, qualitatively, the scattering amplitude for the rotational excitation considered as a function of $1/k$ will follow much the same form as the molecular interaction $V_2(r, R_0)$ as a function of r. In particular the minimum due to the zero at $r = r_0$ will occur approximately at a wave number k_{min} where

$$k_{min} = 2/r_0.$$

This will be much smaller for O_2 than for N_2 and the general shape of the cross-sections in these two molecules will be as shown in Fig. 11.50.

This means that the energy range of validity of the Gerjuoy and Stein formula for N_2 is considerably greater than for O_2. While, at very low electron energies, the cross-section for N_2 is much greater than for O_2 in the ratio of the squares of their quadrupole moments, the rapid rise of the cross-section, due to exposure to the short-range field, occurs at a much lower energy than for N_2. This means that at somewhat greater energies the cross-section for O_2 exceeds that for N_2.

If now we consider the effect of distortion of the incident wave by the spherical part of the molecular field, including the polarization term $-\tfrac{1}{2}\alpha e^2 r^{-4}$ where α is given by (20), we would expect the same general behaviour of the cross-section as discussed above, but because the attraction reduces the wavelength of the p-waves within the atomic field the minimum should occur at a lower energy.

These features are clearly seen by reference to Fig. 11.50, which reproduces the results of calculations carried out by Geltman and Takayanagi.† They chose for the interaction $V(\mathbf{r}, \mathbf{R}_0)$

$$V(\mathbf{r}, \mathbf{R}_0) = \{V_{s0}(r, R_0) - \tfrac{1}{2}\alpha e^2 r^{-4} C_0(r)\} + $$
$$+ [V_{s2}(r, R_0) - \{eqC_1(r)r^{-3} + \tfrac{1}{2}\alpha' e^2 C_2(r)r^{-4}\}]P_2(\hat{\mathbf{r}}.\hat{\mathbf{R}}), \quad (82)$$

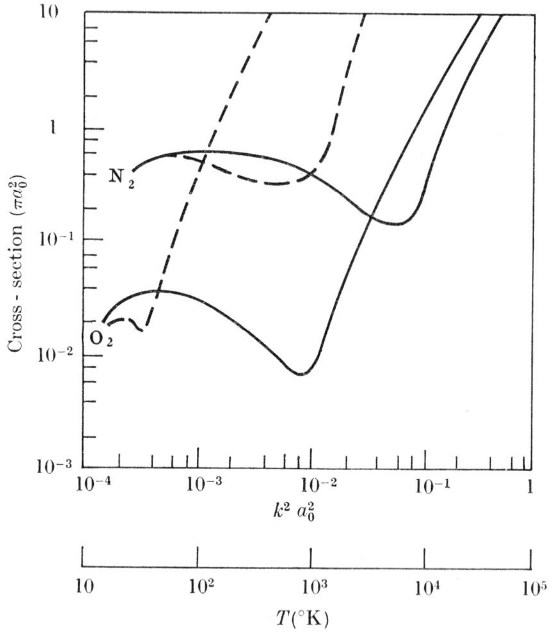

FIG. 11.50. Cross-sections for excitation of the rotational transition $J = 1 \to J = 3$ by electron impact in O_2 and N_2 calculated by Geltman and Takayanagi. —— using Born's approximation. — — — using the distorted wave approximation.

where q, α, and α' are the values at $R = R_0$. The functions $C_0(r)$, $C_1(r)$, $C_2(r)$ are cut-off factors that were taken to be unity for $r \geqslant$ some cut-off distance ρ, but for $r < \rho$ were chosen as

$$C_0(r) = r^4/\rho^4, \quad C_1(r) = r^5/\rho^5, \quad C_2(r) = r^6/\rho^6.$$

The short-range functions V_{s0}, V_{s2} were calculated by assuming that the total short-range interaction is approximately the sum of the fields V_a due to two undistorted atoms at the proper separation. Thus

$$V_s = V_a(r_1) + V_a(r_2), \quad (83)$$

where r_1, r_2 are the distances of the electron from the respective atoms. V_a was calculated from the Hartree–Fock field of the appropriate atom.

† loc. cit., p. 796.

V_{s0} and V_{s2} were then calculated by harmonic analysis of (83). The distorted wave approximation (Chap. 8, § 4.2) was then applied, the distorted p-waves being calculated by electronic solution of the appropriate Schrödinger equation and the cut-off distance ρ chosen so that the elastic scattering given by the central terms in (82) agreed as well as possible with observation.

It will be seen from Fig. 11.50 that the effect of distortion is quite marked for both N_2 and O_2. According to the Born approximation the

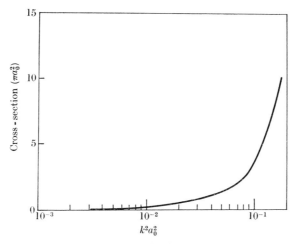

Fig. 11.51. Cross-section for excitation of the rotational transition $J = 0 \rightarrow J = 2$ by electron impact in H_2, calculated by Geltman and Takayanagi using the distorted wave approximation.

nitrogen cross-section would exceed that for oxygen for $ka_0 \leqslant 0\cdot25$, which corresponds to electrons of energy corresponding to the mean thermal energy at 6000 °K. This is inconsistent with the experimental results. However, when distortion is included the oxygen cross-section rises rapidly above that for nitrogen for $ka_0 \geqslant 0\cdot03$, corresponding to electrons at a temperature below 100 °K. Although the quantitative accuracy of the results shown in Fig. 11.50 is not high they nevertheless provide an explanation of the behaviour of the rotational excitation cross-sections for N_2 and O_2. In particular, the large apparent quadrupole moment for O_2 is due to the fact that, in the energy region of importance in the experiments, rotational excitation in O_2 arises from the short-range non-central interaction and not from the long-range quadrupole term at all. For N_2, on the other hand, in the same energy range the latter term is still dominant.

A distorted wave calculation has also been carried out by Mjolsness and Sampson[†] for N_2. They were concerned with distortion due to the polarization rather than to the short-range field and their results are in general agreement with those of Takayanagi and Geltman.

One further point arises for O_2. Whereas for N_2 and H_2 the ground electronic state is a singlet so that there are no complications due to coupling with electron spin, for O_2 the ground state is a triplet and coupling occurs between the electron spin and the molecular rotation. Takayanagi and Geltman[‡] have shown that this coupling has no important effect on the rotational excitation by electron impact.

Hydrogen possesses a positive quadrupole moment so that the cross-section for rotational excitation should vary smoothly with electron energy. This was confirmed by Geltman and Takayanagi[‡] who carried out a distorted wave calculation in exactly the same way as for N_2 and O_2. Fig. 11.51 illustrates their results. It is likely that the departures observed from the formula (26) for H_2 and D_2 are due to distortion but it is difficult to confirm this in detail.[§]

9. Rotational excitation of molecular positive ions

Formula (26) needs modification, even within the limits of Born's approximation, if the molecule is charged. If this is so the electron waves are modified by the long-range Coulomb field, an effect that must be included even if other distorting forces are ignored.

Stabler[||] has allowed for this Coulomb distortion for excitation of molecular positive ions. He finds that, for the same quadrupole moment and energy levels, the cross-section $Q_J^{J'}$ for the ion exceeds that for the neutral molecule by a factor

$$12\pi^2(\ln 2 - \tfrac{2}{3})/kk'a_0^2,$$

where k, k' are the initial and final electron wave numbers respectively.

This is a large factor for slow electrons, becoming infinite at the threshold—for the positive ion the cross-section starts at a finite value in contrast to the zero initial value for the neutral molecule.

Sampson[††] has considered the importance of induced polarization and distortion for these collisions.

  † MJOLSNESS, R. C. and SAMPSON, D. H., *Phys. Rev.* **140** (1965) A1466.
  ‡ loc. cit., p. 796.
  § See TAKAYANAGI, K. and GELTMAN, S., *Phys. Lett.* **13** (1964) 135.
  ‖ STABLER, R. C., *Phys. Rev.* **131** (1963) 679.
  †† SAMPSON, D. H., ibid. **137** (1965) A4.

COLLISIONS OF ELECTRONS WITH MOLECULES—ELECTRONIC EXCITATION, IONIZATION, AND ATTACHMENT—GENERAL THEORETICAL CONSIDERATIONS AND EXPERIMENTAL METHODS

1. Introductory—Quantum states of diatomic molecules

IN Chapter 11 we have already pointed out that the apparently very complicated problem of determining the quantum states of a molecule, even if diatomic, is much simplified by the relatively great mass of the nuclei compared with the electrons. As a result, in all molecules the electrons are much more mobile than the nuclei and are usually able to adjust themselves to changing nuclear motion without sufficient disturbance to produce a transition in their own state of motion. For any fixed nuclear separation R of a molecule AB we can derive a set of electronic energy levels $\epsilon_n(R)$ ($n = 1, 2,...$) with corresponding wave-functions $\psi_n(\mathbf{r}, \mathbf{R})$, $\mathbf{r}$ representing the aggregate of electronic coordinates relative to the centre of mass of the nuclei. In the limit of infinitely large nuclear separation R, $\epsilon_n(R)$ will simply become the sum of the energies of two definite states, either of the atoms A and B, of the ions A^+ and B^-, or of the ions A^- and B^+. It is important to note, however, that more than one molecular electronic level may tend to the same limit at infinite separation. Thus, while the energy of the ground electronic level of a molecule will tend to that of the two atoms in their ground states, there may be other molecular levels which tend to this same limit.

When we allow the nuclei to move, we now assume that the electronic motion adjusts itself so that, when the nuclear separation is R, the energy of the system due to the electron motion is $\epsilon_n(R)$. The nuclei, of charge $Z_1 e$, $Z_2 e$ respectively, can then be regarded as moving in a field of force of potential $\eta_n(R) = \epsilon_n(R) + Z_1 Z_2 e^2/R$. The plot of $\eta_n(R)$ against R is therefore referred to as the potential energy curve for the molecule in the nth electronic state. Its form determines the nature of the nuclear motion. The two most important cases which arise are illustrated in Fig. 12.1.

In Fig. 12.1 (a) the potential energy curve has a minimum at $R = R_0$, which is therefore an equilibrium separation for the nuclei. As $R \to \infty$ the curve tends asymptotically to a point representing the sum of the energies U'_A, U''_B of the two states A', B'' of the atoms A, B. If we now suppose the nuclei moving under the influence of this potential, the quantum states of the nuclear motion will consist of a discrete series of vibrational levels such as at ab, cd in Fig. 12.1 (a), converging to a limit

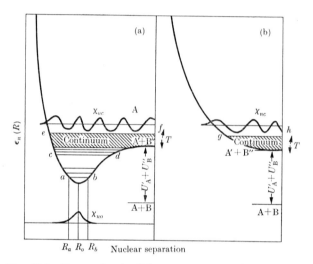

Fig. 12.1. Types of potential energy curves and vibrational wave functions for a diatomic molecule. (a) Typical curve for a bound state. (b) Typical curve for a repulsive state.

at the electronic energy possessed by the system when at infinite nuclear separation. Above this limit there will be a continuum of unclosed nuclear states representing a dissociated molecule in which the nuclei have various amounts of kinetic energy of relative motion.

These nuclear levels are illustrated in Fig. 12.1 (a). The total energy of the molecule in the particular electronic state corresponding to a will be given by

$$\epsilon_n(R_0) + v_{ns},$$

where v_{ns} is the energy of the nuclear motion.

Corresponding to this energy the nuclear wave function will have the form

$$\Psi_{ns}(\mathbf{R}, \mathbf{r}) = \psi_n(\mathbf{R}, \mathbf{r})\chi_{ns}(\mathbf{R}). \tag{1}$$

Here $\psi_n(\mathbf{R}, \mathbf{r})$ is the electronic wave function defined for each nuclear separation and χ_{ns} is the wave function for the nuclear motion. The wave function χ_{n0} for the lowest vibrational state is very small in regions outside the classically allowed motion, i.e. for values of $R < R_a$ or $> R_b$ in Fig. 12.1 (a), and has the general form illustrated in that figure. For higher vibrational levels the vibrational wave functions χ_{ns} are also small outside the classically allowed region but have s nodes within that region.

A state lying within the continuum of nuclear levels no longer represents a stable molecule, but the wave function can still be written approximately in the form (1). The function χ_{nc} representing the nuclear motion in this case will consist of modulated plane waves in the classically allowed region, falling rapidly to zero at closer nuclear separations, as illustrated diagrammatically in Fig. 12.1 (a).

A state in the nuclear continuum represented in Fig. 12.1 (a) by ef corresponds to two atoms A′, B″ moving with relative kinetic energy T. If, in particular, the potential energy curve is that for the ground electronic level, the two atoms will be in their normal states. It is important to note also that, if the level of the curve for infinite nuclear separation is the energy of two ions A$^+$, B$^-$, a state such as ef would correspond to these two ions moving with the relative kinetic energy T.

In the second type of potential energy curve, illustrated in Fig. 12.1 (b), there is no minimum so that an effective repulsive force exists between the atoms (AB, A′B″, A$^-$B$^+$, or A$^+$B$^-$) at all separations. No stationary nuclear states and no stable molecules can result under these conditions— the electronic state is said to be repulsive. The state of affairs is much the same as for curves of type (a) when the nuclear motion lies in the continuum, so the wave function can be written again in the form (1) with χ_{nc} as illustrated in Fig. 12.1 (b). A state represented by gh in Fig. 12.1 (b) has an exactly similar significance to that of ef in Fig. 12.1 (a).

The same considerations apply to molecular ions such as AB$^+$ or AB$^-$. The only difference is that at infinite nuclear separation the potential energy curves tend asymptotically, for AB$^+$, to normal and excited states of A and B$^+$ or A$^+$ and B, and for AB$^-$ to those of A and B$^-$ or A$^-$ and B.

For stable molecules the ground electronic state gives a potential energy curve of type (a). There are molecules such as He$_2$ in which the ground electronic state gives a repulsive curve of type (b) whereas some of the excited electronic levels give stable curves. At best such molecules can only be metastable since eventually they must make a transition to the ground repulsive state.

In the above discussion we have made no mention of molecular rotation. Each of the vibrational levels we have indicated in Fig. 12.1 (a) is in reality composed of rotational levels the presence of which can lead to many important energy transfers within a molecule. As we shall have no occasion to discuss these effects in special detail the reader is referred to books on molecular spectra such as Kronig, *Band spectra and molecular structure*† or Herzberg, *Molecular spectra and molecular structure*‡ for detailed discussion of molecular energy levels.

It must be remembered, furthermore, that the above treatment, useful as it has been found to be, is, nevertheless, an approximation. The effects neglected, involving the interaction of electronic and nuclear motion, can be regarded as small internal perturbations capable of producing transitions between different electronic states defined as above. Important instances of this will be discussed in later sections.

2. Enumeration of electronic states of diatomic molecules

The problem of enumerating the electronic states of diatomic molecules is clearly a more difficult one than for atoms. We may nevertheless follow an analogous procedure. The first step is to consider the states in terms of electron configurations, regarding each electron as moving in the two-centre field due to the nuclei and the mean interaction of the other electrons. The specification of the individual two-centre orbitals is more difficult than for atoms but may be solved by considering the limiting forms taken by the orbital in the limits of infinitely great (separated atom) and vanishing (united atom) nuclear separation. It proves possible to relate the specification quite closely to the attractive or repulsive character of each orbital, an important step in relating the ultimate specification to the form of the corresponding potential energy curve.

Having specified the various configurations it is necessary to consider the terms arising from the configuration. The total spin and the axial components of electronic angular momentum may be used as approximate quantum numbers. The possible values of these numbers may be obtained again in terms of the particular configuration and related to the total spin and angular momentum quantum numbers of the separated and united atom limits to which the term tends.

Finally, to consider the fine structure, we must take account of the interaction of electronic spin and orbital angular momentum with nuclear rotation.

We now discuss in more detail the specification of electron configurations in diatomic molecules.

† Cambridge University Press, 1930.   ‡ Prentice Hall, New York, 1939.

2.1. *Electron configurations in diatomic molecules*

For atoms the orbitals comprising an electron configuration are based on the forms of those known exactly for the simplest atom, that of hydrogen, which contains one electron only. The simplest diatomic molecule, the ion H_2^+, also contains only a single electron and it is possible to solve exactly the Schrödinger equation for the motion of this electron in the field of the two protons at any fixed distance apart. At first sight this would seem to offer a convenient basis on which to specify electron configurations at least of homonuclear molecules. However, the two-centre wave functions are so much more complicated than the hydrogen atom functions that it is not easy to use them as the basis of any scheme. Moreover, their variation as a function of nuclear separation differs in some important ways from that of orbitals for more complicated molecules. Instead we approach the problem of specifying the individual orbitals as follows.

We consider first the stationary states of an electron in a general two-centres field. As there is no longer spherical symmetry the total orbital angular momentum is not a constant of the motion. Instead we may only take as constant the component of orbital angular momentum parallel to the nuclear axis. This component will have as possible values $\lambda \hbar$ where $\lambda = 0, \pm 1, \pm 2,\ldots$. States in which $\lambda = 0, \pm 1, \pm 2, \pm 3,\ldots$ are dubbed $\sigma, \pi, \delta, \phi,\ldots$ states respectively corresponding in Greek to the $s, p, d,\ldots$ notation for atomic states with orbital angular momentum quantum numbers $0, 1, 2,\ldots$. The spin angular momentum is again an approximate constant of motion as for atoms.

One further accurate specification is possible for molecules with nuclei of the same charge. An electronic state will be either symmetric or antisymmetric with respect to interchange of the nuclei. These two alternatives are distinguished by the suffixes g (for 'gerade') and u (for 'ungerade') respectively.

To distinguish the orbitals completely we must introduce two further identifications, replacing the total and azimuthal quantum numbers for the one-centre case. The most useful method of doing this is to relate the molecular orbital to either or both of the limits to which it tends for infinitely separated and united atoms respectively. The actual orbitals in the normal molecule will be intermediate between these two. If the molecule is firmly bound the united atom limit may be nearest the actual state, a situation which prevails for most of the diatomic hydrides. Otherwise the separated atom limit is likely to prove more useful. A special advantage of this method is that it gives an idea of whether the addition of an electron in a particular orbital is likely to lead to a weakening or a strengthening of the molecular binding. If the orbital is such that its occupation leads to an increase of binding energy it is said to be a *bonding* orbital, if to a decrease an *anti-bonding* orbital, while if it has little effect on the binding it is referred to as *inactive* or *non-bonding*. It may happen that a particular orbital is antibonding at certain nuclear separations and bonding at others. The effect of its occupation depends then on the nuclear separation and this may make a considerable difference to the estimated binding energy (cf. N_2).

We now trace the way in which the molecular orbitals arising from orbitals of a pair of separated like atoms go over into those of the united atoms. Consider a particular orbital of one of the atoms. The presence of the other atom at a great distance will perturb this through the application of an electric field along the nuclear axis. As a result a degenerate p orbital will be split into a σ- and two π-states. An s orbital on the other hand will give only a σ-state. Similarly a p orbital of the united atom will split into σ and π orbitals as the atoms are slightly separated. We shall adopt the usual notation. A $\sigma_g\,np$ orbital is an orbital with $\lambda = 0$, symmetrical in the nuclei, arising from an np orbital of the separated atom. On the other hand, $np\sigma$ is an orbital with $\lambda = 0$, symmetrical in the nuclei, arising from an np orbital of the united atom. In this case the nuclear symmetry is fixed by the azimuthal quantum number.

To follow the behaviour as the nuclear separation varies we construct a diagram as in Fig. 12.2 (a) in which the energy of a given orbital is traced as a function of this separation. We are guided by the fact that the curves for levels with the same values of λ cannot cross. This follows from the fact that, if two orbitals are such that a transition can occur between them through interaction with the nuclei, then their interaction must keep the corresponding potential energy curves a finite distance apart (see also § 8 of Chap. 18). On similar grounds it follows that curves with the same g or u symmetry can never intersect. Because of this a g orbital must merge into an orbital of the united atom with l even, a u into one with l odd.

Thus the $1s$ orbitals of two separated atoms give rise to $\sigma_g\,1s$ and $\sigma_u\,1s$ orbitals. These orbitals will tend to the lowest accessible orbitals of the united atom with $\lambda = 0$ and the same nuclear symmetry. Hence the $\sigma_g\,1s$ goes over to $1s\sigma$ but $\sigma_u\,1s$ must go to the much higher $2p\sigma$ orbital. A similar situation arises with the $2s$ orbitals of the separated atoms. $\sigma_g\,2s$ goes to $2s\sigma$ but $\sigma_u\,2s$ is 'promoted' to $3p\sigma$. Again, of the orbitals arising from $2p$ orbitals of the separated atom, $\sigma_g 2p$ is 'promoted' to $3s\sigma$, $\pi_g 2p$ to $3d\sigma$ and $\sigma_u 2p$ to $4p\sigma$ and only $\pi_u 2p$ tends to a 2-quantum state $2p\pi$ of the united atom. Since 'promotion' involves a marked increase in the energy of the orbital at very small nuclear separations it

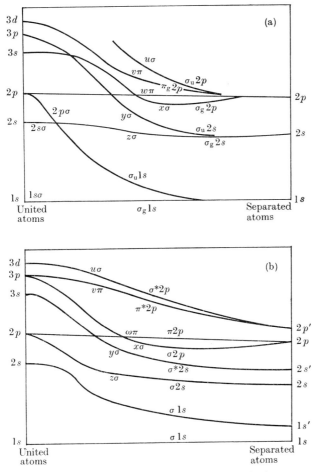

FIG. 12.2. Correlation of molecular orbitals to united atom and separated atom limits. (a) Homonuclear case. (b) Heteronuclear case.

usually means also an increase at intermediate distances. Hence orbitals in which promotion occurs in the united atom limits are usually anti-bonding. The energy order of the orbitals for the less complicated homo-nuclear diatomic molecules is fairly definite so Mulliken has introduced special symbols to distinguish them and to stress the fact that, at the equilibrium separation in the molecule, the orbital is not accurately represented by either the separated or unit atom limit.

Orbitals arising from separated atoms in $1s$ states are virtually inactive in binding atoms with 2-quantum shells, so we omit them from further consideration. The order and designation of the other low-lying orbitals for most diatomic molecules are then as given in Table 12.1, the more firmly bound orbitals being at the top. In the last column of

TABLE 12.1

Orbital	United atom limit		Separated atom limit		Nature	Number of states per orbital
Mulliken notn.	Equal nuclei	Unequal nuclei	Equal nuclei	Unequal nuclei		
$z\sigma$	$2s\sigma$	$2p\sigma$	$\sigma_g\,2s$	$\sigma 2s$	Bonding	2
$y\sigma$	$3p\sigma$	$3s\sigma$	$\sigma_u\,2s$	σ^*2s	Anti-bonding	2
$x\sigma$	$3s\sigma$	$3p\sigma$	$\sigma_g\,2p$	$\sigma 2p$	Anti-bonding (large sepn.) Bonding (small sepn.)	2
$w\pi$	$2p\pi$	$2p\pi$	$\pi_u\,2p$	$\pi 2p$	Bonding	4
$v\pi$	$3d\pi$	$3p\pi$	$\pi_g\,2p$	π^*2p	Anti-bonding	4
$u\sigma$	$4p\sigma$	$3d\sigma$	$\sigma_u\,2p$	σ^*2p	Anti-bonding	2

the table is given the number of electrons that can occupy a particular orbital without violation of the Pauli principle. In deriving these numbers it must be remembered that, for each value of λ, except $\lambda = 0$, there are two quantum states. Although we have placed the orbitals in a definite order of increasing energy this order may be slightly departed from in any particular molecule.

As an illustration of the use of these orbitals we may write down the lowest configurations for N_2 and O_2. In the former molecule there are 14 electrons and in the latter 16. Assigning these to the lowest accessible orbitals we have the configurations

$$(1s)^4\,(z\sigma)^2\,(y\sigma)^2\,(x\sigma)^2\,(w\pi)^4, \qquad N_2;$$
$$(1s)^4\,(z\sigma)^2\,(y\sigma)^2\,(x\sigma)^2\,(w\pi)^4\,(v\pi)^2, \quad O_2.$$

In the sense that all the available places in the occupied orbitals are filled in N_2 it can be regarded as having a closed shell structure. This is not so for O_2 as there are four places available in the outermost $v\pi$ orbital and only two occupied.

Another point of interest in these configurations is to note that in N_2 there are six bonding electrons, two antibonding, and two intermediate, while in O_2 there are two additional antibonding electrons. This is reflected in a much lower dissociation energy for O_2 as against N_2.

Mulliken has defined the number of bonds in diatomic molecules as half the excess of bonding over antibonding electrons. On this basis there is one less bond in O_2 than in N_2. Actually, at the equilibrium separations in these molecules the $x\sigma$ orbitals are bonding rather than antibonding and, allowing for this, there is a double bond in O_2 and a triple bond in N_2 in agreement with the usual chemical assignments.

The above considerations are still approximately valid even if the molecule is heteronuclear provided it is composed of atoms which are not too dissimilar. A little care must be taken in establishing detailed correlations because the distinction in terms of nuclear symmetry is no longer valid and the crossing rule now operates between terms which, in the limit of identical atoms, would have opposing nuclear symmetries. Fig. 12.2 (b) represents a typical correlation scheme. Orbitals arising from different separated atoms are distinguished by the presence or absence of a star as in σ^*2p, $\sigma2p$. Mulliken's scheme as outlined in Table 12.1 still remains valid, the separated and united atom correlations for the heteronuclear case being as indicated.

We may now assign the ground configurations of CO and NO. The former is isoelectronic with N_2 so has the same ground configuration. This explains the great similarity in properties of these two molecules. NO falls between N_2 and O_2 so that the $v\pi$ orbital contains but one electron. It therefore has one less antibonding electron than O_2 and, as expected, its binding energy falls between that of N_2 and of O_2.

When the atoms are very dissimilar as in the diatomic hydrides the above scheme is no longer applicable. The best approximation for these hydrides is to represent the configuration in terms of the united atom limit.

2.2. *Terms arising from a given electron configuration*

Since the effect of the axially symmetrical field between the atoms in a molecule is stronger than that between the electrons no quantum number analogous to the atomic L exists. Instead we have a quantum number Λ, which refers to the component of total electronic orbital angular momentum in the direction of the nuclear axis. It is the algebraic sum of the λ values of the individual orbitals. Capital Greek letters Σ, Π, Δ,... are used to distinguish terms in which $\Lambda = 0$, ± 1, ± 2,....

As spin-orbit coupling is weak for light molecules just as for light atoms, the total spin quantum number S is again a good quantum number. It is derived from the individual electronic spin quantum numbers s in exactly the same way as for atoms. The value of $2S+1$, the so-called multiplicity of the term is then indicated as a left-hand index as for atoms. Thus a $^2\Pi$ state is one with $S = \frac{1}{2}$, $\Lambda = \pm 1$.

For molecules in which the nuclei have the same charge a distinction may also be made between terms symmetrical or antisymmetrical in the nuclei. Just as for single orbitals these are distinguished by the right-hand suffixes g and u respectively. It is clear that a term arising from a configuration will be a g or u term according as the configuration contains an even or odd number in u orbitals.

Finally, Σ states of all molecules can be split into two types depending on the symmetry of the wave function with respect to reflection in any plane through the nuclear axis. If it is unchanged on reflection we have a Σ^+ state, if altered in sign a Σ^- state.

As an illustration consider the terms that can arise from the ground configuration of O_2. The closed shells contribute 0 to Λ and S. From the doubly-occupied outer $v\pi$ orbitals we may have $\Lambda = 0, \pm 2$, the Σ state being doubly degenerate, and $S = 0, 1$. When $\Lambda = \pm 2$ the Pauli principle only permits $S = 0$, giving rise to two $^1\Delta$ states. The Σ^- state is antisymmetric in the two electrons so for it $S = 1$ to make the complete electronic wave function (including spin) antisymmetric. Similarly the Σ^+ state is symmetric in the two electrons so for it $S = 0$. We thus have a $^3\Sigma_g^-$, a $^1\Sigma_g^+$, and two $^1\Delta_g$ terms arising from the configuration.

The nitrogen molecule has a closed-shell ground configuration giving rise only to a $^1\Sigma_g$ term. For NO with only one electron in the outer orbital the ground configuration gives rise only to a $^1\Pi_g$ term.

Table 12.2 summarizes the terms that can arise from different configurations both for equivalent and non-equivalent electrons.

2.3. *Molecular terms from terms of separated atoms*

We now consider the molecular terms that can arise from given terms in the separated atom limit, assuming Russell–Saunders coupling in the atomic states.

We consider first heteronuclear molecules. Let the total orbital and spin quantum numbers of the atomic state be L_1, S_1 and L_2, S_2 respectively. When the two atoms interact with sufficient strength neither L_1 and L_2 remain good quantum numbers but the quantum number specifying the total axial component of orbital angular momentum remains good. Since from the two atomic states the separate values of

TABLE 12.2

Terms arising from different molecular configurations

Configuration	Terms
Non-equivalent electrons	
σ	$^2\Sigma^+$
π	$^2\Pi$
$\sigma\sigma$	$^1\Sigma^+$, $^3\Sigma^+$
$\sigma\pi$	$^1\Pi$, $^3\Pi$
$\pi\pi$	$^1\Sigma^+$, $^3\Sigma^+$, $^1\Sigma^-$, $^3\Sigma^-$, $^1\Delta$, $^3\Delta$
$\sigma\sigma\sigma$	$^2\Sigma^+$, $^2\Sigma^+$, $^4\Sigma^+$
$\sigma\sigma\pi$	$^2\Pi$, $^2\Pi$, $^4\Pi$
$\sigma\pi\pi$	$^2\Sigma(2)$, $^4\Sigma^+$, $^2\Sigma^-(2)$, $^4\Sigma^-$, $^2\Delta(2)$, $^4\Delta$
$\pi\pi\pi$	$^2\Pi(6)$, $^4\Pi(3)$, $^2\Phi(2)$, $^4\Phi$
Equivalent electrons	
σ^2	$^1\Sigma^+$
π^2	$^1\Sigma^+$, $^3\Sigma^-$, $^1\Delta$
π^3	$^2\Pi$
π^4	$^1\Sigma^+$
Equivalent and non-equivalent electrons	
$\pi^2\sigma$	$^2\Sigma^+$, $^2\Sigma^-$, $^2\Delta$, $^4\Sigma^-$
$\pi^2\pi$	$^2\Pi(3)$, $^2\Phi$, $^4\Pi$
$\pi^2\sigma\sigma$	$^1\Sigma^+$, $^1\Sigma^-$, $^1\Delta$, $^3\Sigma^+$, $^3\Sigma^-(2)$, $^3\Delta$, $^5\Sigma^-$
$\pi^2\sigma\pi$	$^1\Pi(3)$, $^1\Phi$, $^3\Pi(4)$, $^3\Phi$, $^5\Pi$
$\pi^2\pi\pi$	$^1\Sigma^+(3)$, $^1\Sigma^-(3)$, $^1\Delta(4)$, $^1\Gamma$, $^3\Sigma^+(4)$, $^3\Sigma^-(4)$, $^3\Delta(5)$, $^3\Gamma$, $^5\Sigma^+$, $^5\Sigma^-$, $^5\Delta$
$\pi^2\pi^2$	$^1\Sigma^+(3)$, $^1\Sigma^-$, $^1\Delta(2)$, $^1\Gamma$, $^3\Sigma^+(2)$, $^3\Sigma^-(2)$, $^3\Delta(2)$, $^5\Sigma^+$
$\pi^3\sigma$	$^1\Pi$, $^3\Pi$
$\pi^3\pi$	$^1\Sigma^+$, $^1\Sigma^-$, $^1\Delta$, $^3\Sigma^+$, $^3\Sigma^-$, $^3\Delta$
$\pi^3\sigma\sigma$	$^2\Pi(2)$, $^4\Pi$
$\pi^3\pi^2$	$^2\Pi(3)$, $^2\Phi$, $^4\Pi$
$\pi^3\pi^3$	$^1\Sigma^+$, $^1\Sigma^-$, $^1\Delta$, $^3\Sigma^+$, $^3\Sigma^-$, $^3\Delta$
$\pi^3\pi^2\sigma$	$^1\Pi(3)$, $^1\Phi$, $^3\Pi(4)$, $^3\Phi$, $^5\Pi$
$\pi^3\pi^3\sigma$	$^2\Sigma^+(2)$, $^2\Sigma^-(2)$, $^2\Delta(2)$, $^4\Sigma^+$, $^4\Sigma^-$, $^4\Delta$

M_L range from $-L_1$ to $+L_1$ and $-L_2$ to $+L_2$ respectively it follows that $\sum M_L$ ranges from $-(L_1+L_2)$ to $+(L_1+L_2)$ thus specifying the possible Λ values, as follows, supposing $L_1 > L_2$,

$$L_1+L_2, L_1+L_2-1,..., \quad \Pi, \Sigma^+$$
$$L_1+L_2-1,..., \quad \Pi, \Sigma^-$$

$$\cdot \quad \cdot \quad \cdot \quad \cdot \quad \cdot$$

$$L_1-L_2,..., \quad \Pi, \Sigma^+ \text{ or } \Sigma^-$$

The assignment of the odd or even characters to the Σ states is determined by the fact that the last Σ state is Σ^+ or Σ^- according as $L_1+L_2+\sum_i l_i + \sum_j l_j$ is even or odd, l_i and l_j being the azimuthal quantum numbers of the ith and jth electron respectively in the respective atomic configurations.

The possible values of S for the molecular states are obtained in the usual way from the vector composition rule. They are

$$S = S_1 \sim S_2, S_1 \sim S_2 + 1, ..., S_1 + S_2 - 1, S_1 + S_2.$$

As an example, from the ground states $(1s)^2(2s)^2(2p)^2\ ^3P$ of carbon and $(1s)^2(2s)^2(2p)^4\ ^3P$ of oxygen the possible values for Λ give rise to Δ, Π, and Σ^+ terms. The possible values of S are 2, 1, 0 so the

TABLE 12.3

Molecular electronic states resulting from given states of the separated (unlike) atoms

States of separated atoms	Molecular states
$S_g + S_g,\ S_u + S_u$	Σ^+
$S_g + S_u$	Σ^-
$S_g + P_g$ or $S_u + P_u$	Σ^-, Π
$S_g + P_u$ or $S_u + P_g$	Σ^+, Π
$P_g + P_g$ or $P_u + P_u$	$\Sigma^+(2)$, Σ^-, $\Pi(2)$, Δ
$P_g + P_u$	Σ^+, $\Sigma^-(2)$, $\Pi(2)$, Δ

molecular terms of CO which can arise are $^5\Delta$, $^5\Pi$, $^5\Sigma^+$, $^3\Delta$, $^3\Pi$, $^3\Sigma^+$, $^1\Delta$, $^1\Pi$, $^1\Sigma^+$. Table 12.3 summarizes the terms that can arise from different pairs of atomic states for unlike atoms, for a number of important cases.

We next consider homonuclear molecules. If the two atoms are in different atomic states the number of states is doubled as compared with the case when the atoms are different—one of these states is a g state, the other a u. Otherwise there is no difference from the heteronuclear case. For two similar atoms in the same state with $L_1 = L_2$, $S_1 = S_2$ the Pauli principle must be taken into account and we have the following possible values depending on whether $S = 0, 1, ..., 2S_1$, is even or odd.

If S is even
$$(2L_1)_g, (2L_1 - 1)_g, ...,\quad \Pi_g, \Sigma_g^+$$
$$(2L_1 - 1)_u, ...,\quad \Pi_u, \Sigma_u^-$$
$$\cdot \qquad \cdot \qquad \cdot$$
$$..., \quad \Pi_u, \Sigma_u^-$$
$$..., \quad \Sigma_g^+$$

If S is odd
$$(2L_1)_u, (2L_1 - 1)_u, ...,\quad \Pi_u, \Sigma_u^+$$
$$(2L_1 - 1)_u, ...,\quad \Pi_g, \Sigma_g^-$$
$$\cdot \qquad \cdot \qquad \cdot$$
$$..., \quad \Pi_g, \Sigma_g^-$$
$$..., \quad \Sigma_u^+$$

As an example, for two O atoms in the ground $(1s)^2(2s)^2(2p)^4$ ^3P state with $L_1 = 1$, $S_1 = 1$ we have $^5\Delta_g$, $^5\Pi_g$, $^5\Sigma_g^+$, $^5\Pi_u$, $^5\Sigma_u^-$, $^5\Sigma_g^+$, $^3\Delta_u$, $^3\Pi_u$, $^3\Sigma_u^+$, $^3\Pi_g$, $^3\Sigma_g^-$, $^3\Sigma_u^+$, $^1\Delta_g$, $^1\Pi_g$, $^1\Sigma_g^+$, $^1\Pi_u$, $^1\Sigma_u^-$, $^1\Sigma_g^+$, molecular terms of O_2.

Table 12.4 summarizes the terms that can arise from states of two like atoms for a number of important cases.

TABLE 12.4

Molecular electronic states resulting from identical states of the separated (like) atoms

States of separated atoms	Molecular states
$^1S + {}^1S$	$^1\Sigma_g^+$
$^2S + {}^2S$	$^1\Sigma_g^+$, $^3\Sigma_u^+$
$^3S + {}^3S$	$^1\Sigma_g^+$, $^3\Sigma_u^+$, $^5\Sigma_g^+$
$^4S + {}^4S$	$^1\Sigma_g^+$, $^3\Sigma_u^+$, $^5\Sigma_g^+$, $^7\Sigma_u^+$
$^1P + {}^1P$	$^1\Sigma_g^+(2)$, $^1\Sigma_u^-$, $^1\Pi_g$, $^1\Pi_u$, $^1\Delta_g$
$^2P + {}^2P$	As for $^1P + {}^1P$ with, in addition, $^3\Sigma_u^+(2)$, $^3\Sigma_g^-$, $^3\Pi_g$, $^3\Pi_u$, $^3\Delta_u$
$^3P + {}^3P$	As for $^2P + {}^2P$ with, in addition, $^5\Sigma_g^+(2)$, $^5\Sigma_u^-$, $^5\Pi_g$, $^5\Pi_u$, $^5\Delta_g$

2.4. *Molecular terms in the united atom limit*

We now consider the molecular terms that can arise from a state of a united atom with given L and S values. The S value for the molecule is the same as for the united atom and the possible values of Λ are the same as the M_L values for the united atom, viz. $\pm L$, $\pm(L-1)$,..., 0. The Σ state is Σ^+ or Σ^- according as $L + \sum_i l_i$ is even or odd, l_i being the azimuthal quantum number of the ith electron in the united atom.

2.5. *Electronic states of negative molecular ions*

In Chapter 9 we have discussed the importance of excited states of negative atomic ions for the determination of fine-structure resonance features in the elastic and inelastic scattering of electrons by the corresponding atoms. So far as atomic collision phenomena are concerned the effective states, at present identified, are those in which an electron is bound to an excited state of the neutral atom. Such states are unstable towards autodetachment but possess a lifetime at least of order 10^{-12} to 10^{-13} s, which is considerably greater than the mean time taken for an electron of a few eV energy to traverse the atom. Because of the short lifetime the energies of the states are not sharply defined, the uncertainty being of order $\hbar/\tau$, where τ is the lifetime. With τ between 10^{-12} and 10^{-13} s it is of order 0·01–0·001 eV.

Provided certain selection rules are satisfied, an incident electron with energy falling within the energy range covered by one of these states can be captured temporarily and then re-emitted through auto-detachment, either with the initial energy, leaving the target atom in its initial state, or with reduced energy leaving the atom in an excited state. Such capture and re-emission processes are made manifest through the occurrence of typical resonance peaks and troughs in the variation of collision cross-sections with energy at energies within 0·01 to 0·001 eV or so of the unstable negative ion states.

It has also been pointed out in Chapter 9, § 9 that unstable states with lifetimes appreciably longer than the travel time of an electron across an atom can conceivably arise in a physically different way. The states we have been discussing above only exist through the fact that the target atom possesses structure and is not just a source of a scattering field. However, even if it were just such a source there are circumstances in which the quantized motion of the electron in that field could lead to an appreciable delay in passage through the field. Thus we have the example discussed in Chapter 9, § 9 of a potential energy of interaction, attractive at short distances and then changing to a repulsion at larger distances so that a potential barrier exists. Provided the parameters defining the strength of the interaction are suitable, for certain incident electron energies penetration of the barrier will be readily effected and the electron captured for an appreciable length of time before leaking out again through the barrier. The electron energies at which this occurs are known as virtual states because of their finite lifetimes. These may be quite long depending on the relative strengths of the attractive and repulsive parts of the interaction. Even if the potential is entirely attractive the time spent by an electron within the field will vary with electron energy and in certain energy ranges may exceed considerably the normal transit time for an electron to cover the distance over which the field is appreciable. As discussed in Chapter 9, § 9 we may consider a virtual state as existing at an electron energy for which the phase shift, for the appropriate angular momentum, passes through an odd integral multiple of $\frac{1}{2}\pi$. For typical atomic fields the maxima in the elastic cross-section at such energies have widths of order 1–2 eV at least, corresponding to electron delay times within the field of order 3–6×10^{-16} s, which is equal to the time taken for a 1-eV electron to travel 2 to 4×10^{-8} cm. This is not much larger than atomic dimensions so it is not very useful to consider these broad features as resonance phenomena. However, for some applications to molecular phenomena, including particularly dissociative attachment, even quite a brief residence of an incident electron in the neighbourhood of the molecule may be important. Furthermore, it is well to remember that for incident electrons with finite orbital angular momentum an effective potential barrier is introduced by the centrifugal force. Even though this does not seem to lead to virtual states of width much smaller than 1 eV in atomic collisions there may be cases in molecular collisions in which such states arise. Indeed the interpretation of the experimental evidence for N_2 seems to depend on the existence of such a virtual state.

Although we are not dealing with structureless sources of field the virtual states we have just been discussing correspond to states of negative atomic ions lying not far above the ground state of the neutral atom. If we need to take into account virtual states we must also allow for the presence of similar states lying just above the different excited states of the atoms. Following the notation of Chapter 9, § 9 we refer to the resonance states in which an electron is bound to an excited atom as of Type I, those which are virtual states in the field of the ground state or of an excited state as of Type II.

In applying these considerations to molecules we need to consider the variation of the energy and lifetimes of the resonance states with nuclear separation.

The effective potential energy in which the nuclei move in such states is complex with a negative imaginary component proportional at each nuclear separation to the lifetime of the state. At a separation for which the state is of Type I the lifetimes will normally be 10^{-13} s or longer. This is considerably greater than the vibrational period in a molecule such as N_2 ($1 \cdot 5 \times 10^{-14}$ s) and vibrational structure would therefore be expected to show up in a resonance process involving this state at the appropriate nuclear separation. A state of Type II can usually be expected to last too short a time for vibrational structure to show up when the state alone is involved but there may be exceptions of which N_2 could be one (see Chap. 13, § 3.7).

In constructing potential energy curves (including the imaginary component) for states of negative molecular ions it is useful to consider the limits to which the state tends when the nuclear separation tends to zero and to infinity. These limits may correspond to a stable state of a negative atomic ion or to a Type I or Type II resonance state. The nature of the state in the united atom limit may be different from that in the separated atom limit. Also, because of the small energy differences involved it does not follow that a state that tends to a Type I atomic

resonance state at both limits will remain of this Type I at all intermediate nuclear separations.

Some of the possibilities are illustrated in Fig. 12.3. In Fig. 12.3 (a),
(b), (c), and (d), the separated atom limit for the negative ion state

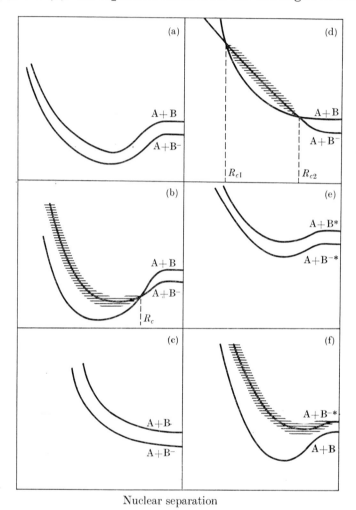

Nuclear separation

FIG. 12.3. Illustrating potential energy curves for electronic states of
negative molecular ions.

of AB$^-$ is one in which the atom A and the ion B$^-$ are in their ground
states, B having a positive electron affinity. Whereas in Fig. 12.3 (a)
the negative molecular ion is stable at all nuclear separations, in Fig.
12.3 (b), for nuclear separations $> R_c$ the molecular ion state is stable

but at smaller separations it is of Type II as indicated by the shaded region which represents the level width.

Fig. 12.3 (c) and (d) show pairs of repulsive curves that tend to the same separated atom limits as in Fig. 12.3 (a) and (b). In (c) the curve for the negative ion always lies below that for the neutral molecule so that the negative ion state is of Type I. On the other hand, in Fig. 12.3 (d) the curves for the neutral and negative ion states cross twice at $R = R_{c1}$ and R_{c2}. Thus, in this case, although in both the united and separated atom limits the negative ion state is of Type I, for nuclear separations R such that $R_{c1} \leqslant R \leqslant R_{c2}$ it is of Type II.

Fig. 12.3 (e) shows a situation similar to that which gives the fine resonances in atomic states. The negative molecular ion state is of Type I at all nuclear separations, tending in the separated atom limit to an atom A in its ground state and a negative ion B^- in a Type I resonance excited state.

Finally in Fig. 12.3 (f) the negative ion state, which tends to a Type II state of the ion B^- in the separated atom limit, is of Type II at all nuclear separations.

It will be realized that, because the negative ion states lie close to states of the neutral molecule, we must expect that interactions between the corresponding potential energy curves will occur frequently and it is only in the lowest states that comparatively simple situations involving Type I states, as shown in Fig. 12.3 (e), are likely to arise.

3. Qualitative description of processes associated with electronic transitions in diatomic molecules

3.1. *Selection rules*

In order that an electronic transition in a diatomic molecule be associated with an electrical dipole moment, i.e. that it is an optically allowed transition, the following rules must be obeyed:

$$\Delta\Lambda = 0, \pm 1, \quad \Sigma^- \leftrightarrow \Sigma^-, \quad \Sigma^+ \leftrightarrow \Sigma^+, \quad \text{but not } \Sigma^- \leftrightarrow \Sigma^+.$$

In addition, for molecules in which the nuclear charges are equal, $u \leftrightarrow g$ is allowed but $u \leftrightarrow u$, $g \leftrightarrow g$ are forbidden.

3.2. *The Franck–Condon principle*

An electronic transition in a molecule may result from the influence of an internal perturbation or of an external one, as in electron impact or in absorption or emission of radiation. The question that immediately arises concerns the way the nuclear separation behaves in the transition. The answer to this follows again from the great ratio of nuclear to electronic mass and is summarized in the Franck–Condon principle, which states that, in an electronic transition, the nuclear separation and velocity of relative nuclear motion alter to a negligible extent—the transition takes place so quickly that the nuclei have not time to move an appreciable distance. This leads to a number of possible consequences of an electronic transition that depend on the shapes of the potential energy curves of the initial and final electronic states. These are best studied in terms of potential energy diagrams, as illustrated in Fig. 12.4 for upward electronic transitions in a molecule AB.

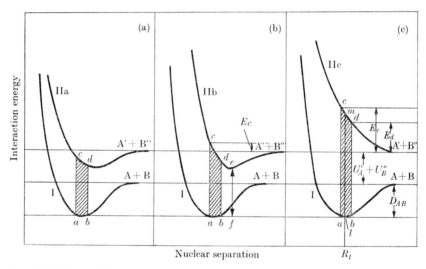

Fig. 12.4. Electronic transitions in molecules from a given initial state to three final states having different potential energy curves, illustrating the consequences of the Franck–Condon principle.

In all three cases represented in Fig. 12.4, curve I is the potential energy curve for the initial electronic state, while curves IIa, IIb, and IIc represent three distinct possibilities for the potential energy curve of the upper state. The nuclear separation in the ground vibrational level will effectively lie between the limits a and b in all cases. Hence, according to the Franck–Condon principle, it must still lie within these limits after the transition. Referring to Fig. 12.4 the final state of the molecule will therefore be represented by points lying between c and d on the upper curves. The three cases that are illustrated then correspond to the following consequences.

Case (a). The final state always lies within the region of the discrete vibrational levels of the upper potential energy curve. The transition always results then in a stable electronically excited molecule possessing also some degree of vibrational excitation.

Case (b). The region in which the final state must lie includes some part of the continuum as well as some discrete vibrational levels of the upper potential energy curve. A certain proportion of the transitions will therefore lead to dissociation of the molecule while others will produce stable excited molecules.

When dissociation occurs the molecule will split into two atoms A′, B″ (or AB or A^+B^- or A^-B^+, depending on the limit of the potential energy curve for large R) with relative kinetic energy ranging from 0 to E_c in Fig. 12.4 (b).

Case (c). The final state lies always within the continuum of nuclear levels. In this case dissociation of the molecule accompanies all transitions from the lower to the upper electronic state. The relative kinetic energy of the atoms or ions into which the molecule dissociates will lie between E_c and E_d in Fig. 12.4 (c).

3.3. Ionization of a molecule

A particular case of an upward electronic transition in a molecule occurs when it is ionized. The same considerations apply as in the general discussion above, except that the upper potential energy curve corresponds to an electronic state of the molecular ion AB^+. A transition of type (a) will thus produce a stable AB^+ ion, one of type (b) either a stable ion or dissociation into a neutral atom and an atomic ion with relative kinetic energy ranging from 0 to some value E_c, and one of type (c) a neutral atom and atomic ion with relative kinetic energy between E_d and E_c.

The ionization energy of a molecule may be defined as the difference between the energy of the ground state of the molecule and molecular ion. This energy may, however, bear no simple relation to the energy required to produce the ion from the neutral molecule. Mulliken has therefore introduced the *vertical ionization energy* as a more useful quantity. This is the minimum energy necessary to remove an electron from the normal molecule without change of nuclear separation. (Referring to Fig. 12.4, in which the upper state is taken to be one of AB^+, the vertical ionization energy is represented by bd in all three cases.) According to the Franck–Condon principle this will be the usual state of affairs in any actual transition. As the equilibrium separation of normal molecule and molecular ion need not be the same it is obvious that the vertical ionization energy will not normally be the difference in energy of the normal states of molecule and molecular ion (thus in Fig. 12.4 (b), while bd represents the vertical ionization energy, ef represents the difference in the energy of the normal states).

It must be remembered that the above treatment is only an approximation, though a very good one, so there may be a small but finite chance of ionization of the molecule if it receives less than the vertical ionization energy, i.e. in Fig. 12.4 (b) a transition from b to e is theoretically possible, but would be associated with a very low probability. This may result in a certain indefiniteness in molecular ionization potentials, depending on the sensitivity of the detecting apparatus. Owing, however, to the small chance of finding the nuclei at a separation represented by f in Fig. 12.4 (b), the probability of ionization appreciably below the vertical ionization potential is usually negligible.

3.4. *Energetic relations in dissociative transitions*

Consider a transition from the ground state of a molecule AB to an upper electronic state. Take as the zero of potential energy the energy of the two normal atoms A, B. If D_{AB} is the dissociation energy of AB, W_{min} the minimum energy of relative motion of the resulting atoms $A'B''$ (or ions A^+B^- or A^-B^+) after the transition, and U'_A, U''_B the excitation energy of these atoms then, according to the Franck–Condon principle, the minimum energy E_{min} necessary to produce the transition is given by

$$E_{min} = U'_A + U''_B + D_{AB} + W_{min}, \qquad (2)$$

as may be seen from Fig. 12.4 (c) in which $W_{min} = E_d$.

This formula applies equally well to ionization of the molecule, leading to production, say, of excited atoms A' and ions B^+. Its importance lies in the possibility of obtaining information about D_{AB} or the nature of the products A', B^+ by measurement of E_{min} and W. In any such measurement it is usually only possible to determine the kinetic energy W^+ of the B^+. If M_A, M_B are the respective masses of the atoms A and B, then it follows from the conservation of momentum that

$$W = (1 + M_B/M_A)W^+. \qquad (3)$$

Applications of these relations will be described in §§ 1.5, 1.6, 3.5, 4.3, 4.4, 5.2, 6, 7, 9, 10, 11.2, and 11.3 of Chapter 13.

3.4.1. *Energy distribution of atoms or ions in a dissociative transition.* In an electronic transition that can lead to dissociation, the resulting atoms or ions may have relative kinetic energy lying within a finite range. It is of interest and importance to have some means of estimating the probability that the energy will have any particular value in this range. This probability is determined mainly by the chance that the nuclear separation should have any particular value in the classically allowed range. Thus, referring to Fig. 12.4 (c), the chance that the transition be from l to m is determined mainly by the probability that the nuclear separation in the initial state be R_l. This probability is given by the value of $\chi_{00}^2(R_l)$, χ_{00} being the vibrational wave function appearing in (1). Using this function as weighting factor, the energy distributions of the atoms or ions resulting from transitions of the respective types (b) and(c) take the forms illustrated in Fig. 12.5. A characteristic difference

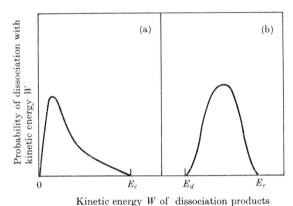

Kinetic energy W of dissociation products

FIG. 12.5. Kinetic energy distribution of products of dissociation by electron impact: (a) when the final state has a potential energy curve similar to IIb of Fig. 12.4 (b); (b) when the final state has a potential energy curve similar to IIc of Fig. 12.4 (c).

may be noted in that, in the former case, illustrated in Fig. 12.5 (a), where atoms or ions with zero kinetic energy may be produced, the distribution curve falls off more or less sharply on the low-energy side, whereas in case (c), where the atoms or ions formed have always some finite kinetic energy, the distribution curve is more nearly symmetrical as illustrated in Fig. 12.5 (b).

3.5. *Downward transitions*

Similar considerations to the above apply to downward transitions. These may result in production of a stable molecule or lead to dissociation, depending on the nature of the potential energy curve for the lower state and the nuclear separation involved.

3.6. *The formation of negative ions from molecules by electron impact*

There are two ways in which stable negative ions may be formed as a result of electron impact with a molecule, one in which the electron is captured and the other in which it simply breaks up the molecule into a positive and negative ion. We shall consider these possibilities separately.

3.6.1. *Negative ion formation by electron capture.* If an electron is captured by a neutral molecule a transition can be regarded as taking place between two electronic levels of the negative molecular ion. In the initial state one of the electrons occupies an unbound orbital and

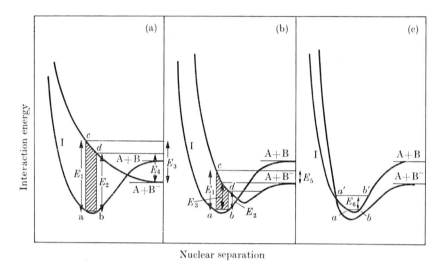

Fig. 12.6. Potential energy curves illustrating three possible ways in which negative ions may be formed from a molecule AB by electron capture.

the potential energy curve is just that of the neutral molecule in its initial state. The position of the upper potential energy curve relative to the lower determines the consequences of the capture. Three situations may arise, as illustrated by the potential energy curves of Fig. 12.6.

In all three cases (*a*), (*b*), and (*c*) curve I is the potential energy curve of the initial state of the neutral molecule AB while the second curve in each case represents that for the state of the negative molecular ion into which the electron is captured. At infinite nuclear separation these upper curves tend to the energy of certain states of A and B^-. If, as is often the case, these are the normal states, then the upper curve will tend, at infinite separation, to an energy value less than curve I by the electron affinity of B. We have not indicated the nature of the molecular negative ion state. It may be stable, of Type I or of Type II, at different nuclear separations (see § 2.5 and Fig. 12.3).

In cases (a) and (b), according to the Franck–Condon principle, the transitions will be confined within the shaded areas $abdc$, i.e. the final state will lie on the upper potential energy curve between c and d.

In case (a) the final state will therefore have an energy in excess of that at infinite nuclear separation.

This can lead to break-up of the negative molecular ion first formed into an atom A and ion B^- with total kinetic energy lying between E_3 and E_4. The threshold energy will be nearly equal to E_2 and the probability of the capture will fall rapidly again when the electron energy exceeds E_1. Whether or not the initial electron capture will lead to dissociative attachment depends on the balance between the rate at which the atom and ion begin to separate and the chance that the system will revert to a state of the neutral molecule through autodetachment.

Thus if $\hbar/\Gamma(R)$ is the mean lifetime against autodetachment when the nuclear separation is R, the fraction of molecular ions which would persist to a time T after formation, if they remained at a distance R apart would be $\exp\{-\Gamma(R)T/\hbar\}$. If the ions are first formed at a nuclear separation R_0, and they move apart with velocity of separation $v(R)$, the number remaining, when at a separation R, will be

$$
\exp\left\{-\frac{1}{\hbar}\int_{R_0}^{R}\frac{\Gamma(R')}{v(R')}\,dR'\right\}. \tag{4}
$$

Referring to Fig. 12.3 (b) we see that once the nuclear separation passes R_c, autodetachment can no longer occur. The cross-section for dissociative attachment is then given by

$$
Q_{\mathrm{da}} = Q_c \exp\left\{-\frac{1}{\hbar}\int_{R_0}^{R_c}\frac{\Gamma(R')}{v(R')}\,dR'\right\}, \tag{5}
$$

where Q_c is the cross-section for formation of the negative ion complex.

It is clear that Q_{da} will depend very much on whether the molecular negative ion state is of Type I or Type II (see § 2.5).

If it is possible to carry out experiments using different isotopes† an estimate may be made of the order of magnitude of the width $\Gamma(R)$ that will at least indicate the nature of the ionic state concerned. This

† HOLSTEIN, T. D., Conference on Gaseous Electronics, Schenectady, N.Y. 1951; DEMKOV, Y. N., Zh. éksp. teor. Fiz. 46 (1964) 1126; Soviet Phys. JETP 19 (1964) 762; Phys. Lett. 15 (1965) 235; BARDSLEY, J. N., HERZENBERG, A., and MANDL, F., Atomic collision processes, ed. McDOWELL, M. R. C., p. 415 (North Holland, Amsterdam, 1964); Proc. phys. Soc. 89 (1966) 321.

is possible because the velocity $v(R')$ is given by

$$\frac{1}{2}\frac{M_1 M_2}{M_1+M_2}v^2(R) = W(R),$$

where $W(R)$ is the height of the ionic curve at R above the limit at infinite separation. It is usual to write

$$Q_{da} = Q_c \exp\{-\bar{\Gamma}\tau/\hbar\}, \tag{6}$$

where τ is the time taken for A and B^- to separate from R_0 to R_s. For different isotopes τ is proportional to $M_1 M_2/(M_1+M_2)$ while Q_c and $\bar{\Gamma}$ are unchanged.

Bardsley, Herzenberg, and Mandl† have also been able to express Q_c in terms of the partial width $\Gamma_0(R)$ for decay of the ionic state to the ground state of the neutral molecule. $\Gamma_0(R)$ will only be the same as $\Gamma(R)$ if the negative molecular ion in the state concerned can only suffer autodetachment by reverting to the ground state of AB. The expression obtained for Q_c is

$$Q_c = \frac{4\pi}{k^2}\frac{2S_c+1}{2(2S_0+1)}\frac{\Gamma_0(R_0)}{\Delta E}\exp[-\{(E-E_c(\bar{R}_0))^2-\tfrac{1}{4}\Gamma^2(\bar{R}_0)\}/\Delta E^2]. \tag{7}$$

Here k is the wave number of the incident electron, S_c and S_0 are the respective total spin quantum numbers of the initial and final molecular states, ΔE is the range of energy over which capture can take place through the operation of the Pauli principle, and R_0 is the nuclear separation in the upper state into which the electron with its particular energy E_c is captured. The exponential factor essentially incorporates the Franck–Condon principle, $\bar{R}_0$ being a mean value of R_0 averaged over the Franck–Condon region.

Having determined Γ approximately from the isotope effect, Γ_0 may be estimated from (7) and the observed variation of Q_{da} with electron energy. We shall illustrate the application of this analysis to H_2 and O_2 in §§ 1.6 and 4.4 respectively of Chapter 13.

In case (b), some of the possible final states fall where the energy is less than that at infinite separation. Dissociation will only occur if the electron energy lies between E_1 and E_3 and the kinetic energy shared between A and B^- will range correspondingly from zero to E_5. Transitions produced by capture of electrons with energies between E_2 and E_3 will produce a vibrationally excited molecular ion AB^-. Unless such an ion loses its surplus energy in some other way first, it will release the captured

† loc. cit., p. 823.

electron and revert to the neutral molecule in the ground state by a process the reverse of that which led to its formation. Alternative ways of getting rid of the surplus energy, leading to a stable molecule AB⁻, are by radiation or by a superelastic impact. Radiation through vibrational transitions occurs very slowly and can usually be ignored, but stabilization by impact may be important if the pressure is not too low.

The rate of formation of stable AB⁻ by electron capture followed by collision stabilization will be proportional to the pressure at low pressures but, owing to saturation, may become independent of pressure at higher pressures. Thus, if τ be the average time taken for a vibrationally excited molecule AB⁻ to be relieved of its excess energy by impact with a gas molecule and θ the time before spontaneous dissociation of the molecule, $1/\tau$ is proportional to the gas pressure, θ independent of it. If, at time $t = 0$, we have an excited molecule AB⁻, the chance that it will not have dissociated in time t will be $e^{-t/\theta}$. The chance that it will be relieved of its excess energy by impact in a time between t and $t+dt$ is $e^{-t/\tau}\,dt/\tau$, so the total probability of the ion being stabilized by collision before it breaks up will be

$$\rho = \int_0^\infty \exp\left\{-t\left(\frac{1}{\tau}+\frac{1}{\theta}\right)\right\} dt/\tau$$
$$= \theta/(\theta+\tau)$$
$$= p/(p+p'),$$

where p is the gas pressure and p' a critical pressure for which $\theta = \tau$. Hence if p is large enough ρ becomes independent of it. Because of this result it is unsafe to assume that if the rate of formation of a particular molecular negative ion is independent of the pressure in a particular pressure range it will remain so at still lower pressure.

Case (c) represents a situation in which collision stabilization must occur if a stable molecular negative ion is to result. With the potential energy curves related as in Fig. 12.6 (c) no transition can occur with appreciable probability unless the neutral molecule is first excited to a vibrational level such as $a'b'$. Once excited to this state the chance of capture of an electron will be high. The following sequence of events may therefore be imagined. The incident electron first excites the neutral molecule to the vibrational level $a'b'$ and is thereupon captured into an excited vibrational state of the molecule AB⁻. Left to itself this molecule will break up again by the inverse process, but it can be stabilized by impact in the same way as the excited molecules AB⁻ discussed under case (*b*). This process will only be possible if the electrons have energies very close to E_6 of Fig. 12.6 (c).

A slight variant of case (*c*) would be if the incident electron first excited the molecule to a new electronic state with such vibrational excitation that the electron could then be captured into a vibrationally excited level of some electronic state of AB^-.

Summarizing these possibilities we have:

Case (*a*). Electrons with energies between E_1 and E_2 are captured and molecular dissociation results, giving atoms A and ions B^- with kinetic energies lying between E_3 and E_4. This process we call dissociative attachment.

Case (*b*). Electrons with energies between E_1 and E_3 are captured, again leading to molecular dissociation into atoms A and ions B^- with kinetic energies lying between 0 and E_5. Electrons with energies between E_2 and E_3 are also captured, giving rise to vibrationally excited molecular ions AB^-. In the absence of collision stabilization, i.e. at low pressures, these ions will be transitory and in a short time of order 10^{-14} s will revert to the initial neutral molecule with release of an electron. At higher pressures, however, stable molecular ions AB^- will be formed.

Case (*c*). Electrons with energies in the neighbourhood of E_6 are captured giving rise to vibrationally excited molecular ions AB^-, which will be transitory unless the pressure is high enough for stabilization to occur.

Normally the energy ranges E_1, E_2, etc., will be of the order of a few electron volts and would be expected to be somewhat larger in case (*a*) owing to the steepness of the repulsive part of the potential energy curve. It will be seen also that E_2, the minimum electron energy necessary to produce dissociative attachment, cannot be zero unless the electron affinity of B is greater than the dissociation energy D_{AB} of the ground state of the molecule AB. Except for the halogen molecules this is rarely true, so negative ions formed by capture of very slow electrons will normally be molecular ions AB^- formed as described above. Since they will only be stabilized if the pressure is sufficiently high, it is unlikely that they will be formed in any appreciable quantity in low-pressure experiments.

3.6.2. *Determination of dissociation energies and electron affinities.* The relation (2) applies to dissociative attachment as well as to other collisions in which dissociation occurs. The sum $U'_A + U''_B$ in (2) becomes $-E_a$, where E_a is the electron affinity of B, if both the products A and

B^- are in their ground states. Hence, from an experimental study of the relation between the energy of the incident electron and the kinetic energy of the ion B^-, either the dissociation energy of AB or the electron affinity of B may be determined. It must be remembered, however, that the products may be excited states and there may be doubt about their identification. Further difficulties involved in an experimental programme of this kind are discussed in § 7.

3.6.3. *Dissociation into positive and negative ions—polar dissociation.* A molecule AB may be excited by electron impact into an electronic state that dissociates, not into neutral atoms, but into ions A^+ and B^-, either in their normal or excited states. The process only differs essentially from other excitation processes leading to dissociation in the nature of the dissociation products. Once more the relation between the electron energy E and the kinetic energy W of the resulting ions is of the form (2). If the ions A^+ and B^- are in their normal states $U'_A + U''_B$ becomes $I - E_a$, the difference between the ionization energy I of A and the electron affinity E_a of B.

Apart from the nature of the dissociation products, the difference between this process of negative ion production, which we call polar dissociation, and that involving capture is manifest in the variation of probability with electron energy. The capture process is important only in a narrow energy range of a few electron volts, whereas the probability of the non-capture one will vary with energy in much the same way as any electronic excitation process, viz. will increase to a maximum at an energy a few times the threshold value and then fall off gradually for increasing energies (see Chap. 4, Fig. 4.19). Examples of non-capture production of negative ions will be discussed in Chapter 13, §§ 4.4, 5.2, 6, and 11.3.

4. Theoretical description of electronic transitions in a diatomic molecule

4.1. *Excitation of a stable electronic state*

It is instructive and useful to consider in more theoretical detail the excitation of a higher stable electronic state of a molecule by electron impact allowing for the existence of vibrational and rotational states. We shall confine our analysis to the case of a diatomic molecule and assume at first the validity of Born's first approximation (Chap. 7, § 3).

A molecular state is characterized by a set of quantum numbers n that specify the state of electronic motion relative to the nuclei, the vibrational quantum number v, and the rotational quantum numbers

J, M, it being supposed that the component of nuclear rotation about the nuclear axis vanishes—there is no difficulty in generalizing the analysis to allow for cases in which this is not so. To a good approximation the wave function for the state can be written in the form (see Chap. 11 (2))

$$\Psi_{nvJM} = \psi_n(\mathbf{r}_i, \mathbf{R})\chi_{nv}(R)\rho_{JM}(\Theta, \Phi), \tag{8}$$

where $\mathbf{r}_i$ denotes the coordinates of the electrons relative to the nuclei, R the nuclear separation, and Θ, Φ the polar angles of the nuclear axis relative to an axis fixed in space.

We consider now the excitation of a transition from the state $nvJM$ to one $n'v'J'M'$ by impact of an electron with initial and final wave vectors $\mathbf{k}$, $\mathbf{k}'$ respectively. According to Born's first approximation the differential cross-section for this process is given by

$$I_{nvJM}^{n'v'J'M'}\,\mathrm{d}\omega = \frac{4\pi^2 m^2 e^4}{h^4}\frac{k'}{k}\times$$

$$\times\left|\iiint\sum_i\frac{1}{|\mathbf{r}_i-\mathbf{r}|}e^{i(\mathbf{k}-\mathbf{k}').\mathbf{r}}\Psi_{nvJM}\Psi^*_{n'v'J'M'}\,\mathrm{d}\mathbf{R}\,\mathrm{d}\mathbf{r}\prod_i\mathrm{d}\mathbf{r}_i\right|^2\mathrm{d}\omega, \tag{9}$$

$\mathbf{r}$ denoting the coordinates of the incident electron relative to the nuclei.

Using the relation

$$\int\frac{e^{i\mathbf{k}.\mathbf{r}}}{|\mathbf{r}-\mathbf{r}_i|}\,\mathrm{d}\mathbf{r} = \frac{4\pi}{k^2}e^{i\mathbf{k}.\mathbf{r}_i} \tag{10}$$

and substituting the form (8) for the molecular wave function we obtain

$$I_{nvJM}^{n'v'J'M'} = \left|\int\chi_{nv}(R)\chi_{n'v'}(R)G(R,\Theta,\Phi)\rho_{JM}(\Theta,\Phi)\rho^*_{J'M'}(\Theta,\Phi)\,\mathrm{d}\mathbf{R}\right|^2, \tag{11}$$

where

$$G(R,\Theta,\Phi)$$

$$= \frac{8\pi^2 m e^2}{h^2}\left(\frac{k'}{k}\right)^{\frac{1}{2}}|\mathbf{k}-\mathbf{k}'|^{-2}\int\psi_n(\mathbf{r}_i,\mathbf{R})\sum_i e^{i(\mathbf{k}-\mathbf{k}').\mathbf{r}_i}\psi^*_n(\mathbf{r}_i,\mathbf{R})\prod_i\mathrm{d}\mathbf{r}_i. \tag{12}$$

G is just the amplitude that would be given by Born's approximation for a transition in which the nuclear coordinates were held fixed at R, Θ, Φ.

The Franck–Condon principle simply expresses the fact that for given $\chi_{nv}(R)$ the product $\chi_{nv}\chi_{n'v'}$ will only be large when v' is such that the maxima of the two functions coincide as closely as possible. In that case, since χ_{nv} has a maximum at $R = R_0$, the initial equilibrium nuclear separation, and the electronic wave functions are slowly varying functions of R,

$$I_{nvJM}^{n'v'J'M'} \simeq \left|\int\chi_{nv}(R)\chi_{n'v'}(R)G(R_0,\Theta,\Phi)\rho_{JM}(\Theta,\Phi)\rho^*_{J'M'}(\Theta,\Phi)\,\mathrm{d}\mathbf{R}\right|^2. \tag{13}$$

Furthermore, under these conditions it is a good approximation to suppose that the rotational wave numbers are independent of the electronic and vibrational quantum numbers.

We are now in a position to calculate the differential cross-section summed over all final vibrational and rotational states provided we can make the further approximation that k' is independent of v', J', M', for all significant values of these quantum numbers. This will certainly be valid if the separation between the excited states of vibration and rotation are small compared with the final energy of the colliding electron. Since

$$\sum_{J'M'} \left| \int \lambda(\Theta, \Phi) \rho_{J'M'}(\Theta, \Phi)\, d\Omega \right|^2 = \int |\lambda(\Theta, \Phi)|^2\, d\Omega, \tag{14 a}$$

$$\sum_{v'} \left| \int S(R) \chi_{n'v'}(R)\, dR \right|^2 = \int |S(R)|^2\, dR, \tag{14 b}$$

we have

$$I_{nv}^{n'} = \sum_{v'} \sum_{J'M'} I_{nvJM}^{n'v'J'M'} = \iint |G(R, \Theta, \Phi) \rho_{JM}(\Theta, \Phi) \chi_{nv}(R)|^2\, d\Omega\, dR \tag{15 a}$$

$$\simeq \int |G(R_0, \Theta, \Phi) \rho_{JM}(\Theta, \Phi)|^2\, d\Omega. \tag{15 b}$$

If the initial rotational wave function is spherically symmetrical the differential cross-section summed over the final states of nuclear motion is the average, over all orientations of the nuclear axis, of that calculated for the electronic excitation with the nuclei fixed at the equilibrium separation. The same result holds if an average is taken over all initial rotational states supposed of equal *a priori* probability.

For some purposes the differential cross-section for excitation of separate vibrational levels of the upper state may be required. In that case we sum over the final rotational states only to obtain

$$I_{nv}^{n'v'} = \sum_{J'M'} I_{nvJM}^{n'v'J'M'} = \int \left| \int \chi_{nv}(R) \chi_{n'v'}(R) G(R, \Theta, \Phi) \rho_{JM}(\Theta, \Phi)\, dR \right|^2 d\Omega. \tag{16}$$

It is often a good approximation in this case to assume that $G(R, \Theta, \Phi)$ varies so gradually with R that

$$I_{nv}^{n'v'} \simeq (p_{nv}^{n'v'})^2 \int |G(R_0, \Theta, \Phi) \rho_{JM}(\Theta, \Phi)|^2\, d\Omega, \tag{17}$$

where

$$p_{nv}^{n'v'} = \int \chi_{nv}(R) \chi_{n'v'}(R)\, dR. \tag{18}$$

The cross-section for a particular vibrational transition would then be proportional to the square of the overlap integral of the corresponding vibrational wave function, as is often assumed.

If Born's approximation is not valid then in general (16) and (17) will still remain correct but $G(R, \Theta, \Phi)$ will no longer be given by (12).

In the evaluation of overlap integrals of the form (18) a good approximation to the vibrational wave functions $\chi_{nv}(R)$ may often be obtained by fitting the potential curve in the attractive region analytically by

$$V(R)-V(R_0) = C\{1-e^{-a(R-R_0)}\}^2, \qquad (19)$$

where R_0 is the equilibrium separation (see Chap. 10 (16)). The Schrödinger equation for the vibrational motion in this potential may be solved analytically. This procedure, which is often used, was first suggested by Morse.†

If the electronic energy of the final state is above that required for dissociation as in Fig. 12.4 (c), the relative motion of the dissociating fragment will not be quantized, i.e. the final wave function $\chi_{n'v'}$, will not be a discrete but a continuum function $\chi_{n'w}$, where W is the kinetic energy of relative motion of the product fragments at infinite separation. If $\chi_{n'w}$ is normalized per unit energy range then the distribution of the relative energy of the product fragments will be given by (16) with $\chi_{n'v'}$ replaced by $\chi_{n'w}$.

It is often a good approximation to represent $\chi_{n'w}$ as a delta function

$$\chi_{n'w} = S\delta(R-R_c),$$

where R_c is the nuclear separation at the classical turning point for the final relative motion, i.e. it is the separation at which the nuclear potential energy in the final state is equal to W. S is a normalizing factor which varies slowly with W. The overlap integral then becomes simply

$$S^2\chi_{nv}^2(R_c).$$

This is often known as the reflection approximation, as it amounts to reflecting the ground state wave function in the potential energy curve of the upper state.

4.2. *Angular distribution of products of molecular dissociation*

In the preceding section we discussed excitation of a stable electronic state of a molecule by electron impact. We now consider the case in which the excited state is unstable so that the molecule breaks up into atoms or into a pair of positive and negative ions. Experimental information on the rates of these processes often involves observation of the dissociation products, and absolute values for the total cross-sections are obtained on the assumption that their angular distribution is isotropic. In fact this is by no means so in general. This may be seen both by a general physical argument and by a detailed analysis.

Thus the time during which dissociation takes place is short compared with the period of molecular rotation, so that the products will move

† MORSE, P. M., *Phys. Rev.* **34** (1929) 57.

off very nearly in the direction of the vibration just before impact. It follows that any dependence of the probability of the electronic transition on the direction of the incident electron will be reflected in an anisotropic distribution of the relative directions of motion of the dissociation products.

In terms of the analysis of § 4.1 the differential cross-section for a collision in which the products of dissociation move off with a relative direction of motion within the solid angle $d\Omega$, about the direction (λ, μ), will be given by

$$I^{n'}_{nvJM}(\lambda, \mu) = \left| \int G(R_0, \Theta, \Phi)\chi_{nv}(R)\rho_{JM}(\Theta, \Phi)F(\lambda, \mu; R, \Theta, \Phi) \, d\mathbf{R} \right|^2,$$

where $F(\lambda, \mu; R, \Theta, \Phi)$ is the wave function for the unbound relative motion of the dissociation products.† It will have the asymptotic form

$$F(\lambda, \mu; R, \Theta, \Phi) \sim e^{i\mathbf{\varkappa}.\mathbf{R}} + R^{-1}e^{i\kappa R}f(\lambda, \mu; \Theta, \Phi), \tag{20}$$

$\mathbf{\varkappa}$ being a vector in the direction (λ, μ) of magnitude determined by the energy of relative motion of the dissociation products. F may be expanded in the form

$$F(\lambda, \mu; R, \Theta, \Phi) = \sum_s (2s+1)F_s(R)P_s(\cos\vartheta), \tag{21}$$

where
$$\cos\vartheta = \cos\lambda\cos\Theta + \sin\lambda\sin\Theta\cos(\Phi - \mu). \tag{22}$$

Suppose, for example, that G depends on Θ so that

$$G(R_0, \Theta, \Phi) = \sum_p (2p+1)G_p(R_0)P_p(\cos\Theta). \tag{23}$$

If the excitation occurs from the ground rotational state so that $\rho_{JM} = 1$ we have now

$$I^{n'}_{nvJM} \simeq \left| \sum G_p(R_0)(2p+1) \int \chi_{nv}(R)F_p(R)R^2 \, dR \, P_p(\cos\lambda) \right|^2. \tag{24}$$

In particular, if the electronic transition is optically allowed we may, by using the same expansion as (41) of Chapter 7, show that

$$\frac{\hbar^2}{2me^2}\left(\frac{k}{k'}\right)^{\frac{1}{2}} G(R_0, \Theta, \Phi) = K^{-1}\{X\cos\Theta\cos\Phi + Y\sin\Theta\sin\Phi + Z\cos\Theta\}, \tag{25}$$

where $K = |\mathbf{k} - \mathbf{k}'|$ and (Θ, Φ) are polar angles referred to the direction of K as axis. X, Y, Z are the components of the dipole moment associated with the transition, resolved along orthogonal axes x, y, z with z along the axis of the molecule.

If $X = Y = 0$, which will be the case for Σ–Σ transitions, we find from (25) that $I^{n'}_{nvJM}$ varies as $\cos^2\lambda$. To convert this into a distribution relative to the direction $\hat{\mathbf{k}}$ of the incident electron beam we note that, if with respect to $\hat{\mathbf{k}}$ as axis the polar angles of K are (η, χ) and of the direction of relative motion of the products of dissociation (θ, ϕ), then

$$\cos\lambda = \cos\theta\cos\eta + \sin\theta\sin\eta\cos(\phi - \chi).$$

We require the average of $\cos^2\lambda$ over all χ giving for the angular distribution in terms of θ
$$\cos^2\theta\cos^2\bar{\eta} + \tfrac{1}{2}\sin^2\theta\sin^2\bar{\eta},$$

where $\bar{\eta}$ is the mean value of η.

Close to the threshold $\bar{\eta}$ tends to zero and the distribution has a

---

† This function replaces $\chi_{n'v'}\rho^*_{J'M'}$ in (11) when dissociation occurs.

maximum at $\theta = 0$ and π. As the electron energy increases $\bar\eta$ increases and the minimum at $\theta = \frac{1}{2}\pi$ becomes shallower until when $\tan\eta = \sqrt{2}$, $\theta = 54{\cdot}7°$, the distribution becomes isotropic. For larger η a maximum develops at $\theta = \frac{1}{2}\pi$.

If $Z = 0$, which is so for Σ--Π transitions, $I^{n'}_{nvJM}$ varies as $\sin^2\lambda$ and in terms of θ the distribution becomes

$$\cos^2\theta\sin^2\bar\eta + \tfrac{1}{2}\sin^2\theta(1+\cos^2\bar\eta).$$

The distribution as a function of electron energy goes through the same sequence as for Σ–Σ transitions but in the reverse order. Near the threshold the distribution has a maximum at $\theta = \frac{1}{2}\pi$, which flattens as $\bar\eta$ increases to a completely isotropic distribution at $\theta = 54{\cdot}7°$ and then develops a gradually deepening maximum at $\theta = \frac{1}{2}\pi$.

When ionization as well as excitation occurs allowance must be made for the direction of ejection of the second electron. As, however, this electron will tend to be ejected in the direction of $-\mathbf{K}$ to conserve momentum (see Chap. 7, § 5.6.2) similar considerations apply, somewhat less definitely, as for excitation. A detailed experimental investigation by Dunn and Kieffer of the angular distribution of the protons resulting from dissociative ionization of H_2 is given in Chapter 13, § 1.5.2.

As far as processes involving dissociation with capture are concerned we note that the incident plane wave is symmetric with respect to all rotations about $\mathbf{k}$ and with respect to reflections in planes containing $\mathbf{k}$. For a given direction of the nuclear axis of the target molecule the initial molecular state may possess certain symmetries with respect to these operations. Transition to a particular final state will only then be possible if it possesses these same symmetries.

Proceeding on these lines Dunn† considered the transitions, for homonuclear diatomic molecules within sets of states Σ_g^+, Σ_g^-, Σ_u^+, Σ_u^-, Π_g, Π_u, Δ_u, Δ_g. He finds that transitions can occur with the nuclear axis parallel to the incident beam for the following cases only:

$$\Sigma_g^+\text{–}\Sigma_g^+,\ \ \Sigma_g^+\text{–}\Sigma_u^+,\ \ \Sigma_g^-\text{–}\Sigma_g^-,\ \ \Sigma_g^-\text{–}\Sigma_u^-,\ \ \Sigma_u^+\text{–}\Sigma_u^+,\ \ \Sigma_u^-\text{–}\Sigma_u^-,$$

$$\Pi_g\text{–}\Pi_g,\ \ \Pi_g\text{–}\Pi_u,\ \ \Pi_u\text{–}\Pi_u,\ \ \Delta_g\text{–}\Delta_g,\ \ \Delta_g\text{–}\Delta_u,\ \ \Delta_u\text{–}\Delta_u.$$

and with the nuclear axis perpendicular to the incident beam for the following cases only:

$$\Sigma_g^+\text{–}\Sigma_g^+,\ \ \Sigma_g^+\text{–}\Pi_u,\ \ \Sigma_g^+\text{–}\Delta_g,\ \ \Sigma_g^-\text{–}\Sigma_g^-,\ \ \Sigma_g^-\text{–}\Pi_u,\ \ \Sigma_g^-\text{–}\Delta_g,$$

$$\Sigma_u^+\text{–}\Sigma_u^+,\ \ \Sigma_u^+\text{–}\Pi_g,\ \ \Sigma_u^+\text{–}\Delta_u,\ \ \Sigma_u^-\text{–}\Sigma_u^-,\ \ \Sigma_u^-\text{–}\Pi_g,\ \ \Sigma_u^-\text{–}\Delta_u,$$

$$\Pi_g\text{–}\Pi_g,\ \ \Pi_g\text{–}\Delta_u,\ \ \Pi_u\text{–}\Pi_u,\ \ \Delta_g\text{–}\Delta_g,\ \ \Delta_u\text{–}\Delta_u.$$

† Dunn, G. H., *Phys. Rev. Lett.* **8** (1962) 62.

The latter results are of particular interest in relation to the Lozier method for studying ionization and attachment phenomena (see § 7.3.1), for in this method only ions, positive or negative, which are produced in a direction normal to the electron beam are observed.

For heteronuclear molecules the corresponding results are, for transitions with nuclear axis parallel:

$$\Sigma^+-\Sigma^+, \ \Sigma^--\Sigma^-, \ \Pi-\Pi, \ \Delta-\Delta,$$

and with axis perpendicular:

$$\Sigma^+-\Sigma^+, \ \Sigma^+-\Pi, \ \Sigma^+-\Delta, \ \Sigma^--\Sigma^-, \ \Sigma^--\Pi, \ \Sigma^--\Delta, \ \Pi-\Pi, \ \Pi-\Delta, \ \Delta-\Delta.$$

These considerations show that, in general, the angular distribution of the dissociated products will not be isotropic.

This possibility was first raised by Sasaki and Nakao† in 1935 as a result of their observations of the angular distribution of 845 and 105-eV protons resulting from dissociative ionization of H_2 (see Chap. 13, § 1.5). They attempted‡ to justify these results theoretically by applying Born's approximation to calculate the angular distribution of atoms resulting from excitation of the unstable $^3\Sigma_u^+$ state of H_2 (see Chap. 13, § 1.2) by 14·8-eV electrons, but the validity of the approximation for such slow collisions is very doubtful (see Chap. 7, § 5.6). However, many years later, Kerner§ calculated the distribution of protons resulting from excitation of H_2^+ to the unstable $^2\Sigma_u^+$ state (see Chap. 13, § 2.2) by electrons of such energies that Born's approximation is valid. This falls within the optically-allowed case discussed above and gives the $\cos^2\lambda$ distribution. The importance of the anisotropic distribution for many types of experiment was first stressed by Dunn‖ who also, in collaboration with Kieffer†† made the first thorough experimental investigation of such a distribution (see Chap. 13, § 1.5.2).

5. Diffraction of fast inelastically scattered electrons by molecules

5.1. *Theoretical considerations*

In Chapter 10, § 1 we discussed the diffraction of fast elastically scattered electrons by molecules. If the electrons lose energy in the course of the collision the scattering is usually referred to as incoherent but this does not apply if electrons that have suffered a particular energy loss are separately observed. Already in Chapter 5, § 5.3.2 and Chapter 8,

---

† SASAKI, N. and NAKAO, T., *Proc. imp. Acad. Japan* **11** (1935) 138; **17** (1941) 75.
‡ Ibid. 413.   § KERNER, E. H., *Phys. Rev.* **92** (1953) 1441.
‖ DUNN, G. H., *Phys. Rev. Lett.* **8** (1962) 62.
†† DUNN, G. H. and KIEFFER, L. J., *Phys. Rev.* **132** (1963) 2109.

§ 7.2 we have discussed the angular distributions of electrons scattered by atoms after suffering a definite energy loss. In such cases maxima and minima, rather similar to those observed in elastic scattering, are observed and can be understood theoretically. These maxima and minima arise from coherence of waves scattered inelastically, but with a definite wavelength, from different parts of the atomic field. We should expect then that diffraction effects due to coherence of electron waves scattered inelastically, with a definite final wavelength, by different atoms in a molecule should also be observable.

Consider for illustration the simplest case of an inelastic collision between an electron and a homonuclear diatomic molecule. For simplicity we shall also assume that in the molecular excitation one active electron distinguished by the suffix 1 is concerned. Using the same approximation as that employed in § 4.1 the wave functions Ψ_i, Ψ_f for the initial and final electronic states may now be written

$$\Psi_i = \frac{1}{\sqrt{2}}\{\phi_i(\mathbf{r}_{1A})\pm\phi_i(\mathbf{r}_{1B})\}\psi_c(\mathbf{r}_c), \qquad (26\,a)$$

$$\Psi_f = \frac{1}{\sqrt{2}}\{\phi_f(\mathbf{r}_{1A})\pm\phi_f(\mathbf{r}_{1B})\}\psi_c(\mathbf{r}_c). \qquad (26\,b)$$

Here ϕ_i, ϕ_f are atomic orbitals and $\mathbf{r}_{1A}$, $\mathbf{r}_{1B}$ are the coordinates of the active electron relative to the nuclei A, B, respectively, $\mathbf{r}_c$ denotes the coordinates of the core electrons, and ψ_c the wave function for these electrons. Thus $\mathbf{r}_{1B} = \mathbf{r}_{1A}+\mathbf{R}$, where $\mathbf{R}$ is the vector separation of the nuclei (see Fig. 12.7).

We consider a collision with a fast electron in which the nuclear separation is fixed at its equilibrium value and the orientation of the axis is also fixed.

According to Born's first approximation the differential cross-section for the excitation by electrons of initial and final wave vectors $\mathbf{k}$, $\mathbf{k}'$ respectively, where $\mathbf{k}-\mathbf{k}' = \mathbf{K}$, is given by

$$I(\theta)\,d\omega = \frac{4\pi^2m^2e^4}{h^4}\left|\tfrac{1}{2}\int\int\frac{e^{i\mathbf{K}.\mathbf{r}}}{|\mathbf{r}-\mathbf{r}_1|}\{\phi_i(\mathbf{r}_{1A})\pm\phi_i(\mathbf{r}_{1B})\}\times\right.$$
$$\left.\times\{\phi_f^*(\mathbf{r}_{1A})\pm\phi_f(\mathbf{r}_{1B})\}\,d\mathbf{r}d\mathbf{r}_1\right|^2 d\omega. \qquad (27)$$

Again making the same approximation as in Chapter 10, § 1.1, neglecting overlap between the atomic orbitals on the two centres we have, using (10),

$$I(\theta)\,d\omega = \frac{4\pi^2m^2e^4}{K^4h^4}\left|\int e^{i\mathbf{K}.\mathbf{r}_A}\phi_i(\mathbf{r}_{1A})\phi_f^*(\mathbf{r}_{1A})\,d\mathbf{r}_{1A}\right|^2 \left.\begin{matrix}\cos^2\\\sin^2\end{matrix}\right\}\mathbf{K}.\mathbf{R}\,d\omega. \qquad (28)$$

The outside factor, depending on the orientation of the nuclear axis is either $\cos^2 \mathbf{K.R}$ or $\sin^2 \mathbf{K.R}$ according as the nuclear symmetry is unchanged or changed in the transition. Averaging over all orientations of the nuclear axis as in Chapter 10, § 1.1 we have

$$I(\theta) \, d\omega = 2I_a(\theta)\left(1 \pm \frac{\sin KR}{KR}\right) d\omega, \tag{29}$$

where $I_a \, d\omega$ is the differential cross-section for excitation of a single atom. The $\pm$ signs in the bracket that represents the diffraction factor are to be taken according as the initial and final electronic states do or do not possess the same symmetry. At high energies and not too small angles of scattering $K = 2k \sin \frac{1}{2}\theta$ so we see that, when there is no change of nuclear symmetry, the molecular diffraction factor is nearly the same as for elastic scattering but that the

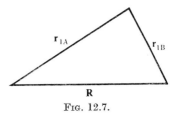

FIG. 12.7.

diffraction pattern is 'inverted', as it were, when there is a change of nuclear symmetry. Karle[†] has extended the analysis to cover excitation of polyatomic molecules such as CCl_4, C_2H_2, etc.

5.2. *Experimental confirmation*

These considerations were first discussed by Massey and Mohr[‡] in 1932 but did not receive experimental confirmation until 1961 when Swick and Karle[§] investigated the scattering of electrons of 20-keV energy by bromine molecules.

They used an instrument first developed by I. and J. Karle[||] in which the velocity analyser was of the type introduced by Möllenstedt[††] (see Chap. 5, § 5.2.3). Fig. 12.8 illustrates the general arrangement.

Electrons from the gun were accelerated to 20–40 keV and focused by the magnetic lenses 1 and 2 to produce a beam less than 0·1 mm in diameter with an energy spread less than 1 eV. The beam could be cut off by means of the upper electrostatic deflecting plates.

The scattering gas could be admitted from a nozzle in bursts of less than 0·1 s duration, the flow being controlled by a plunger valve activated by a solenoid. Opening of the solenoid was synchronized with

† KARLE, J., *J. chem. Phys.* **35** (1961) 963.
‡ MASSEY, H. S. W. and MOHR, C. B. O., *Proc. R. Soc.* **A135** (1932) 258.
§ SWICK, D. A. and KARLE, J., *J. chem. Phys.* **35** (1961) 2257.
|| KARLE, I. L. and KARLE, J., ibid. **17** (1949) 1052.
†† MÖLLENSTEDT, G., *Optik* **5** (1949) 499.

grounding of the upper deflexion plates, which allowed the beam to enter the scattering chamber.

After passage through the chamber the beam, including the scattered electrons, passed through the wide aperture magnetic lens 3, which had the effect of compressing the diffraction pattern so that more of the diffracted electrons were able to enter the velocity analyser.

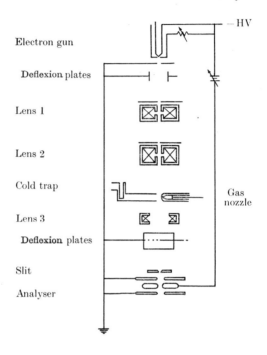

FIG. 12.8. Illustrating the arrangement used by Swick and Karle for investigating the diffraction patterns for electrons scattered elastically or inelastically (with definite energy loss) by gas molecules.

The analyser consisted of a fine slit, a few microns wide and 5 cm long, above an electrostatic cylindrical lens that consisted of two grounded electrodes parallel to and on opposite sides of an electrode maintained at or near to cathode potential. All three electrodes had rectangular apertures of dimensions roughly 1×5 cm. As mentioned in Chapter 5, § 5.2.3 a lens of this type has a very high dispersion at points of the axis. Calibration of the energy scale was made by photographing the displacement of the electron beam, in the absence of scattering from the gas jets, produced by applying to the central electrode a small potential difference relative to the cathode. Fig. 12.9 is an example of a typical microphotometer record obtained in this way for successive applications of an additional potential of 1 V.

The choice of the range of scattering angles of electrons to be analysed could be varied by means of the lower deflexion plates. Before injection of gas the apparatus was normally evacuated to 10^{-5} torr.

The pattern on the photographic plates when a scattering experiment was carried out consisted of images, formed by the cylindrical lens, of a diameter of the diffraction pattern as selected by the slit. These

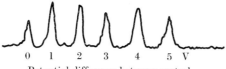

Potential difference between central
electrode and cathode

FIG. 12.9. Microphotometric record illustrating the method of energy calibration in the experiments of Swick and Karle on the diffraction of fast electrons by gas molecules.

images take the form of a series of arcs one for each energy value (or range of values) in the diffracted beam. The angular distribution of the scattering for a given energy range was obtained by tracing along an arc with a microphotometer. Angular calibration was carried out by comparison with a known pattern such as the elastic pattern for CCl_4 (see Chap. 10, Fig. 10.2).

Fig. 12.10 illustrates photometer traces of the elastic and some of the inelastic patterns found in early experiments for CCl_4 and C_6H_6. The presence of diffraction effects in the patterns for electrons scattered after suffering a definite energy loss is clearly seen. For both molecules, however, the peaks in the inelastic patterns occur at nearly the same angle of scattering as for the corresponding elastic pattern. Quite different results were obtained for bromine, for which molecule Swick and Karle observed the elastic pattern and that for electrons that had lost $7\cdot4\pm0\cdot2$ eV in exciting an optically allowed transition. After removing the form factors $(I_a(\theta))$ in (29) they obtained the results shown in Fig. 12.11. The inversion of the pattern for the inelastic case is clearly seen and is to be expected because, if the transition excited is optically allowed, it must involve a change of nuclear symmetry in the electronic state. The minimum in the elastic case and the maximum in the inelastic occur at very nearly the same angle which is close to that expected from the known nuclear separation in Br_2.

6. Interpretation of resonance effects in collisions of electrons with neutral molecules

The chief additional feature that must be introduced in dealing with resonance effects in collisions with neutral molecules is the existence of vibrational structure if the lifetime of the intermediate negative ion is longer than the vibrational period. If this is not so there will be no

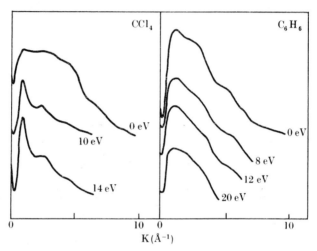

Fig. 12.10. Microphotometric records of diffraction patterns for electrons scattered elastically and inelastically with fixed energy loss, in CCl_4 and C_6H_6. The energy loss is indicated on each curve.

time for even a cycle of vibration to develop. On the other hand, if the opposite conditions apply, a complex pattern of vibrational structure will be interwoven among the fine structure characteristic of a resonance effect as in atomic collisions.

Thus suppose that the target molecule is initially in a state with electronic quantum numbers denoted by 0 and vibrational quantum number v. The incident electron is captured to form a molecular negative ion in the electronic state n, say. If the lifetime of the ionic state is long compared with the vibrational period it will also be possible to distinguish the vibrational states of this ion by an appropriate quantum number u. Thus if E_n is the energy of the electronic level the resonance energies will be given by

$$E_{nu} = E_n + (u + \tfrac{1}{2})h\nu_n,$$

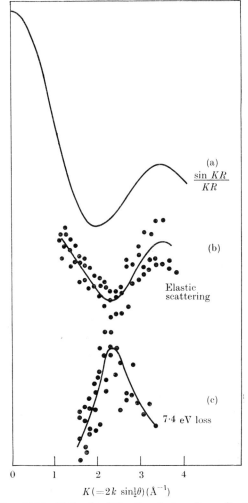

Fig. 12.11. Molecular diffraction factors for scattering of 20-keV electrons by bromine molecules. These have been obtained by subtraction of the atomic scattering background. (a) The function $\sin KR/KR$, where R is the nuclear separation in Br_2 and $K = 2k \sin \frac{1}{2}\theta$ (see (29)). (b) Observed diffraction factor for elastically scattered electrons. (c) Observed diffraction factor for electrons scattered after suffering an energy loss of 7·4 eV.

it being assumed that the vibrations in the ionic state are effectively simple harmonic with fundamental frequency ν_n. The corresponding level widths we denote by Γ_{nu}. E_{nv} will, of course, depend on the nuclear separation, but its range is limited by the Franck–Condon principle and it will be assumed to have an appropriate mean value averaged over the Franck–Condon region.

This complex will break up through autodetachment, an electron being emitted. Provided the energy relations are satisfied the residual neutral molecule may be left in the ground or an excited state. Even if the energy is insufficient for the molecule to be left in an excited electronic state it may be left with more or less quanta of vibration than initially. In this general way we can understand qualitatively such results as those of Schulz and others on the excitation of vibration in N_2 and CO by electron impact (see Chap. 11, § 3 and Figs. 11.5 and 11.7).

Chen† has analysed this situation in detail. Without following through the details of this work we can describe his results in the following way. Consider the amplitude for the transition

$$0v \to nu \to 0v',$$

in which the ultimate result is a change in the vibrational quantum number from v to v' through the intermediate ionic state which possesses u vibrational quanta.

According to the analysis of Chapter 9, p. 606, the amplitude is given by

$$f^u_{vv'} = \frac{\gamma^{0v}_{nu}\gamma^{0v'}_{nu}}{E - E_{nu} + \frac{1}{2}i\Gamma_{nu}}, \tag{30}$$

where
$$|\gamma^{0v}_{nu}|^2 = \Gamma^{0v}_{nu}, \quad |\gamma^{0v'}_{nu}|^2 = \Gamma^{0v'}_{nu} \tag{31}$$

and
$$\Gamma_{nu} = \sum_v \Gamma^{0v}_{nu}. \tag{32}$$

Thus Γ^{0v}_{nu} is the partial width of the state nu due to break-up leaving the target molecule in the state $0v$. Γ_{nu}, the total width of the state nu, is the sum of the partial widths over all such final states.

Again in terms of the analysis of Chapter 9, § 1.2,

$$\Gamma^{0v}_{nu} = 2\pi\{\langle \Psi_{nu}|H|\Psi^{0v}_c \rangle\}^2. \tag{33}$$

Ψ_{nu} is the wave function for the complex in the state nu, which can be written approximately in the form

$$\Psi_{nu} = \psi_n(\mathbf{r}_m, \mathbf{r}, \mathbf{R})\chi_{nu}(R), \tag{34}$$

where $\mathbf{r}_m$ denotes the coordinates of the electrons in the target molecule, and $\mathbf{r}$ that of the incident electron. ψ_n is the electronic and χ_{nu} the vibrational wave function.

Ψ^{0v}_c denotes the wave function of the initial state, which can be written approximately

$$\Psi^{0v}_c = \omega_0(\mathbf{r}_m, \mathbf{R})F^v_0(\mathbf{r}, \mathbf{R})\chi_{0v}(R). \tag{35}$$

† CHEN, J. C. Y., J. chem. Phys. **40** (1964) 3507, 3513.

Here ω_0 is the electronic wave function of the ground state of the neutral molecule, χ_{0v} the initial vibrational wave function of that molecule, and F_0^v is the wave function, appropriately normalized, of the incident electron. We indicate a dependence of this latter wave function on v because the final vibrational state determines the energy of the free electron. Thus $\Psi_c^{0v'}$ differs from Ψ_c^{0v} in the replacement of $F_0^v(\mathbf{r}, \mathbf{R})$ by $F_0^{v'}(\mathbf{r}, \mathbf{R})$ and χ_{0v} by $\chi_{0v'}$.

H is the Hamiltonian for the total system.

We may now write approximately

$$\langle \Psi_{nu} | H | \Psi_c^{0v} \rangle \simeq \beta_v \int \chi_{nu} \chi_{0v} \, dR$$

$$= \beta_v \langle \chi_{nu} \chi_{0v} \rangle, \tag{36}$$

where
$$\beta_v = \iint \psi_m(\mathbf{r}_m, \mathbf{r}, \overline{\mathbf{R}}) H \omega_0(\mathbf{r}_m, \overline{\mathbf{R}}) F_0^v(\mathbf{r}, \overline{\mathbf{R}}) \, d\mathbf{r}_m \, d\mathbf{r} \tag{37}$$

and $\overline{\mathbf{R}}$ is a mean of $\mathbf{R}$ over the Franck–Condon transition region. Thus the amplitude (32) is given by

$$\frac{\beta_v \beta_{v'} \langle \chi_{nu} \chi_{0v} \rangle \langle \chi_{nu} \chi_{0v'} \rangle}{E - E_n - (u + \frac{1}{2}) h\nu_n + \frac{1}{2} i\Gamma_{nu}} \, \exp\{i\lambda_{vv'}^u\}, \tag{38}$$

where $\lambda_{vv'}^u$ is a phase that we have not yet determined.

The total amplitude for the transition $v \to v'$ will now be given by

$$f_{vv'} = \sum_u f_{vv'}^u$$

$$= \beta_v \beta_{v'} \sum_u \frac{\langle \chi_{nu} \chi_{0v} \rangle \langle \chi_{nu} \chi_{0v'} \rangle}{E - E_n - (u + \frac{1}{2}) h\nu_n + \frac{1}{2} i\Gamma_{nu}} \, \exp\{i\lambda_{vv'}^u\}, \tag{39}$$

where
$$\Gamma_{nu} = \sum_v \Gamma_{nu}^{0v} = \sum_v \beta_v^2 \langle \chi_{nu} \chi_{0v} \rangle^2. \tag{40}$$

In Chapter 9, § 1.2 the phase shift $\lambda_{vv'}^u$ is determined simply by the phases of the wave functions F_0^v, $F_0^{v'}$ and is independent of the intermediate complex. We may therefore assume that $\lambda_{vv'}^u$ is independent of u and may be taken outside the sum in (39).

As we are concerned only with the cross-section

$$Q_{vv'} = \frac{\pi}{k_v'^2} |f_{vv'}|^2 = \frac{\pi \beta_v^2 \beta_{v'}^2}{k_v'^2} \left| \frac{\sum_u \langle \chi_{nu} \chi_{0v} \rangle \langle \chi_{nu} \chi_{0v'} \rangle}{E - E_n - (u + \frac{1}{2}) h\nu_n + \frac{1}{2} i\Gamma_{nu}} \right|^2, \tag{41}$$

a knowledge of $\lambda_{vv'}^u$ is not then required.

The amplitude contributed by the resonance to the elastic scattering is given simply by f_{vv}.

If $\Gamma_{nu} \ll h\nu_n$ the variation of the cross-section with electron energy for each vibrational transition will consist of a series of resonance peaks approximately separated by the vibrational quantum energy $h\nu_n$ and occurring at the same locations for each transition. When Γ_{nu} and $h\nu_n$

are of the same order the pattern will be more complicated for the contributions from the different terms will overlap and the final result will depend on the vibrational overlap integrals.

If the overlap is very marked the analysis needs revision† as the separate states of the complex are no longer distinguishable.

There is no difficulty in extending this analysis to the case in which the target molecule may be left in an excited electronic state.

An account is given of the application of these results to the discussion of resonant effects in the elastic and inelastic electron scattering by H_2 and by N_2 in Chapter 13, § 1.3 and § 3.7 respectively.

7. Experimental methods for studying inelastic collisions that involve electronic transitions

7.1. *Introduction—special requirements and problems which arise in studying molecular collisions*

The inelastic collisions of electrons with a great variety of molecules have been studied using a number of different techniques. The kind of data obtained and the general nature and limitations of the technique employed to obtain them can be briefly summarized as follows.

(*a*) *Dissociation into neutral fragments by electron impact.* The major problem here is the detection of the neutral products. Illustrations of the methods used which involve chemical detectors will be described in Chapter 13, § 1.2.

(*b*) *Appearance potentials and cross-sections for excitation of discrete electronic states.* Similar techniques may be used as for atoms. These may be optical, as in Chapter 4, § 1, if the excited state combines optically with the ground state, or electrical as in Chapter 4, § 2 and Chapter 5, § 2.

(*c*) *Absolute cross-sections for ionization.* Similar apparatus may be used for this purpose as for the corresponding study of the ionization cross-sections of atoms described in Chapter 3, § 2. For accurate determination (better than a few per cent) account must be taken of the fact that some of the ions may be produced with considerable kinetic energy. To secure a high efficiency of collection of these ions such strong collecting fields are required that serious disturbance may be introduced elsewhere in the apparatus. If this occurs one possible way out is to use pulsed fields so that ions are only collected at intervals during which the ionizing electron beam is turned off. A second possibility is to use a

† See MOTT, N. F. and MASSEY, H. S. W., *The theory of atomic collisions*, 3rd edn, Chap. xiii, § 4.2 (Clarendon Press, Oxford, 1965).

crossed-beam technique in which the target atoms comprise an atomic beam of some keV energy produced by neutralization of an ion beam through charge exchange. In this case the kinetic energy of the ion fragments is small compared with the translational energy of the beam atoms and introduces no special problem. On the other hand, the problem of obtaining a beam of fast atoms effectively all in their ground states is a difficult one (see p. 849).

(d) *Appearance potentials and nature, relative abundance, and kinetic energy of the different ionized products.* The chief methods used are the mass spectrograph and the ion energy analyser of the type introduced by Lozier. Difficulties in determining appearance potentials, already considerable in dealing with ionization of atoms (see Chap. 3, § 2.4.4) are more severe in the case of molecules due to the greater number of states available to the final products. The fact that frequently ions may be formed with kinetic energies of the order of a few eV is a further complication that makes it difficult to ensure a high collection efficiency for all products and can lead to unreliable and indeed very inaccurate data about relative abundances.

Chantry and Schulz† have also pointed out an additional difficulty in determining appearance potentials for production of ions with kinetic energy. This arises from the thermal motion of the target molecules. The problem is exactly similar to that of calculating the energy distribution of positive ions relative to an earth satellite moving through the ionosphere with a velocity large compared with the r.m.s. velocity of the ions.‡ If E_0 is the energy of an ion of mass M_i emitted from a collision with a molecule of mass M at rest, then the fraction of ions emitted with energy lying between E and $E+\mathrm{d}E$ is given by

$$(4\pi\beta\kappa TE_0)^{-\frac{1}{2}}\exp\left\{-\frac{1}{\beta\kappa T}(E^{\frac{1}{2}}-E_0^{\frac{1}{2}})^2\right\}\mathrm{d}E, \tag{42}$$

where $\beta = M_i/M$.

This distribution has a maximum at $E = E_0$ but its width at half maximum is $(11\cdot0\beta\kappa TE_0)^{\frac{1}{2}}$. Thus for $T = 350\ °\mathrm{K}$ and $E_0 = 2$ eV this width is as great as $0\cdot56$ eV so that a considerable number of ions are produced with kinetic energy greater than E_0. This is particularly serious in those experiments in which the direction of the ions on entering the energy-analysing retarding field is not well-defined, for it was customary to regard the maximum kinetic energy possessed by

† CHANTRY, P. J. and SCHULZ, G. J., *Phys. Rev. Lett.* **12** (1964) 449.

‡ See, for example, MASSEY, H. S. W., *Space Physics*, pp. 147–9 (Cambridge University Press, 1964).

observed ions to be the correct one to take for appearance potential determinations. With a broad distribution of the type (42) this is clearly not valid.

Consider an experiment carried out in which the gas molecules are at a temperature T °K. If W_i is the initial kinetic energy of ions produced by electron impact with molecules at rest, the minimum retarding potential V_r^m necessary to stop ions produced in the gas will exceed $W_i e$ by $\Delta W_i e$, where ΔW_i is proportional to the half-width of the distribution (42). Thus we may write

$$eV_r^m = W_i + \alpha(\beta \kappa T W_i)^{\frac{1}{2}}, \tag{43}$$

where α depends on the experimental procedure for determining V_r^m from a retarding potential curve. The usual method is to extrapolate linearly to the axis of zero current. Chantry and Schulz showed that in a typical case this amounted to taking $\alpha = 2 \cdot 0$.

In an experiment on dissociative attachment we have

$$W_i = (1-\beta)\{E_e - (D-A)\} \tag{44}$$

as in (2) and (3) (see also § 6.2), E_e being the kinetic energy of the incident electrons, D the dissociation energy of the molecule, and A the electron affinity of the negative ion produced. Using (43) and (44) with $\alpha = 2$ we have

$$eV_r^m/(1-\beta) = W + \{4\beta \kappa T/(1-\beta)\}^{\frac{1}{2}} W^{\frac{1}{2}}, \tag{45}$$

where

$$W = E_e - (D-A).$$

Fig. 12.12 illustrates the relation (45) for two cases. In the first the reaction is

$$O_2 + e \rightarrow O + O^-, \tag{46}$$

for which $\beta = 1/12$, and in the second

$$H_2O + e \rightarrow H^- + OH, \tag{47}$$

for which $\beta = 1/18$. It will be seen that for the former the departure from the simple linear relation

$$eV_r^m/(1-\beta) = W \tag{48}$$

previously assumed is quite marked. Experimental results obtained by Schulz are shown and it is clear that linear extrapolation on the assumption of the validity of (48) leads to an error of the order 0·5 eV in the determination of $D-A$. In the second case (47) the error is much smaller because $\beta \ll 1$.

A further factor which must be taken into account, both in studying with relative abundance and appearance potentials, is the likelihood that the angular distribution of the productions will often be anisotropic with respect to the direction of the existing electron beam.

(e) *Negative ion formation.* This has been studied for impact of electrons with energies greater than a few eV by using homogeneous electron beams either with the mass spectrograph or the Lozier method as in (d).

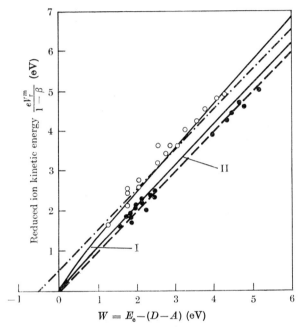

$$W = E_e - (D - A) \text{ (eV)}$$

Fig. 12.12. Illustrating the effect of temperature motion of the target molecules on the relation between the apparent maximum kinetic energy of the negative ions and the kinetic energy of the electrons which produce them by attachment. V_r is the retarding potential necessary to stop the ions and $W = E_e - (D - A)$, where E_e is the electron kinetic energy, D the dissociation energy of a molecule, and A the electron affinity of the negative ion. I. calculated for O^- from dissociative attachment to O_2. II. calculated for H^- from dissociative attachment to H_2. — — — calculated ignoring temperature motion. ○ ○ ○ typical observations for O_2. — · — · — usual extrapolation to obtain $D - A$ for O_2. ● ● ● typical observations for H_2O.

For lower energy electrons it is necessary to work with a swarm of electrons either in experiments in which the electrons drift under the action of a uniform electric field or by microwave probing of afterglows in the gas under investigation.

(f) *Derivation of cross-sections from analysis of transport coefficients of electrons in the gas under consideration.* Just as for atoms (Chap. 2, § 7) a trial-and-error procedure may be used to derive information about the momentum-loss cross-section and the excitation and ionization cross-sections from measurements of drift velocity, characteristic energy, and excitation and ionization coefficients as a function of F/p for electrons drifting through the gas at pressure p under the action of an electric field F.

We shall now describe the techniques that are particularly applicable or specially required for the investigation of electron collisions with molecules which involve electronic excitation.

7.2. Measurement of absolute (apparent) ionization cross-sections

It has been pointed out under (c) above that for accurate measurement of apparent ionization cross-sections (see Chap. 3, § 2.1) certain specific problems arise when dealing with the ionization of molecules. These are due to the possibility of producing ions of considerable kinetic energy (several eV) through molecular dissociation.

Particular attention was paid to this problem by Rapp, Englander-Golden, and Briglia† and by Schram, de Heer, van der Wiel, and Kiste-maker,‡ who were careful to check that the collected current of positive ions was fully saturated. Fig. 12.13 illustrates the arrangement used by the former authors, depending on the principles discussed in detail in Chapter 3.

The electron beam from the cathode K, after acceleration and colli-mation by passage between holes in the electrodes 1, 2, and 3, passed successively through the collision chamber C and an electron collector shield ECS to be collected on the plate ECP contained within the collector cylinder ECC. A magnetic field of order 500 gauss was applied along the electron beam direction.

Positive ions produced by the beam in passing through the collision chamber were collected on the plate IC flanked by two guard-rings G. These electrodes were maintained at a negative potential $V_{\parallel}$ relative to a parallel system of collector plates and guard-rings placed sym-metrically on the other side of the electron beam.

The electrodes were of a nonmagnetic copper–nickel alloy (advance metal) and the insulators of alumina. The centre holes in plates 1, 2, and 3 were respectively 0·05, 0·014, and 0·030 inches in diameter. The electron-collector shield ECS had slits $0·86 \times 0·1$ in and the cylinder

† Rapp, D., Englander-Golden, P., and Briglia, D. D., J. chem. Phys. 42 (1965) 4081; Rapp, D. and Englander-Golden, P., ibid. 43 (1965) 1464.

‡ Schram, B. L., de Heer, F. J., van der Wiel, M. J., and Kistemaker, J., Physica 31 (1965) 94.

ECC a slit 0.86×0.036 in, in each case parallel to the parallel ion collector plates, which were 0.40 in apart.

A typical arrangement of voltages for the study of ionization by electrons of 100-eV energy was as follows

$$V_C = -100 \text{ V}, \quad V_{IC} = 0, \quad V_{\parallel} = 10 \text{ V}, \quad V_B = V_{ECS} = V_{ECC} = 5 \text{ V}.$$
$$V_{ECP} = +505 \text{ V}.$$

Although in such a case electrons are incident on the collector with 600-eV energy and any secondary electrons are retarded by only 500 V, the number of secondaries that are produced with energies greater than

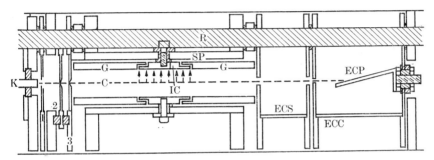

FIG. 12.13. Arrangement of the apparatus used by Rapp, Englander-Golden, and Briglia for determining absolute cross-sections for ionization of molecules by electron impact.

500 eV is small as judged from observations of the variation of collected electron current with V_{ECP}. During an experiment V_{ECP} could not be made too positive because positive ions formed by electrons in this neighbourhood of ECP are accelerated towards the collision chamber. With $V_{ECP} = 100$ V these ions contributed about $\frac{1}{2}$ per cent to the current collected on IC.

The voltage $V_{\parallel}$ between the parallel plates in the collision chamber must be large enough to collect all the positive ions produced by the electron beam over a path length equal to the length of the ion collector plate IC. For ionization of atomic gases, in which the ions are produced with very small kinetic energy, saturation occurs when $V_{\parallel}$ is only a few volts and the ions pursue nearly straight paths normal to the collector plate. With molecular gases, however, collection of the energetic ions requires $V_{\parallel}$ to be much larger. Moreover, the paths of the ions are no longer linear and saturation occurs when the number of ions formed in the region below the collector plate IC, which are collected on a guard-ring, is equal to the number formed below a guard-ring but collected on IC. Because of this the guard-rings must be long enough to make this

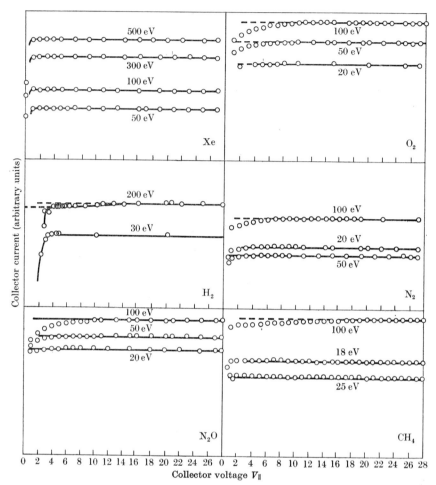

Fig. 12.14. Positive ion collector current as a function of collector voltage $V_{\parallel}$ for different gases and different electron energies, observed by Englander-Golden and Rapp.

balance possible. As an example a proton formed in H_2 with 10-eV initial kinetic energy and moving parallel to the collector plate can travel 0·6 in before collection if $V_{\parallel}$ is 25 V. In the experiments the guard-rings were about 1 inch in length.

Fig. 12.14 illustrates a number of typical ion current characteristics obtained with different gases at different electron energies. It will be seen that for an atomic gas such as xenon saturation occurs when $V_{\parallel}$ is greater than 1 V, irrespective of electron energy. For H_2, on the other hand, at least 4 V is required to produce saturation when the energy of the incident electrons is 30 eV and as much as 16 V for 200-eV electrons.

Similar features are seen for the other molecular gases, CH_4 and 100-eV electrons presenting an extreme example. The actual values for the saturation draw-out field strengths in the experiments for the different molecular gases are given in Table 12.5. For the rare gases 5 V/cm proved adequate.

TABLE 12.5

Field strengths in V/cm used to saturate collection of energetic positive ions produced in dissociative ionization of molecular gases

Gas	H_2	D_2	N_2	O_2	NO	CO	N_2O	CO_2	CH_4
$V_\parallel$ (V)	25	25	15	25	21	15	20	20	26

In the early experiments of Tate and Smith† (Chap. 3, § 2) the collecting efficiency for positive ions of several eV kinetic energy was probably considerably smaller than unity. Nevertheless, comparison of their results with those of later authors, who were careful to saturate the positive ion collection, does not reveal any features that are not present in a similar comparison for atomic gases (see Chap. 3, § 2.5). In many cases the relative probability of producing energetic ions is quite small, so inadequate collection of these ions introduces relatively small errors in the derived absolute cross-section.

Petersen‡ has suggested a method which in principle avoids any difficulties due to production of energetic ions, but which has its own problems. The method is very similar to the crossed-beam methods described in Chapter 3, § 2.3, the additional feature being the substitution for the thermal beam of an energetic molecular beam produced by charge exchange (see Vol. IV). In comparison with the already large kinetic energy of the target molecules the kinetic energy of the ions produced in dissociative ionization is small and makes little difference to the efficiency of analysis of the ions produced.

Apart from the background effects common to crossed-beam experiments that may be eliminated in the usual way (Chap. 3, § 2.3) there is the additional complication that a small fraction of ions will be present in the molecular beam due to ionization of residual gas by the fast molecules. These ions will interact with the electron beam to give a background signal. Perhaps the most important problem, however, is that of ensuring that the energetic molecular beam is effectively free from metastable species produced in the charge transfer process.

† TATE, J. T. and SMITH, P. T., *Phys. Rev.* **39** (1932) 270.

‡ PETERSEN, J. R., *Atomic collision processes*, ed. McDOWELL, M. R. C., p. 465 (North Holland, Amsterdam, 1964).

Preliminary results obtained by this method for atomic gases may be checked against those of Tate and Smith and of other investigators given in Chapter 3, § 2.5. The apparent ionization cross-sections obtained do not agree very well, especially at electron energies above 100 eV. Thus for argon Petersen obtains an apparent cross-section about 50 per cent greater than Smith's for 500-eV electrons. For neon the discrepancy is smaller. The ratios of double to single ionization cross-sections also disagree by factors of 2 or so with the results of Bleakney (Chap. 3, § 2.5.2). In view of these discrepancies applications to the study of molecular ionization must be treated with reserve until the validity of the technique is more thoroughly established.

7.3. *Determination of appearance potentials and the nature, abundance, and relative kinetic energy of the different ionized products*

The first experimental evidence of the formation of ions with kinetic energies of a few eV, by ionization of hydrogen molecules, was noted by Bleakney and Tate[†] and, in more detail, by Bleakney[‡] in 1930. They used an apparatus of the type described in Chapter 3, § 2.2 in which ions produced by an electron beam were accelerated to a slit through which they passed into a magnetic analysing chamber. The ion current peaks observed in the collector after passing through this chamber were quite sharp under certain conditions of electron energy but in others were found to be rather diffuse, indicating a considerable spread in initial velocity of the ions. This work was rendered more precise by Hagstrum and Tate[§] and has formed the basis of many later investigations. The disadvantage of this technique is that the efficiency of collection through the exit slit falls quite rapidly with the kinetic energy of the ions so that it is unsuitable for the study of ion energy distributions. An alternative method introduced by Lozier[||] a few months after Bleakney's first experiment partly overcomes these difficulties but provides no means of mass analysis. It has been used effectively for studying the production of negative as well as positive ions. We shall first describe Lozier's method and then comment on the use of the mass spectrograph.

7.3.1. *Lozier's method.* The apparatus used by Lozier, which is of a type employed in a considerable variety of experiments in molecular gases, is illustrated in Fig. 12.15. The principle is a very direct one. An electron beam of homogeneous energy eV_e is fired through the gas at a

† BLEAKNEY, W. and TATE, J. T., *Phys. Rev.* **35** (1930) 658.
‡ BLEAKNEY, W., ibid. **35** (1930) 1180.
§ HAGSTRUM, H. D. and TATE, J. T., ibid. **59** (1941) 354.
|| LOZIER, W. W., *Phys. Rev.* **36** (1930) 1285 and 1417.

pressure of the order 2×10^{-4} torr, a magnetic field parallel to the beam, of between 100 and 150 gauss, preventing lateral spread. Of the ions produced by the beam, only those moving in a direction perpendicular to it can move through slots cut in the enclosing metal cylinder. Velocity analysis of these ions may then be carried out by the retarding potential method as they have unidirectional velocities. Referring to Fig. 12.15, K is the tungsten filament source of the electron beam, which was accelerated through the holes in the diaphragm D by a potential difference V_e between K and D. After passage through the collision chamber the beam passed through the holes in the diaphragm E and

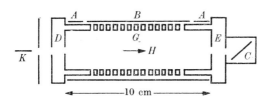

FIG. 12.15. Lozier's apparatus for measuring the velocity of ions formed by electron impact.

was collected at C. A holding potential of 175 V was applied between C and E to ensure complete collection. Ions formed in the chamber moved to the walls of the enclosing circular cylinder G in which were cut a number of slots running round the circular boundary of the cylinder. These permitted the passage of ions, moving radially, into the space outside G. They could be collected on the enclosing cylinder B for which the parts A acted as guard-rings. A retarding potential V_R could be applied between G and B, so making possible a velocity analysis of the positive ions.

The apparatus was of copper, sealed in a Pyrex glass tube, and could be baked out at 400 °C. The cylinder was 10 cm long and the slots in G were 0·5 mm wide and 3 mm thick. Electron beam currents of from 5 to 8 μA were used. Corrections were applied to V_e for the initial electron velocities and to the observed voltage of the positive ions for the effect of the magnetic field on the positive ion trajectories.

Referring to (2) and (3) it will be seen that if $E_{\min}$ is the minimum electron energy at which ions of energy W^+ can be produced in a particular process

$$E_{\min} = \frac{1}{\beta} W^+ + W_0,$$

where

$$\beta = (M - M^+)/M,$$

M^+ being the mass of the ion and M that of the initial target molecule. The constant W_0 is the sum of the excitation energies of the ion (including the ionization energy) and neutral molecular fragment and the dissociation energy of the molecule.

Lozier therefore introduced the technique of plotting E_{min} as a function of W^+. Extrapolation to zero then gave W_0. Unfortunately the broadening of the ion energy distribution due to the thermal motion of the target molecule as discussed in § 7.1 is likely to introduce an important error of the order 0·5 eV in the determination of W_0 in this way. This is less important when β is small and when the identification of the process rather than accurate determination of appearance potentials is all that is required (see, for example, § 7.1 and Fig. 12.12). It also does not arise if the electron energy required to produce ions of zero kinetic energy can be *directly* observed.

No additional error is introduced due to anisotropy of the distribution of ion momentum unless the chance of ejection at right angles to the electron beam is relatively very small. This is because the system of slots selects out for retarding potential analysis only those ions moving in this direction and no blurring occurs due to ions entering the field at an angle.

On the other hand, as shown by Tozer,† the collection efficiency, because of the slots, is not independent of the energy of the ions. This may be avoided by choosing the collecting potential to be inversely proportional to the ion kinetic energy but this is often inconvenient or impossible to apply over the whole energy range.

Lozier's method has been used extensively in the past for studying the energy distribution of positive ions produced by electron impact and for determining appearance potentials, much of which has been done without realization of the importance of the temperature motion of the target molecules in certain cases or of the variation of collective efficiency with ion energy. It has also been used for studying the production of negative ions and this application will be discussed further in § 7.4.

7.3.2. *Use of the mass spectrograph.* A mass spectrograph of some kind must be used to identify the products of a reaction between an electron beam and gas molecules. In typical experiments the arrangement is essentially similar to that in a total ionization tube in which an electron beam passes through the gas under investigation between two parallel plates. Positive ions produced in the gas by the electron beam are attracted to one of the plates by application of an electric

† Tozer, B. A., *J. Electron. Control* **4** (1958) 149.

field between them. The additional feature is the extraction of a sample of these ions through a slit cut in the collector plate. The ions issuing through the slit are then accelerated through a second slit into the mass analyser which may be one of several types.

The appearance potentials for ions of different kinds may be determined directly from the variation of the analysed current of the ions in the usual way but there are complications in dealing with ions which

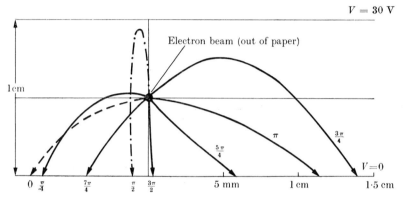

FIG. 12.16. Trajectories of protons formed by electron impact with 8-eV kinetic energy and various initial directions of motion in the plane of the diagram. The beam is perpendicular to the paper and passes parallel to and midway between two electrodes 1 cm apart across which a potential of 30 V is applied.

are formed with initial kinetic energy of a few eV. This is because the collecting power of the slit depends quite strongly on this kinetic energy. As an illustration of this† we show in Fig. 12.16 a series of trajectories for protons of 8-eV kinetic energy produced by an electron beam passing symmetrically between two plates 1 cm apart between which a potential difference of 30 V is applied. It is assumed also that a magnetic field of 150 gauss is applied parallel to the electron beam. Protons produced with initial velocities in different directions within the plane perpendicular to the electron beam reach the collector plate at such widely separated points that it is clearly out of the question to extract a major fraction of the ions through a fine slit in the collector plate. Relative abundance measurements for different ions can clearly not be made with a mass spectrograph if the ions possess appreciable kinetic energy.

Hagstrum and Tate‡ discussed the effect of ion kinetic energy on the shape of the peak corresponding to the ion in the mass spectrum. They

† RAPP, D., ENGLANDER-GOLDEN, P., and BRIGLIA, D. D., *J. chem. Phys.* **42** (1965) 4081.

‡ HAGSTRUM, H. D. and TATE, J. T., *Phys. Rev.* **59** (1941) 354.

examined the trajectories of the ions through a 180° mass analyser with geometry as shown in Fig. 12.17. The slits S_1 and S_2 were of width 0·037 and 0·022 cm respectively and S_6 was adjustable from 0 to 0·050 cm. The collecting field between P and P_1 was about 30 V/cm, the radius of the semicircular path of the collected ions was 20 cm, and the defining slit of the electron beam had dimensions 0·01 × 0·5 cm.

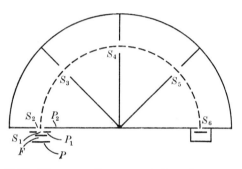

FIG. 12.17. Diagrammatic representation of mass spectrograph used to study the products of ionization by electron impact.

The collecting efficiency of the spectrometer as a function of ion kinetic energy was calculated from the ion trajectories neglecting space charge and the magnetic field. It is shown in Fig. 12.18 (b). Apart from the spread in the shape due to the ions possessing an initial energy distribution there will also be a spread due to instrumental effects. Thus for ions formed with zero initial kinetic energy the peak shape will be as shown in Fig. 12.18 (a). Corresponding to three typical shapes of the energy distribution as shown in Fig. 12.18 (c), (e), and (g) respectively we obtain, by combination of the spread due to collector efficiency variation and to instrumental effects, corresponding peak shapes as shown in (d), (f), and (h). These are sufficiently different and characteristic to make it possible to derive some information about the initial ion energy distribution from a study of the peak shapes.

A great deal of experimental work, aimed at the determination of the nature and appearance potentials of ions formed by electron impact with different molecules, has been carried out. If the ions are formed with little initial kinetic energy there is no difficulty in interpretation, but in many experiments little attention has been paid to the examination of peak shapes so that inadequate information about the initial ion kinetic energies has been forthcoming. For this reason the relative abundances of different ions formed by electrons of fixed energy as given from mass analysis must be regarded very often as unreliable. As in other experiments the same complications arise from the thermal motion of the target molecules when the ions are formed with kinetic energy of the order of a few eV. Despite these limitations a surprising

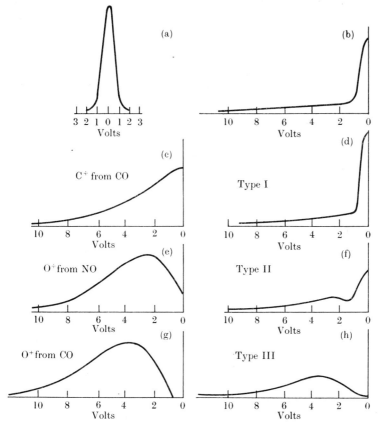

Fig. 12.18. Dependence of peak shape in a mass spectrograph on the kinetic energy with which ions are formed. (a) Symmetrical peak shape produced in a mass spectrograph by an ion formed with no initial energy. (b) Estimated collection-efficiency curve for the mass spectrograph of Hagstrum and Tate. (c), (e), and (g) some typical distributions in energy of ions formed by electron impact. (d), (f), and (h) show the corresponding observed peak shapes.

degree of agreement with thermochemical data has been obtained from analysis of data on appearance potentials obtained using mass analysis.

7.3.3. *Crossed-beam methods for studying near-threshold behaviour of the ionization cross-section.* The crossed-beam technique introduced by McGowan and his associates that has been described in Chapter 3, § 2.4.5 is immediately applicable to the study of the ionization cross-section

as a function of electron energy close to the threshold. Applications to
H_2, N_2, and O_2 are described in Chapter 13, §§ 1.4.2, 3.4.2, and 4.3.2
respectively.

7.3.4. *Absolute measurement of cross-sections for production of
energetic ions.* The problem of determining absolute cross-sections for
production of positive ions with initial kinetic energy of several eV is
difficult because the efficiency of collection of such ions after passing
through a slit system as, for example, in a mass spectrograph, varies
with the ion energy. However, in 1930 Bleakney[†] showed how the total
cross-section for production of energetic ions could be obtained with an
apparatus designed to measure total ionization. In the usual mode of
operation the ion collector plate is negative with respect to the beam
so as to saturate positive ion collection. Bleakney reversed the field
on this plate and so collected only ions which were sufficiently energetic
to cross the retarding field. In this way the use of slits is avoided.

This method has been applied by Rapp, Englander-Golden, and
Briglia[‡] using the apparatus discussed in § 7.2 and illustrated diagrammatically in Fig. 12.13. Molecular ions, formed with only thermal
energies, could be prevented from reaching the ion collector plate when
it was maintained at a potential of 0·5 V above that of the parallel plate.
This was established by carrying out experiments with electrons of
energy less than that of the threshold for dissociative ionization. The
field was then maintained while experiments were carried out at higher
electron energies. Ions with kinetic energy $> 0·25$ eV could then reach
the positive collector plate. With such a small field the motion of
energetic ions is not much affected but the fraction of energetic ions
collected depends on their angular distribution on formation, relative
to the electron beam. Using a wide collector plate the effect of an
anisotropic distribution is minimized and it was estimated from trajectory calculations that about 40 per cent of ions with kinetic energy
$> 0·25$ eV were collected.

Results obtained in this way for a number of molecular gases are
described in Chapter 13. A check on these results can be made[‡] with a
180° mass spectrograph of the type used by Bleakney. In this instrument the electron gas and mass analyser are both contained inside a
long solenoid. The usual mode of operation involves acceleration of
the ions out of the region between the parallel plates, through a slit

† BLEAKNEY, W., *Phys. Rev.* **35** (1930) 1180.
‡ RAPP, D., ENGLANDER-GOLDEN, P., and BRIGLIA, D. D., *J. chem. Phys.* **42** (1965)
4081.

into the mass analyser. However, by reversing the ion extraction field the instrument can be used to measure total ionization. By working at an electron energy below the threshold for dissociative ionization the ratio of molecular ion current as measured by the mass analyser to the total ion current can be obtained. This gives the collection efficiency C of the analyser for the molecular ions. At a higher electron energy the ratio of molecular ion current i_m in the mass analyser to the total ion current i_t will fall and this can only be due to the production of ions through molecular dissociation. Thus

$$i_m/i_t < C$$

and
$$i_m/C = i_t^m,$$

where i_t^m is the contribution to the total ion current at the higher electron energy, due to molecular ions. The contribution to this current due to dissociative ionization is then

$$i_t^d = i_t - i_t^m.$$

A further estimate may be made by examination of the ion collector saturation characteristics such as shown in Fig. 12.14, which indicates clearly, for sufficiently high electron energies, that an appreciable fraction of energetic ions is present.

7.4. *Study of negative ion formation by beam methods*

The Lozier method (§ 7.3.1) may be applied without difficulty to the determination of appearance potentials for negative ions though the same complications arise from the temperature motion of the target molecules. It has also been used for the observation of the relative yield of negative ions at different electron energies and for the measurement of the energy distribution of the ions. For these purposes it is not very satisfactory as the collection efficiency varies with the ion kinetic energy as mentioned in § 7.3.1. To correct for this it is necessary to vary the collecting voltage inversely as the ion kinetic energy, a procedure often inconvenient and difficult in practice, which was not applied in earlier measurements. It is even less satisfactory for accurate determination of cross-sections for negative ion production, because ions with initial directions of motion that make small angles with the direction of the electron beam are not collected.

For these reasons, in recent years, experimenters have studied negative ion formation using total ion-collecting devices in which the slots in Lozier's apparatus are replaced by grids. Thus Fig. 12.19 illustrates

H

the arrangement of apparatus used by Schulz† in which the aim is to collect effectively all the ions produced. The electron gun was equipped for use with the retarding potential difference method (Chap. 3, § 2.4.2) so that observations could be made with electrons with a narrow range of energies.

Asundi, Craggs, and Kurepa‡ have used a total ionization tube similar to that of Tate and Smith for measurement of positive ion cross-sections,

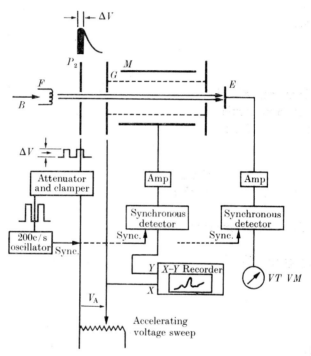

Fig. 12.19. Apparatus used by Schulz for measurement of total cross-sections for negative ion production.

while the apparatus shown schematically in Fig. 12.13 and described in § 7.2 has been adapted by Rapp and Briglia§ for the measurement of total cross-sections for negative ion production. To do this effectively account must be taken of certain factors that make it more difficult to measure these cross-sections for negative than for positive ions.

The principal difference is that, at energies above the respective thresholds, the production cross-sections for negative ions are of order

† Schulz, G. J., *J. chem. Phys.* **33** (1960) 1661.
‡ Asundi, R. K., Craggs, J. D., and Kurepa, M. V., *Proc. phys. Soc.* **82** (1963) 967.
§ Rapp, D. and Briglia, D. D., *J. chem. Phys.* **43** (1965) 1480.

10^3 times smaller than for positive ions. Also the cross-sections for dissociative attachment (see Fig. 12.5) are of a resonance type with a breadth of order 1–2 eV as compared with the typical shape of the cross-section–energy-curves for positive ion production (see Chap. 3, Fig. 3.10). Because of these differences the following specific problems arise when using the equipment for accurate measurement of negative ion production cross-sections.

(i) Because of the small cross-sections, the effect of positive ions formed in front of the electron collector plate (see Fig. 12.13) being accelerated back into the ionization region is more important than when studying positive ions. Care must therefore be taken to operate the electron collector at the lowest positive potential compatible with effectively complete collection of the electron beam.

(ii) Because of the peaked shape of the variation of the dissociative attachment cross-section with electron energy (see § 3.6.1) errors due to energy spread in the electron beam that may arise from thermal sources, space charge, or field fluctuations are more significant.

(iii) If large extraction fields are required to saturate the negative ion collection there is a danger that scattered or secondary electrons formed from or by the main beam with little axial velocity will reach the negative ion collector despite the axial magnetic field. This presents a real problem when the electron energy is above the threshold for dissociative ionization as it is then necessary to apply high extraction fields to prevent energetic positive ions from reaching the negative ion collector. Occurrence of the effect is indicated by a failure to saturate the collected negative current. In practice it was found that, for electron energies less than about 60 eV, a compromise choice of extraction voltage could be made, which, while large enough to prevent collection of energetic positive ions and saturate negative ion collection, was nevertheless not so large as to begin to collect electrons.

This is illustrated in Fig. 12.20, which reproduces a number of negative-ion current collection characteristics in CO taken for different energies of the electron beam. For 9·9-eV electrons saturation is reached with a voltage as low as 2 V across the plates. When this rises above 8 V the collected current begins to fall. This is a consequence of the energy spread across the electron beam due to the transverse field, the effect of which is magnified because the cross-section for dissociative attachment in CO is a maximum for 9·9-eV electrons, as referred to in (ii).

As the incident electron energy increases to 30 eV the initial current collected is positive, due to collection of the energetic positive ions from dissociative ionization. However, as the extraction field increases the current reverses in sign and is saturated at 17 V. A similar result is found for 50-eV electrons except that the current remains positive until 8 V is applied. By the time the electron energy has increased to 100 eV it is no longer possible to produce saturation as scattered and secondary electrons are being collected before the energetic positive ion are all completely repelled.

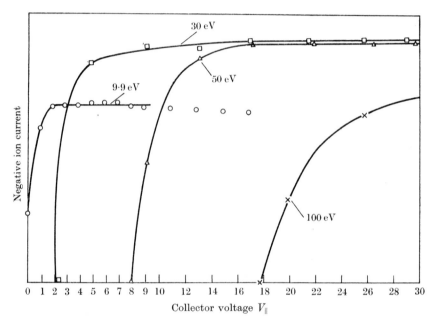

FIG. 12.20. Negative ion collector current as a function of collector voltage $V_{\parallel}$ observed in CO with electron beams of different energies (Rapp and Briglia).

Table 12.6 lists the negative ion extraction fields used by Rapp and Briglia for different gases and different electron energy ranges.

(iv) The small cross-sections for dissociative attachment for some gases make it especially important to reduce the concentration of impurities that have relatively large cross-sections. In the experiments of Rapp and Briglia the actual vacuum system could be baked out and the background pressure was 10^{-4} of the working pressure.

In carrying out a systematic series of attachment cross-section measurements in a number of molecular gases and vapours, Rapp et al. determined the absolute cross-section $Q^-(E_1)$ for negative ion formation by electrons of a particular energy E_1 by comparing the saturation negative ion current $i^-(E_1)$, obtained with the extraction field positive, to $i^+(E_2)$, the saturation positive ion current obtained for electrons of a chosen energy E_2, with the sense of the extraction field reversed. Then if $Q^+(E_2)$ is the total apparent cross-section for ionization by electrons of energy E_2 determined by earlier measurements (see § 7.2),

$$Q^-(E_1) = \{i^-(E_1)/i^+(E_2)\}Q^+(E_2),$$

provided the electron beam current and gas pressure remain unchanged.

TABLE 12.6

Negative ion extraction fields used in the experiments of Rapp and Briglia

Gas	Electron energy range (eV)	Extraction field (V/cm)
CO_2	0–12	4
	12–60	25
CO	0–15	5
	15–60	26
O_2	0–12	5
	12–60	26
N_2O	0–6	3
	6–60	30
NO	0–14	4
H_2, D_2, HD	0–12·8	24
	12·8–18	10
SF_6	0–10	2

Table 12.7 gives a comparison of results obtained by Rapp and Briglia[†] with those of Schulz[‡] and of Asundi, Craggs, and Kurepa[§] using the apparatus referred to above. Some results obtained by Buchel'nikova[||] are also included. She has carried out a number of measurements of cross-sections for negative ion formation using apparatus based essentially on the same principles.

It will be seen that reasonably good agreement exists for O_2, CO, and both the peaks in CO_2. For SF_6 the peak is so narrow (see Chap. 13, § 11.1) that the results obtained are extremely sensitive to the electron energy distribution. In fact, as described in Chapter 13, § 11.1, the shape of the observed production cross-section–energy-curve in SF_6 is often used to determine the actual energy distribution in a low-energy electron beam. Rapp and Briglia estimate the full width at half maximum of this distribution in their experiments to be 0·2 eV.

[†] loc. cit., p. 858.
[‡] SCHULZ, G. J., *Phys. Rev.* **128** (1962) 178.   [§] loc. cit., p. 858.
[||] BUCHEL'NIKOVA, I. S., *Zh. éksp. teor. Fiz.* **35** (1958), 1119; *Soviet Phys. JETP* **35**(8) (1959) 783.

In Buchel'nikova's experiments the energy spread was similar but, by insertion of a diaphragm in the path of the electron beam to which a suitable potential could be applied, she was able to cut off electrons with energies less than a chosen value so that the resultant distribution was effectively exponential. This being so it was possible to unfold the distribution from the cross-section data.

TABLE 12.7

Comparison of maximum cross-sections for dissociative attachment in various gases as measured by different investigators

Gas	Investigators	Electron energy at peak (eV)	Max. cross-section (πa_0^2)	Peak width at $\frac{1}{2}$ max (eV)
O_2	RB	6·5	0·016	2·1
	S	6·7	0·015	2·2
	ACK	6·5	0·015	2·0
	B	6·3	0·015	1·9
CO	RB	9·9	0·0023	1·3
	S	10·1	0·0018	1·4
	ACK	10·1	0·0027	1·3
CO_2 (1)	RB	8·1	0·0049	1·1
	S	8·2	0·0051	1·1
	ACK	8·0	0·0049	1·3
CO_2 (2)	RB	4·3	0·0017	0·9
	S	4·5	0·0017	0·9
	ACK	4·5$_5$	0·0017	1·0
SF_6	RB	0·1	2·4	0·2
	B	0·0	5·7	0·5

RB, Rapp and Briglia; S, Schulz; ACK, Asundi, Craggs, and Kurepa; B, Buchel'nikova.

Although the use of a mass analyser to select the negative ions precludes the determination of absolute or relative production cross-sections because of the dependence of the collecting efficiency on the ion kinetic energy as discussed in § 7.3.2, it offers the only means for identifying the ionized products arising from reactions with electrons of a particular energy. Appearance potentials, subject to the usual complication by the thermal motion of the target molecules, may also be obtained.

Detailed results of investigations for different gases using these techniques are described in Chapter 13.

7.5. *Study of negative ion formation using electron swarms—the attachment coefficient*

Consider a stream of electrons drifting through a gas at pressure p under the influence of a uniform electric field F. The mean energy of the electron stream, which is a function of F/p at a fixed gas temperature T, can be determined from experiments of the type described in Chapter 2, § 2. In general there will be a chance α_a that an electron of the stream will attach to a gas molecule in drifting unit distance in the direction of the field. The loss δI of electron current in passing a distance δx in this direction, due to this cause, will be given by

$$\delta I = -I\alpha_a\,\delta x.$$

Integrating this, we have for the ratio of the electron current in the stream at two points x_1, x_2

$$I_2/I_1 = e^{-\alpha_a(x_1-x_2)}. \tag{49}$$

By observation of this ratio, α_a, which is known as the attachment coefficient, may be determined.

To see the relation between α_a and the attachment cross-section we note that, in passing a distance δx in the direction of the field, the actual distance traversed by the electron is not δx but $c\,\delta x/u$, where c is the random and u the drift velocity. Hence, if N is the number of molecules/cm^3 we have

$$\alpha_a = N\bar{Q}_a c/u, \tag{50}$$

where $\bar{Q}_a$ is a mean cross-section for attachment. To unfold the electron energy distribution an indirect analysis on the same lines as that described in Chapter 11, § 7 may be applied.

We now proceed to discuss the methods that have been used for the measurement of α_a.

7.5.1. *The method of the electron filter.* To apply the formula (49) in practice a device must be introduced for measuring the fraction of the beam current which is carried by electrons, at any particular plane perpendicular to the field. This may be done by insertion, at the plane concerned, of an electron filter of the type described in Chapter 2, § 3.1. The filter consists of a grid between alternate wires of which a high-frequency alternating electric field may be applied. By suitable adjustment of this field all the electrons in the stream that reach the filter may be swept to one or other grid wire, leaving the comparatively heavy negative ions unaffected. The decrease in beam current on passage through the filter may then be taken as a measure of its electron content just before reaching the filter.

The way in which this method was adapted by Bradbury[†] in his extensive series of measurements of α_a is illustrated schematically in Fig. 12.21. Electrons generated photoelectrically from the plate A

† BRADBURY, N. E., *Phys. Rev.* **44** (1933) 883.

were allowed to diffuse under the action of an electric field between this plate and the plate P. The field was maintained uniform by the guard-rings B, C, D, E. One half of each of the guard-rings B and D consisted of a grid of fine platinum wires (G, G' in Fig. 12.21), which could be slid into position to intercept the electron stream. An alternating potential of frequency 4–15 Mc/s could be applied across the grid wires by an oscillator O.

The apparatus was enclosed in a Pyrex tube and could be baked out at 200 °C, so that the background pressure could be reduced to 10^{-6} torr. Working pressures ranged from 3 to 90 torr.

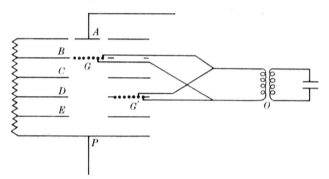

FIG. 12.21. Apparatus employing the electron filter used by Bradbury to measure attachment coefficients.

The measurements made were of the current reaching P, first with one grid in place with and without the alternating field applied, and then with the second grid in place. From these I_2/I_1 and hence α_a could be determined. Later measurements have been made with essentially the same design as that used by Bradbury (see Chap. 13, §§ 4.4, 7.2, 8).

7.5.2. *Pulse methods.* A variety of pulse methods have been introduced. We first describe one due to Doehring,† which has been extensively developed and used by Chanin, Phelps, and Biondi.‡ The principle is as follows.

A pulse of electrons liberated from a photocathode is allowed to drift a distance d through the gas, in a uniform electric field, to a collector electrode. During their passage some of the electrons will form negative ions in collision with gas molecules. These ions will drift about 1000 times more slowly than the electrons so that, after an electron pulse is collected, there will arrive a further delayed current due to the ions.

† DOEHRING, A., *Z. Naturf.* **7a** (1952) 253.

‡ CHANIN, L. M., PHELPS, A. V., and BIONDI, M. A., *Phys. Rev. Lett.* **2** (1959) 344; *Phys. Rev.* **128** (1962) 219.

The ion current received at a time t after the pulse will have been generated at a distance $u_i t$ from the electrode where u_i is the ion drift velocity. At this distance the electron concentration in the pulse will have been reduced from its initial value $n_e(0)$ to $n_e(x) = n_e(0)e^{-\alpha_a x}$

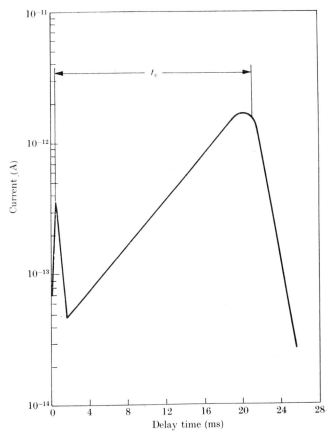

FIG. 12.22. Form of the signal received at the collector electrode in the drift tube used by Chanin, Phelps, and Biondi.

where x is the distance from the photocathode. Since $x = d - u_i t$, it follows that the negative ion current will vary as $e^{\alpha_a u_i t}$. Hence, by observing this current as a function of time $\alpha_a u_i$ may be obtained. Fig. 12.22 illustrates the general form of the signal received at the collector electrode. The exponential rise terminates in a sharp cut-off. This occurs when the time is that required for a negative ion to drift the full distance d. Thus if t_c is the cut-off time $u_i = d/t_c$. Hence both α_a and u_i may be obtained.

Alternatively α_a may be obtained from the ratio of the total electron current to the total ion current received at the collector. The number of negative ions produced at a distance between x and $x+\delta x$ from the photocathode is given by

$$\delta n_i = \alpha_a\, n_e(x)\, \delta x = \alpha_a\, n_e(0) e^{-\alpha x}\, \delta x. \tag{51}$$

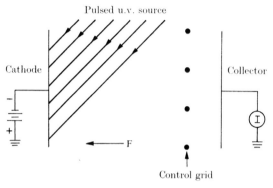

FIG. 12.23. Illustrating the arrangement of electrodes in the drift tube used by Chanin, Phelps, and Biondi.

The total number of negative ions received at the collector is then given by

$$n_i = n_e(0)\alpha_a \int_0^d e^{-\alpha_a x}\, dx$$

$$= n_e(0)(1-e^{-\alpha_a d}) \tag{52}$$

and $\qquad n_e(d)/n_i = e^{-\alpha_a d}/(1-e^{-\alpha_a d}). \tag{53}$

This method is perhaps less satisfactory than the first because it supposes that the shutter can be made equally efficient for both electrons and negative ions.

Fig. 12.23 illustrates a typical arrangement of electrodes in a drift tube used by Chanin, Phelps, and Biondi.† The electron pulses were produced by focusing ultra-violet light from a pulsed mercury discharge lamp on to the cathode. The grid was of essentially the same type as that used by Bradbury (see § 7.5.1). It consisted of a series of wires 0·0025 in diameter in which the alternate wires were electrically connected. A bias was normally applied to the grid to prevent electrons and ions from passing through it. To open the gate, rectangular voltage pulses were applied to each half of the grid so that the field between alter-

† loc. cit., p. 864.

nate wires was reduced to zero. The time of opening of the grid relative to the light pulse could be varied so the current to the collector could be measured at any desired time interval after the emission of the electron pulse.

Two drift distances 2·54 and 10·16 cm were used. The repetition rate at which measurements were conducted varied from about 10 c/s at low F/p to 200 c/s at high F/p with duty cycles of about 5 per cent. Currents between 10^{-11} and 10^{-15} A were used.

Very good vacuum conditions were maintained. Thus the drift tube was mounted on a standard ultra-high vacuum handling system.

Measurements could be carried out at 77 °K by immersing the whole tube, except for the fused silica window, in a liquid nitrogen bath. The vacuum system was baked out at 300 °C for about 16 h before each set of measurements. Immediately after bake-out the residual gas pressure was as low as 5×10^{-9} torr, rising at about 5×10^{-10} torr/min.

It is assumed in the above analysis that no appreciable detachment of electrons from the negative ions occurs due to collisions with the gas molecules. To check this, Chanin, Phelps and Biondi, in their first experiments, at temperatures of 300 °K and below and pressures up to 54 torr, operated the control grid as an electron filter (see Chap. 2, § 3.1) so that it would collect electrons but not negative ions. No evidence was found of a delayed electron current due to electrons detached from negative ions in transit. However, in experiments at higher temperatures and pressures such currents were observed. Pack and Phelps† were then able by suitable analysis of the electron and total current received to determine both the attachment coefficient and the rates of collision detachment. A detailed account of their method and results is given in Chapter 19, § 4.

Bortner and Hurst‡ have developed a pulse method that is particularly suitable for the determination of attachment coefficients for electrons drifting in a non-attaching gas containing a small partial pressure of gas that attaches readily. The principle of the method, which depends on pulse height measurements, is as follows.

Consider the region between two parallel plates between which a steady potential difference is applied. Suppose N_0 electrons are formed at a distance d from the positive electrode. If V_0 is the change of potential due to passage of a single electron from the point of formation to the

† PHELPS, A. V. and PACK, J. L., *Phys. Rev. Lett.* **6** (1961) 111; PACK, J. L. and PHELPS, A. V., *J. chem. Phys.* **44** (1966) 1870.
‡ BORTNER, T. E. and HURST, G. S., *Hlth. Phys.* **1** (1958) 39.

electrode, the change which will be observed when the electrons have travelled a distance x will be given by

$$N_0 V_0 \frac{x}{d}$$

provided no loss of electron by attachment occurs. Since we may relate x to the time t since production of the electrons by

$$t = x/u,$$

where u is the drift velocity of the electrons in the field, the time variation of the voltage pulse due to formation of the N_0 electrons will be as

$$N_0 V_0 t/t_0 \quad (0 \leqslant t \leqslant t_0),$$

where $t_0 = d/u$.

If, however, attachment can occur the numbers of free electrons at a distance x from the plane of formation will be no longer N_0 but

$$N(x) = N_0 e^{-\alpha_a x},$$

where α_a is the attachment coefficient. The change dV in electrode potential due to the work done in moving $N(x)$ electrons a further distance δx is

$$N_0 e^{-\alpha_a x} V_0 \, \delta x/d.$$

Hence the change in electrode potential after the initially produced electrons have moved a distance x towards the positive electrode will be

$$V(x) = \frac{N_0 V_0}{d} \int_0^x e^{-\alpha_a x} \, dx = \frac{N_0 V_0}{\alpha_a d} (1 - e^{-\alpha_a x}),$$

and the time variation of the voltage pulse will accordingly be

$$V(t) = \frac{N_0 V_0}{\alpha_a d} (1 - e^{-\alpha_a \, dt/t_0}). \tag{54}$$

Suppose now that this pulse is observed by a linear pulse amplifier the response of which to a function with a step at $t = 0$ is given by

$$V_a(t) = \frac{t}{t_1} e^{-t/t_1}, \tag{55}$$

where t_1 is the time constant of the amplifier. The output from the amplifier due to the pulse (54) at a time $t_1 > t_0$ will therefore be

$$V_a(t_1) = \int_0^{t_0} \frac{dV(t)}{dt} V_a(t_1 - t) \, dt. \tag{56}$$

This integration may be carried out readily, using the forms (54) and

(55) for $V(t)$ and $V_a(t)$ respectively and the resultant expression for $V_a(t_1)$ has a maximum when $t_1 = t_{max}$. The pulse height $V_a(t_{max})$ is a function of α_a and u. If u is known α_a may be obtained from observation of $V_a(t_{max})$. Fig. 12.24 illustrates the variation of pulse height with $\alpha_a d$ for various values of t_0/t_1.

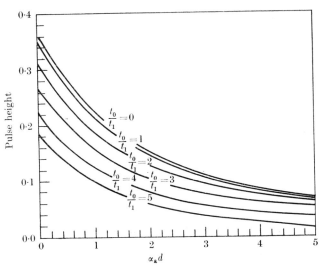

FIG. 12.24. Variation of pulse height with $\alpha_a d$ for various values of t_0/t_1 in the method of Bortner and Hurst for measuring attachment coefficients.

Applying these principles in practice, Bortner and Hurst produced the electrons by alpha particles from a ^{239}Pu source, collimated so as to produce the particles in a plane parallel to the collecting electrode. The source to collector distance was normally 6·0 cm, though it could be varied. The time constant t_1 of the linear pulse amplifier used was 35 μs and the shape of the response followed the form (55) out to 50 microseconds.

It was found convenient to measure the electron drift velocity u in the same apparatus. This was done by the method due to Bortner, Hurst, and Stone described in Chapter 2, § 3.3, in which the electrons are produced by an alpha-radioactive source in a plane-geometry chamber. There is a real advantage in measuring the drift velocity in the same gas as the effect of any non-attaching impurities in the drift velocity will be allowed for.

Apart from ordinary precautions about gas purity it was found essential to remove water vapour for which purpose either P_2O_5 or anhydrone was used.

As a check on the performance of the equipment the pulse height was measured as a function of F/p in CO_2 dehydrated by passage through anhydrone and further purified with a cold trap. In this state of purity CO_2 does not attach electrons to an appreciable extent for F/p less than $2 \cdot 0$ V cm^{-1} torr^{-1}. Fig. 12.25 shows the agreement between the predicted and observed variation of pulse height with F/p on the assumption that $\alpha_a = 0$ and the drift velocities are as measured.

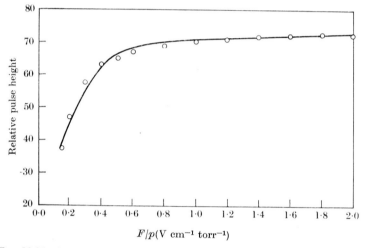

FIG. 12.25. Comparison of observed and calculated variations of pulse height with F/p for pure CO_2. ○ observed (Bortner and Hurst). —— calculated.

Since it depends on the measurement of electron pulse heights this method is especially suitable for weakly attaching gases or for gas mixtures in which the attaching component is in small relative concentration.

7.5.3. *Diffusion methods.* In Chapter 2, § 2 we described how the characteristic energy of a swarm of electrons drifting in gas under the action of a uniform electric field F could be determined by a diffusion method originally due to Townsend and Bailey. A swarm of electrons entered a drift space through a narrow aperture. As the swarm drifted along the field direction it spread laterally by diffusion. The extent of this spread after drifting a certain distance was measured by comparing the current received at a central electrode and at outer enclosing electrodes. It was then possible, by applying a suitable analysis in terms of diffusion theory, to determine the characteristic energy. Later refinements have introduced simpler geometrical conditions so as to render the analysis less complicated.

This method is not applicable in its simplest form when attachment occurs in the gas because the sideways diffusion of the negative ions is much less marked than that of the electrons. However, by carrying out experiments at two different drift distances it is possible to determine both the characteristic energy and the attachment coefficient.

Such a technique was first employed by Bailey[†] who used a somewhat modified geometry illustrated in Fig. 12.26.

Electrons generated photoelectrically from a plate A were allowed to drift under the action of a uniform electric field F maintained between the parallel plates A, B, C, D, E, parallel and superposed slits being cut in B, C, D to allow the passage of drift current to the plate E. Some current was collected on the plates B, C, D due to transverse diffusion, and the measurement of this formed an important part of the method.

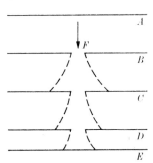

FIG. 12.26. Illustrating the principle of Bailey's method for measuring attachment coefficients.

The separations of the plates B, C and C, D were maintained equal throughout the experiment although both were changed in the course of the observations. These consisted of measurements of the currents to the plates C, D, E at two sets of homologous values of plate separation x, gas pressure p, and electric field F. To show how α_a was obtained we introduce the following symbols:

S, the ratio of the current passing through a slit to the total current arriving on the plane containing the opening. Suffixes 0, 1, 2, 3 refer to plates B, C, D, E respectively;

R_e, the fraction of electrons, falling on a plane, which pass through the slit;

R, the corresponding fraction for ions;

r, the corresponding fraction for ions formed between plates B and C, or C and D;

ϵ_k, the characteristic energy;

n_0, N_0, the number of electrons and ions respectively passing through the slit in plate B;

n_1, N_1; n_2, N_2, the corresponding quantities for plates C and D respectively.

† BAILEY, V. A., *Phil. Mag.* **50** (1925) 825.

Then

$$S_1 = \frac{n_1 + N_1}{n_0 + N_0}, \qquad S_1 S_2 = \frac{n_2 + N_2}{n_0 + N_0},$$

$$n_1 = n_0 R_e e^{-\alpha_a x}, \qquad n_2 = n_1 R_e e^{-\alpha_a x},$$

$$N_1 = N_0 R + r(1 - e^{-\alpha_a x})n_0, \qquad N_2 = N_1 R + r(1 - e^{-\alpha_a x})n_1. \quad (57\,\mathrm{a})$$

Elimination of r gives

$$R_e e^{-\alpha_a x} = \frac{S_1(R - S_2)}{R - S_1} = a_1, \quad \text{say.} \qquad (57\,\mathrm{b})$$

R_e and R are known, from diffusion theory, as functions of $F/\epsilon_k x$ but ϵ_k, which may be taken as κT, where T is the gas temperature, for the negative ions, is not known for the electrons. However, since ϵ_k is a function of F/p, R_e is unaltered if F, p, and x are all increased in the same ratio n. Under these conditions

$$R_e e^{-n\alpha_a^n x} = a_n.$$

If the attachment probability per collision is independent of the pressure, $\alpha_a^n = n\alpha_a$ and

$$\alpha_a = \frac{1}{x(n^2 - 1)} \ln \frac{a_1}{a_n}. \qquad (58)$$

This gives α_a in terms of directly measurable quantities apart from R, which may be calculated reliably from diffusion theory.

In a typical apparatus of this type the electrode plates B, C, D consisted of rings over which silver foil was tightly stretched. The slits cut in the foil were 4 mm wide and their distance apart was 4 cm. A system of guard-rings, 2 cm apart, maintained a uniform field between the plates. The currents measured were of the order 10^{-12} A.

Recent modifications of the technique have depended on the choice of geometrical conditions so as to render simpler the analysis that yields the measured current ratio S. Huxley[†] analysed the problem for a circular geometry essentially similar to that used by Crompton and Elford[‡] for the measurement of characteristic energies and described in Chapter 2, § 2. His results were valid provided these energies are much greater than the mean energy of a gas molecule. The analysis was later extended by Hurst and Huxley[§] so as to avoid this limitation.

If the source is a small hole in the cathode and the swarm is entirely interrupted by an anode consisting of a central disk and two annuli, Huxley showed that, provided the ratio k of the mean electron energy to that of the gas molecules is $\gg 1$, the ratio S of the current received

† HUXLEY, L. G. H., *Aust. J. Phys.* **12** (1959) 171.
‡ CROMPTON, R. W. and ELFORD, M. T., *Proc. phys. Soc.* **74** (1959) 497.
§ HURST, C. A. and HUXLEY, L. G. H., *Aust. J. Phys.* **13** (1960) 21.

by the inner annulus to that received by both annuli is given by

$$1-S = f(b,h,\lambda,\mu)/f(a,h,\lambda,\mu), \tag{59}$$

where

$$f(x,h,\lambda,\mu) = (h/d)\exp(\lambda h - \mu d) + (\lambda h \alpha_a/\mu) \int_0^1 \exp(\lambda h s) \times$$

$$\times \left[\exp\left\{ -\mu h \left(\frac{x^2}{h^2} + s^2 \right)^{\frac{1}{2}} \right\} - \exp\left\{ -\mu h \left(\frac{x^2}{h^2} + \overline{2-s^2} \right)^{\frac{1}{2}} \right\} \right] ds,$$

and

$$d = (x^2 + h^2)^{\frac{1}{2}}.$$

Here h is the length of the diffusion space, $2\lambda = u/D$ the ratio of the drift velocity to the diffusion coefficient for the electrons,

$$\mu^2 = \lambda^2 + 2\lambda\alpha \quad \text{and} \quad \alpha = \alpha_a - \alpha_i,$$

where α_i is the ionization coefficient that must be included if the mean energy of the swarm is high enough to produce appreciable ionization in the drift space.

If measurements are made of S for two suitable values of h and α_i is known, λ and α_a may both be determined. The procedure adopted by Huxley, Crompton, and Bagot† was to determine μ from the observed ratio with the shorter value of h, assuming α_a and α to be zero. This approximation for μ was then substituted in the observed ratio for the larger separation and a first approximation to α_a determined. This iterative procedure was then continued until accurate solutions were found for λ and α_a. By using an apparatus in which the chamber length could be varied continuously from 1 to 10 cm by an induction motor drive Huxley *et al.* checked the consistency of the method by obtaining α_a for $F/p = 5$ V cm^{-1} torr^{-1} with four different combinations of pressure p and chamber length h.

In later experiments‡ the geometry used is illustrated in Fig. 12.27, the ratio S now referring to the ratio of the current received by the annular section A_2 to the sum of the current received by the sections A_2 and A_3, the range for h being adjustable as before.

7.5.4. *Attachment coefficients from measurement of current in a prebreakdown discharge.* Consider the passage of electron current between two plane parallel electrodes at a separation d in a gas at pressure p. According to Townsend's original theory, if α_i is the ionization coefficient for electrons drifting in the gas under the influence of a uniform electric

† HUXLEY, L. G. H., CROMPTON, R. W., and BAGOT, C. H., ibid. **12** (1959) 303.
‡ CROMPTON, R. W. and JORY, R. L., *Aust. J. Phys.* **15** (1962) 451; REES, J. A., ibid. **18** (1965) 41; CROMPTON, R. W., REES, J. A., and JORY, R. L., ibid. **18** (1965) 541.

field arising from a voltage difference V between the plates, then the current i arriving at the anode is given by

$$i = i_0 e^{\alpha_i d}, \qquad (60)$$

where i is the initial electron current leaving the cathode.

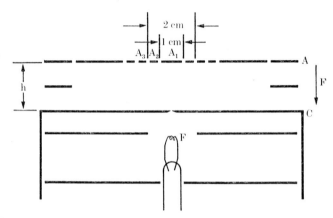

FIG. 12.27. Illustrating the geometry of the apparatus used by Crompton and Jory for measuring attachment coefficients.

This theory assumes that no loss of electrons occurs by attachment in the gas. If, however, there is an attachment coefficient α_a for electrons in the stream drifting from cathode to anode, then it is easy to show that (60) must be modified to give

$$i = i_0\{\alpha_i/(\alpha_i-\alpha_a)\}\exp\{(\alpha_i-\alpha_a)/d\}-i_0\,\alpha_a/(\alpha_i-\alpha_a). \qquad (61)$$

In deriving this relation it is supposed that in the attachment process no positive ions are simultaneously produced, i.e. that the process is of the type

$$AB+e \to A+B^-.$$

As described in Chapter 5, § 3.1.1 α_i is usually obtained by plotting $\ln(i/i_0)$ as a function of d. This plot will be linear if no attachment occurs but there will be departures from linearity if $\alpha_a \neq 0$. Thus Fig. 12.28 illustrates plots obtained by Harrison and Geballe[†] for oxygen at $11 \cdot 2$ torr pressure and different values of F/p. The non-linearity of the lines in the plot is marked at lower F/p because under such conditions α_a is relatively larger.

By appropriate choice of α_i and α_a a set of curves such as shown in Fig. 12.28 may be fitted thereby giving values of both coefficients. This method is useful at fairly large values of F/p at which α_i and α_a are comparable and also for gases with large values of α_a.

† HARRISON, M. A. and GEBALLE, R., *Phys. Rev.* **91** (1953) 1.

In typical experiments carried out by Harrison and Geballe[†] the electrodes were polished copper discs 9 cm in diameter. The separation could be adjusted to any distance between 0 and 4 cm using a precision screw with 40 threads per in. The envelope was of Pyrex, coated internally with Dag to prevent surface charging. Both the vessel

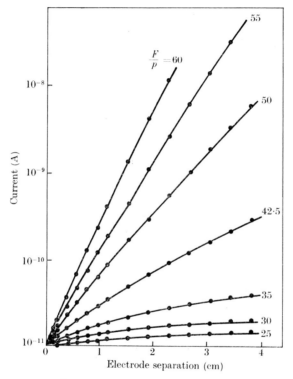

Fig. 12.28. Variation of current with electrode separation at different values of F/p in oxygen, as observed by Harrison and Geballe (indicated in V cm⁻¹ torr⁻¹ on each curve).

and electrodes were baked at temperatures exceeding 400 °C *in vacuo* and in an atmosphere of hydrogen for several hours. Experiments were only begun if the vacuum conditions were such that a pressure less than 10^{-5} torr could be maintained for several hours in the sealed-off vessel.

7.5.5. *Attachment rates from microwave probing of discharge afterglows.* We have already described in Chapter 2 how the measurement of the microwave conductivity of a discharge afterglow may be used to determine the momentum-transfer cross-sections for collisions of

† Ibid.

electrons of nearly thermal energy with gas atoms and molecules. There are many other applications of the technique of microwave probing of afterglows. An account of the technique for measuring the variation of electron concentration with time in an afterglow is described in Chapters 2, § 5.1 and 19, § 3.2.

Electrons are lost from an afterglow due to diffusion to the walls of the container, to recombination with ions in the gas and to attachment. If we ignore any reactions which repopulate the electrons, a dangerous assumption in some cases, then we may sort out the contribution from these three processes as follows.

As explained in Chapter 19, § 3.2 loss by diffusion involves an exponential decrease with time, the exponent varying inversely as the gas pressure. Attachment also produces an exponential rate of decrease but in this case the exponent is proportional to the gas pressure. Finally, recombination, if dominant, leads to a quite different rate of loss. Thus if η is the recombination coefficient so that n_e is the electron concentration

$$\frac{\mathrm{d}n_e}{\mathrm{d}t} = -\eta n_e^2,$$

then on integration we have that, if $n_e(0)$ is the concentration at $t = 0$,

$$\frac{1}{n_e(0)} = \frac{1}{n_e(t)} - \eta t.$$

It is therefore easy to determine whether recombination is important and, if not, attachment loss may be distinguished from that due to diffusion by investigating the variation with gas pressure. It is usually possible to choose the working range of pressure so that attachment, if it arises, is dominant. Details of the method of investigation are given in Chapter 19, § 3.2 with particular reference to the measurement of the diffusion coefficient of positive ions.

The great uncertainty of this method is the assumption that no repopulation of electrons is occurring. There are so many unusual species in an afterglow that many possibilities of electron production exist. We shall have occasion in Chapter 13, § 4.4.4 to discuss the results obtained for the attachment rate of slow electrons to oxygen molecules in which the apparent rates given from afterglow observations in pure oxygen are very much lower than those measured by other methods. In these cases the situation may sometimes be clarified by working with a mixture in different proportions of the gas under investigation with a second gas such as helium or nitrogen which does not form negative ions (see Chap. 13, § 4.4.4).

13

COLLISIONS OF ELECTRONS WITH MOLECULES—ELECTRONIC EXCITATION, IONIZATION, AND ATTACHMENT— APPLICATIONS TO SPECIFIC MOLECULES

In the preceding chapter we have given a general theoretical account of the various aspects of electronic transitions in molecules that lead to excitation of discrete states, to ionization with or without dissociation, and to attachment. An account was also given of the techniques employed for the experimental study of these phenomena. We now discuss applications to specific molecules. As a rough guiding rule we shall confine the choice of molecules to a selection in which none contains more than five atoms, concentrating mainly on the simplest and most important diatomic molecules.

We begin with neutral molecular hydrogen rather than the ion H_2^+ because, although the latter is in some respects the simplest molecule, it is difficult to study experimentally. Moreover, since H_2^+ ions must be produced initially from H_2 the interpretation of experimental data involving impact with such ions depends on knowledge of the processes that occur when the neutral molecule is ionized.

For each molecule considered we shall discuss first the excitation of states which are either stable excited states of the molecule or which dissociate into neutral fragments. This will be followed by an account of the production of positive ions from the molecule by electron impact. The next section will deal with processes involving negative ions. This will discuss the role, if any, played by such ions as intermediate states as well as the production of negative ions by dissociative attachment or polar dissociation. Finally, analysis of swarm data at mean electron energies high enough for electronic transitions to be important will be discussed in terms of assumed cross-sections for the relevant inelastic processes, based on measurements carried out with electrons of nearly homogeneous energy.

1. Molecular hydrogen (including deuterium and hydrogen deuteride)

The potential energy curves of the low electronic states of the hydrogen molecule and molecular ion H_2^+ are known rather more fully than for any other molecules. They exemplify almost all the effects discussed in Chapter 12, § 3 so that the effects of electron impact in producing electronic excitation with a variety of possible consequences have been quite extensively investigated experimentally. Good confirmation, both of the predicted forms of the potential energy curves and of the consequences of the excitation derived from use of the Franck–Condon principle has been provided. We shall therefore devote some space to a detailed description both of the expected effects and the experiments which have confirmed them.

1.1. *Predicted effects*

Fig. 13.1 illustrates the potential energy curves for the more interesting electronic states of the hydrogen molecule, which lie within 20 eV of the ground state, while Fig. 13.2 illustrates on a reduced scale some states up to 50 eV above that state, including particularly the most interesting states of the ionized molecule H_2^+.

Referring first to Fig. 13.1, curve I is that for the ground electronic state of the molecule with equilibrium nuclear separation 0·76 Å and dissociation energy 4·4 eV. In the ground vibrational state the classical range of the nuclear separation is between M and N. This ground electronic state is a singlet ($X\,^1\Sigma_g^+$). The lowest triplet state ($b\,^3\Sigma_u^+$) is a repulsive one, represented by curve II of Fig. 13.1. In the limit of large nuclear separation the molecule in this state tends to two normal hydrogen atoms, just as for the ground state. This repulsive state is of special interest because its existence was predicted by Heitler and London† at the introduction of their quantum theory of valency. The result that an electron pair (two electrons with opposite spins) gives rise to a bonding effect, whereas two electrons of parallel spins tend rather to oppose bonding, was generalized by them to a theory of the electron pair bond, now extensively employed in quantum chemistry. Curves III and IV represent two well-known stable singlet excited states—the so-called $B\,(^1\Sigma_u^+)$ and $C\,(^1\Pi_u)$ states respectively. Curve V represents the lowest stable triplet state $a\,(^3\Sigma_g^+)$ and curve VI the ground $^2\Sigma_g^+$ state of the molecular ion H_2^+.

For many purposes it is useful to have available the limits to which these states tend as the nuclear separation tends to zero (united atom limit) and to infinity (separated atom limit) (see Chap. 12, §§ 2.3–4). These are as follows:

† HEITLER, W. and LONDON, F., *Z. Phys.* **44** (1927) 455.

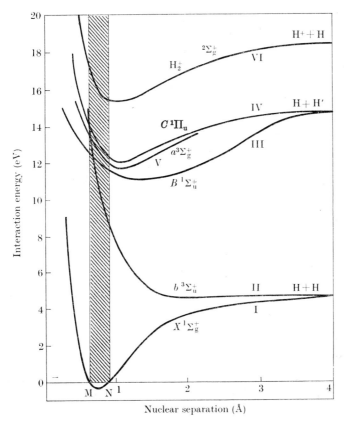

FIG. 13.1. Potential energy curves for electronic states of H_2 and H_2^+ lying within 20 eV of the ground state.

United atom limit	Molecular state	Separated atom limit
He $(1\,^1S)\,(1s\sigma)^2$	$X\,^1\Sigma_g^+$	$(\sigma_g\,1s)^2$
He $(2\,^1P)\,1s\sigma\,2p\sigma$	$B\,^1\Sigma_u^+$	$\sigma_g\,1s\,\sigma_u\,2s$
He $(2\,^1P)\,1s\sigma\,2p\pi$	$C\,^1\Sigma_u\,\Pi_u$	$\sigma_g\,1s\,\pi_u\,2p$
He $(2\,^1S)\,1s\sigma\,2s\sigma$	$E\,^1\Sigma_g^+$	$\sigma_g\,1s\,\sigma_g\,2s$
He $(2\,^3S)\,1s\sigma\,2s\sigma$	$a\,^3\Sigma_g^+$	$\sigma_g\,1s\,\sigma_g\,2s$
He $(2\,^3P)\,1s\sigma\,2p\sigma$	$b\,^3\Sigma_u^+$	$\sigma_g\,1s\,\sigma_u\,1s$

The only case in which the correlation at different nuclear separations offers some complication is that of the B state. It seems that, near the equilibrium separation this state is largely polar in character but does not dissociate into H^- and a proton because of interaction with other states at large nuclear separations.

We have included, in addition to the states previously mentioned, the $E\,^1\Sigma_g^+$ state which is the third singlet state that dissociates into a normal hydrogen atom and one in a 2-quantum excited state.

In contrast to helium, the corresponding two-electron atom, molecular hydrogen has no obvious metastable state. This is because of the molecular states that tend to metastable states of He in the united atom limit, the $b\,^3\Sigma_g^+$ state (tending to $2\,^3S$) is unstable and the $E\,^1\Sigma_g^+$ lies above the $B\,^1\Sigma_u^+$ state to which it may make allowed optical transitions. The B state in its turn may make such transitions to the ground $X\,^1\Sigma_g^+$ state. It appears, however, that a metastable state does exist. This is the $c\,^3\Pi_u$ state with no vibrational excitation. Optically allowed transitions can be made from this state to a $^3\Sigma_g$ but not a $^3\Sigma_u$ state. The energy difference between the $a\,^3\Sigma_g$ state and the $c\,^3\Pi_u$ state near their minima is very small so that, whereas the $c\,^3\Pi_u$ ($v=0$) level lies below $a\,^3\Sigma_g$ ($v=0$), $c\,^3\Pi_u$ ($v=1$) lies above. It follows that the $c\,^3\Pi_u$ ($v=0$) state can only radiate through a magnetic dipole transition to the $b\,^3\Sigma_u$ state towards which its lifetime will be of order 10^{-3} s. There should therefore be little difficulty in observing these molecules in an experimental apparatus of ordinary dimensions.

Confirmation of the existence of the metastable state, with an excitation threshold of $11\cdot9\pm0\cdot25$ eV was provided by the experiments of Lichten.† Using a radio-frequency resonance technique he was able to detect the presence of metastable molecules in a hydrogen molecular beam that had been bombarded by electrons with energy in the range $11\cdot9$–25 eV. The metastable molecules were aligned by passage through a magnetic field and the energy separation of their Zeeman substates determined by inducing resonance transitions between them. These transitions were detected by changes in the trajectories of the metastable molecules in an inhomogeneous magnetic field.

Just as for atoms we would expect that triplet electronic states will only be excited with appreciable probability by slow electrons. From the correlations with the united atom limit shown above we would expect the cross-sections for excitation of the B and C states to be generally similar to those for the 2^1P state of helium (Chap. 4, § 1.4.2.1, Chap. 7, § 5.6.7) both as regards magnitude and variation with electron energy. They should therefore be the most strongly excited states at high electron energies. By the same argument the cross-sections for excitation of the E state should resemble those for the 2^1S state of helium (Chap. 4, § 1.4.2.1; Chap. 7, § 5.6.9). It will therefore be expected to be

† LICHTEN, W., *Phys. Rev.* **120** (1960) 848.

considerably smaller than for the B and C states and fall off a little faster at high energies. We shall return to the discussion of these cross-sections for electrons of medium and high energy in § 1.3. Meanwhile we consider the effects that are to be expected due to impact of slow electrons, including particularly the threshold phenomena.

According to the Franck–Condon principle (Chap. 12, § 3.2), transitions from the ground state due to electron impact must occur vertically within the shaded region in Fig. 13.1. We would therefore expect the following effects as the energy of the exciting electrons increases.

No appreciable electronic excitation should occur until the electron energy reaches 8·8 eV when excitation of the repulsive $^3\Sigma_u^+$ triplet state begins. This will result in dissociation of the molecule into two normal atoms each with 2·2 eV kinetic energy. Further increase will make possible excitation of the B and C states, which will be followed by emission of ultra-violet light as radiative transitions occur from these states back to the ground state. At 11·8 eV excitation of the stable triplet state (curve V) can occur. This will result also in dissociation, for radiative transitions can take place from this state only to the lower triplet state, which is repulsive. Apart from the possibility of excitation of other electronic levels of the molecule, the next result of increasing the electron energy will be to produce stable H_2^+ ions when it exceeds the vertical ionization energy 15·4 eV.

To follow the further effects of increasing the electron energy it is convenient to refer to Fig. 13.2. Here curves I and II are those for the normal $^1\Sigma_g$ state of the molecule and $^2\Sigma_g$ of the molecular ion H_2^+. Curve III is that for a repulsive $^2\Sigma_u$ state of H_2^+ dissociating into normal H and H^+. The existence of this state is of interest in connection with the other current theory used in quantum chemistry, that known as the molecular orbital method. Instead of thinking in terms of electron pairs, the properties of the molecule are built up in terms of those of the individual electrons. Whether a particular electron contributes a bonding or antibonding effect depends on the symmetry of its wave function in the nuclear coordinates. This is exemplified in the simplest form in the two curves II and III of Fig. 13.2 for H_2^+, for in II the wave function of the single electron in the molecular ion is symmetric in the nuclear coordinates, in III antisymmetric. Finally, curve IV represents the Coulomb repulsion of two protons.

As in Fig. 13.1 the probable transitions must take place vertically within the shaded area. Hence, when the electron energy reaches 28 eV, excitation of the unstable state of H_2^+, represented by curve III, begins.

This will produce a hydrogen atom and a proton, each with 5 eV kinetic energy. Increased energy of the incident electrons can increase this energy up to about 7 eV by making possible excitation from initial nuclear separations nearer to M in Fig. 13.2. Eventually, when the bombarding energy reaches 46 eV, excitation to H_2^{++}, i.e. to two protons, can occur. These protons will have energies ranging from 7 to 10 eV.

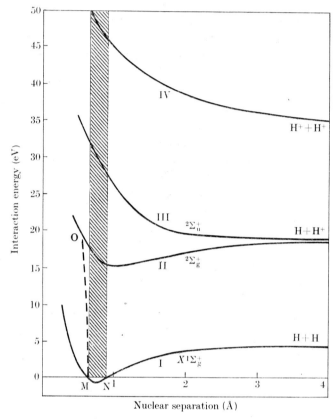

FIG. 13.2. Potential energy curves for higher energy states of H_2^+.

It will be noted, however, that, although all transitions to the $^2\Sigma_g$ level of H_2^+ that correspond to the Franck–Condon principle lead to stable H_2^+, a transition such as MO in Fig. 13.2 would lead to H_2^+ excited above dissociation. This transition would represent a comparatively weak violation of the Franck–Condon principle and might be expected to occur to a detectable extent for the lightest molecule H_2. The onset potential is 18 eV and the protons produced will have a very small kinetic energy—the smaller this kinetic energy the more nearly is the Franck–

Condon principle obeyed. It should therefore be possible to distinguish proton production by this transition from that due to excitation of the unstable state, not only by the onset potential but by the proton energy (see § 1.5.2).

A summary of these effects is provided in Table 13.1.

TABLE 13.1

Onset energy (eV)	State excited	Predicted effects
8·8	$H_2\ b\,^3\Sigma_u$ (lowest triplet state)	Dissociation into normal H atoms with 2·2 eV kinetic energy.
11·5	$H_2\ B\,^1\Sigma_u^+$	Ultra-violet radiation due to radiative transitions from B to ground state.
11·8	$H_2\ a\,^3\Sigma_g$	Dissociation into normal H atoms accompanied by emission of continuous spectrum due to radiative transitions from $a\,^3\Sigma_g$ to the repulsive $b\,^3\Sigma_u$ state.
12·6	$H_2\ C\,^1\Pi_u$	Ultra-violet radiation due to radiative transitions from C to ground state.
15·6	$H^+\ ^2\Sigma_g$ (ground state)	Ionization without dissociation.
18	$H_2^+\ ^2\Sigma_g$	Production of slow protons from dissociation into normal H and H^+ (transition weakly violating the Franck–Condon principle).
28	$H_2^+\ ^2\Sigma_u$	Dissociation into normal H and H^+, each with 5 eV kinetic energy.
46	H_2^{++} (repulsive)	Dissociation into two protons with 10 eV kinetic energy.

We shall now consider these processes separately in detail.

1.2. *Dissociation into normal H atoms due to electronic excitation of triplet states*

1.2.1. *Beam experiments.* It was remarked as early as 1925 by Blackett and Franck† that the intensity of excitation of the Balmer lines of atomic hydrogen by passage of an electron stream through hydrogen at low pressures was approximately proportional to the pressure. This indicated that dissociation resulted directly from impact of a molecule with an electron and was not mainly due to a series of processes such as:

$$H_2 + e \rightarrow H_2^+ + e,$$
$$H_2^+ + H_2 \rightarrow H_3^+ + H.$$

In 1927 the work of Glockler, Baxter, and Dalton‡ and of Hughes and

---

† BLACKETT, P. M. S. and FRANCK, J., *Z. Phys.* **34** (1925) 389.
‡ GLOCKLER, G., BAXTER, W., and DALTON, R., *J. Am. chem. Soc.* **49** (1927) 58.

Skellet† established the existence of a direct dissociative collision process

$$H_2 + e \rightarrow H + H + e,$$

and their results were confirmed two years later by Dorsch and Kallmann.‡

The principle of the experimental method used was the same in all three cases. An electron beam of definite and controllable energy was fired through gaseous hydrogen and the production of atomic hydrogen detected in some way. The method of detection differed in the three investigations. The two earlier groups of workers used a stationary mass of hydrogen at a pressure of the order 0·1 torr and dissociation became manifest by a falling pressure. In the method of Glockler *et al.* the pressure fall was secured by oxidizing the inner walls of the copper cylinder containing the hydrogen. Any atomic hydrogen formed reduced this oxide, forming water vapour that was condensed on a liquid-air trap. Hughes and Skellet froze the atomic hydrogen out directly on the glass walls of the hydrogen container, which was a glass tube coated internally with a conducting film of evaporated tungsten and immersed in liquid air. In both experiments the pressure fall was observed with a hot wire gauge. Both found that the onset potential for dissociation was between 11·4 and 11·5 eV. As this is more than 4 eV less than the ionization energy of the molecule it is clear that the phenomenon is in no way dependent on ionization. In addition Hughes and Skellet made a special study of the variation of dissociation rate with pressure from which they derived strong evidence in favour of the process being a primary one.

The method used by Dorsch and Kallmann differed from that of the earlier workers in that the dissociation was produced in a stream of hydrogen gas and the presence of atomic hydrogen detected by blackening of a deposit of lead chloride due to the reaction

$$PbCl + H \rightarrow HCl + Pb.$$

This increased the sensitiveness of detection and indications of the occurrence of dissociation by electrons with energies somewhat below 11·5 eV (down to 8·0 eV) were observed. The effectiveness of the electron beam in producing dissociation was found to increase considerably as the energy rose from 10 eV up to the ionization energy.

An interval of nearly forty years elapsed before further attempts were made to use chemical techniques as a means for detecting quantitatively the production of normal hydrogen atoms by impact dissociation.

† HUGHES, A. L. and SKELLET, A. M., *Phys. Rev.* **30** (1927) 11.
‡ DORSCH, K. and KALLMANN, H., *Z. Phys.* **53** (1929) 80.

This much later work due to Corrigan† has provided the most direct evidence as yet available about the magnitude of the cross-section and its energy variation.

The principle of these experiments is simple. An electron beam of controlled energy is fired through molecular hydrogen contained in a vessel the inside walls of which are coated with molybdenum trioxide, a substance that adsorbs hydrogen atoms with practically 100 per cent

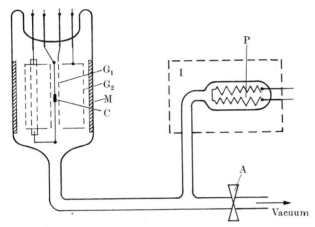

FIG. 13.3. Illustrating the arrangement of apparatus in Corrigan's experiments on the dissociation of H_2 by electron impact.

efficiency. If the electron beam is maintained the pressure in the vessel gradually falls due to dissociation and adsorption of the dissociated products. The rate of change of pressure can be related to the cross-section for dissociation in terms of the electron current strength, the path length of the electrons on the gas, and the volume and pressure of the main gas. Although the principle is simple in practice the experiment is difficult to carry out effectively.

Fig. 13.3 illustrates the general arrangement of Corrigan's apparatus. The collision chamber contained a three-electrode structure of which C is the oxide-coated platinum filament which on heating provided the electrons. The filament was coated only over 3 mm of its length so as to reduce end effects. G_1 and G_2 are two grids which, during the main experiment, were maintained at the same potential. G_1 was closely woven of nickel wire of internal diameter 2 mm surrounding the filament while G_2 was of much larger diameter (22 mm) and wound with fine nickel wire so as to have a high transparency (about 90 per cent).

† CORRIGAN, S. J. B., *J. Chem. Phys.* **43** (1965) 4381.

Electrons from the heated cathode were accelerated through G_1 by a potential V which could be varied from a few V to about 100 V. Almost all dissociative collisions occurred in the space between G_1 and G_2 through which the electrons passed with nearly constant energy. Hydrogen atoms produced in the space diffused through the nearly transparent grid G_2 to reach the molybdenum-trioxide coating M where they were trapped.

The electron energy scale was calibrated by carrying out a retarding potential analysis between G_2 and G_1.

The pressure in the collision chamber was measured by Pirani gauge immersed in an air bath to reduce the effects of drifts due to temperature changes. The nickel filament of the gauge was connected in a bridge circuit and balance obtained with a certain potential across the bridge which was measured by a voltmeter. This was calibrated against a standard McLeod gauge. As the experiment proceeded the potential V_b across the bridge was maintained constant and the change of pressure recorded through the out-of-balance current I flowing across the bridge. Special provision was made for maintaining V_b constant so that no contribution was made to I due to variation of V_b. The sensitivity was such that pressure changes somewhat less than 10^{-7} torr/s could be detected.

If Ω is the total volume occupied by the main gas the rate of loss of H_2 molecules from the gas is given by

$$\frac{dn}{dt} = L\Omega \frac{dp}{dt},$$

if L is the number of gas molecules/cm^3 at a pressure of 1 torr and p is measured in torr. Allowing for incomplete trapping efficiency of the MoO$_3$ layer, the total rate of dissociation is given by $\eta^{-1} dn/dt$ if η is the fraction of H atoms produced which are trapped. Hence

$$\frac{L\Omega}{\eta} \frac{dp}{dt} = lpLQ_{\text{diss}} \frac{dn_e}{dt},$$

where dn_e/dt is the number of electrons per second passing through the chamber through which their path length is l and Q_{diss} is the cross-section for a dissociative collision. If i is the electron current in amperes

$$\frac{dn_e}{dt} = 6{\cdot}2 \times 10^{18} i,$$

so

$$Q_{\text{diss}} = 1{\cdot}6 \times 10^{-19} \frac{\Omega}{\eta l i} \frac{1}{p} \frac{dp}{dt}, \tag{1}$$

the pressure unit now being immaterial.

Of the quantities occurring in (1) Ω can be measured without difficulty and l determined with reasonable accuracy. The current i should be as large as possible consistent with avoidance of space charge effects and was chosen to be about $1–4 \times 10^{-15}$ A. It was shown in earlier experiments by Melville and Robb† that the trapping efficiency of fresh MoO_3 is effectively unity. However, some exposure of hydrogen atoms to metal parts is unavoidable and from geometrical considerations about 20 per cent of the atoms could be expected to make first contact with such parts rather than with the MoO_3 coating. It seems therefore that η can be taken to lie between 0·8 and 1·0.

The working pressure of the main gas was in the range 5 to 20×10^{-3} torr. It is essential for the success of the experiment that the background pressure should be held constant so that the observed change with time could be ascribed to impact dissociation. To this end all ground stop-cocks were eliminated and the whole chamber baked out thoroughly before an experiment. A possible source of variable pressure was the hot cathode. To check this, at the beginning and end of the experiment dp/dt was measured with the gas at operating pressure and the cathode heated to provide the working electron current but no accelerating potential applied between C and G_1. When conditions had settled down the beam was accelerated and further measurements taken for a suitable interval. The accelerating voltage was again switched off and the background of pressure observed once more. Data were only admitted when the evidence from such series of observations showed that this background rate was either nearly zero or small and constant enough to be subtracted off. In the course of this work it was found that an indirectly heated filament was much less satisfactory from the point of view of constancy of background pressure.

Checks were made to verify that, for fixed accelerating potential, Q_{diss} obtained from (1) was independent of current and gas pressure.

Fig. 13.4 illustrates the results obtained for the dissociation cross-section. Below the ionization threshold for H_2 (15·4 eV) there is no ambiguity in the interpretation of these data but at higher electron energies the question arises as to whether a further contribution to pressure fall arises from clean-up of positive ions. Thus Corrigan points out that, at the highest energies observed, the apparent $Q_{\overline{\text{diss}}}$ is closely equal to the measured ionization cross-section of H_2 (see § 1.4.1). As dissociation into neutral atoms arises from the excitation of a triplet electronic state the true Q_{diss} could be expected to fall rapidly with

† MELVILLE, H. W. and ROBB, J. C., *Proc. R. Soc.* A**196** (1949) 445.

electron energy after passing through a maximum at an energy not far above the threshold. If the measured ionization cross-section of H_2 is subtracted from the apparent Q_{diss} in Fig. 13.4 the resulting curve has the form expected for singlet–triplet excitation (see Fig. 4.19).

Despite this, which assumes that in some way all positive ions are cleaned up from the main gas, there is considerable doubt about the interpretation. Thus Corrigan found no effect on the apparent Q_{diss}

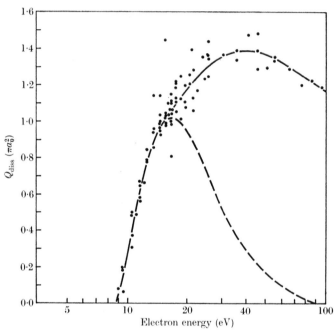

FIG. 13.4. Apparent cross-section, Q_{diss}, for dissociation of H_2 into neutral H atoms by electron impact as observed by Corrigan. • Experimental points. —— best fit to experimental points. — — — estimated true Q_{diss} for electron energies beyond the ionization threshold.

when a potential was applied to the trioxide layer to prevent positive ions from reaching it. This is not conclusive because the clean-up may occur through ionic reactions in the gas or on electrode surfaces leading to neutral H atoms which would then be trapped in the trioxide layer.

The threshold for production of H atoms was found to be $8 \cdot 8 \pm 0 \cdot 2$ eV, very close to the expected value for excitation of the unstable state (see Fig. 13.1 and Table 13.1). Even assuming the validity of the subtraction of the total ionization cross-section, Q_{diss} is quite large at its maximum, being nearly πa_0^2.

Evidence that slow electrons may suffer an energy loss of 9·5 eV, not far from the dissociation threshold in hydrogen, has been given by Jones and Whiddington[†] using the apparatus described in Chapter 5, § 4.1 in which a velocity analysis was made of an initially homogeneous electron beam after passage, without appreciable deviation, through the gas. They found that the proportion of electrons suffering this loss was a maximum at about 16-eV incident energy and fell off very rapidly at high energies. At its maximum the probability of the loss was 0·25 of that due to the excitation of the $C^1\Pi_u$ state (see § 1.1), requiring 12·6 eV. Electrons in producing the latter excitation (which is of the optically allowed type, see Chap. 7, § 5.6.7) suffer little deviation, whereas those producing the triplet excitations are distributed much more nearly uniformly after impact (see Chap. 7, Tables 7.3 and 7.4). Hence, in experiments such as those of Jones and Whiddington, a much larger proportion of collisions in which C-state excitation occurs will be observed than for those leading to triplet excitation. The observations are therefore not incompatible with a cross-section as large as that estimated from Corrigan's experiments.

The same energy loss has also been detected by Ramien[‡] using the Hertz diffusion method described in Chapter 5, § 2.2. The loss first became apparent for electrons with initial energy a little over 9 eV and became more probable as this energy increased to 11·7 eV, the maximum used in the experiments. At this energy the cross-section was about $\frac{1}{40}$ of the total collision cross-section ($12\pi a_0^2$, see Fig. 10.21) giving Q_{diss} as about $0·3\pi a_0^2$, somewhat smaller than that observed by Corrigan.

Finally, Kruithof and Ornstein,[§] in the course of an investigation of the optical excitation functions of certain hydrogen molecular and atomic lines (see Chap. 4, § 1), obtained a curve giving the variation of the probability of impact dissociation of the molecule as a function of electron energy in the range 14–50 eV. This curve exhibits a steep rise as the energy falls from 18 eV just as would be expected if the dissociation is due to excitation of a triplet state of the molecule.

The method used to obtain the dissociation curve was as follows. Dissociation proceeds by electron impact and recombination takes place on the walls of the experimental chamber so that, in equilibrium at pressure p, if m is the degree of dissociation,

$$Ap\frac{1-m}{1+m} = p^2\frac{4m^2}{(1+m)^2},\tag{2}$$

† Jones, H. and Whiddington, R., *Phil. Mag.* 6 (1928) 889.
‡ Ramien, H., *Z. Phys.* 70 (1931) 353.
§ Kruithof, A. A. and Ornstein, L. S., *Physica*, 2 (1935) 611.

I

where A is proportional to the rate of dissociation per molecule by impact. From (2),

$$1 + 4p/A = 1/m^2.$$

By observing the variation of the intensity of a molecular line with pressure, m and hence A, may be obtained at each electron energy.

1.2.2. *Swarm experiments.* Further evidence concerning the dissociative triplet excitation has been obtained from experiments in which electrons generated in a discharge with a distribution of velocities have been used to provide the excitation, the number η of dissociations produced per unit energy input being measured. The chief complication in the interpretation of these experiments is that an important contribution to atom production may come from the fast reaction (cf. Chap. 19, § 3.5)

$$H_2^+ + H_2 \to H_3^+ + H. \tag{3}$$

In the absence of these complications if $f(E)\,dE$ is the number of electrons with energies between E and $E + dE$, the number of dissociations produced per second is

$$N \int_0^\infty Q_{\text{diss}}(E) v f(E) \, dE,$$

where v is the velocity of an electron of energy E, N is the number of molecules/cm³, and Q_{diss} is the cross-section for dissociative collisions. In this time the work done on the electron swarm will be eFu, where u is the drift velocity and F the electric field strength. η is therefore given by

$$\eta = \frac{N}{eFu} \int_0^\infty Q_{\text{diss}}(E) v f(E) \, dE. \tag{4}$$

As $f(E)$ and u are determined by the ratio F/p, where p is the gas pressure, and N is proportional to p, it follows that η will be a function of F/p, independent of p for fixed F/p.

If (3) is important, since its rate depends on the square of the pressure, η, as measured, will increase with p for fixed F/p. Reactions of H_3^+ ions on the walls may also complicate the pressure variation. Unfortunately, no check has been made, in any experiments, of the pressure variation of η for fixed F/p, so doubt remains about the interpretation of the results. From this point of view they should be more reliable the lower the mean electron energy (small F/p).

In Poole's experiments,[†] hydrogen was allowed to stream at an

---

† POOLE, H. G., *Proc. R. Soc.* A**163** (1937) 404, 415, 424.

adjustable rate through a tube in which a striated glow discharge was maintained. Atomic hydrogen, formed by the electrons in the positive column of the discharge, passed, together with the outflowing molecular hydrogen, into a calorimeter in which it recombined, the heat of recombination being measured. From this and the known binding energy of H_2 the rate at which atomic hydrogen reached the calorimeter could be derived. The power input into the positive column was obtained from the discharge-tube current and the potential drop down the column. From these measurements a quantity μ was obtained giving the number of dissociated molecules received as atoms at the calorimeter per unit of energy supplied to the positive column. This is not equal to the quantity η appearing in (4), for the degree of dissociation in the positive column was considerable and atoms were lost by wall recombination before reaching the calorimeter. To derive η, experiments were conducted at a fixed pressure, for a series of values of hydrogen flow rates U and power input W. In the limit of vanishing W and infinitely large U the proportion of atomic hydrogen tends to vanish and there is no time for wall recombination to occur before the calorimeter is reached. Hence, by extrapolating μ to vanishing W and $1/U$, η could be found at each pressure.

The general arrangement of the apparatus is illustrated in Fig. 13.5. Hydrogen passed through the capillary flowmeter F from which it was saturated with water vapour. It entered the discharge tube at two points K and L in the neighbourhood of the electrodes E and passed out through the side tube A to the calorimeter J and thence to the pump. By means of a device for varying the pumping speed it was possible to adjust conditions to maintain a constant pressure at different flow rates. The glass walls of the discharge tube and calorimeter and of the side tube were coated with metaphosphoric acid to poison them for recombination. To secure permanent effectiveness of this coating the water-vapour content of the hydrogen was important. Although recombination was not reduced to a negligible value by this arrangement it remained constant to a degree unattainable otherwise. The calorimeter consisted of a glass tube enclosing a concentric platinum-coated copper tube through which water passed at an adjustable rate. The gas from the positive column entered through a centrally placed side tube and the hydrogen atoms recombined on the walls of the copper tube. In equilibrium the rate of heat input due to the recombination could be determined from the water flow and the temperature difference between inlet and outlet measured by two thermocouples. The glass tube of the calorimeter was of 25 mm bore and 42 cm long, while the copper tube was of external diameter 5·5 mm, thickness 0·75 mm, and 65 cm long. Water-flow rates from 4 to 55 cm³/min were employed.

The discharge-tube voltage was 5000 V d.c., the inlet hydrogen pressure varied from 0·4 to 1·35 torr, the flow rates from 700 to 6000 cm³/s, and the discharge currents up to 100 mA. The results obtained are illustrated in Fig. 13.6 as a function of F/p.

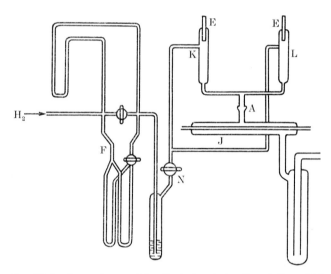

FIG. 13.5. Apparatus used by Poole to study dissociation of H_2 by electron impact.

The energy efficiency of dissociation of H_2 has also been measured by Corrigan and von Engel† and by Shaw.‡ The former authors measured the rate of production of hydrogen atoms by an r.f. electrodeless discharge in a cylindrical glass vessel. This was done by partially coating the inner surface of the vessel with molybdenum oxide MoO_3 (see previous section). The rate of production of atoms is then proportional to the rate of decrease of pressure in the vessel and the constant of proportionality may be obtained in terms of the fraction of surface covered by the oxide and the volume of the vessel. Knowing the average discharge current carried over a half cycle and the r.m.s. value of the applied field, the dissociation coefficient α_d could be obtained as a function of the ratio of equivalent d.c. field strength (the r.m.s. field) F

† CORRIGAN, S. J. B. and VON ENGEL, A., *Proc. R. Soc.* A245 (1958) 335.

‡ SHAW, T. M., *J. chem. Phys.* 30 (1959) 1366.

to pressure p. The energy efficiency is then simply given by $(\alpha_d/p)/(F/p)$. Measurement was made at two r.f. frequencies of 5·1 and 2·5 Mc/s and the results obtained are illustrated in Fig. 13.6. The agreement with Poole is not unsatisfactory at the lower values of F/p but there is wide disagreement near $F/p = 100$ V cm^{-1} torr^{-1}.

Shaw studied the production of atomic hydrogen in a microwave gas discharge at a frequency of 3100 Mc/s. Hydrogen was pumped at low pressure continuously through a quartz tube about 70 cm long and 0·8 cm inside diameter, and the discharge was confined to a region about

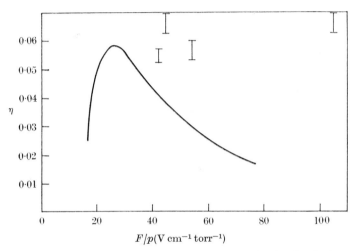

FIG. 13.6. Energy efficiency η ($= \alpha_d/F$) of atom production for electrons in hydrogen. —— observed by Poole. ⊥ observed by Corrigan and von Engel.

5 cm long near one end of the tube. To reduce loss of atoms by recombination the inner wall of the tube was coated with a mixture of dimethyldichlorosilane and methyltrichlorosilane. The number of atoms flowing out from the discharge was measured by means of an electron spin-resonance spectrometer and an atom recombination calorimeter. The energy efficiency was derived from the rate of production of hydrogen atoms and the power dissipated in the discharge. A maximum of $6·5 \times 10^{-2}$ dissociations per eV was obtained. This agrees well with the corresponding value $5·8 \times 10^{-2}$ obtained by Poole (see Fig. 13.6).

1.2.3. *Theoretical calculations.* The first calculations of Q_{diss}, or rather of the cross-section for excitation of the $b\,^3\Sigma_u^+$ unstable state, were carried out using the Born–Oppenheimer approximation (Chap. 7, § 5.4.1) and gave quite unsatisfactory results, exceeding by a considerable factor the maximum possible values (see Chap. 7, § 5.5). Edelstein† has obtained interesting results by applying the distorted wave method (Chap. 8, § 4.2), which proved to be reasonably satisfactory for the calculation of the cross-sections for the excitation of the 2^3S state of helium (see Chap. 8, § 6.2). Because of the very considerable additional complexity introduced by the two-centre molecular field, rather more drastic approximations had to be made so that the numerical evaluation could be completed. For the wave function of the ground $X\,{}^1\Sigma_g^+$ state the self-consistent separable form of Coulson‡ was used and for the $b\,{}^3\Sigma_u^+$ state the two-parameter form of the variational function used by James and Coolidge.§ The wave function of the incident electron wave was taken to be that determined by Massey and Ridley‖ (see Chap. 10, § 3.3), while that for the final wave was calculated by a similar variational approximation.

Fig. 13.7 illustrates the cross-section finally obtained by Edelstein. It seems to be of about the right order of magnitude and is the most complicated case to which the distorted wave method has yet been applied.

1.2.4. *The continuous spectrum of molecular hydrogen.* It remains to consider the other manifestation of the excitation of the $a\,{}^3\Sigma_g$ state (see Table 13.1)—the continuous spectrum of molecular hydrogen. Considerable success has been attained in understanding the details of this spectrum even though it is not possible to predict the absolute value of the cross-section for excitation of the initial state concerned.

It has been shown that a consequence of the excitation of the $a\,{}^3\Sigma_g$ state of molecular hydrogen should be the emission of a continuous spectrum due to transitions from the $a\,{}^3\Sigma_g$ state to the unstable $b\,{}^3\Sigma_u$ state. The existence of a continuum, stretching approximately over the wavelength range from 4000 to 2000 Å, in the emission spectrum of molecular hydrogen had long been known and, until Winans and Stueckelberg†† provided the correct explanation in 1928, aroused much speculation as to its origin. Fig. 13.8 illustrates three spectrograms obtained by Vencov‡‡ on which the continuum appears clearly. Since 1928 the intensity distribution in the continuum under controlled excitation conditions has been

---

† EDELSTEIN, L. A., *Nature, Lond.* **182** (1958) 932.
‡ COULSON, C. A., *Proc. Camb. phil. Soc. math. phys. Sci.* **34** (1938) 204.
§ JAMES, H. M. and COOLIDGE, A. S., *J. chem. Phys.* **1** (1933) 825.
‖ MASSEY, H. S. W. and RIDLEY, R. O., *Proc. phys. Soc.* A69 (1956) 659.
†† WINANS, J. G. and STUECKELBERG, E. C. G., *Proc. natn. Acad. Sci. U.S.A.* **14** (1928) 867. See also FINKELNBURG, W., *Z. Phys.* **62** (1930) 624.
‡‡ VENCOV, S., *Annls Phys.* **15** (1931) 131.

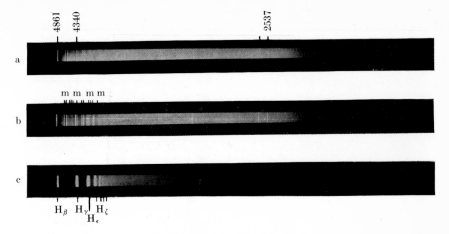

Fig. 13.8. The H_2 continuous spectrum, photographed by Vencov. The exciting electron energy is least in (a), which shows the continuous molecular spectrum. In (c), a higher electron energy favours production of the continuous atomic spectrum. Case (b) is intermediate.

studied in very considerable detail.† In particular, in 1931 Finkelnburg and Weizel‡ showed that the threshold electron energy for excitation of the spectrum was 11·8 eV, in good agreement with the predicted value.

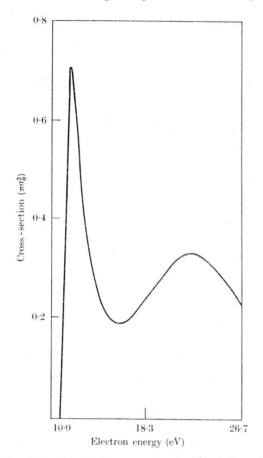

FIG. 13.7. Calculated cross-section for dissociation of hydrogen due to impact excitation of the $X\,^1\Sigma_g^+ - b\,^3\Sigma_u^+$ transition.

James, Coolidge, and Present§ have also worked out the theory of the intensity distribution, and good agreement has been obtained with experiment, particularly with the observations of Coolidge.‖

† HUKUMOTO, Y., *Tohoku Univ. Sci. Rep.* **19** (1930) 773; CHALONGE, D., *C.r. hebd. Séanc. Acad. Sci. Paris,* **192** (1931) 1551; *Annls Phys.* **1** (1934) 123; FINKELNBURG, W., *Phys. Z.* **34** (1933) 529.
‡ FINKELNBURG, W. and WEIZEL, W., *Z. Phys.* **68** (1931) 577.
§ JAMES, H. M., COOLIDGE, A. S., and PRESENT, R. D., *J. chem. Phys.* **4** (1936) 187.
‖ COOLIDGE, A. S., *Phys. Rev.* **65** (1944) 236.

A complete theory of the intensity distribution involves not only the calculation of the relative probability of a radiative transition from each vibrational level of the upper $^3\Sigma_g^+$ state to different states of unquantized nuclear motion associated with the unstable lower $^3\Sigma_u^+$ state, but also the determination of the relative population of excited molecules in each vibrational state. The second part is the more difficult to carry out accurately for any given experimental conditions so, as explained below, Coolidge[†] arranged the conditions of his experiments to minimize the contribution from excited vibrational levels of the $^3\Sigma_g^+$ state.

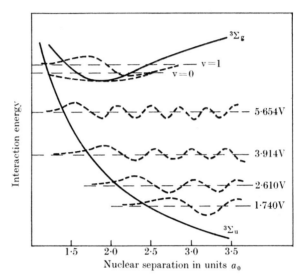

Fig. 13.9. Potential-energy curves of the $^3\Sigma_u$ and $^3\Sigma_g$ states of H_2 as calculated by Coolidge, James, and Present. The continuous spectrum is produced by transitions $^3\Sigma_u$–$^3\Sigma_g$. The broken lines show the wave functions for the two lowest vibrational states of the $^3\Sigma_g$ level and for four states of different kinetic energy associated with the repulsive $^3\Sigma_u$ level.

The first part of the calculation may be carried out in the usual way. The relative probabilities of different transitions are determined by the overlap of the appropriate nuclear wave functions, of the form illustrated in Fig. 13.9.

The functions may be calculated from the known potential-energy curves for the two electronic states. Fig. 13.9 illustrates the results obtained for the relative probability of transitions from the ground vibrational level of the upper $^3\Sigma_g^+$ state to the lower level, involving emission of radiation over the wavelength range shown. Similar curves have been obtained by James and Coolidge[‡] for other initial vibrational levels of the $^3\Sigma_g^+$ state. To complete the calculation they suppose that the excitation function for any particular level of the upper state increases linearly with the excess of the electron energy above that just sufficient to produce the transition and that the constant of proportionality varies, for different vibra-

† Coolidge, A. S., *Phys. Rev.* **65** (1944) 236.
‡ James, H. M. and Coolidge, A. S., ibid. **55** (1939) 184.

tional levels of the upper $^3\Sigma_g^+$ state, as the square of the overlap integral between the nuclear wave functions of the initial state (the ground vibrational state of the $^1\Sigma_g^+$ ground electronic level), and of the final state (see Chap. 12 (17)).

To test the results of this theory in detail Coolidge excited the spectrum in streaming hydrogen at a pressure of 0·015 torr by a 25-μA beam of electrons of nearly homogeneous energy, emitted from an indirectly heated oxide-coated cathode. Elaborate precautions were taken to secure clean conditions. The radiation was analysed by photographic comparison with a recalibrated standard mercury arc and the use of a small Hilger quartz spectrograph, great care being taken to ensure accurate photometry in spite of the extremely weak source. To minimize the effects of inaccurate estimation of the relative population of different vibrational levels of the upper state, an extensive series of measurements was made in which the mean electron energy was very close to the vertical excitation energy of the ground vibrational level, 11·72 eV. Correction was then made for the small amount of excitation of higher vibrational levels due to the energy spread of the electron beam, determined from measurement of the filament temperature (1100 °K). The close agreement obtained between theory and observation, under such conditions, is illustrated in Fig. 13.10.

This work has also been extended to deuterium, for which slightly different results are expected. While the electronic potential energy curves are the same as for hydrogen, the vibrational states are changed by the increase in nuclear mass. The accord between theory and experiment is equally close in this case.

There seems little doubt from this and earlier work that the continuous spectrum does arise in the manner suggested and the close verification of the theory is gratifying. Much less attention has been paid to the investigation of the excitation function for the $^3\Sigma_g^+$ state for different electron energies. As a multiplicity change is involved the general form of this function should be typical of such transitions (see previous section). Apart from some early observations of Finkelnburg,[†] which favoured this, the only detailed experimental work that has been carried out on this aspect is due to Lunt, Meek, and Smith[‡] using the electron swarm method. They observed the relative energy efficiency η_{ex} of excitation of the continuous spectrum, as a function of F/p, for a steady discharge in streaming hydrogen. This is related to the cross-section $Q_{ex}(E)$ for excitation of the upper state by electrons of energy E in the same way as is η to Q_{diss} in (4). The observed variation is consistent with an excitation function of the characteristic form for an inter-combination transition but, in the absence of a satisfactory theory of Q_{ex} and of complete knowledge of the velocity distribution of the exciting electrons (see, however, § 1.7), no very detailed knowledge of the variation of Q_{ex} with electron energy may be derived.

† FINKELNBURG, W., Z. Phys. 62 (1930) 624.
‡ LUNT, R. W., MEEK, C. A., and SMITH, E. C. W., Proc. R. Soc. A158 (1937) 729.

1.3. *Excitation of other discrete states*

1.3.1. *Excitation by electrons of medium to high energy.* Energy loss spectra in hydrogen have been observed for electrons with incident

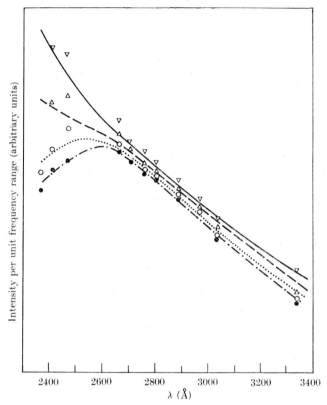

Fig. 13.10. Comparison of calculated and observed distribution of intensity in the H_2 continuous spectrum excited by electron impact. The vertical excitation potential is $11 \cdot 72$ eV. The curves show the calculated intensity distribution for electrons of energy Δ_0 eV above the vertical excitation potential —— $\Delta_0 = 0 \cdot 5$; ———— $\Delta_0 = 0 \cdot 3$; ..... $\Delta_0 = 0 \cdot 1$; —·—·— $\Delta_0 = -0 \cdot 1$. The points show the observed intensity distributions for electrons corresponding to different values of Δ_0: ▽, $0 \cdot 55$; △, $0 \cdot 35$; ○, $0 \cdot 15$; ●, $-0 \cdot 05$. The possibility of excitation for negative Δ_0 arises from the small but finite spread in electron energies due to the filament temperature (1100 °K).

energies in the range 400–600 eV by Lassettre and Francis[†] and by Lassettre and Jones[‡] and for much more energetic electrons (25 keV) by Geiger.[§] The techniques used were those which are already described in Chapter 5, § 4.1.

† Lassettre, E. N. and Francis, S. A., *J. chem. Phys.* **40** (1964) 1208.
‡ Lassettre, E. N. and Jones, E. A., ibid. 1222.
§ Geiger, J., *Z. Phys.* **181** (1964) 413.

Fig. 13.11 shows the shape of the spectra observed both for medium and high energy electrons. In these experiments the energy distribution of the incident electron is of the order 1 eV wide and it will be seen that at both ranges of incident electron energy a single peak is observed in the energy loss spectrum at 12·7 eV. Since the respective thresholds for

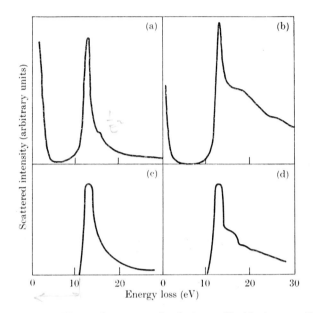

FIG. 13.11. Energy-loss spectra for electrons of incident energy E scattered at small angles θ in H_2. (a) $E = 25$ keV, $\theta < 2·5 \times 10^{-4}$ rad (Geiger). (b) $E = 25$ keV, $\theta = 1·9 \times 10^{-2}$ rad (Geiger). (c) $E = 390$ eV, $\theta = 0°$ (Lassettre and Jones). (d) $E = 390$ eV, $\theta = 5°$ (Lassettre and Jones).

excitation of the B and C states are 11·4 and 12·7 eV it seems that both states contribute to the energy loss in the first peak—it must be remembered that as there will be an underlying vibrational structure, masked by the incident electron energy spread, the contribution from each electronic excitation will extend over an appreciable energy range.

To investigate this structure Geiger carried out experiments at a very small scattering angle ($< 3 \times 10^{-4}$ rad) using very high resolution (0·04 eV). The technique by which this resolution was obtained is described in Chapter 5, § 5.2.3 and its application to observe the excitation of vibration in CO_2 and other molecules is discussed in Chapter 11, § 2. Fig. 13.12 reproduces an energy loss spectrum obtained in this way, showing the vibrational structure in detail. The relation of the peaks to the vibrational levels of the B and C states is shown. Levels associated with the $D(3p\pi\,{}^1\Pi_u)$ state are also included. From a spectrum such as this the separate contributions from the B, C, and D states may be analysed out.

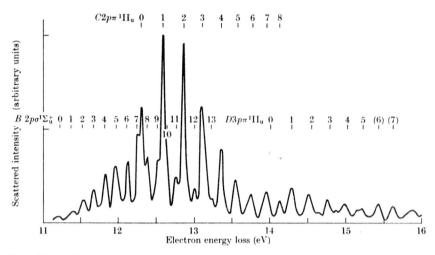

Fig. 13.12. Energy-loss spectra for electrons of incident energy 25 keV scattered through angles $< 3 \times 10^{-4}$ rad in H_2 (Geiger, using high-energy resolution).

The relative probability of excitation to a level with vibrational quantum number v' of a particular electronic state n from the ground state of H_2 is closely proportional to

$$p_{00}^{nv'} = N\left\{\int \chi_{00}(R)\,\chi_{nv'}(R)\,\mathrm{d}R\right\}^2,$$

where $\chi_{00}(R)$, $\chi_{nv'}(R)$ are the respective vibrational wave functions (see Chap. 12, § 4.1). N is taken so that

$$\sum_{v'} p_{00}^{nv'} = 1.$$

We give in Table 13.2 a comparison between calculated values of $p_{00}^{nv'}$ for the B and C states and those derived from energy loss spectra such as that shown in Fig. 13.12. For both states the theoretical values were obtained on the assumption that the potential energy curve for the appropriate electronic state could be represented by a Morse function

$$V(R) = D[\{e^{-a(R-R_0)} - 1\}^2 - 1],$$

where D is the dissociation energy and R_0 the equilibrium separation. It is seen from Table 13.2 that this gives good results for the C but not the B state.

Finally, the rate of the intensity of excitation of the B, C, and D bands by electrons scattered through angles less than 3×10^{-4} rad may be obtained from the high resolution spectra as $1:1 \cdot 22:0 \cdot 16$.

Geiger then made use of the low resolution spectra, which were obtained over a wider angular range, in order to determine the oscillator strengths for the three transitions. From the high resolution spectra it was estimated that all but 25 per cent of the contribution to the $12 \cdot 6$ peak observed at low resolution came from excitation of the B and C states.

TABLE 13.2

v'		0	1	2	3	4	5	6
$p_{00}^{nv'}$ obs	$\}\, B$	0·009	0·016	0·033	0·051	0·072	0·086	0·091
$p_{00}^{nv'}$ calc		0·014	0·059	0·147	0·213	0·225	0·171	0·095
$p_{00}^{nv'}$ obs	$\}\, C$	0·143	0·232	0·232	0·145	0·093	0·054	0·035
$p_{00}^{nv'}$ calc		0·153	0·241	0·223	0·168	0·099	0·057	0·028

The intensity of the inelastic scattering giving rise to $B+C$ excitation was normalized by comparison with the elastic which had already been normalized to agree with theory (see Chap. 10, § 1.7.3). Fig. 13.13 shows the differential cross-section obtained in this way for excitation of the combined B and C states. On the same diagram are shown the results of theoretical calculations by Roscoe,[†] using Born's approximation and approximate wave functions first suggested and used by MacDonald.[‡] The agreement is not unsatisfactory.

The normalized differential cross-section was then used to obtain the generalized oscillator strengths (Chap. 7, § 5.2.4) for the two combined transitions or functions of the electron momentum change $K\hbar$. Extrapolation to zero K (cf. Chap. 7, § 5.6.7) then gave the combined optical oscillator strengths for the $B+C$ transitions as $0 \cdot 57 \pm 0 \cdot 08$. From the ratio of the intensities of excitation of the separate states derived from the high resolution spectra as described above the optical oscillator strengths for the B, C, and D states were obtained.

These results are compared with a number of theoretical calculations in Table 13.3. It will be seen that, considering the difficulty of carrying out accurate calculations for such a complicated problem, the agreement is not unsatisfactory. It must be remembered that in Geiger's analysis

† ROSCOE, R., *Phil. Mag.* **31** (1941) 349.
‡ MACDONALD, J. K. L., *Proc. R. Soc.* A**136** (1932) 528.

allowance was made for excitation of the E state, which has a threshold very close to that of C. This could make an error of order 10 per cent.

Roscoe has calculated cross-sections for excitation of the $E({}^1\Sigma_g^+)$ state. His results for this transition and also for excitation of the B, C, and D states are given in Table 13.4.

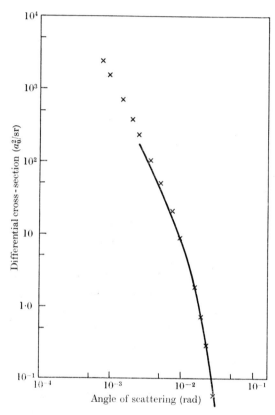

Fig. 13.13. Comparison of observed and calculated differential cross-sections for inelastic scattering of electrons after exciting the B and C states of H_2. ✕ observed, Geiger. —— calculated, Roscoe.

1.3.2. *Excitation of discrete states by low energy electrons.* The excitation function for the metastable state of H_2 has been observed by Olmsted, Newton, and Street,† using the technique described in Chapter 4, § 2, and by Lichten‡ in the course of his experiments on the detection and identification of the state (see § 1.1). Their results are shown in

† OLMSTED, J., NEWTON, A. S., and STREET, K., *J. chem. Phys.* **42** (1965) 2321.
‡ LICHTEN, W., *Phys. Rev.* **120** (1960) 848.

Fig. 13.14. It will be seen that they do not agree very well. Lichten's function rises much more slowly to a peak at about 15·5 eV and then falls off quite rapidly in contrast with the flat maximum found by Olmsted *et al.* It is possible that this latter result is really due to spurious

TABLE 13.3

Oscillator strengths for transitions from the B, C, and D states of H_2 to the ground state

	B	C	D
Observed (Geiger)[†]	0·26±0·04	0·31±0·04	0·049±0·007
Calculated (Roscoe)[‡]	0·23	0·19	0·036
(Mulliken and Ricke)[§]	0·24	0·38	0·052
(Shull)[‖]	0·18	0·42	
(Ehrenson and Phillipson)[††]	0·27		
(Peek and Lassettre)[‡‡]	0·28	0·28	

signals arising from photons emitted from excited upper states. Such an effect would not occur in Lichten's experiments as the metastable concentration was monitored by radio-frequency resonance methods. In any case, in interpreting the observations it must be realized that for

TABLE 13.4

Cross-sections for excitation of B, C, D, and E states of H_2, calculated by Born's first approximation

Electron energy (eV)	Cross-section in units πa_0^2			
	B	C	D	E
100	0·27	0·21	0·03$_5$	0·08
200	0·16	0·13$_5$	0·02$_5$	0·04
300	0·14	0·10$_5$	0·02	0·03

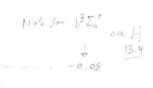

Note for $^3\Sigma_u^+$
see H_j
13.4
∼0.08

energies above the threshold for excitation of upper states that combine optically with the metastable $^3\Pi_u$ state, the latter state will be populated by cascade as well as by direct excitation.

Kruithof and Ornstein[§§] have measured the optical excitation functions

† loc. cit.   ‡ loc. cit.
§ MULLIKEN, R. S. and RICKE, C. A., *Rep. Prog. Phys.* **8** (1941) 231.
‖ SHULL, H., *J. chem. Phys.* **20** (1952) 18.
†† EHRENSON, S. and PHILLIPSON, P. E., *J. chem. Phys.* **34** (1961) 1224.
‡‡ PEEK, J. M. and LASSETTRE, E. N., ibid. **38** (1963) 2392.
§§ KRUITHOF, A. A. and ORNSTEIN, L. S., *Physica* **2** (1935) 611.

of the lines $\lambda = 4634$ Å and 4617 Å that arise from the $3d\sigma\,^1\Sigma_g^+ \to B$ and $4p\pi\,^3\Pi_u \to a\,^3\Sigma_g^+$ transitions. Allowance was made for the variation in the degree of dissociation of the hydrogen with electron energy (see § 1.2.1 (2)). They found as expected that the shapes of the excitation functions were similar to those for excitation of the corresponding state of the united atom.

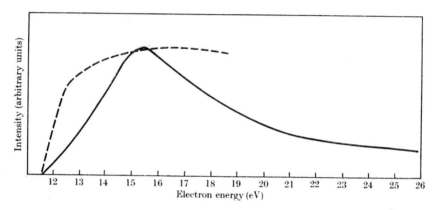

Fig. 13.14. Observed excitation functions for the metastable state of H_2. —— Lichten. — — — Olmsted, Newton, and Street.

1.3.3. *The photon-production coefficient.* A further source of evidence comes from observations of the production of ultra-violet photons by an electron swarm diffusing as usual in a gas at pressure p under the influence of a uniform electric field F. We may define an ultra-violet photon production coefficient α_{ph} in an exactly similar way to the ionization and dissociation coefficients previously discussed (see Chap. 5, § 3), i.e. α_{ph} is the number of quanta produced per electron per cm path in the direction of the field. Just as for the other coefficients α_{ph} will be a function of F/p.

If N_0 is the number of electrons emitted per second from the cathode, the number of quanta emitted per second from an element of the electron swarm measured in the direction of the field at a distance x from the cathode will be given by

$$\Delta\phi = \alpha_{ph}\,N_0\,e^{\alpha_i x}\,\Delta x,$$

where α_i is the ionization coefficient. The current received at the anode will be given by

$$J = eN_0\,e^{\alpha_i d}, \qquad (5)$$

where d is the electrode operation, e the electronic charge. Hence

$$\Delta\phi = \alpha_{ph}(J/e)e^{\alpha_i(x-d)}\,\Delta x. \qquad (6)$$

By measuring $\Delta\phi/\Delta x$ as a function of x, for fixed J and F/p, α_i may be obtained. Measurement of J and d then gives α_{ph}.

Alternatively, the total photon flux ϕ from the whole length of the swarm may be measured, giving

$$\alpha_{ph}/\alpha_i = (\phi e/J)/(1-e^{-\alpha_i d}). \qquad (7)$$

For $\alpha_i d \geqslant 1$ the value of α_{ph}/α_i will not depend much on α_i.

Measurement of α_{ph} for hydrogen is of particular interest because ultra-violet quanta are almost certainly mainly emitted in the bands which arise from excitation of the B and C states. There will also be a contribution from excitation of the continuous spectrum, discussed in § 1.2.4, but there is evidence that this is weak compared with that of these bands.

In the experiments of Corrigan and von Engel† the ultra-violet radiation in the range from 500 to 3000 Å was observed through fluorescent conversion by sodium salicylate crystals to visible (blue) radiation, the intensity of which was measured by means of a photomultiplier. This procedure avoided the use of a vacuum spectrograph.

Fig. 13.15 illustrates the general arrangement of their apparatus which was designed to measure $\Delta\phi/\Delta x$. The electrons were produced photoelectrically by a beam of ultra-violet light from a mercury lamp, incident normally through a hole in the anode An on to a polished zinc mirror forming part of the cathode K. In this way light not absorbed in the zinc was reflected back and did not enter the photon-detector system. To maintain a uniform field the hole was crossed by fine wires and the cathode surrounded by a guard-ring.

The detector probe was constructed as follows. A microcrystalline layer L of sodium salicylate was deposited on the flat end of a transparent silicon rod Q which guided the blue fluorescent light to a photomultiplier P. To confine the portion viewed of the electron swarm to a narrow section an aperture system A was introduced between the layer L and the swarm. It admitted light within a fan-shaped solid angle, the plane of which was perpendicular to the direction of the electric field. In this direction the half-angle of admission was 9° while in the plane of the fan it was greater than 30° so as to screen light from the full width of the swarm. Radiation of wavelength less than 1800 Å could be prevented from reaching the layer L by introduction of a quartz filter F between it and the aperture A.

The detector system was mounted on a bellows B so that its distance

† CORRIGAN, S. J. B. and VON ENGEL, A., *Proc. R. Soc.* A245 (1958) 335.

from the swarm could be varied and a check made as to whether absorption of the ultra-violet light was occurring in the gas. By means of a magnet operating a screw from outside, the electrode separation could be varied and the electrode assembly could be moved bodily past the detector system so that the variation of the emission along the length of the swarm could be investigated.

The electrodes were circular of diameter 8 cm and the maximum separation was 3·5 cm. The largest current used was about 10^{-8} A.

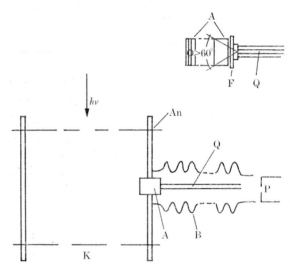

FIG. 13.15. Arrangement of apparatus used by Corrigan and von Engel to measure photon production coefficients for electrons drifting in H_2.

The fluorescent detector was calibrated using an auxiliary photomultiplier sensitive to wavelengths down to 2100 Å. An absolute measurement of the cathode sensitivity of this tube was made for the mercury resonance line at 2537 Å as well as the gain over a voltage range (200–850 V) through which it increased from 20 to 2×10^6. This then permitted absolute measurement of light fluxes of the order 10^6 to 10^7 quanta $cm^{-2} s^{-1}$ and was used to determine the over-all efficiency of the fluorescent detector for the 2537-Å line. This was sufficient because the quantum efficiency of the conversion by the sodium salicylate is independent of wavelength over the range of interest.

To establish that the observed radiation arises from single collisions between electrons and gas molecules the proportionality of the observed photon flux to the current was checked.

The semilogarithmic plot in Fig. 13.16 shows that the observed flux does indeed grow exponentially with distance from the cathode. From this plot the ionization coefficient is determined. α_{ph} is obtained using (7), and is shown in Fig. 13.17.

When the quartz filter F was introduced so as to exclude radiation with wavelength less than 1800 Å the photon flux was reduced to at

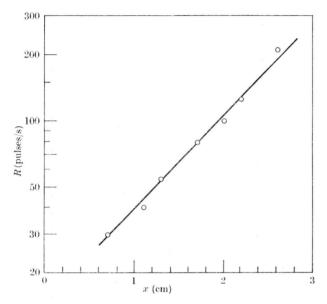

FIG. 13.16. Photon counting rate R as a function of the distance x from the cathode in the experiments of Corrigan and von Engel.

least 0·2 of its original value. This shows that at least 80 per cent of the radiation is shorter than that radiated in the continuous spectrum (see § 1.2.4) and strengthens the assumption that most of the quanta observed arise from transitions from the B and C states.

Measurement of α_{ph} have also been made by Legler[†] using the fluorescent conversion detector technique but measuring the total flux from the main region by (7). He found that for pressures greater than 20 torr α_{ph}/α_i varied with the pressure approximately according to the law

$$\alpha_{ph}/\alpha_i = (\alpha_{ph}/\alpha_i)_0(1+p/p_0)^{-1},$$

where $p_0 = 20$ torr. This effect was presumably due to the occurrence of collisions of the second kind in the gas which deactivate excited

† LEGLER, W., *Z. Phys.* **173** (1963) 169.

molecules before they can emit radiation (see Chap. 18, § 2). By extrapolation to zero pressure Legler obtained the results shown in Fig. 13.17, which agree well with those of Corrigan and von Engel who worked always with pressures < 12 torr.

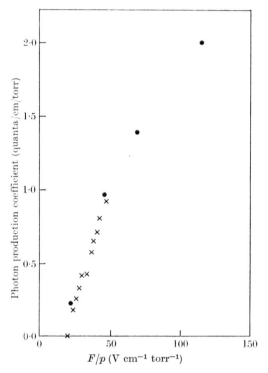

Fig. 13.17. Photon production coefficients for electrons drifting in H_2. ● observed, Corrigan and von Engel. × observed, Legler.

Comparison with the dissociation coefficients α_d (see § 1.2.2 and Fig. 13.6) shows that α_{ph} is about three times smaller. Since the contribution to α_{ph} from the continuous spectrum is less than 20 per cent it follows that the contribution to the dissociation rate from transitions to the $a\,^3\Sigma_g^+$ state must be less than 7 per cent of the total, which must therefore mainly arise directly from excitation of the unstable $b\,^3\Sigma_u^+$ state.

These data on photon production coefficients will be taken account of in the analysis in § 1.7 of all swarm results in H_2 which depend on electronic excitation.

1.4. *Ionization of* H_2, D_2, *and* HD

1.4.1. *Total ionization cross-sections.* In Fig. 13.18 the results of observations of total cross-section for ionization of H_2 and D_2 made by different investigators† are shown. The techniques involved have been described in Chapter 3 (see Fig. 3.10 for comparable data for ionization of rare gas atoms).

The chief problem in obtaining accurate data is that of collecting efficiently the energetic protons that are produced by dissociative ionization. As described in Chapter 12, § 7.2, Rapp and Englander-Golden took special precautions to ensure saturation in the collection of these ions. The agreement of their results for H_2 with the much earlier ones of Tate and Smith is surprisingly good. At first sight it might seem that the lower values obtained by Schram, de Heer, van der Wiel, and Kistemaker at the higher energies are due to incomplete collection of energetic ions but this seems unlikely because the same discrepancy appears in comparison of data for the rare gases (see Fig. 3.10). The much higher values obtained by Harrison are not easy to explain.

It will be seen from comparison of the results in Fig. 13.18 (a) and (b) that, as would be expected, there is no evidence of any difference between the total cross-sections of H_2 and of D_2.

1.4.2. *Behaviour of the cross-section near the threshold for* H_2^+ *production.* We are now concerned with the process

$$H_2(X\,^1\Sigma_g^+)+e \to H_2^+(^2\Sigma_g^+)+2e.$$

At ordinary temperatures the $H_2(X\,^1\Sigma_g^+)$ target molecule will be in its ground vibrational state so that, according to (17) of Chapter 12, the probability of excitation to a final state n with vibrational quantum number v' is proportional to

$$N(p_{00}^{nv'})^2, \tag{8}$$

where
$$p_{00}^{nv'} = \int \chi_{nv'}(R)\chi_{00}(R)\,\mathrm{d}R. \tag{9}$$

Here χ_{00} is the ground vibrational wave function for the $^1\Sigma_g$ state of H_2 and $\chi_{nv'}$ that for the level of vibrational quantum number v' associated with the $^2\Sigma_g^+$ ground state of H_2^+ (see Fig. 13.39).

The form taken by the factor N at the threshold is still a matter for

† TATE, J. T. and SMITH, P. T., *Phys. Rev.* **39** (1932) 270; RAPP, D. and ENGLANDER-GOLDEN, P., *J. chem. Phys.* **43** (1965) 1464; SCHRAM, B. L., DE HEER, F. J., VAN DER WIEL, M. J., and KISTEMAKER, J., *Physica* **31** (1964) 94; HARRISON, H., *The experimental determination of ionization cross-sections of gases under electron impact,* Table 4 (Catholic University of America, Washington, 1956).

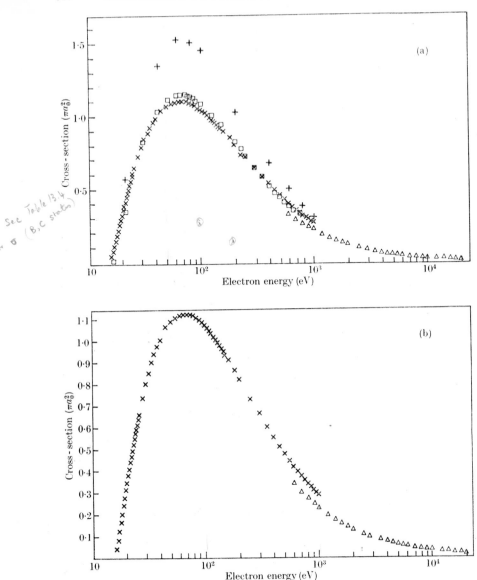

See Table 13.4 for σ (B,C states)

FIG. 13.18. Observed total ionization cross-sections for (a) H_2, (b) D_2. □ Tate, Smith; + Harrison; × Rapp, Golden; △ Schram, de Heer, van der Wiel, Kistemaker.

theoretical argument but can probably be written

$$N = A(E - E_t)^s, \qquad (10)$$

where E is the electron energy and E_t the threshold energy, A a constant, and s an index close to if not actually unity (see Chap. 9, § 10.8).

The overlap integrals have been calculated by several authors,[†] for H_2, HD, and D_2. For the vibrational wave function χ_{00} for the ground state all have used the Morse approximation and most have used the same approximation for the final vibrational wave functions $\chi_{nv'}$. In Table 13.5 we give the results for H_2 and D_2 obtained by Dunn who calculated $\chi_{nv'}$ from the accurate potential-energy curve for the $^2\Sigma_g^+$ state of H_2^+ obtained by Bates, Ledsham, and Stewart (see pp. 941–2). We also include for comparison results obtained by Wacks who used the Morse approximation in all cases. The results are expressed in the form of relative values of $(p_{00}^{nv'})^2$ taking $(p_{00}^{n2})^2$ as unity.

TABLE 13.5

Relative transition probabilities to different vibrational levels (v') of $H_2^+({}^2\Sigma_g^+)$ in impact ionization of ground state H_2, HD, and $D_2(X^1\Sigma_g^+)$ molecules

v'	0	1	2	3	4	5	6	7	8
H_2									
Calc. (Wacks)	0·4706	0·8938	1·0000	0·8678	0·6508	0·4461	0·2889	0·1812	0·1118
Calc. (Dunn)	0·5089	0·9090	1·0000	0·8851	0·6972	0·5139	0·3646	0·2535	0·1745
Obs. (Kerwin–Marmet)	0·46	0·77	1·00	0·55	0·35				
HD									
Calc. (Wacks)	0·3619	0·7919	1·0000	0·9625	0·7875	0·5780	0·3982	0·2611	0·1667
D_2									
Calc. (Dunn)	0·2693	0·6809	1·0000	1·1258	1·0832	0·9428	0·7678	0·5980	0·4522
Calc. (Wacks)	0·2465	0·6625	1·0000	1·1229	1·0507	0·8680	0·6576	0·4681	0·3181

Experimental values for these overlap integrals are best obtained from observation of the energy distribution of electrons ejected by photo-ionization of H_2 as described in Chapter 14, § 8.2. Evidence from electron impact experiments is at present somewhat confused.

Kerwin and Marmet[‡] observed the production of H_2^+ ions by electrons of very homogeneous energy near the threshold. They incorporated their low-energy electrostatic electron selector (Chap. 3, § 2.4.1), which has an energy spread of only 0·03 eV, into the source of a mass spectrometer. Typical ionization curves which they obtained are shown in Fig. 13.19 and kinks arising from excitation of successive vibrational levels may be seen. The appearance potential comes out to be $15·37\pm0·05$ eV

† KRAUSS, M. and KROPF, A., *J. chem. Phys.* 26 (1957) 1776; WACKS, M. E., *J. Res. natn. Bur. Stand.* A68 (1964) 631; HALMAN, M. and LAULICHT, I., *J. chem. Phys.* 43 (1965) 1503; DUNN, G. H., ibid. 44 (1966) 2592.

‡ MARMET, P. and KERWIN, L., *Can. J. Phys.* 38 (1960) 972; KERWIN, L., MARMET, P., and CLARKE, P., ibid. 39 (1961) 1240.

very close to the accepted value 15·427 eV. From analysis of the observed curve in terms of (8), the separations of the vibrational levels of H_2^+ as well as the relative values of $(p_{00}^{nv'})^2$ could be obtained. Comparison between the observed and calculated energy separations is given

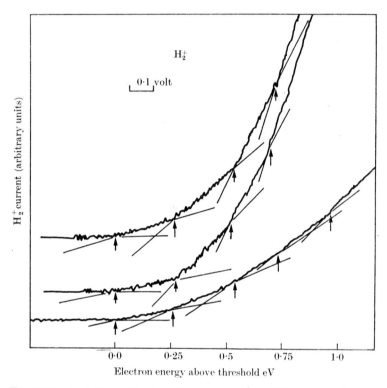

FIG. 13.19. Variation of the ionization cross-section of H_2 near the threshold as observed by Marmet and Kerwin.

in Table 13.6 and is seen to be quite satisfactory. For the relative transition probabilities given in Table 13.5, the agreement is not so close, particularly for the upper observed levels ($v' = 3$ and 4).

However, in a careful study of the form of the cross-section as it rises from the threshold, Briglia and Rapp,† using their total ionization apparatus (see Chap. 12, § 7.2 and Fig. 12.13) with an electron source operated by a phase-sensitive modulated r.p.d. method (Chap. 3, § 2.4.2), were quite unable to observe any evidence at all of changes of slope in the ionization curve near threshold. Thus Fig. 13.20 reproduces one of

† BRIGLIA, D. D. and RAPP, D., *Phys. Rev. Lett.* **14** (1965) 245.

their typical results for H_2^+. The curve is accurately linear from the threshold to 20 eV and is in no way different in shape from the corresponding curve for He^+ shown in the same Fig. 13.20.

TABLE 13.6

Comparison of observed and calculated energy separation of various vibrational levels of H_2^+

	Energy separations (eV)					
Vibnl. levels	0–1	1–2	2–3	3–4	4–5	5–6
Calculated	0·269	0·254	0·238	0·223	0·208	0·192
Observed	0·272	0·263	0·233	0·237	0·21	0·20

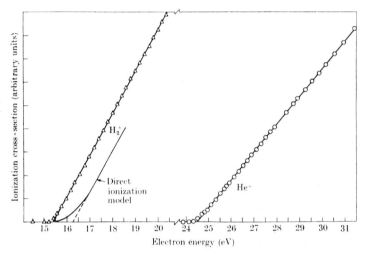

FIG. 13.20. Variation of the ionization cross-section of H_2 near the threshold as observed by Rapp and Briglia. The curve labelled 'Direct ionization model' is that calculated from (8) allowing for the electron velocity distribution. Results for He are also shown for comparison.

The situation has been further confused by the observations of McGowan, Fineman, Clarke, and Hanson† made with the technique described in Chapter 3, § 2.4.5. They used an electrostatic analyser of the same type as did Marmet and Kerwin, to produce an electron beam homogeneous to about 0·03 eV in energy. Instead of exciting a volume of gas in a chamber by passing the electron beam through it, they crossed the electron beam with a molecular beam. Their results are illustrated in Fig. 13.21.

† McGowan, J. W., Fineman, M. A., Clarke, E. M., and Hanson, H. P., *Phys. Rev.* **167** (1968) 52.

In contrast to Briglia and Rapp they find a curve that is not straight but, on the other hand, shows no sign of discontinuities of slope. Thus in Fig. 13.21 the form of the cross-section to be expected if the process is one of direct ionization to different vibrational states of H_2^+ is shown.

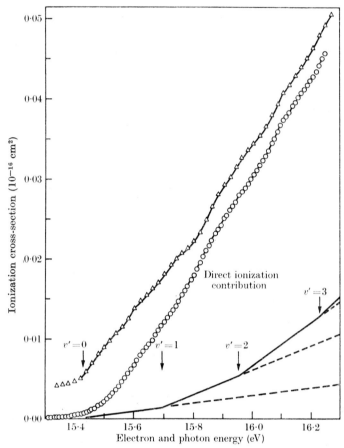

FIG. 13.21. Variation of the ionization cross-section of H_2 near the threshold as observed by McGowan, Fineman, Clarke, and Hanson, ○ experimental points. The integrated photo-ionization cross-section observed by Dibeler, Reese, and Krauss is shown for comparison, △ experimental points. The estimated contribution from direct ionization is also shown.

This has been derived from (8), (9), and (10) with $s = 1$ and A chosen so that for electrons with energies 3 eV above the threshold the slope of the ionization curve agrees with that observed by Rapp and Briglia who made absolute measurements. It will be seen that, if these assumptions are justified, only a fraction of the observed ionization could be ascribed to direct production of H_2^+. The remainder would presumably come from auto-ionization. In support of the reality of the small kinks noticeable in the ionization curve of Fig. 13.21 McGowan *et al.* point out that if $Q_i^{ph}(\nu)$ is the cross-section for photo-ionization of H_2 by radiation of frequency ν (see Chap. 14, § 8.2), the integrated cross-section

$$ h \int_{\nu_0}^{\nu} Q_i^{ph}(\nu) \, d\nu, $$

where ν_0 is the threshold frequency and $\nu = E/h$ where E is the incident electron energy in Fig. 13.21, varies with E in much the same way as does the impact-ionization cross-section.

This may be seen by reference to Fig. 13.21 in which the integrated photo-ionization cross-section derived from the observations of Dibeler, Reese, and Krauss[†] is shown. The similarity is more marked if the derivative of the electron-impact ionization cross-section is compared with the photo-ionization cross-section, the kinks then appearing as peaks.

It seems clear from analysis of the photo-ionization data, which makes it possible to distinguish auto-ionization from direct processes (Chap. 14, § 5.5.4.1), that the kinks in the integrated cross-section curve are indeed due to auto-ionization. This lends strong support to the same interpretation of the kinks in the electron ionization curve. On this basis the percentage of the total ionization contributed by auto-ionization would fall gradually from 87 per cent very close to the threshold to 60 per cent at 16·5 eV and 40 per cent at 17·4 eV.

Two problems remain unanswered. Why did Briglia and Rapp find evidence neither of auto-ionization nor vibrational structure and why did Marmet and Kerwin find the latter but not the former? We must, at the time of writing, let the matter rest in this unsatisfactory state.

1.4.3. *Energy loss spectra beyond the ionization threshold.* Information about the variation of the probability of ionization with energy of the scattered electron obtained from observations of energy loss spectra is discussed in Chapter 5, § 4. We have already referred to the observations by Lassettre and Francis[‡] and by Lassettre and Jones[§] of energy loss spectra up to the ionization threshold (15·4 eV) in H_2, but in the course of these experiments spectra were taken out to 29·7 eV. From these results generalized differential oscillator strengths were derived and thence, by extrapolation in the usual way (Chap. 7, § 5.6.7), the differential optical oscillator strengths that may then be compared

† DIBELER, V. H., REESE, R. M., and KRAUSS, M., *J. chem. Phys.* **42** (1965) 2045.
‡ loc. cit., p. 898.   § loc. cit., p. 898.

with results obtained from optical absorption measurements in the ultra-violet.

Such a comparison is given in Table 13.7, the optical observations being those of Po Lee and Weissler† (Chap. 14, § 8.2). The agreement is satisfactory.

TABLE 13.7

Comparison of differential optical oscillator strengths for H_2

Energy loss (eV)	Differential oscillator strength/eV	
	From electron impact	From optical absorption
20·15	0·064	0·061–0·064
20·55	0·059	0·061
21·29	0·053	0·047
22·33	0·049	0·051
22·95	0·049	0·047
26·03	0·037	0·042

The sum of the optical oscillator strengths for H_2 should be equal to 2, the number of atomic electrons (see Chap. 7, § 5.2.2). From the extrapolated impact data the value is 1·80. The discrepancy may be partly due to the need for extrapolation of the impact results to energy losses above 29·7 eV which was done linearly up to 38·7 eV but no further.

A still further check is based on the expression of the refractive index μ for radiation of frequency ν in terms of the optical oscillator strengths. Thus, when $\mu \simeq 1$ (see Chap. 7, eqns. (83) and (84),

$$\mu - 1 = 2\pi N\alpha/V, \qquad (11\,a)$$

where

$$\alpha = \frac{e^2\hbar^2}{m}\left(\sum + \int\right)\frac{f_i}{E_i^2 - h^2\nu^2}. \qquad (11\,b)$$

Here N and V are the Avogadro number and molar volume respectively and f_i is the optical oscillator strength for the transition involving energy E_i. The integral applies over the continuous spectrum.

In Table 13.8 comparison is made between the values of α derived from observed refractive indices with those calculated from (11 b) using the extrapolated oscillator strength derived from their energy loss spectra by Lassettre and Francis.‡ The agreement is moderate.

1.5. *Dissociative ionization*

We have already pointed out in Chapter 12, § 7.3 how Bleakney§ detected the production of protons formed by dissociative ionization of

† Po LEE and WIESSLER, G. L., *Astrophys. J.* **115** (1952) 570.
‡ loc. cit., p. 898.   § BLEAKNEY, W., *Phys. Rev.* **35** (1930) 1180.

hydrogen. Energetic protons were distinguished from protons of thermal energy by the differences in the peak shapes observed in the mass analysis. In this way Bleakney determined appearance potentials for the two processes,

$$H_2 + e \rightarrow H_2^+ (^2\Sigma_u) + 2e \rightarrow H + H^+ + 2e, \tag{12 a}$$

which yields energetic protons (see Fig. 13.2) and

$$H_2 + e \rightarrow H_2^+ (^2\Sigma_g) + 2e \rightarrow H + H^+ + 2e, \tag{12 b}$$

which yields slow protons (see Fig. 13.2).

See sec. 1.4.?
p. 909

TABLE 13.8

Wavelength (Å)	α (in 10^{-24} cm³) (see (11))	
	From refractive index	From energy loss spectra
6564	0·8213	0·808
5876	0·8267	0·812
5461	0·8302	0·815
4358	0·8702	0·827

Discrimination between the protons in terms of peak shapes in mass spectra was further exploited by Hagstrum and Tate† (Chap. 12, § 7.3.2).

In between these investigations Lozier‡ made his first measurements for hydrogen using the apparatus described in Chapter 12, § 7.3.1. For the process (12 a) the appearance potential for ions of kinetic energy W^+ is given by

$$E_{\min} = 2W^+ + W_0,$$

where W_0, the sum of the dissociation energy of H_2 and the ionization energy of H, $\simeq$ 18 eV.

Similarly, for the process

$$H_2 + e \rightarrow H^+ + H^+ + 3e, \tag{13}$$

W_0 is greater by the ionization energy of H.

Lozier observed $E_{\min}$ as a function of W^+ for both these transitions. This was done by fixing the retarding potential V_r between the grid and collecting cylinder so that only ions with kinetic energy exceeding eV_r could be collected. The current to the collector was then observed as a function of electron energy. No ion current was observed until the electron energy became equal to $E_{\min}$ for the process (12 a) after which it increased smoothly until it reached a certain higher value beyond

† HAGSTRUM, H. D. and TATE, J. T., ibid. **59** (1941) 354.
‡ LOZIER, W. W., ibid. **36** (1930) 1285 and 1417.

which the slope of the curve changed. This marked the onset of the process (13).

Fig. 13.22 illustrates the comparison between the observed and predicted relations between E_{min} and W^+ for the two processes. Although there are a number of reasons why these observations could not now be regarded as precise they leave no doubt of the interpretation of the results in terms of the two processes (12 a) and (13).

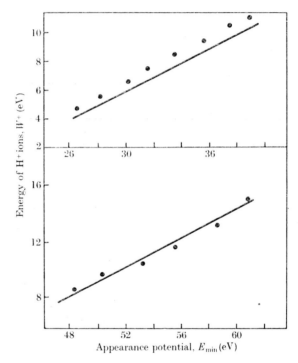

FIG. 13.22. Appearance energy E_i for ions of kinetic energy W^+ produced in the dissociative ionization of H_2 into $H^+ + H$ and $H^+ + H^+$ respectively. ● experimental points. —— predicted curve.

Once the processes have been identified for H_2 there is little interest in determining appearance potentials with precision because they can be predicted with certainty. Much more attention has been paid to the difficult problems of determining the absolute cross-sections for production of protons by different processes and of measuring their energy distributions. A further problem is that of determining the angular distribution of proton production relative to the direction of the electron beam.

1.5.1. *Absolute magnitude of the cross-section for production of protons.* Because of the difficulty of ensuring that the collecting efficiency of any experimental device for protons is independent of proton energy, the only measurements of absolute cross-sections to which any quantitative significance may be attached are those of Rapp, Englander-Golden, and Briglia,† using the technique described in Chapter 12, § 7.3.4, in which they used their total ionization apparatus to collect all ions with kinetic energy greater than a suitably chosen value, slower ions being stopped by a retarding potential on the collector.

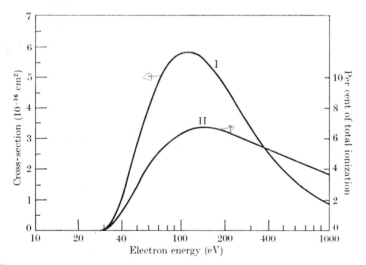

FIG. 13.23. Cross-sections for production of ions with kinetic energy greater than 2·5 eV by electron impact in H_2. I, absolute cross-section. II, percentage of total ionization cross-section.

In hydrogen they collected ions with kinetic energy $\geqslant 2\cdot5$ eV, which should include all protons except those produced by (12 b). Fig. 13.23 illustrates the derived cross-sections for production of these protons as a function of incident electron energy. It will be seen also from the figure that the percentage of the total ionization cross-section that can be ascribed to energetic proton production is only a little over 6 per cent at the maximum.

The absolute value of the cross-section for production of slow protons by the process (12 b) is much more difficult to determine. As these protons are only produced through violation of the Franck–Condon principle the cross-section for their formation should be small. In principle the

† RAPP, D., ENGLANDER-GOLDEN, P., and BRIGLIA, D. D., *J. chem. Phys.* **42** (1965) 4081.

relative cross-section compared to that for production of H_2^+ ions could be calculated approximately from the formula (8).

First we note, in the notation of that formula, that

$$\left(\Sigma + \int\right)(p_{00}^{nv'})^2 = \left(\Sigma + \int\right)\left\{\int \chi_{nv'}(R)\chi_{00}(R)\,\mathrm{d}R\right\}^2$$

$$= \int \{\chi_{00}(R)\}^2\,\mathrm{d}R$$

$$= 1.$$

Here we have denoted the sum over all discrete final vibrational levels by Σ and the integral over all unquantized vibrational states by $\int$. Now if $Q(H^+)$, $Q(H_2^+)$ are the respective cross-sections for production of H^+ and H_2^+ ions

$$Q(H^+)/\{Q(H^+) + Q(H_2^+)\} = \int (p_{00}^{nv'})^2 \Big/ \left(\Sigma + \int\right)(p_{00}^{nv'})^2$$

$$= 1 - \Sigma\,(p_{00}^{nv'})^2. \tag{14}$$

If the sum over all discrete final vibrational states can be evaluated then the required ratio of cross-sections may be obtained. Unfortunately as this ratio is small the evaluation of $\Sigma\,(p_{00}^{nv'})^2$ must be very accurate.

Dunn,† using vibrational wave functions calculated as described in § 1.4.2, finds $\Sigma\,(p_{00}^{nv'})^2 = 0.9714$ and 0.9937 for H_2 and D_2 respectively, giving 2·8 per cent and 0·63 per cent as the percentage fraction of ionization processes that lead to production of slow H^+ and D^+ ions from their respective parent homonuclear molecules.

Experiments designed to measure the rates of the cross-sections for production of atomic and molecular ions through the process (12 b) do not suffer from kinetic energy discrimination between the ions because both possess nearly thermal kinetic energies. Other sources of discrimination cannot be excluded when the intensities of atomic and molecular ion currents are compared by mass analysis, but they are probably not so important as to change the order of magnitude of the results.

Schaeffer and Hastings‡ measured the cross-section ratios for H_2, D_2, and T_2 using a mass spectrometer of conventional 60° sector type with a 6-in radius of curvature and 0·5-mm slit width. The results depended to some extent on the voltage applied to accelerate the ions into the spectrometer. At the highest accelerating voltage employed (4200 V) the percentage production by 30-V electrons of atomic as compared with diatomic ions was found to be 1·3, 0·7, and 0·36 per cent for H_2, D_2, and T_2 respectively. These percentages changed to 1·1, 0·57, and 0·31 per cent

† Dunn, G. H., *J. chem. Phys.* **44** (1966) 2592.
‡ Schaeffer, O. A. and Hastings, J. M., ibid. **18** (1950) 1048.

when the accelerating voltage was reduced to 2500 V but the ratio of the percentages for different molecules remained much the same. In certain measurements by Hipple,† using an 180° mass spectrometer, the percentage production of atomic ions from H_2 and D_2 by 23-V electrons was found to be 0·96 and 0·34 per cent respectively, while Bauer and Beach‡ found 0·58 per cent for H_2 and 0·38 per cent for D_2.

The spread of these results is an indication of the difficulty of making accurate measurement of abundances. Having regard to this there is no inconsistency with the calculated ratios for H_2 and D_2.

1.5.2. *Energy and angular distributions of protons from dissociative ionization of* H_2

1.5.2.1. *Experimental method.* Problems of energy discrimination have beset the interpretation of many of the experimental results on the energy distribution of energetic protons resulting from dissociative ionization in H_2—so much so that Stevenson§ in 1960 considered that the Franck–Condon principle (Chap. 12, § 3.2) did not hold. Relatively little effort had been made during this period to observe the angular distribution of the energetic protons apart from the experiments of Sasaki and Nakao‖ for protons with energies between 5 and 9 eV produced by impact of electrons of 100 eV energy. In 1963 Dunn and Kieffer†† combined a thorough experimental investigation of the angular distributions of the protons with measurements of the energy distribution, in which they paid special attention to the avoidance of energy discrimination.

The general arrangement of their experiments is shown in Fig. 13.24. The electron gun G was attached to a cylindrical scattering chamber S so as to fire an electron beam radially through the chamber to be collected in a Faraday cup C also attached to, though insulated from, the chamber. Ions formed by impact of the electron beam within the chamber drifted out through a slot into an aperture tube system L_1, L_2, L_3, L_4 and thence into the spectrometer A, which was of the 60° magnetic sector type. Those ions that had a chosen momentum were focused on the exit slit E_1. They then passed through two grids D and E_2, normally at the potential of the spectrometer, and then were accelerated through a further grid F, which was at a negative potential of about 150 V, to fall on to the cathode

---

† Hipple, J. A., private communication to Schaeffer and Hastings.
‡ Bauer, N. and Beach, J. Y., *J. chem. Phys.* **15** (1947) 150.
§ Stevenson, D. P., *J. Am. chem. Soc.* **82** (1960) 5961.
‖ Sasaki, N. and Nakao, N., *Proc. imp. Acad. Japan* **11** (1935) 138; **17** (1941) 71.
†† Dunn, G. H. and Kieffer, L. J., *Phys. Rev.* **132** (1963) 2109.

of a magnetic electron multiplier M. To obtain measurements of the
proton current as a function of the angle θ between the direction of
motion of the proton and of the electron beam, the scattering chamber S,
carrying the electron gun and collector, could be rotated about its axis.

The apparatus was contained in a glass bell-jar vacuum system and
the background pressure was around 2–3×10^{-7} torr. During an experiment commercial hydrogen was leaked into the system continuously so

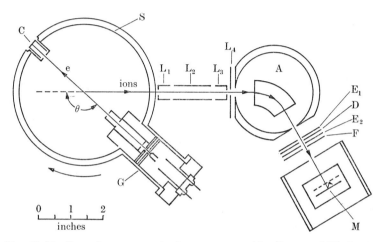

Fig. 13.24. General arrangement of apparatus used by Dunn and Kieffer for
observing angular and energy distributions of energetic protons produced by
electron impact dissociation of H_2.

that, with the mercury diffusion pump operating, the chamber pressure
was about 4×10^{-5} torr. Impurities were monitored by operating the
sector magnet as a mass analyser and were observed to contribute a
small constant background current over the pressure range (about three
decades) within which the proton signal was proportional to the pressure.

The electron beam, which usually was of strength about $10\ \mu\text{A}$, had
a divergence angle less than $3°$ and a diameter between 1 and 3 mm.
Retarding-potential analysis at the collector C indicated an energy half-
width in the beam of about 1 eV.

Apart from the exit slit E_1, which was 1 mm width, all other defining
apertures in L_1, L_3, L_4, D, E_2, and F, were of width 3 mm. The analyser
constant α, which is such that the magnetic field B required to focus ions
of mass M and energy W is given by

$$B = \alpha(MW)^{\frac{1}{2}},$$

was determined by accelerating H_2^+ ions through a known voltage and measuring the value of B required to focus them. By carrying out this calibration first with quite high accelerating potentials and then at progressively lower values the effective contact potential ΔV could also be determined. With surfaces coated with an overlay of Aqua-dag over gold plating ΔV was less than 0·1 V. The scattering chamber and drift region were magnetically shielded so that the maximum field within them was about 50 milligauss.

Three types of measurements were carried out. To measure the angular distributions of the protons the electron energy E and magnetic analyser field B were kept fixed and the ion current measured as a function of θ. This was carried out for ten values of E in the range 35–1500 eV and magnetic fields that selected protons of energies 3·7, 8·6, and 11·8 eV.

Next, to measure the energy distribution of the protons E and θ were kept fixed and ion currents measured for different values of the magnetic analyser field B. Finally, keeping θ and B fixed, the ion current was measured as a function of the electron energy E.

In all these measurements the protons were exposed to *no* accelerating or electrostatically focusing fields until they were accelerated through the grid F into the multiplier.

The magnetic electron multiplier was run at a gain of 10^6–10^7 giving measured output currents from 10^{-13} to 10^{-10} A.

1.5.2.2. *Discussion of observations of angular distributions of energetic protons.* Fig. 13.25 illustrates observed angular distributions for 8·6-eV protons for different incident electron energies E. It will be seen that, at low energies, the distribution has a minimum near 90°. This minimum becomes shallower as E increases till the distribution becomes isotropic. At higher energies it exhibits a flat maximum at 90°.

These features can be understood in terms of the theory outlined in Chapter 12, § 4.2. According to that theory, for a dissociative transition between Σ states in which two neutral fragments are produced, the angular distribution of the direction of relative motion of the fragments is given by

$$A\{\cos^2\theta\,\cos^2\bar{\eta} + \tfrac{1}{2}\sin^2\theta\,\sin^2\bar{\eta}\}, \tag{15}$$

where $\bar{\eta}$ is the mean angle between the incident direction of the colliding electron and the direction of the change of momentum $K\hbar$ suffered by the electron in the dissociating collision. A is a factor independent of θ but dependent on the electron energy. It was also pointed out that, even if one of the fragments produced is a positive ion, the most probable direction of motion of the secondary electron is opposite to that of **K** (see Chap. 7, § 5.6.2) so that (15) still applies to a good approximation. This being so the variation of the angular distribution with electron energy is as observed. Furthermore $\bar{\eta}$ may be determined from the experimental data as a function of electron energy for protons of different kinetic energy W^+. Results obtained for protons of these different energies are as shown in Fig. 13.26. At high electron energies $\bar{\eta}$ tends towards 60°. This should be apparent in observations of the angular distribution

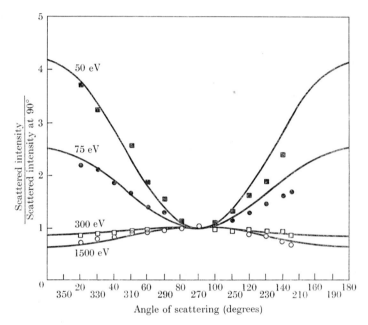

Fig. 13.25. Angular distribution of 8·6-eV protons produced by impact dissociation of H_2 by electrons of different incident energies indicated on the separate curves. The intensity scale is normalized to unity at 90° for all curves.

of ejected electrons arising from ionization of H_2 by energetic electrons. Evidence in general support of this is provided by the observations by Mohr and Nicoll† shown in Chapter 5, § 5.3.2.

1.5.2.3. *Observations of the energy distribution of energetic protons.* Fig. 13.27 illustrates the energy distribution of protons produced in dissociative ionization of H_2 by impact of 75-eV electrons, as observed by Dunn and Kieffer. According to the theory the distribution should be given by (8), (9), and (10), bearing in mind that the final wave function of nuclear motion is now in the continuum, corresponding as it does to

† Mohr, C. B. O. and Nicoll, F. H., *Proc. R. Soc.* **A144** (1934) 596.

a dissociating state. With the reflection approximation of Chapter 12, § 4.1, p. 830 the overlap integral (9) now becomes

$$\chi_{00}^2(R_c),$$

where R_c is the nuclear separation at the classical turning point for the relative motion of proton and hydrogen atom in the final state and χ_{00} is the wave function for the lowest vibrational level of the ground state of H_2.

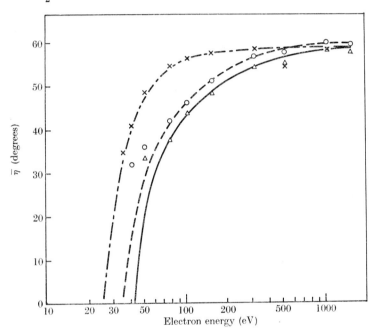

FIG. 13.26. The mean angle $\bar{\eta}$ between the momentum vector of the incident electron and the vector change of momentum which it suffers in the ionizing collision. ✕ —·—·— 3·7 eV protons. ○ — — 8·6-eV protons. △ —— 11·8-eV protons.

Assuming a linear law for the rise in the ionization cross-section with energy above the threshold, the energy distribution of the ejected protons is determined by

$$(E - E_0 - 2W^+)\chi_{00}^2(R_c), \tag{16}$$

where E is the electron energy, E_0 the threshold energy for producing ions of zero kinetic energy, and W^+ the kinetic energy of the protons. R_c is the nuclear separation in the final state at which the internuclear potential energy is $2W^+$ above that, at infinite separation, of a proton and normal hydrogen atom. It may be calculated from the exact potential energy curve for the $^2\Sigma_g^+$ state of H_2^+ obtained by Bates, Ledsham, and Stewart.† The wave function χ_{00} may be represented as a simple harmonic oscillator function with constants adjusted to fit the properties of the ground vibrational state of H_2.‡

At sufficiently high electron energies the dependence of the linear factor on W^+ is small. In Fig. 13.27 the distribution given by $\chi_{00}^2(R_c)$ is shown for comparison with that observed by Dunn and Kieffer. The agreement is not unsatisfactory in that the maximum occurs at the correct energy and the variation at higher energies is closely similar.

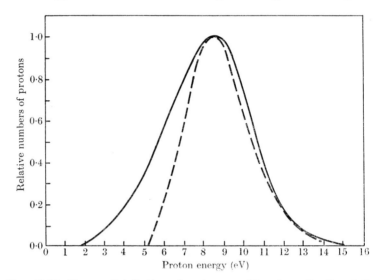

FIG. 13.27. Energy distribution of protons resulting from the dissociative ionization of H_2 by 75-eV electrons. —— observed. — — — calculated.

On the low energy side the calculated distribution falls off more rapidly than the observed but this may be partly due to the use of the reflection approximation in which the final wave function for the nuclear motion is replaced by a delta function (see Chap. 12, § 4.1, p. 830).

As the electron energy is reduced the linear factor in (16) becomes significant and the energy W_m^+ at which the peak production of protons occurs gradually shifts to lower energies. Fig. 13.28 shows the variation of W_m^+ with electron energy E observed at different angles θ compared with the variation predicted from (16) for the total cross-section. According to (15) and (16) the energy distribution of ions produced with velocity in the direction θ is given by

$$(E - E_0 - 2W^+)\chi_{00}^2(R_c)/D(E, W^+, \theta), \tag{17}$$

† BATES, D. R., LEDSHAM, K., and STEWART, A. L., Phil. Trans. R. Soc. A246 (1953) 215.   ‡ STEVENSON, D. P., J. Am. chem. Soc. 82 (1960) 5961.

where $$D(E, W^+, \theta) = \frac{1+\tfrac{1}{3}P(E, W^+)}{1+P(E, W^+)\cos^2\theta},$$

and $$P(E, W^+) = 2\cot^2\bar{\eta}(E, W^+) - 1.$$

We note that $D = 1$ when $\cos^2\theta = \tfrac{1}{3}$, i.e. $\theta = 54\cdot7°$.

From the values of $\bar{\eta}$ derived from analysis of the observation of angular distributions of protons for different values of E and W^+ (see

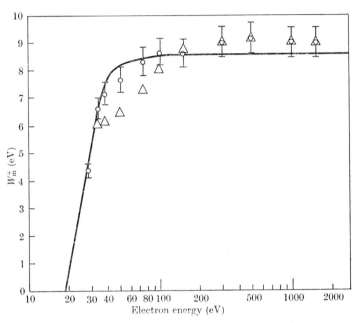

FIG. 13.28. The energy W_m^+ at which the proton-energy distribution has a maximum, as a function of electron energy. ○ observed for $\theta = 30°$ and $40°$. △ observed for $\theta = 90°$. —— calculated assuming an isotropic proton distribution.

Fig. 13.25) D can be calculated and hence from (17) W_m^+ as a function of θ. Fig. 13.29 shows $W_m^+(E, \theta)/W_m^+(E, 54\cdot7°)$ for $\theta = 90°$ and $30°$. It will be seen that these results are consistent with the dependence on θ of the variation of $W_m^+(E, \theta)$ with E.

1.5.2.4. *Production of protons as a function of electron energy.* Finally in Fig. 13.30 we show how the variation with electron energy of the cross-section for production of protons of a given kinetic energy depends on the direction of motion of the observed protons. The importance of allowing for anisotropy in the proton distribution in experiments in which total collection of the protons is not achieved, is clear.

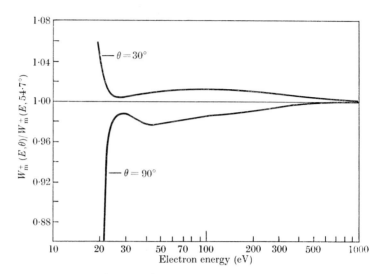

FIG. 13.29. Ratio $W_m^+(E, \theta)/W_m^+(E, 54 \cdot 7°)$ of the energies at which the peak of the proton energy distribution occurs for protons observed at an angle θ to that for protons observed at $54 \cdot 7°$, as a function of electron energy E, for $\theta = 30°$ and $90°$.

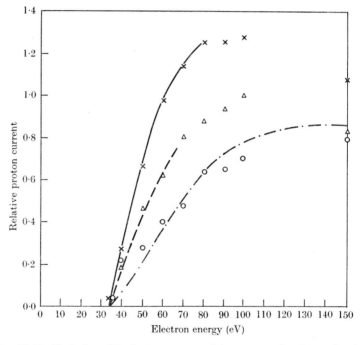

FIG. 13.30. Variation with electron energy of the cross-section for production of protons of kinetic energy $8 \cdot 6$ eV, observed in different directions. $\times$————$\times$ $\theta = 30°$, $\triangle$— — —$\triangle$ $\theta = 54 \cdot 7°$, $\bigcirc$ —·—·— $\bigcirc$ $\theta = 90°$.

1.6. *Production of negative ions in* H_2, *HD, and* D_2

1.6.1. *The first experiments in* H_2. The cross-sections for negative ion production by electron impact in H_2 are very small. Lozier[†] in 1930 applied his technique to hydrogen and was able to detect two peaks in the negative ion currents at electron energies of 6·6 and 8·8 eV. These peaks were so weak that Lozier attributed them to impurities and confirmed this by showing that when water vapour was admitted the peaks grew in magnitude while still occurring at the same energies. In 1958 Khvostenko and Dukel'skii[‡] used a mass spectrometer to study ion

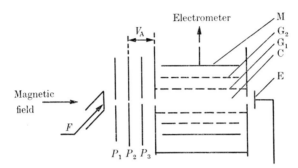

FIG. 13.31. Schematic diagram of the apparatus used by Schulz for studying dissociative attachment in H_2.

production in H_2. They observed a peak current of H^- ions at 7·2 eV, which was present when either H_2 or H_2O was present, but in addition observed the H^- current out to electron energies of 38 eV and observed a further peak at 14·5 eV.

As the cross-section for H^- production from H_2 is clearly very small it is necessary to use ultra-high vacuum techniques if accurate measurements of the current arising from H_2 alone are to be made. Schulz[§] applied the method that he developed to study the negative ion production. It is essentially similar to that described in Chapter 12, § 7.4 but with some modifications. Fig. 13.31 shows a schematic diagram (cf. Fig. 12.19). The electron gun used the retarding potential difference method in the standard way. To reduce contact potentials all parts of the tube were gold-plated but, in addition, the voltage scale was calibrated from the threshold value 15·56 V for production of positive ions from H_2.

The vacuum system used a copper trap. With the whole system baked

† LOZIER, W. W., *Phys. Rev.* **36** (1930) 1417.
‡ KHVOSTENKO, V. I. and DUKEL'SKII, V. M., *Zh. éksp. teor. Fiz.* **33** (1957) 851; *Soviet Phys. JETP* **6** (1958) 657.
§ SCHULZ, G. J., *Phys. Rev.* **113** (1959) 816.

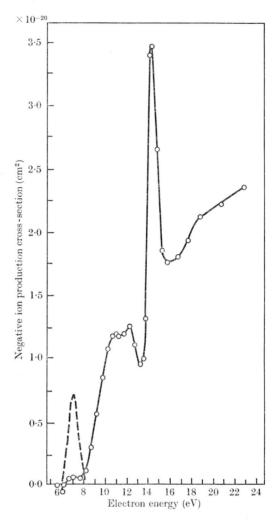

FIG. 13.32. Variation with electron energy of negative ion
production by electron impact in H_2 as observed by Schulz.
— — — additional production when the reagent grade
hydrogen was introduced with no liquid air in the trap,
presumably due to production from water vapour.

at 400 °C a background pressure of 2×10^{-10} torr could be obtained
without use of a liquid air trap, but such a trap was employed so as to
minimize the partial presence of water vapour built up through reactions
occurring in the tube.

Fig. 13.32 illustrates the results obtained by Schulz of the negative
ion production when the liquid air trap was in operation. Without this,

a further peak, shown as a broken line, developed at an electron energy of 6·8 eV, its magnitude depending on the purity of the hydrogen gas admitted to the system. It is almost certainly due to water vapour as in the earlier experiments (cf. Fig. 13.106). Otherwise the negative ion currents were found to be proportional to the H_2 pressure and the electron beam current. The general form of the variation agrees with that of Khvostenko and Dukel'skii.

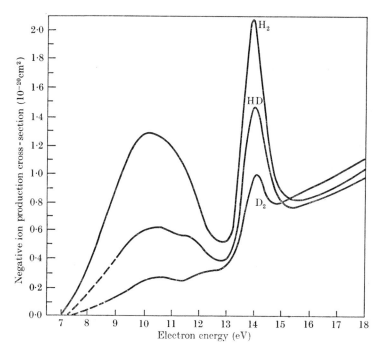

FIG. 13.33. Cross-sections for negative ion formation in H_2, HD, and D_2 as a function of electron energy, as observed by Rapp, Sharp, and Briglia.

The processes which lead to H^- production around 10- and 14·2-eV electron energy are certainly due to electron capture

$$H_2 + e \rightarrow H + H^-.$$

If the hydrogen atom is produced in the ground state the appearance potential according to Chapter 12 (2) for H^- ions of zero kinetic energy would be $D(H_2) - A(H)$, where $D(H_2)$ is the dissociation energy (4·46 eV) for H_2 and $A(H)$ is the electron affinity (0·76 eV) of atomic hydrogen. This gives 3·7 eV, which agrees, remarkably enough, with the value obtained by applying the linear extrapolation technique. The minimum kinetic energy of the ions produced is about 2·5 eV so that, according to the formula (42) of Chapter 12, the half-width of the energy spread due to the temperature motion of the target molecules is about 0·3 eV. This is allowing for the fact that the gas was maintained at the temperature of liquid nitrogen (77 °K) to freeze out impurities.

The peak at 14·2 eV is probably due to production of a hydrogen atom in a 2-quantum excited state. For ions of zero kinetic energy the appearance potential would be 13·8 eV, which is consistent with the observations, assuming that the ions observed do not possess much kinetic energy.

Polar dissociation into H^+ and H^- can occur at electron energies greater than 17·2 eV $(= D(H_2) - A(H) + I(H)$, where $I(H)$ is the ionization energy of H). It seems probable that the rising part of the observed curve shown in Fig. 13.32 about this energy is due to processes of this kind.

1.6.2. *The isotope effect.* A few years after Schulz obtained the above discussed results, Rapp, Sharp, and Briglia† applied their technique described in Chapter 12, § 7.4 to measure cross-sections for production of negative ions not only in H_2 but also in the isotopic molecules HD and D_2. As expected the ions formed in the electron energy range 7–12·8 eV possessed appreciable kinetic energy and, to saturate their collection, an extraction field of 24 V/cm was required as compared with 10 V/cm above 12·8 eV.

Fig. 13.33 shows the results they obtained for the three molecules. For H_2 there is general but not detailed agreement with the observations of Schulz. For the broad peak near 10 eV Rapp *et al.* obtain an absolute cross-section quite close to that determined by Schulz, but for the peak at 14 eV they obtain a somewhat smaller cross-section ($2·1 \times 10^{-20}$ cm² as compared with $3·5 \times 10^{-20}$ cm²). In any case the cross-sections are very low compared with corresponding cross-sections for negative ion production in other gases (O_2 § 4.4, CO_2 § 8, H_2O § 7, SF_6 § 11.1, I_2 § 11.2). For this reason, when a strong source of H^- ions is required as for the tandem van der Graaf proton accelerator, water vapour rather than H_2 is used as the target gas (see § 7).

A striking feature of the results shown in Fig. 13.33 is the very pronounced difference between the cross-sections for the different isotopic molecules. In all cases the cross-section falls off as the mass of the molecule increases.

An even more pronounced isotope effect was found by Schulz and

† RAPP, D., SHARP, T. E., and BRIGLIA, D. D., *Phys. Rev. Lett.* **14** (1965) 533.

Asundi.† They noted that in early experiments in H_2 there was some evidence of a negative ion current at an electron energy of about 3·7 eV, and concentrated attention on the verification or otherwise of the reality of this current as due to H^- formed from H_2. In order to be able to work at relatively high pressures of H_2 they used a differentially pumped collision chamber. Ions produced by the passage of the electron beam through the chamber were expelled from the chamber by an electric field of about 2 V/cm and analysed in a magnetic mass spectrometer.

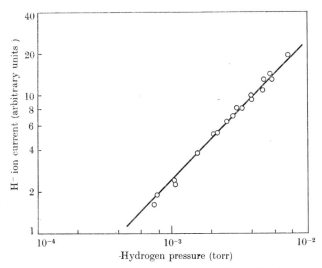

FIG. 13.34. Pressure dependence of the peak involving H^- formation from dissociative attachment in H_2 by impact of electrons of 3·75 eV energy, as observed by Schulz and Asundi.

Pressures up to 5×10^{-2} torr could be introduced into the collision chamber without destroying the linearity of the variation of ion current with gas pressure. The retarding potential difference method was used to reduce the energy spread of the electron beam for experiments in H_2 but not when the attachment cross-sections are very small as for D^- produced in HD and D_2.

Using this apparatus Schulz and Asundi were able to establish the existence of a process of dissociative attachment in H_2 with an appearance potential of $3·75 \pm 0·07$ eV. The linearity of the variation of H^- ion current produced in this process with pressure of H_2 was established, as may be seen from typical observations shown in Fig. 13.34. The observed variation of the cross-section with electron energy for H_2, obtained using

† SCHULZ, G. J. and ASUNDI, R. K., ibid. **15** (1965) 946.

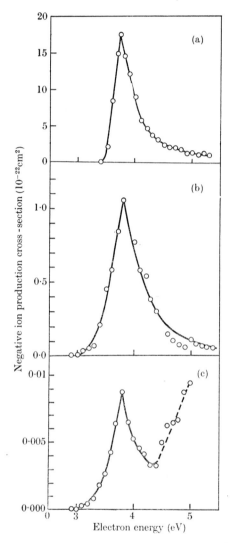

FIG. 13.35. Cross-sections for negative ion formation in H_2, HD, and D_2 by low energy electrons, as observed by Schulz and Asundi. (a) H^- from H_2. (b) D^- from HD. (c) D^- from D_2.

the retarding potential difference method, is shown in Fig. 13.35 (a). It will be seen that the peak cross-section is only $1\cdot8\times10^{-21}$ cm^2. Small as this is, the cross-section for D_2 shown in Fig. 13.35 (c) is nearly two orders of magnitude smaller. That for production of D^- from HD, shown in Fig. 13.35 (b) is intermediate in magnitude.

1.6.3. *Theoretical interpretation of results.*† A general theoretical discussion of the dissociative attachment from the theoretical point of view has been given in Chapter 12, § 3.6.1. It was shown that the cross-section Q_{da} may be written in the form

$$Q_{da} = Q_c \exp\left\{ -\frac{1}{\hbar} \int_{R_0}^{R_s} \frac{\Gamma(R')}{v(R')} \, dR' \right\}. \tag{18}$$

Here Q_c is the cross-section for initial capture of the electron that occurs when the nuclear separation in the molecule is R_0. $\hbar/\Gamma(R')$ is the lifetime of the molecular ion formed against autodetachment when the nuclear separation is R', and $v(R')$ is the velocity of separation of atomic ion and neutral atom when at the same separation. R_s is the nuclear separation beyond which the system is stable against autodetachment.

It is convenient to write (18) in the form

$$Q_{da} = Q_c \exp\{-\overline{\Gamma}\tau/\hbar\}, \tag{19}$$

where $\overline{\Gamma}$ is a mean value of Γ and τ is the time taken for the atom and atomic ion to separate from R_0 to R_s. For the different isotopic molecules Q_c and $\overline{\Gamma}$ will remain constant but τ will be proportional to

$$\{M_1 M_2/(M_1+M_2)\}^{\frac{1}{2}}$$

where M_1 and M_2 are the masses of the constituent atoms. We then have

$$\frac{Q_{da}(H_2)}{Q_{da}(HD)} = \frac{1\cdot6\times10^{-21}}{2\cdot1\times10^{-22}} = \exp\left[\frac{\overline{\Gamma}\tau}{\hbar}\{(1\cdot33)^{\frac{1}{2}}-1\}\right], \tag{20}$$

$$\frac{Q_{da}(H_2)}{Q_{da}(D_2)} = \frac{1\cdot6\times10^{-21}}{8\times10^{-24}} = \exp\left[\frac{\overline{\Gamma}\tau}{\hbar}\{2^{\frac{1}{2}}-1\}\right]. \tag{21}$$

From (20) $\overline{\Gamma}\tau = 8\cdot7\times10^{-15}$ eV s and from (21) $8\cdot4\times10^{-15}$ eV s so that the interpretation is remarkably consistent.

The separation time τ can be expected to be the order 10^{-14} s giving $\overline{\Gamma}$ as about 1 eV. Also on substitution of the numerical value for $\overline{\Gamma}\tau$ in (19) we find

$$Q_c = 9\times10^{-16} \text{ cm}^2,$$

which is not unreasonable.

The large value found for $\overline{\Gamma}$ shows that, in the notation of Chapter 12, § 2.5 the responsible molecular negative ion state must be of Type II and the relation of the potential energy curve to that for the ground state is in general terms as illustrated in Fig. 12.3 (b).

A similar analysis carried out for the process with peaks near 10 eV

† See BARDSLEY, J. N., HERZENBERG, A., and MANDL, F., *Proc. phys. Soc.* **89** (1966) 321.

gives $\bar{\Gamma}\tau$ between 2·5 and 3×10^{-15} eV s so $\bar{\Gamma}\simeq0\cdot2$ to 0·3 eV. While smaller than for the low energy process this is still so large as to require that, in the Franck–Condon region, the upper state concerned is also of Type II. Much the same conclusion is arrived at for the process with peak near 14 eV for which $\bar{\Gamma}\tau$ comes out to be a little less than

$$2\times10^{-15}\text{ eV s.}$$

Fig. 13.36 illustrates the probable disposition of the potential energy curves for H_2 and H_2^-, which give rise to the dissociative attachment

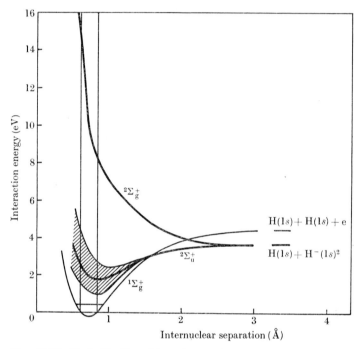

FIG. 13.36. Probable disposition of potential curves for states of H_2^- responsible for dissociative attachment.

processes observed. The ground state of H_2^-, which is a $(1s\,\sigma_g)^2(2p\,\sigma_u)$ $^2\Sigma_u^+$ state, is stable at nuclear separations greater than about 1·5 Å, tending at large nuclear separations to a normal H atom and an H^- ion in its stable ground state. At smaller separations it is unstable towards autodetachment with a short lifetime as it is then essentially a one-body resonance state.

The $(1s\,\sigma_g)(2p\,\sigma_u)^2\,^2\Sigma_g^+$ state of H_2^- is also stable towards autodetachment at sufficiently large nuclear separations. It is also relatively stable in the Franck–Condon region towards autodetachment to the ground state of H_2. To understand its short lifetime as judged from the isotope effect we must suppose that, in the Franck–Condon region it lies above the $b^3\Sigma_u$ repulsive state of H_2, so making a Type II state in this region. In the united atom limit the $^2\Sigma_g^+$ state tends to the $(1s)(2p)^2\,^2D$ doubly excited state of He^-, which lies above the $(1s)(2p)\,2^3P$ state of He to which the $b^3\Sigma_u$ state tends in the same limit.

Additional support for this interpretation is based on an estimate† of the cross-section for the initial capture of the incident electron into the $^2\Sigma_g^+$ state, using (7) of Chapter 12. This gives for the contribution Γ_0 to the level width in the Franck–Condon region due to autodetachment to the ground state of H_2 only 0·004 eV. Thus if autodetachment to the $b^3\Sigma_u$ state were not possible no isotope effect would be observable and the $^2\Sigma_g^+$ state would be of Type I.

As regards the 14·2-eV peak, the appearance potential for ions of zero kinetic energy is not inconsistent with the assumption that the state responsible dissociates into a hydrogen atom in a 2-quantum excited state and H^- in its ground state. This would require the appearance potential to be 13·85 eV. Since ions of zero kinetic energy are produced the potential energy curve for the state, which may be the $(1s\,\sigma_g)(2p\,\pi_u)^2\,^2\Sigma_g$ state, is presumably attractive over part of the Franck–Condon region shown in Fig. 13.36. It is not clear why such a state would be of Type II over this region.

Account must be taken of the fact that a Type I state of H_2^- certainly lies in the same energy range. This is because the observations of Kuyatt, Mielczarek, and Simpson‡ described in Chapter 10, § 3.5.1 have established the existence of fine resonances in the transmission of electrons with energies in the range 11·6–13·3 eV. These resonances showed a clear vibrational structure with separations close to those of the corresponding vibrational levels of the $(1s\,\sigma_g)(2p\,\pi_u)\,C^1\Pi_u$ state of H_2. The H_2^- state concerned must be of Type I. Possible identifications have been discussed by Taylor and Williams§ and by Taylor, Nazaroff, and Golebiewski.‖

1.7. *Analysis of swarm experiments in H_2 and D_2 in which electronic excitation is important*

In Chapter 11, § 7 we described how use of the Boltzmann transport equation made it possible, by a trial-and-error procedure, to obtain

---

† BARDSLEY, J. N., HERZENBERG, A., and MANDL, T., *Proc. phys. Soc.* **89** (1966) 321.
‡ KUYATT, C. E., MIELCZAREK, S. R., and SIMPSON, J. A., *Phys. Rev. Lett.* **12** (1964) 293.   § TAYLOR, H. S. and WILLIAMS, J. K., *J. chem. Phys.* **42** (1965) 4063.
‖ TAYLOR, H. S., NAZAROFF, G. V., and GOLEBIEWSKI, A., ibid. **45** (1966) 2872.

information about momentum-loss and rotational and vibrational excitation cross-sections of molecules for impact with slow electrons. For H_2 and D_2 in particular we described in § 7.1 of Chapter 11 the results of such an analysis for drifting electron swarms for which the characteristic energy ϵ_k (Chap. 2, § 2) was < 1.0 eV. At such low energies the influence of energy loss processes due to electronic transitions is negligible. We now discuss the extension of this analysis to higher characteristic energies ϵ_k.

One difficulty in carrying out this analysis is that hardly any reliable measurements of ϵ_k as a function of the ratio F/p of electric field strength F to gas pressure p, in this range, are available. It is necessary then to rely on measurement of the ionization, dissociation, and photon excitation coefficients as functions of F/p (or F/N, where N is the number of molecules per cm³). Furthermore, it is no longer possible to separate clearly the influence of elastic and inelastic collisions in terms of the two effective collision frequencies ν_m and ν_u defined in (75) and (76) of Chapter 11. An effectively separate check on the assumed momentum-loss cross-section Q_d is therefore not available and the inelastic cross-sections giving the best fit depend on Q_d.

Fig. 13.37 illustrates the cross-sections assumed by Engelhardt and Phelps.† The momentum-transfer cross-section Q_d was taken from total cross-section measurement‡ (see Fig. 10.21) and the ionization cross-section as given by Tate and Smith§ (see Fig. 13.18). The cross-section Q_{diss} for dissociation into neutral atoms was taken to be consistent with that of Ramien‖ with a threshold at 8·85 eV and to have a shape typical of excitation of an intercombination transition. Because of the paucity of information about the cross-sections for optically allowed discrete transitions these were lumped together in a 'photon' excitation cross-section Q_{ph} with a threshold at 12 eV. In carrying out the analysis to fit the observed ionization coefficient α_i/N for H_2† Q_d and Q_{diss} were kept constant while Q_{ph} and the vibrational excitation cross-section Q_v were varied. In the case of Q_v the variation only referred to the cross-section at electron energies above 2 eV so as to leave the result of the analysis for characteristic energies below 1 eV unaffected. Having obtained a good fit for α_i/N for H_2 the vibrational cross-section Q_v was varied for deuterium until a good fit was found with the observed α_i/N††

† ENGELHARDT, A. G. and PHELPS, A. V., *Phys. Rev.* **131** (1963) 2115.
‡ BRODE, R. B., *Rev. mod. Phys.* **5** (1935) 257.
§ TATE, J. T. and SMITH, P. T., *Phys. Rev.* **39** (1932) 270.
‖ RAMIEN, H., *Z. Phys.* **70** (1931) 353.
†† ROSE, D. J., *Phys. Rev.* **104** (1956) 273; FROMMHOLD, L., *Z. Phys.* **160** (1960) 554.

for this gas also. The cross-sections Q_v for the two molecules are shown in Fig. 13.37.

It will be seen from Fig. 13.38 that in this way a very good fit is obtained for both gases. Engelhardt and Phelps also calculated the excitation coefficient α_e/N for H_2 and compared this with the dissociation coefficient α_{diss}/N derived from Poole's observations (§ 1.2.2). This they

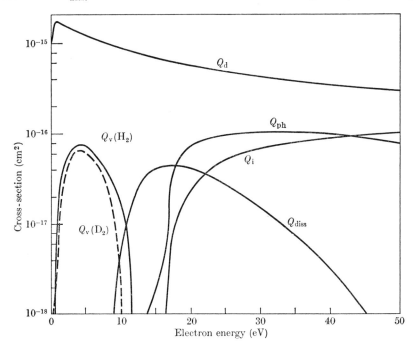

FIG. 13.37. Cross-sections assumed by Engelhardt and Phelps in their analysis of swarm data in H_2 and D_2 for electrons of characteristic energy greater than 1 eV. Q_d momentum-transfer cross-section, Q_i ionization cross-section, Q_{diss} dissociation cross-section, Q_{ph} photon excitation cross-section, Q_v vibrational excitation cross-section (——— H_2, ——— D_2).

justified because the contribution to α_e/N from their photon production cross-section Q_{ph} was only about 10 per cent. However, the observed photon production coefficient α_{ph}/N is about 30 per cent of the observed α_{diss}/N (see § 1.2.2 and Figs. 13.6 and 13.17) so that a good deal still remains to be done, both in the provision of much more experimental data on excitation cross-sections and on further swarm experiments, before a satisfactory consistent analysis is available.

It is of interest to note that the peak of the assumed dissociation cross-section Q_{diss} is a little more than one-half of that assumed by Corrigan,[†]

† loc. cit. p. 885.

while not inconsistent, allowing for the uncertainties, with the dissociation coefficients measured by Poole and others† (§ 1.2.2).

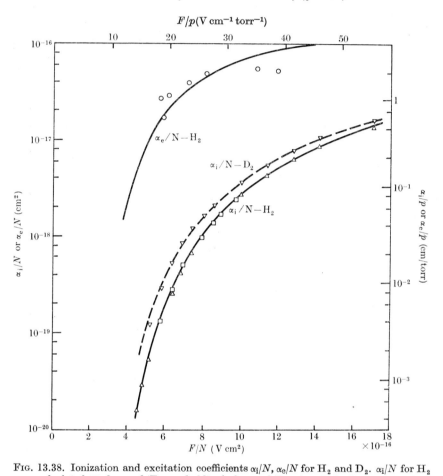

Fig. 13.38. Ionization and excitation coefficients α_i/N, α_e/N for H_2 and D_2. α_i/N for H_2 —— calculated, □ observed (Frommhold). △ observed (Rose). α_i/N for D_2 — — — calculated, ▽ observed (Rose). α_e/N for H_2 —— calculated. α_{diss}/N for H_2 ○ observed (Poole).

References: Rose, D. J., *Phys. Rev.* **104** (1956) 273; Frommhold, L., *Z. Phys.* **160** (1960) 554; Poole, H. G., loc. cit., p. 890.

2. Excitation and dissociation of H_2^+ by electron impact

From the theoretical point of view, the molecular ion H_2^+ is uniquely simple in that it is possible to calculate the allowed electronic energies and wave functions for any fixed nuclear separation to any desired accuracy. On the other hand, the experimental study of collision

† loc. cit., pp. 890, 892.

processes involving the ion is very difficult. Apart from anything else the preparation of H_2^+ ions in a particular vibrational state is not yet possible, and the best that can be done is to work with an ion beam with a known distribution among the various allowed vibrational levels. Even when such a source is available difficulties arise similar to those encountered in the measurement of ionization and excitation cross-sections for impact of electrons with positive atomic ions (see Chap. 3, § 3). Just as these latter difficulties have been largely overcome, measurements are now available of the cross-section for dissociation by electron impact of H_2^+ ions with a fairly well-known initial distribution of vibrational energy. Calculations of the cross-section have been carried out using Born's approximation and quite accurate molecular wave functions.

2.1. *The electronic states and potential energy curves for* H_2^+

The potential curves for various electronic states of H_2^+, calculated accurately by Bates *et al.*† from solution of the two-centre wave equation, are shown in Fig. 13.39.

The ground $1s\sigma_g$ state is attractive and accommodates nineteen vibrational levels. On the other hand, the second, $2p\sigma_u$ state, which also tends in the separated atom limit to a proton and a normal H atom, is repulsive. Excitation of the $2p\sigma_u$ state will therefore lead to dissociation of the molecule into H and H^+ with a relative kinetic energy depending on the initial vibrational level occupied in the $1s\sigma_g$ state.

Of the states that tend in the separated atom limit to an H atom in a 2-quantum state the $2s\sigma_g$ is repulsive. The $3d\sigma_g$ is attractive at quite large nuclear separations that are unlikely to be reached in an impact-excited transition from the ground state unless the initial vibrational quantum number is greater than eleven. In any case, even if reached, a radiative transition to the repulsive $2p\sigma_u$ state will soon occur, leading once more to dissociation. There remains the $2p\pi_u$ state, which is very weakly attractive but only at such large nuclear separations that there is very little chance of excitation from the ground state, even when very highly excited vibrationally. Apart from this remote possibility it can be assumed that all excitation will result in dissociation.

We also include in Fig. 13.39 the curve corresponding to Coulomb repulsion between two protons that will represent the situation when ionizing collisions occur, removing the single molecular electron.

† BATES, D. R., LEDSHAM, K., and STEWART, A. L., *Phil. Trans. R. Soc.* A246 (1953) 215.

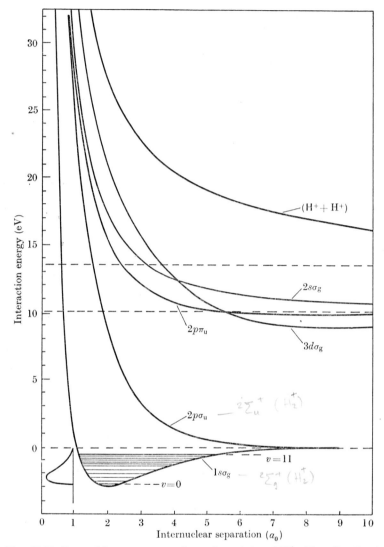

Fig. 13.39. Potential energy curves for various states of H_2^+. The vibrational levels associated with the ground electronic state are shown together with the relative probability of population of the different levels when the H_2^+ is produced by electron impact excitation from the ground state of H_2.

2.2. Calculation of excitation (dissociation) cross-sections

According to the analysis of Chapter 12, § 4.1 (16) the differential cross-section for excitation of the $n'v'$ state from the ground $0v$ state is given by $I_{0v}^{n'v'}\,\mathrm{d}\omega$ where

$$I_{0v}^{n'v'} = \iint \left| \int \chi_{0v}(R)\,\chi_{n'v'}(R)\,G_0^{n'v'}(R,\Theta,\Phi)\,\mathrm{d}R \right|^2 \sin\Theta\,\mathrm{d}\Theta\mathrm{d}\Phi. \quad (22)$$

Here $\chi_{0v}(R)$, $\chi_{n'v'}(R)$ are the respective initial and final vibrational wave functions, and $G_0^{n'v'}(R,\Theta,\Phi)$ is the matrix element according to Born's approximation for the electronic transition at fixed nuclear separation R and orientation angles Θ, Φ of the nuclear axis. Thus if $\mathbf{k}$, $\mathbf{k}'$ are the initial and final wave vectors for the electron motion and $\mathbf{K} = \mathbf{k}-\mathbf{k}'$,

$$G_0^{n'v'}(R,\Theta,\Phi) = (8\pi^2 me^2/h^2)(k'/k)^{\frac{1}{2}}K^{-2}\int \psi_0(\mathbf{r},\mathbf{R})e^{i\mathbf{K}.\mathbf{r}}\psi_n^*(\mathbf{r},\mathbf{R})\,d\mathbf{r}, \quad (23)$$

ψ_0, ψ_n being the initial and final electronic wave functions. It is assumed here that there is an equal *a priori* probability of all initial rotational states. $G_0^{n'v'}$ depends on v' only through the energy ΔE involved in the transition. In many cases this dependence is small so that, if the differential cross-section is summed over all final vibrational states,

$$I_{0v}^{n'} = \sum_{v'} I_{0v}^{n'v'} = \int \chi_{0v}^2(R)|\overline{G_0^{n'}(R)}|^2\,dR, \quad (24\,a)$$

where
$$\overline{|G_0^{n'}(R)|^2} = \int |G_0^{n'}(R,\Theta,\Phi)|^2\sin\Theta\,d\Theta d\Phi. \quad (24\,b)$$

The only difficulty in applying this result is that, if ΔE varies appreciably with v' it is hard to decide what mean value should be used in the calculation of $G_0^{n'}$. An alternative approximate procedure has been used by Peek[†] in which this difficulty does not arise.

In the cases in which we are concerned the final vibrational state is one of a continuum and may be represented to a good approximation by a suitably normalized delta function so

$$\chi_{n'v'}(R) = N_{v'}\,\delta(R-R_c), \quad (25)$$

where R_c is the classical closest distance of approach at the appropriate final energy of relative motion of the separating nuclei. Peek then obtains (23) once more but with ΔE quite definitely defined as the energy separation between the ground vibrational level and that of the final electronic state at separation R.

The first calculations of excitation cross-sections were carried out by Kerner[‡] and by Ivash.[§] Both considered the $1s\sigma$–$2p\sigma$ transition and used approximate electronic wave functions. They were further limited in that Kerner considered an excitation to a specific vibrational and rotational state while Ivash discussed only transitions from the ground vibrational state. Much more elaborate calculations have been carried out by Peek.[||] He used accurate electronic wave functions, as given by

† PEEK, J. M., *Phys. Rev.* **140** (1965) A11.
‡ KERNER, E. H., ibid. **92** (1953) 1441.
§ IVASH, E. V., ibid. **112** (1958) 155.
|| PEEK, J. M., ibid. **134** (1964) A877; **140** (1965) A11.

Bates *et al.*,† considered several electronic transitions, and examined the dependence of the cross-sections on the initial vibrational states.

If $v = 0$ initially and $\overline{|G_0^{n'}(R)|^2}$ varies slowly with R we have from (24 a) that

$$I_{00}^{n'} \simeq \overline{|G_0^{n'}(R_0)|^2}, \qquad (26)$$

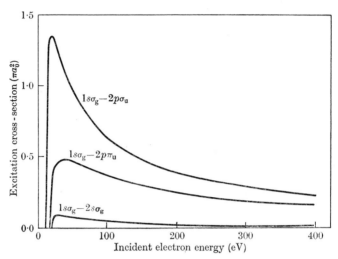

FIG. 13.40. Cross-sections for excitation of electronic transitions to $2p\sigma_u$, $2p\pi_u$, and $2s\sigma_g$ states of H_2^+ from the ground state calculated for the equilibrium nuclear separation $2a_0$.

where R_0 is the equilibrium separation $2\cdot0a_0$ in the ground electronic state. Fig. 13.40 shows the total excitation cross-sections

$$Q_{00}^{n'} = \int I_{00}^{n'} \, d\omega \qquad (27)$$

calculated by Peek‡ from (26) for the three transitions from $1s\sigma_g$ to $2p\sigma_u$, $2p\pi_u$, and $2s\sigma_g$ respectively. It will be seen that, while the cross-section for the $1s\sigma_g$–$2p\sigma_u$ transition is the largest, that for $1s\sigma_g$–$2p\pi_u$, which is also an optically allowed transition, is by no means negligible.

To examine the validity of the approximation (26) and also the importance of the dependence of the cross-section on the initial vibrational states, Fig. 13.41 shows the variation of $Q_{00}^{n'}$ for the $1s\sigma_g$–$2p\sigma_u$ transition with the assumed value of R_0 as calculated by Peek. It will be seen that the dependence on R_0 is quite marked so that (26) cannot be regarded as very reliable. Furthermore, as the classical amplitude of vibration in the state with $v = 3$ lies between $R = 1\cdot4$ and $3\cdot2\,a_0$ we can expect a considerable dependence of $Q_{0v}^{n'}$ on v.

† loc. cit., p. 941.   ‡ loc. cit., p. 943, n. ‖.

Peek therefore carried out detailed calculations for the $1s\sigma_g$–$2p\sigma_u$ transition both for $v = 0$ and $v = 3$, using (22) and (23) with accurate vibrational wave functions given by Cohen, Hiskes, and Riddell.[†] His results are given in Table 13.9 and confirm the expected strong dependence on v.

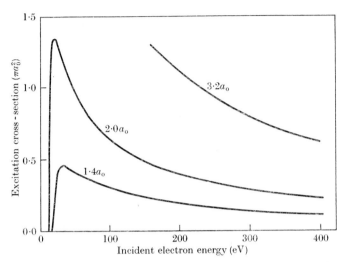

FIG. 13.41. Cross-sections for excitation of the $1s\sigma_g$–$2p\sigma_u$ transition in H_2^+ calculated for three different fixed nuclear separations R as indicated on each curve.

TABLE 13.9

Total cross-section for excitation of the $1s\sigma_g$–$2p\sigma_u$ transition in H_2^+ by electron impact, for two different values of the initial vibrational quantum number v

Incident electron energy (eV)	50	100	200	300	400
Cross-section (πa_0^2), $v = 0$	1·09	0·698	0·431	0·320	0·248
$v = 3$	1·92	1·17	0·691	0·502	0·383

Experimental measurement of total dissociation cross-sections will refer to H_2^+ ions with a wide initial distribution of vibrational states so that it is necessary to carry out the calculations for a wide range of values of v. For this purpose $\overline{|G_0^{n'}(R)|^2}$ must be calculated out to large R.

Remembering that $\overline{|G_0^{n'}(R)|^2}\,d\omega$ is the differential cross-section averaged over all orientations of the molecular axis for the 0–n transition, assuming that the nuclear separation remains fixed at R, we have,

† COHEN, S., HISKES, J. R., and RIDDELL, R. J., *Phys. Rev.* **119** (1960) 1025.

from (44 b) of Chapter 7 that

$$Q_0^{n'}(R) = \int |G_0^{n'}(R)|^2 \, d\omega$$

$$\simeq \frac{4\pi m^2 e^4}{k^2 \hbar^4} \overline{|z_{0n'}|^2} \ln \frac{2mv^2}{E_n - E_0}, \tag{28}$$

where $v = k\hbar/m$ is the electron velocity, $\overline{|z_{0n'}|^2}$ is written for

$$\frac{1}{4\pi K^2} \int \left| \int \psi_0(\mathbf{r}, \mathbf{R}) \mathbf{K} . \mathbf{r} \psi_{n'}(\mathbf{r}, \mathbf{R}) \, d\mathbf{r} \right|^2 \sin\Theta \, d\Theta d\Phi, \tag{29}$$

$\psi_0(\mathbf{r}, \mathbf{R})$, $\psi_{n'}(\mathbf{r}, \mathbf{R})$ are the initial and final electronic wave functions, and $\mathbf{K}$ is the wave-vector change suffered by the electron in the collision. We may therefore write

$$k^2 a_0^2 Q_0^{n'}(R) = \overline{|z_{0n'}(R)|^2} \{ \alpha \ln \Delta E(R) + \beta \}, \tag{30}$$

where $\Delta E(R) = E_{n'}(R) - E_0(R)$ and α and β are constants for given electron velocity v. One advantage of this procedure is that the calculations may be linked up with that of the photo-ionization cross-section which is determined essentially by $\overline{|z_{0n'}(R)|^2}$. Peek used (30) to interpolate between the three values of R for which $Q_0^{n'}(R)$ was already calculated using exact wave functions. For larger values of R he calculated $Q_0^{n'}(R)$, using approximate molecular wave functions. A correction was then applied by multiplying these approximate cross-sections by a factor $g(R)$ where

$$g(R) = \overline{|z_{0n'}(R)|^2}_{\text{ex}} / \overline{|z_{0n'}(R)|^2}_{\text{app}},$$

i.e. in the ratio of the value of $\overline{|z_{0n'}(R)|^2}$, calculated using exact wave functions to that calculated using the approximate functions. Values of $g(R)$ were available from photodissociation calculations (see Chap. 14, § 8.3).

A similar procedure was carried through for the $2p\pi_u$ transitions.

It follows from the form (30) that, for encounters at not too low an energy, the cross-section can be written in terms of the electron velocity V as

$$(\hbar^2 V^2/e^4) Q_{0v}^{n'} = \{ A_v^{n'} \ln(\hbar V/e^2) + B_v^{n'} \} \pi a_0^2, \tag{31}$$

where transitions that are optically forbidden make no contribution to $A_v^{n'}$. Table 13.10 gives $A_v^{n'}$, $B_v^{n'}$ for $v = 0$ to 18 both for excitation of the $2p\sigma$ and $2p\pi_u$ states.

So far we have calculated the contribution from all initial vibrational levels but only for transitions to the $2p\sigma_u$ and $2p\pi_u$ final electronic states. In any experiments, at least initially, the production of protons from all dissociative transitions, including break-up into two protons, will be observed. Furthermore, in the latter case the contribution must be doubled because both protons will be observed. Thus the observed cross-section will be

$$Q_{\mathrm{obs}} = Q_{\mathrm{disc}} + 2Q_{\mathrm{cont}},$$

where Q_{disc} is the total cross-section for excitation of a discrete electronic state and Q_{cont} that for dissociation into 2 protons.

TABLE 13.10

Coefficients $A_v^{n'}$, $B_v^{n'}$ in the formula (31) for the dissociation cross-section of H_2^+. The units are such that $Q_{0v}^{n'}$ is given in πa_0^2

	$2p\sigma_{\mathrm{u}}$		$2p\pi_{\mathrm{u}}$		All states but $2p\sigma_{\mathrm{u}}$	
	A_v	B_v	A_v	B_v	A_v''	B_v''
0	3·10	1·96	2·79	0·007	3·56	1·48
2	3·84	2·40	3·04	0·111	3·85	1·68
4	4·79	5·48	3·27	0·221	4·15	1·91
6	6·00	8·55	3·49	0·334	4·19	2·20
8	7·61	13·2	3·68	0·448	4·86	2·54
10	9·82	20·6	3·82	0·558	5·28	2·94
12	13·1	33·2	3·94	0·666	5·71	3·42
14	18·4	58·2	4·05	0·778	6·18	3·98
16	30·0	12·6	4·17	0·892	6·72	4·59
18	91·8	78·2	4·29	0·969	10·3	2·94

To estimate the importance of Q_{cont} Peek† used the summation formula

$$\sum_n \left| \int \psi_0(\mathbf{r}, \mathbf{R}) e^{i\mathbf{K}\cdot\mathbf{r}} \psi_n^*(\mathbf{r}, \mathbf{R}) \, d\mathbf{r} \right|^2 = 1, \tag{32}$$

so that

$$\sum_n' \left| \int \psi_0(\mathbf{r}, \mathbf{R}) e^{i\mathbf{K}\cdot\mathbf{r}} \psi_n(\mathbf{r}, \mathbf{R}) \, d\mathbf{r} \right|^2 = 1 - \left| \int |\psi_0(\mathbf{r}, \mathbf{R})|^2 e^{i\mathbf{K}\cdot\mathbf{r}} \, d\mathbf{r} \right|^2, \tag{33}$$

where $\sum_n'$ is a sum over all excited states n.

If we choose a suitable mean energy loss for all states that contribute appreciably to $\sum_n'$ so that a mean $\bar{K}$ can be taken for all, then

$$\sum' \overline{G_0^n(R)}^2 = (64\pi^4 m^2 e^4 / h^4)(k'/k)\bar{K}^{-4}\left\{1 - \left| \int |\psi_0(\mathbf{r}, \mathbf{R})|^2 e^{i\mathbf{K}\cdot\mathbf{r}} \, d\mathbf{r} \right|^2\right\}. \tag{34}$$

This will be a good approximation for high energy impacts for which the most probable energy loss is small compared with the incident energy. Hence, from (34) the sum

$$Q_{\mathrm{disc}} + Q_{\mathrm{cont}}$$

may be calculated with an uncertainty determined by that involved in choosing $\bar{K}$. This was taken to correspond to an excitation energy equal

† PEEK, J. M., *Phys. Rev.* **154** (1967) 52.

to the energy difference between that of the ground $1s\sigma_g$ level and that of two protons.

Writing

$$Q'' = Q_{\text{disc}} + Q_{\text{cont}} - Q(2p\sigma) = \{A'' \ln(\hbar V/e^2) + B''\}\pi a_0^2 e^4/V^2\hbar^2,$$

the values of A'' and B'' may now be obtained and are given in Table 13.10 for each value of the initial vibrational quantum number v. The contribution from Q_{cont} must be less than

$$Q'' - Q(2p\pi_u).$$

It is already clear from comparison of the corresponding values in the last two pairs of columns in Table 13.10 that this is quite small except for large values of v, which, as we shall see in Table 13.11, are relatively unimportant. Application of the theory to calculate cross-sections for comparison with experimental observations will be further discussed in § 2.4.

2.3. *The measurement of dissociation cross-sections*

Measurements of cross-sections for dissociation of H_2^+ ions by electron impact have been made by Dunn and van Zyl[†] and by Dance, Harrison, Rundel, and Smith (A. C. H.)[‡] As pointed out earlier, the H_2^+ ions are produced by electron impact ionization of H_2 and will have a distribution of vibrational energy which is determined by this process of formation.

If the target H_2 molecules are in their ground states the probability of producing an H_2^+ ion in the $^2\Sigma_g^+$ state with a vibrational level of quantum number v' is proportional to

$$N(p_{00}^{n'v'})^2,$$

where

$$p_{00}^{n'v'} = \int \chi_{n'v'}(R)\chi_{00}(R)\,dR.$$

χ_{00} is the ground vibrational wave function for the $^1\Sigma_g^+$ state of H_2 and $\chi_{n'v'}$ that for the $^2\Sigma_g^+$ state of H_2^+ of quantum number v'. N is the factor arising from the electronic wave functions. If the incident electron energy is large compared with the range ΔE of energies required to produce the transition within the Franck–Condon region N is independent of v'. The initial vibrational distribution in the H_2^+ ions is then determined by the factor $(p_{00}^{n'v'})^2$, values of which are given in Table 13.5 In Fig. 13.39 the shape of this distribution is shown and it will be seen that the most probable value of v' is 2.

For the interpretation of measured cross-sections the initial vibrational

† Dunn, G. H. and van Zyl, B., *Phys. Rev.* **154** (1967) 40.
‡ Dance, D. F., Harrison, M. F. A., Rundel, R. D., and Smith, A. C. H., *Proc. Phys. Soc.* **92** (1967) 577.

distribution must be maintained throughout the region in which the
ion beam and the crossed electron beam overlap. Fortunately the lifetime of the different vibrational states towards radiative decay is very
long, the H_2^+ molecule being homonuclear. To avoid disturbance of the
distribution through collision it is necessary to work with low pressures
and high extraction fields in the ion source.

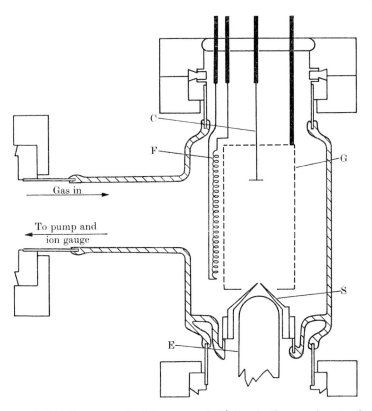

Fig. 13.42. Arrangement of the source of H_2^+ ions in the experiments of
Dunn and van Zyl.

To check that the initial vibrational distribution is as assumed and
that it is preserved throughout the collision region, the observed cross-
section should be independent of source pressure and the voltage distribution in the ion source, over a considerable range of the variables about
the values used to obtain the experimental results. Working at ordinary
temperatures the target H_2 molecules will be almost all in the ground
vibrational state.

Fig. 13.42 illustrates the arrangement of the ion source used by Dunn and van Zyl.† In typical operation if V_F, V_C, V_G, and V_S refer to the voltages applied to the hot filament F, the ion pusher electrode C, the grid G, and the ion exit slit S, $V_C = V_G = V_F+200$ V $= V_S+400$ V. Retarding potential analysis of the energies of the ions produced showed that they were formed in regions in which the electrons possessed energies between 50 and 150 eV. No observable change in the measured cross-sections was found when the potentials were changed so

$$V_C = V_G = V_F+400 \text{ V} = V_S+100 \text{ V},$$

in which case the ions were mainly produced by electrons with energies between 100 and 400 eV. The source pressure was about 10^{-3} torr and used electron currents of about 60 mA.

The ions passing out through the extractor E were formed into a parallel beam by a two-cylinder lens, made into a ribbon beam by passage through a quadrupole lens pair, and mass-analysed in a 60° sector magnetic field. After separation the H_2^+ beam was made parallel once more by a second quadrupole lens pair. It is important before the beam enters the collision region that it should be 'purified' of any H^+ ions produced by dissociation of H_2^+ in gas or surface collisions. To do this the beam, after entering the first chamber, passed through a three-element rectangular cylinder lens to which suitable voltages were applied.

The normal energy of the ion beam was about 10 keV, of strength 1 to 2 μA and of cross-sectional dimensions about 0·15 by 0·65 cm.

Fig. 13.43 shows the general arrangement of the experiments carried out by Dunn and van Zyl to measure the cross-section for dissociation of the H_2^+ ions by electrons of different incident energy. The ion beam was crossed at right angles by an electron beam of about 1 mA and cross-sectional dimensions about 2 cm². Protons produced by dissociative collisions possess about one-half the energy of the primary beam and were separated from the unaffected H_2^+ ions by a 45° parallel-plate electrostatic analyser similar to that described by Yarnold and Bolton‡ and used by Lineberger, Hooper, and McDaniel§ in their experiments on the further ionization of positive atomic ions. In this analyser, fringe field effects are minimized by guard plates in which baffles are introduced along equipotentials, as shown, to prevent stray charged particles from entering the proton collector.

† loc. cit., p. 948

‡ YARNOLD, G. D. and BOLTON, H. C., *J. scient. Instrum.* **26** (1949) 38.

§ LINEBERGER, W. C., HOOPER, J. W., and McDANIEL, E. W., *Phys. Rev.* **141** (1966) 151.

It is necessary, in choosing the apertures in the analyser to allow for the fact that the protons produced will have an energy spread of about 10 eV. Dunn and van Zyl confirmed that they were collecting all the protons by enlarging the apertures by 50 per cent above those used in the main experiments. This had no observable effect on the measured cross-sections.

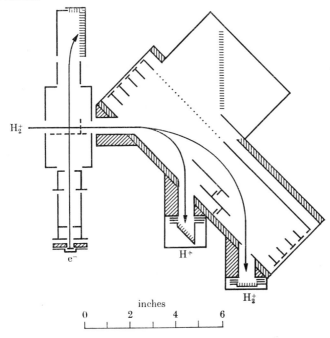

FIG. 13.43. General arrangement of the experiments of Dunn and van Zyl on electron impact dissociation of H_2^+.

The ion collectors were Faraday cups with collecting surfaces covered with metal honeycomb and which included electrodes to suppress secondary electron emission. It was verified that the H_2^+ current measured in the normal collector for these ions remained the same when deflected into the proton collector so that reflection effects were probably unimportant.

As discussed in Chapter 3, § 3.1.1 the dissociation cross-section Q_d is given by

$$Q_d = \frac{hei(H^+)}{i(H_2^+)i(e)} \frac{vV}{(V^2+v^2)^{\frac{1}{2}}} F,\tag{35}$$

where $i(H_2^+)$, $i(e)$ are the total current strengths in the H_2^+ and electron beams respectively and $i(H^+)$ is the collected current of protons. V and v are the respective velocities of the H_2^+ ions and electrons and F is a factor determined by the degree of overlap of the two beams. It is given by

$$F = \frac{\int j_+(z)\,dz \int j_e(z)\,dz}{h \int j_+(z)j_e(z)\,dz}. \tag{36}$$

Here z is measured in the direction mutually perpendicular to both beams and $j_+(z)\,dz$, $j_e(z)\,dz$ are the respective H_2^+ ion and electron currents passing between z and $z+dz$ throughout the region of interaction. h is the height of the ion beam. F was measured as in experiments on further impact ionization of atomic ions (see Chap. 3, § 3.1.1) by using an L-shaped probe.

The wanted signal of protons was about 10^{-15} A as compared with a background signal about 10^3 times larger due to protons produced by collisions with surfaces and with residual gas. By chopping the electron beam at low audio-frequencies and using phase-sensitive detection in the usual way the wanted signal could be distinguished and measured. As pointed out in Chapter 3, § 3.1.1, this method is less effective than with crossed-beam experiments involving one neutral beam as the background signal is likely to exhibit the same modulation as the wanted signal. Such modulation may arise from space-charge effects due to interaction between the two charged beams. This may change the geometry of the ion beam in such a way as to modify the effectiveness of collection of the ions. Any effect of this sort should depend on the energy of the ions and one check on its unimportance was made by showing the measured cross-section to be independent of ion energy over the working range. This was further confirmed by more detailed study of the collection conditions.

A further effect that could give rise to a modulated background signal is the release of adsorbed gas by the electron collector when bombarded by electrons. Before commencing an experiment the collector was heated and simultaneously bombarded by electrons under high vacuum conditions for such a long time that all adsorbed gas was removed. During an experiment gas would then only collect when the electron beam was interrupted and the background modulation due to this effect should increase with the gas pressure. Fig. 13.44 shows a typical observed variation of detected signal with residual gas pressure in the collision chamber. As the working pressure was between 7 and 20×10^{-10} torr the correction is small.

In practice the limitation in the experiments came from noise in the electronics which, for one second integration times, was a little greater than the wanted signal and several times the background proton signal. Long integration times were therefore necessary.

The usual tests were made to check that the cross-section given by (35) was independent of $i(H_2^+)i(e)$, V, v, and F. Fig. 13.45 shows the observed cross-section as a function of electron energy.

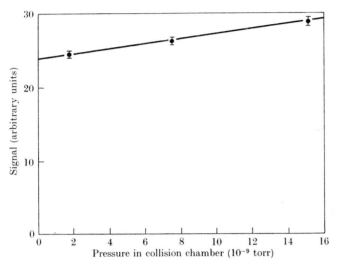

FIG. 13.44. Observed dependence of detected signal on residual gas pressure in the experiments of Dunn and van Zyl.

Dance et al.† used very much the same experimental arrangement as that employed by Dolder, Harrison, and Thonemann‡ for the measurement of the cross-section for ionization of He^+, Ne^+, and N^+ ions by electron impact and which is described in Chapter 3, § 3.1.1. Although they worked with an ion source at a considerably greater pressure than did Dunn and van Zyl (8×10^{-3} torr as compared with 4×10^{-4} torr) they found no dependence of the observed dissociation cross-section on the pressure over a range from 3×10^{-3} to 0·13 torr. This seems to indicate a low rate of vibrational deactivation of H_2^+ ions by collision with normal molecules. It was also verified that the cross-section was independent of the electron accelerating potential in the ion source from 50 to 200 V, the working value being 100 V.

Background effects were reduced to inappreciable proportions by the same techniques as in the experiments of Dolder et al. referred to above.

† loc. cit. p. 948.
‡ DOLDER, K. T., HARRISON, M. F. A., and THONEMANN, P. C., *Proc. R. Soc.* A264 (1961) 367.

As in the experiments of Dunn and van Zyl care was taken to check that the observed cross-section was independent of the ion beam current over the working range (10–20 keV). Background due to pressure changes arising from outgassing of the electron gun and collector were eliminated by the coincidence–anticoincidence procedure described in Chapter 3, § 3.1.1.

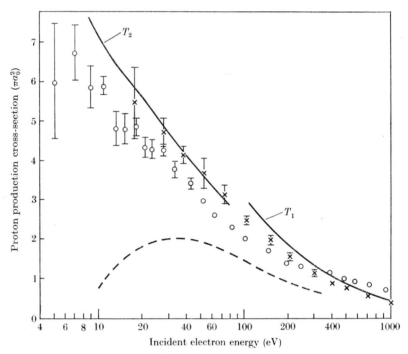

FIG. 13.45. Cross-sections for production of protons by electron impact dissociation of H_2^+ ions. ✖ observed by Dunn and van Zyl. ⊕ observed by Dance, Harrison, Rundel, and Smith. —— T_1 calculated using (38). —— T_2 calculated using (39). — — — calculated assuming all the H_2^+ ions to be in their ground vibrational state.

The H_2^+ beam current (1 μA) and the electron beam current (1 mA) were much the same as in the experiments of Dunn and van Zyl, as was also the background pressure (2×10^{-9} torr).

Fig. 13.45 illustrates the results obtained. The agreement between the two sets of observations is not unsatisfactory considering the difficulty of the experiments. Dance *et al.* seem to have obtained more accurate data at low electron energies and there is some evidence from their observations of the onset of a new process of dissociation setting in at about 13 eV, which might be interpreted as arising from excitation of the $2p\pi_u$ state (see Fig. 13.39).

2.4. *Comparison with theory*

To compare with theoretical predictions the cross-sections calculated by Peek (see § 2.2) for different initial vibrational levels of the H_2^+ molecule must be averaged over an initial distribution corresponding to that of the H_2^+ ions produced by ionization of H_2 molecules in their ground vibrational level. When this is done, using the weighting factor $(p_{00}^{n'v'})^2$ as given in Table 13.5 (see the form of the distribution shown in Fig. 13.39) and writing the weighted mean cross-section for a particular electron velocity V in the form

$$(V^2\hbar^2/e^4)\tilde{Q} = \tilde{A}\ln(V\hbar/e^2)+\tilde{B} \tag{37}$$

we obtain the values of $\tilde{A}$ and $\tilde{B}$ given in Table 13.11.

TABLE 13.11

Constants $\tilde{A}$, $\tilde{B}$ determining the weighted mean cross-sections for excitation of dissociating states of H_2^+, in terms of (37)

Transition	$\tilde{A}$	$\tilde{B}$
$1s\sigma_g$–$2p\sigma_u$	4·97	6·79
$1s\sigma_g$–$2p\pi_u$	3·20	0·20
All transitions except $1s\sigma_g$–$2p\sigma_u$	4·11	1·90

For comparison with the observed results we need not just the sum Q_{diss}^t of the cross-sections that lead to dissociation but $Q_{diss}^t+Q_c$ where Q_c is the cross-section for break-up into two protons. This is because this contribution from the process must be counted twice as both protons are observed.

It will be seen from Table 13.11 that the contribution to $\tilde{A}$ from dissociation into two protons must be less than 0·91. This process cannot therefore contribute more than about 13 per cent to the total cross-section for dissociation, Q_{diss}^t.

A semi-classical calculation of Q_c by Alsmiller† indicates that it is about 10 per cent of the total, which indicates that most of the contribution to Q_{diss}, apart from that arising from the excitation of the $2p\sigma_u$ and $2p\pi_u$ states, is due to break-up into two protons.

For comparison with experiments at high electron energies the summation method of Peek may be used to calculate Q_{diss}^t and if we assume

$$Q_{diss}^t \simeq Q(2p\sigma_u)+Q(2p\pi_u)+Q_c,$$

then

$$Q_{diss}^t+Q_c \simeq 2Q_{diss}^t-Q(2p\sigma_u)-Q(2p\pi_u). \tag{38}$$

† ALSMILLER, R. G., *Oak Ridge Lab. Rep. No. ORNL–3232.*

The results given by (38) are shown as T_1 in Fig. 13.45 for electron energies greater than 122 eV.

For low energies the summation method is not accurate so, for this range, Dunn and van Zyl[†] take

$$Q_{\text{diss}}^{\text{t}} + Q_{\text{c}} \simeq Q(2p\sigma_{\text{u}}) + Q(2p\pi_{\text{u}}) + Q(2s\sigma_{\text{g}}) + 2Q_{\text{c}}', \qquad (39)$$

where Q_{c}' is as calculated by Alsmiller[‡] and the remaining cross-sections by Peek.[§] Results obtained in this way are shown as T_2 in Fig. 13.45 for electron energies from 10 to 80 eV.

Allowing for the uncertainties, both in the experiment and theory, the agreement is good even down to quite low electron energies. This may partly be due to cancellation of two opposing factors. In the theory two approximations are really involved. The correct first-order approximation is really the Coulomb–Born in which the wave functions of the colliding electron are not plane waves but waves distorted by the Coulomb field of a positive charge (see Chap. 7, § 5.6.5). Neglect of this distortion reduces the cross-section at low energies. On the other hand, the first-order Coulomb–Born approximation tends to over-estimate the cross-section at these energies (Chap. 7, § 5.6.5).

It is of interest to note the importance of the initial distribution of vibrational levels in determining the effective cross-section. The cross-section shown in Fig. 13.45, which is calculated on the assumption that the H_2^+ ions are all in their ground vibrational level, bears little relation to the observed results.

3. Nitrogen

Considerable interest attaches to the study of the excitation and ionization effects produced in nitrogen by electron impact. This is partly due to its interest for the interpretation of optical and ionization phenomena in the upper atmosphere—thus bands of N_2 and N_2^+ are prominent features of the auroral spectrum and certain bands of N_2 are also observed in the spectrum of the airglow. Apart from these applications two remarkable afterglow effects occur in nitrogen—the Lewis–Rayleigh afterglow, often referred to as active nitrogen, and the auroral or Kaplan afterglow. The interpretation of these phenomena certainly requires, among other things, a detailed knowledge of electron impact effects in nitrogen gas.

Unlike most gases, no stable negative ions are formed by electron impact in nitrogen—neither N_2^- nor N^- ions are energetically stable.

[†] loc. cit., p. 948.   [‡] loc. cit., p. 955.
[§] PEEK, J. M., *Phys. Rev.* **154** (1967) 52 and **134** (1964) A877.

We shall therefore be concerned only with phenomena that result in excitation (including ionization), but in some cases the formation of intermediate, ephemeral, states of N_2^- is important.

3.1. *The electronic states of N_2 and N_2^+*

Fig. 13.46 shows the energy relations at different nuclear separations of some of the electronic states of N_2 and N_2^+ that are of most importance.

The ground state of N_2 is a $^1\Sigma_g^+$ state further distinguished by the presymbol X. Transitions to this state from the $b^1\Pi_u$ state, of which the lowest vibrational level occurs at 12·8 eV above the ground state, produce the intense Birge–Hopfield bands in the ultra-violet. A Π_g state ($a^1\Pi_g$) lies below the $b^1\Pi_u$ state, but only a weak band system, the Lyman–Birge–Hopfield bands, arises from transitions from this state to the ground state. These bands are weak because a g–g transition violates the dipole selection rules (see Chap. 12, § 3.1) and can only occur through electric quadrupole radiation. The lowest vibrational band of the $a^1\Pi_g$ state lies 9·1 eV above the ground state.

The lowest triplet state $A^3\Sigma_u^+$ lies 6·2 eV above the ground state. Transitions to the ground state are very weak as they are of inter-combination type but they are observed under low pressure conditions such as in the upper atmosphere. They give rise to the Vegard–Kaplan bands.

Two strong band systems arise from transitions between triplet states. They are known as positive bands because they show up prominently in the spectrum of the light emitted from the positive column of a glow discharge in nitrogen. The first positive bands arise from transitions between the $B^3\Pi_g$ state and the $A^3\Sigma_u^+$ that lies 1·15 eV below, while the second positive arises from transitions to the $B^3\Pi_g$ from the $C^3\Pi_u$ that lies 3·7 eV above.

The first negative bands of nitrogen, so called because they are strongly developed in the spectrum of the negative glow of a nitrogen discharge, arise actually from transitions between the $B^2\Sigma_u^+$ state of N_2^+ and the ground $X^2\Sigma_g^+$ state of that ion. To produce the $^2\Sigma_u^+$ state of N_2^+ from the ground $X^1\Sigma_g^+$ state of N_2 requires 18·75 eV, while 15·6 eV is required to produce the $X^2\Sigma_g^+$. An intermediate state of N_2^+, the $A^2\Pi_u$, is the upper state for the Meinel bands that were first discovered in the auroral spectrum though since observed in the laboratory.

In addition to the $A^3\Sigma_u^+$ and $a^1\Pi_g$ states there are a number of other metastable states including $a'^1\Sigma_u^-$ and $B'^3\Sigma_u^-$, as well as others of high multiplicity which are not normally reached in impact phenomena.

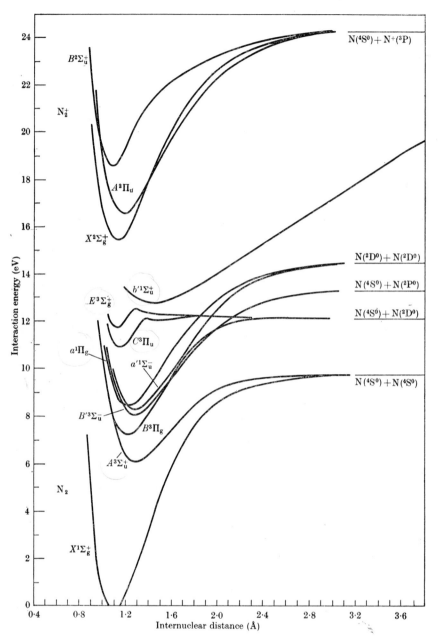

FIG. 13.46. Potential energy curves for some electronic states of N_2 and N_2^+.

Fig. 13.47 (a) shows an excitation spectrum of N_2 obtained by Schulz[†] using the trapped-electron method (Chap. 5, § 2.3) with a well depth of 0·2 V. The peak due to the excitation of the $A^3\Sigma_u^+$ triplet state is strong whereas, while there is clear evidence of excitation of the $b^1\Pi_u$ and $a^1\Pi_g$ states, the peaks are much weaker and less clear. The sharp peak at 11·5 eV (with an onset at 11·2 eV) was identified by Schulz as due to excitation of $C^3\Pi_u$. When the experiment was repeated with the well depth increased to 0·8 V this peak was no longer present, as may be seen by reference to Fig. 13.47 (b).

The energy loss spectra obtained at incident electron energies of order 500 eV show quite a different distribution of relative intensities. Thus in Fig. 13.48 we see a typical spectrum, obtained by Lassettre and Krasnow[‡] for 511-eV electrons scattered at 9·6° in N_2, using the method described in Chapter 5, § 4. There is no longer any trace of excitation of the triplet states but strong excitation, particularly of the $b^1\Pi_u$, and also, though less strongly, of the $a^1\Pi_g$, state. Ionization occurs at 15·9 eV and the discrete loss at 14·01 eV is probably due to excitation of the $^1\Sigma_u^+$ states of N_2, which are upper states for the Worley bands.

Fig. 13.49 shows energy loss spectra between 8·5 and 14 eV at quite low incident energies, 15·7 and 35 eV respectively. These were obtained by Heideman, Kuyatt, and Chamberlain,[§] using the high energy resolution spectrometer developed by Simpson[‖] and described in Chapter 1, § 5.1. The results of their experiments in N_2, as far as elastic scattering and vibrational excitation are concerned, have been described in Chapter 10, § 3.5.2 and Chapter 11, § 3 respectively.

At both incident energies the vibrational structure associated with excitation of the $a^1\Pi_g$ state is well resolved. There is evidence of vibrational structure associated with the $C^3\Pi_u$ excitation at 15·7 eV but not at 35 eV, whereas the $b^1\Pi_u$ excitation behaves in exactly the opposite way. In addition there is a well-defined peak at 13·26 eV that is not yet clearly identified.

Measurements of the excitation functions for production of metastable states have been made by Lichten,[††] and Olmsted, Newton, and Street.[‡‡] In all cases the metastable atoms were produced by cross

† Schulz, G. J., *Phys. Rev.* **116** (1959) 1141.
‡ Lassettre, E. N. and Krasnow, M. E., *J. chem. Phys.* **40** (1964) 1248.
§ Heideman, H. G. M., Kuyatt, C. E., and Chamberlain, G. E., ibid. **44** (1966) 355.
‖ Simpson, J. A., *Rev. scient. Instrum.* **35** (1964) 1698.
†† Lichten, W., *J. chem. Phys.* **26** (1957) 306.
‡‡ Olmsted, J., Newton, A. S., and Street, K., ibid. **42** (1965) 2321.

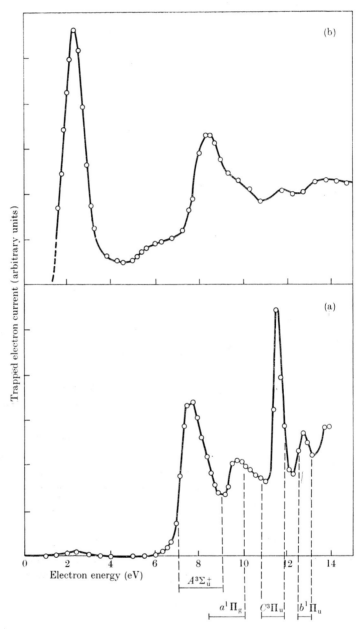

Fig. 13.47. Excitation spectrum of nitrogen obtained by Schulz using the trapped-electron method. (a) with well depth of 0·2 V. (b) with well depth of 0·8 V.

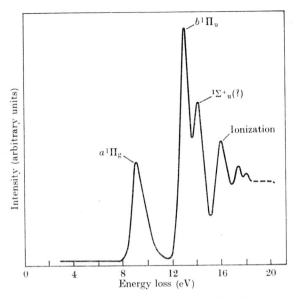

Fig. 13.48. Energy loss spectra of electrons of incident energy 511 eV scattered through 9·6° in N_2 as observed by Lassettre and Krasnow.

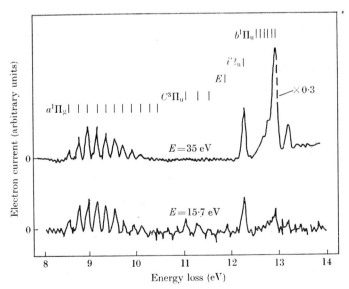

Fig. 13.49. Energy loss spectra (between 8·5 and 14 eV) for electrons of incident energy 15·7 and 35 eV respectively, observed by Heideman, Kuyatt, and Chamberlain.

bombardment of a molecular beam by an electron beam, and detected by electron emission from a metal surface. Fig. 13.50 shows the results obtained by Olmsted *et al.* using the technique described in Chapter 4, § 2. As they operated the electron gun by the retarding potential difference method their energy resolution was high. The striking feature of their results is the presence of a sharp peak with an onset energy of 11·8 eV.

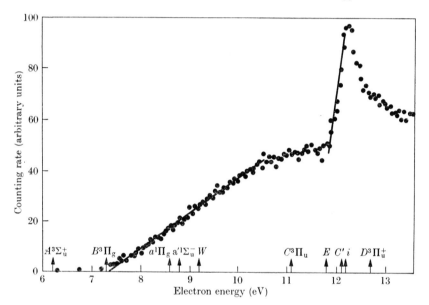

FIG. 13.50. Excitation function for metastable states of N_2 observed by Olmsted, Newton, and Street.

Heideman *et al.*,[†] in their high resolution measurement of the transmission of electrons through nitrogen (see Chap. 10, § 3.5.2 and Fig. 10.35), observed a peak at 11·87 eV which they noted corresponded closely with the threshold for excitation of the $E^3\Sigma_g^+$ state as determined spectroscopically. To examine this further they observed the threshold behaviour of the cross-sections at zero scattering angle for the inelastic process with energy loss 11·87 eV, by sweeping the incident electron energy between 11·6 and 12·8 eV, keeping the energy loss fixed. They obtained the extraordinarily sharp peak shown in Fig. 13.51.

Ehrhardt and Willmann,[‡] using the apparatus already described in Chapter 10, § 3.5.2 observed three energy losses at 11·03, 11·87, and 12·23 eV. The intensities of the transitions concerned, at a scattering angle of 20°, varied with electron energy as shown in Fig. 13.52. No

† loc. cit., p. 959.   ‡ loc. cit., p. 716.

great variation in the shapes of these excitation functions was noted as the angle of scattering was varied. At small angles the 11·87 eV loss was the strongest. This seems certainly to be due to excitation of the $E^3\Sigma_g^+$ state and confirms the sharpness of the excitation function found by Heideman *et al.* at zero scattering angle. The 11·03 eV loss must be ascribed to excitation of $C^3\Pi_u$. In connection with the third loss

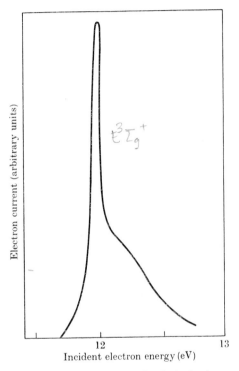

FIG. 13.51. Excitation function for the inelastic process involving an energy loss of 11·87 eV in N_2 as observed by Heideman, Kuyatt, and Chamberlain.

process, at 12·23 eV, it is noteworthy that Čermak, in studying the ionization of various molecules by impact with metastable N_2 molecules, found evidence for a state of N_2 for which the electron excitation function had a maximum at 12·6 eV. This agrees well with the location of the maximum for the 12·23-eV loss shown in Fig. 13.52.

It may well be that the peak identified as due to $C^3\Pi_u$ excitation in the excitation spectrum shown in Fig. 13.47 (a) is really due to $E^3\Sigma_g^+$. This would be consistent with its disappearance when the well depth was increased to 0·8 V, because at this energy above the threshold the

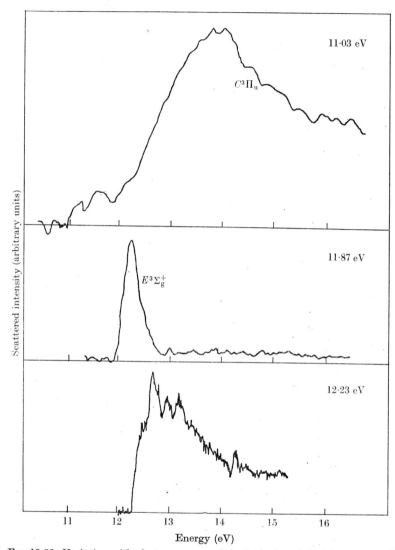

FIG. 13.52. Variation with electron energy of the intensity of electrons scattered through 20° in N_2 after suffering energy losses of 11·03, 11·87, and 12·23 eV respectively, as observed by Ehrhardt and Willmann.

excitation cross-section according to Fig. 13.51 would have fallen by a factor of nearly 10 from its peak value! In the same way it is possible to understand why, in the energy loss spectrum shown in Fig. 13.49 for 15·7-eV incident electrons, excitation of the $C^3\Pi_u$ but not the $E^3\Sigma_g^+$ state is seen. The optical excitation function for the second positive bands, for which the upper state is $C^3\Pi_u$, has been observed by Stewart and

Gabathuler.[†] It is discussed in § 3.2. Meanwhile it can be said that it is not inconsistent with the above interpretation.

There remains the need to explain how the $E^3\Sigma_g^+$ state, which can make an optically allowed transition to the $A^3\Sigma_u^+$ state, has a long enough lifetime to be observed in experiments such as those of Olmsted *et al.* It has, however, been pointed out by Mulliken that, although optically allowed, the transition to the $A^3\Sigma_u^+$ state is a two-electron one. This may reduce the probability enough for $E^3\Sigma_g^+$ excited molecules to live sufficiently long to be recorded.

The reason for the sharp peak close to the threshold in the excitation function of $E^3\Sigma_g^+$ and the absence of vibrational structure is not understood at the time of writing.

It is of interest to note that there is no experimental evidence of discrete energy losses of appreciable probability that are not associated with known stable electronic states. Excitation of unstable states leading to dissociation into neutral atoms is therefore very improbable. Also it does not seem likely from the vibrational development of the Lyman–Birge–Hopfield, Birge–Hopfield, Worley, and Vegard–Kaplan bands that vertical transitions from the ground state will ever reach repulsive portions of the potential energy curves of the upper electronic states concerned. Experiments similar to those of Hughes and Skellet described in § 1.2.1 for hydrogen showed no decrease in pressure in nitrogen until the electron energy exceeded $17 \cdot 8$ eV. As the vertical ionization potential of N_2 is $15 \cdot 7$ V and the onset potential for the reaction

$$N_2 + e \rightarrow N + N^+ + e$$

is $24 \cdot 3$ V (see § 3.5.1) it is likely that the observed effects were associated in some way with chemical activity of N_2^+.

3.2. *Excitation functions for bound electronic states of* N_2

Differential cross-sections and generalized oscillator strengths have been measured for excitation of the $b^1\Pi_u$ and $a^1\Pi_g$ states of N_2 by Lassettre and Krasnow.[‡] The method used was as described in Chapter 5, § 5. For the excitation of a diatomic molecule the peak width, provided the instrumental resolution is high enough, will be determined by the probability distribution of excitation of the vibrational levels of the upper state. As shown in Chapter 12, § 3.2 this is independent of incident electron energy, being determined by the overlap of the vibrational wave function for the upper state with that of the ground state. Lassettre and

† STEWART, D. T. and GABATHULER, E., *Proc. phys. Soc.* **72** (1958) 287.
‡ loc. cit., p. 959.

Krasnow observed that for N_2 the peak widths were roughly twice as large as for an atomic gas such as helium and they verified that the peak shape, as defined by the ratio of peak width to peak height, was independent of electron energy.

Fig. 13.53 illustrates the generalized oscillator strengths $f(K)$ observed for the two transitions as functions of K^2, where K is the magnitude of the change in the wave vector of the incident electron, in atomic units. These have been integrated over the peak and do not refer just to the differential oscillator strength at the peak. Whereas $f(K)$ tends to a finite limit of 0·837 as $K \to 0$ for the $b^1\Pi_u$ state, giving the optical oscillator strength for the $b^1\Pi_u - X^1\Sigma_g$ transition, that for the $a^1\Pi_g$ definitely tends to zero in this limit. This is to be expected because the $a^1\Pi_g - X^1\Sigma_g$ transition does not obey the dipole selection rules.

Also included in Fig. 13.53 are the oscillator strengths observed by Silverman and Lassettre[†] for the transition that occurs at 13·99 eV. For this f tends to a finite value of 0·602 as $K \to 0$, showing that the transition concerned obeys the dipole selection rules. This is consistent with the identification of the peaks concerned with the excitation of the upper $^1\Sigma_u^+$ states of the Worley bands.

Fig. 13.54 illustrates the total cross-sections for excitation of these states derived from the generalized oscillator strengths as in Chapter 7, § 5.6.7.

The optical excitation function for a number of lines in the second positive bands have been measured by Stewart and Gabathuler[‡] using a technique similar to those described in Chapter 4, § 1 for the excitation of atoms. Photoelectric detection was employed. Two methods for reducing background effects were used. In one the light beam was chopped at 200 c/s, so making phase-sensitive detection possible. The other, more sensitive method, depended on direct counting of individual pulses due to electron emission from the photocathodes, with discrimination against smaller pulses due to thermionic emission from the dynodes.

Table 13.12 gives the cross-sections observed for excitation of different vibrational transitions within the band when excited by electrons of energy 35 eV. The intensity of emission of the (v', v'') band will be proportional to

$$ q_{Cv'}^{Bv''} \sum_v g_v p_{Av}^{Cv'}, \tag{40} $$

where $q_{Cv'}^{Bv''}$ is the optical transition probability from the Cv' to Bv'' states. $p_{Av}^{Cv'}$ is the relative probability of population of the Cv' upper level by

† Silverman, S. M. and Lassettre, E. N., *J. chem. Phys.* **42** (1965) 3420.
‡ Stewart, D. T. and Gabathuler, E., *Proc. phys. Soc.* **72** (1958) 287.

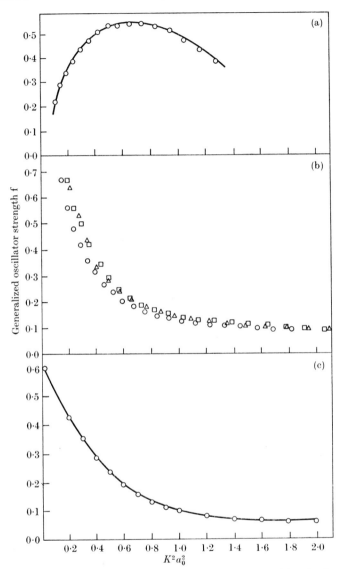

FIG. 13.53. Generalized oscillator strengths for electronic transitions in nitrogen. (a) $X^1\Sigma_g^+ - a^1\Pi_g$ (9·10 eV) derived from observations of Lassettre and Krasnow. (b) $X^1\Sigma_g^+ - b^1\Pi_u$ (12·85 eV) $\bigcirc, \square, \triangle$ derived from observations of Lassettre and Krasnow with electrons of incident energies 417·2, 510·78, and 602·6 eV respectively. (c) $X^1\Sigma_g^+ - {}^1\Sigma_u^+$ (?) (13·99 eV)—derived from observations by Lassettre and Silverman.

impact excitation from the vibrational level v of the ground electronic state. g_v is the fraction of molecules occupying the initial Av level under experimental conditions.

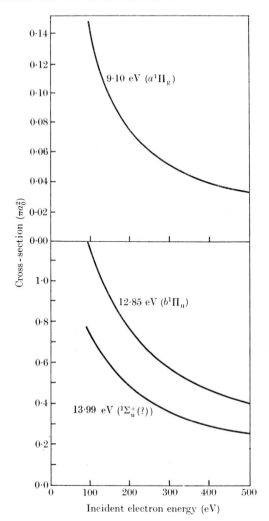

Fig. 13.54. Cross-sections for excitation of the $a^1\Pi_g$, $b^1\Pi_u$, and $^1\Sigma_u^+(?)$ states of N_2 by electron impact calculated using the generalized oscillator strengths illustrated in Fig. 13.53.

If χ_{Av}, $\chi_{Cv'}$, $\chi_{Bv''}$ are the vibrational wave functions for the Av, Cv', Bv'' states respectively then the optical transition probability is proportional to

$$v_{v'v''}^4 \left| \int \chi_{Cv'}(R)\chi_{Bv''}(R) \, dR \right|^2, \qquad (41)$$

where $v_{v'v''}$ is the frequency of the transition concerned. Also, according to the Franck–Condon principle (see Chapter 12, § 3.2),

$$p_{Av}^{Cv'} \propto \left| \int \chi_{Av}(R)\chi_{Cv'}(R) \, dR \right|^2. \qquad (42)$$

Since $\nu_{v'v''}$ varies very little with v' and v'' the intensity is proportional to the product of squares of the Av, Cv' and Cv', Bv'' vibrational overlap integrals. These have been calculated by Bates,[†] Fraser and Jarmain,[‡] and Pillow[§] by using a Morse representation of the potential energy curves of the electronic states concerned. In Table 13.12 the calculated intensity ratios are compared with the observed, the calculated values

TABLE 13.12

Cross-sections $Q_{v'v''}$ for excitation of various bands of the second positive system of N_2 by impact of electrons of energy 35 eV

$(v'v'')$	$Q_{v'v''}$ obs. ($\times 10^{-18}$ cm^2)	$Q_{v'v''}$ calc. ($\times 10^{-18}$ cm^2)	
		I§	II‡
(0, 0)	2·96	2·96	2·96
(0, 2)	0·858	0·946	1·011
(0, 3)	0·269	0·378	0·371
(1, 2)	0·635	0·850	0·908
(1, 3)	0·71	0·870	0·908
(2, 0)	0·71	0·257	0·303
(2, 4)	0·22	0·397	0·334
(3, 3)	0·084	0·066	0·105
(3, 5)	0·05	0·098	
(4, 4)	0·038	0·046	

Calculated values are relative to the cross-section for the (0, 0) band.

being normalized so as to agree with the observed value for the (0, 0) band. It will be seen that there is fair agreement, particularly when allowance is made for the difficulty of calibration of the optical equipment over a wavelength range from 3000 to 4500 Å.

In earlier work on the bands Langstroth[∥] found that the vibrational intensity distribution no longer remained independent of electron energy above 30 eV. Stewart and Gabathuler[††] were careful to check that the light emitted was proportional to both electron beam intensity and gas pressure over the energy range studied (up to 200 eV) but it is by no means certain that at the higher energies secondary processes are not contributing. Fig. 13.55 illustrates their observed excitation function for the (0, 0) band. The relative magnitude at 200 eV is much higher than would be expected theoretically for an intercombination excitation.

† BATES, D. R., *Proc. R. Soc.* A196 (1949) 217.
‡ FRASER, P. A. and JARMAIN, W. R., *Proc. phys. Soc.* A66 (1953) 1145; ibid. 1153.
§ PILLOW, M. E., ibid. A67 (1954) 780.
∥ LANGSTROTH, G. O., *Proc. R. Soc.* A146 (1934) 166.　　　†† loc. cit., p. 965.

It is true that the total spin is a less good quantum number for molecules than for atoms but it would be surprising if it were so poor as is indicated from the relatively slow decrease with energy shown by the curve of Fig. 13.55 (cf. the corresponding problem for excitation of atoms Chap. 4, § 1.4.2.1 and Chap. 7, § 5.6.9).

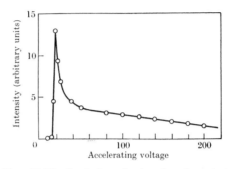

FIG. 13.55. Optical excitation function of the (0, 0) band of the second positive system of N_2, as observed by Stewart and Gabathuler.

Attempts have been made to carry out similar measurements for the second positive bands. These have not been successful because of the difficulty of avoiding secondary effects. Thompson and Williams[†] first noticed that most of the emission in this case came from outside the beams and this was confirmed much later by Stewart.[‡]

3.3. Ultra-violet photon-production coefficient for N_2

The ultra-violet photon production coefficient α_{ph} for a swarm of electrons drifting through nitrogen gas at pressure p under the influence of a uniform electric field F has been measured by Legler[§] by the same method as that used for hydrogen (see § 1.3.3 and Fig. 13.17). He found, as for that gas, that the ratio α_{ph}/α_i, where α_i is the ionization coefficient, varies with pressure according to the law

$$\alpha_{ph}/\alpha_i = (\alpha_{ph}/\alpha_i)_0(1+p/p_0)^{-1},$$

with $p_0 \simeq 60$ torr. Extrapolating to zero pressure he obtained $(\alpha_{ph}/\alpha_i)_0$ as a function of F/p, and using values of α_i observed by Masch[||] derived the values of α_{ph}/p shown in Fig. 13.56.

† THOMPSON, W. and WILLIAMS, S. E., *Proc. R. Soc.* A147 (1934) 583.
‡ STEWART, D. T., *Proc. phys. Soc.* A68 (1955) 404.
§ LEGLER, W., *Z. Phys.* 173 (1963) 169.
|| MASCH, K., *Arch. Electrotech.* 26 (1932) 587.

3.4. *Ionization of* N_2

A number of investigations of the ionization of nitrogen by electron impact have been made. For some time they were pursued with the hope of deciding the then controversial issue of the dissociation energy

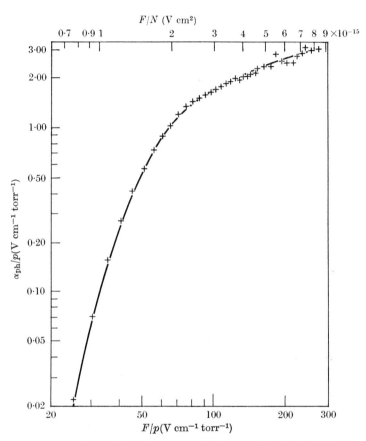

Fig. 13.56. Ultra-violet photon production coefficient for electrons drifting in N_2. + observed by Legler. —— curve best fitting the experimental points.

of the normal molecule. More recently a considerable amount of effort has been directed towards the determination of the cross-section for production of N^+ ions. This is of particular interest in connection with the production of atomic nitrogen—it has been pointed out that no strong process of impact dissociation into neutral atoms occurs.

We begin, however, by discussing ionization processes in which N_2^+ is the sole or at least the main ion concerned.

3.4.1. *Total ionization cross-section.* In Fig. 13.57 the results of observations of total cross-sections for ionization of N_2, made by different investigations, are shown. The techniques involved have been described in Chapter 3, § 2.

In contrast with the results for H_2 (Fig. 13.18 (a)), the cross-sections observed by Rapp and Englander-Golden,[†] who took special precautions

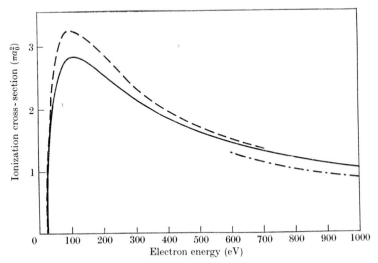

Fig. 13.57. Observed total ionization cross-sections for N_2. —— Rapp and Englander-Golden. — — — Tate and Smith. — · — · — Schram *et al.*

(see Chap. 12, § 7.2) to ensure saturation in the collection of positive ions, fall substantially below those of Tate and Smith[‡] over a considerable energy region in which the maximum cross-section occurs. Below about 30 eV the agreement is good. Just as for H_2 and some atomic gases the cross-sections observed by Schram, de Heer, van der Wiel, and Kistemaker[§] fall somewhat below those of Rapp and Englander-Golden.

3.4.2. *Behaviour of the cross-section near the threshold for* N_2^+ *production.* The first ionization potential leading to production of N_2^+ ions in their ground electronic and vibrational states, was determined as $15·60\pm0·01$ V by Fox and Hickam,[||] using the pulsed retarding potential difference method (see Chap. 3, § 2.4.2). They calibrated their voltage scale using the ionization potential of xenon as standard. Their observed

† Rapp, D. and Englander-Golden, P., *J. chem. Phys.* **43** (1965) 1464.
‡ Tate, J. T. and Smith, P. T., *Phys. Rev.* **39** (1932) 270.
§ Schram, B. L., de Heer, F. J., van der Wiel, M. J., and Kistemaker, J., *Physica* **31** (1965) 94.
|| Fox, R. E. and Hickam, W. M., *J. chem. Phys.* **22** (1954) 2059.

ionization efficiency curve as a function of electron energy near the threshold is shown in Fig. 13.58 (a). Frost and McDowell,† using the same method, obtained very similar results (see Fig. 13.58 (b)), the ionization potential being $15 \cdot 63 \pm 0 \cdot 02$ eV. These results compare well with the spectroscopic value $15 \cdot 58$ eV.

Fox and Hickam, by analysing their ionization efficiency curve in terms of linear intercepts (see Chap. 3, § 2.5.3), found evidence for the onset of new processes at energies $1 \cdot 33 \pm 0 \cdot 04$ and $3 \cdot 24 \pm 0 \cdot 04$ eV above the first ionization potential. These they ascribed to ionization leaving the N_2^+ molecule in the $A^2\Pi_u$ and $B^2\Sigma_u^+$ states respectively. Frost and McDowell (see Fig. 13.58 (b)) also found evidence for the excitation of these states at $1 \cdot 21$ and $3 \cdot 13$ eV above the ionization threshold. Clarke,‡ using the electron velocity selector described in Chapter 3, § 2.4.1, obtained the ionization curve shown in Fig. 13.58 (c). It will be seen that there is a reduction of slope at the point C. He attributes the curvature between A and C as due to excitation of vibrational levels. Ignoring this and extrapolating the linear portions AB and EF to meet at D he obtains what he considers to be the onset potential for excitation of the $A^2\Pi_u$ state as $1 \cdot 0 \pm 0 \cdot 2$ eV above the first ionization potential. This is to be compared with the spectroscopic value $1 \cdot 115$ eV obtained from analysis of the Meinel bands (see p. 957).

Kerwin, Marmet, and Clarke,§ using the same equipment as that with which they detected the excitation of separate vibrational levels of H_2 (see Fig. 13.19 and § 1.4.2), observed the excitation of such levels for N_2. The observed energies, $0 \cdot 278$, $0 \cdot 260$, $0 \cdot 257$ eV respectively associated with the $0 \to 1$, $1 \to 2$, and $2 \to 3$ transitions, are in close agreement with those obtained spectroscopically—$0 \cdot 270$, $0 \cdot 266$, and $0 \cdot 262$ eV respectively.

3.4.3. *Energy loss spectra beyond the excitation threshold.* Silverman and Lassettre‖ have made a thorough experimental study of the probability of energy losses by electrons in nitrogen above the ionization limit. They used apparatus and method very similar to that described in Chapter 5, § 4. Fig. 13.59 illustrates a typical energy loss spectrum showing a considerable amount of structure above the ionization limit. Differential generalized oscillator strengths, defined as in Chapter 7, § 5.2.4, obtained for different energy losses at and above the ionization

† FROST, D. C. and McDOWELL, C. A., *Proc. R. Soc.* A**232** (1955) 227.
‡ CLARKE, E. M., *Can. J. Phys.* **32** (1954) 764.
§ KERWIN, L., MARMET, P., and CLARKE, E. M., *Advances in mass spectrometry*, Vol. 2, p. 522 (Pergamon Press, 1962).
‖ SILVERMAN, S. M. and LASSETTRE, E. N., *J. chem. Phys.* **42** (1965) 3420.

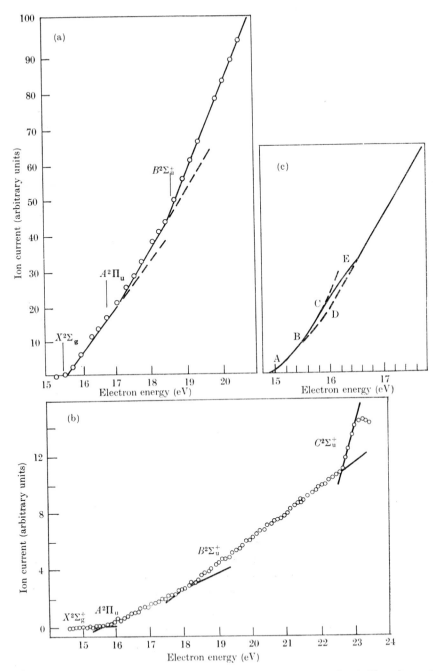

FIG. 13.58. Variation of the ionization cross-section of N$_2$ near the threshold as observed (a) by Fox and Hickam, (b) by Frost and McDowell, (c) by Clarke.

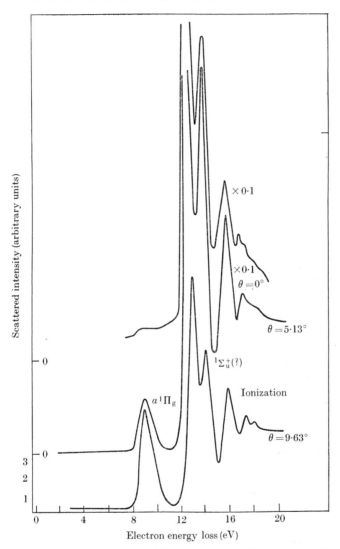

FIG. 13.59. Energy loss spectrum for N_2 as observed by Silverman and Lassettre. The main peak is due to excitation of the $b^1\Pi_u$ state (see Fig. 13.48).

limit, are shown in Fig. 13.60 as functions of K^2, the square of the wave vector change of the incident electron.

By interpolating to zero K^2 the corresponding optical oscillator strengths are obtained. If this is done over a wide energy loss range and combined with the data available from transitions to discrete states (see § 3.2 and Fig. 13.53) the integrated oscillator strength is found to be 11·8. Theoretically this sum should equal the total number of

electrons, 14, less the contribution from inner-shell excitation. As the
latter involves 2 electrons only we would expect the sum calculated
from the observations of Silverman and Lassettre to be close to 12, as
indeed it is.

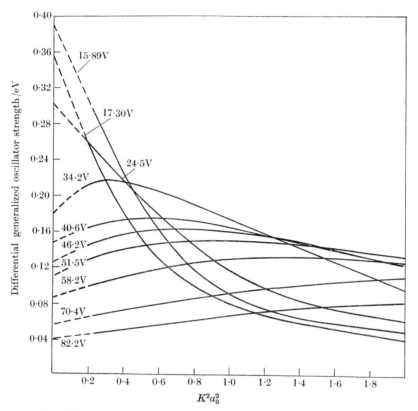

Fig. 13.60. Differential generalized oscillator strengths for transitions excited by electron impact in N_2 at and beyond the ionization limit, as functions of $K^2 a_0^2$.

A further check may be made by comparison with differential optical
oscillator strengths derived from observed absorption coefficients. The
best comparison is in a region of wavelength in which the absorption
coefficient is slowly varying as the oscillator strengths derived from the
impact experiments are averages over a considerable energy range.
These conditions are well satisfied over the spectral region 580–400 Å
(see Chap. 14, § 8.5). The average differential optical oscillator strength
measured over this range is 0·32 electrons per eV, very close to that
from the impact data (0·315 electrons per eV). On the other hand,
calculation of the refractive index μ for radiation of frequency ν from

the formula (11 a) gives $\mu = 3 \cdot 94 \times 10^{-4}$ in the limit of zero frequency. The best value for μ from microwave measurement is $2 \cdot 94 \times 10^{-4}$, which is very substantially smaller. The reason for this discrepancy is not clear, particularly as it does not appear in the corresponding analysis for hydrogen (see § 1.4.3). A similar discrepancy is found, however, for oxygen (see § 4.3.3).

3.4.4. *Direct excitation of* N_2^+ *bands*. The direct impact excitation of the first negative bands of N_2^+ has been studied by several investigators. In all cases the fact that both ionization and excitation occurred in a single collision has been verified by checking the proportionality of intensity to beam current and gas pressure.

Langstroth,[†] who made the first observations in 1934, was particularly concerned with verifying (40)–(42). He found that the intensity ratio of the $0 \rightarrow 1$ and $1 \rightarrow 2$ vibrational transitions was independent of the energy of the bombarding electrons and equal to $7 \cdot 2$, in close agreement both with an early calculated value by Hutchisson[‡] and a later one, $7 \cdot 3$, by Bates.[§]

Fig. 13.61 shows the absolute excitation functions for the $(0, 0)$, $(0, 1)$, and $(0, 2)$ vibrational transitions observed by Stewart.[‖] They are closely similar in shape, as expected, and yield the respective intensity ratios $1 \cdot 0 : 0 \cdot 30 : 0 \cdot 10$ to be compared with $1 \cdot 0 : 0 \cdot 31 : 0 \cdot 072$ calculated by Bates.[§]

Sheridan, Oldenberg, and Carleton[††] and Hayakawa and Nishimura[‡‡] have also measured the cross-section for excitation of the $(0, 0)$ transition up to considerably higher bombarding energies. Comparison of their results with those of Stewart is shown in Fig. 13.62. Kishko[§§] has obtained evidence of a more complicated structure in the excitation function near its maximum.

Stewart[‖‖] has observed the excitation function for the Meinel bands, again excited in a single impact. It is of the form typical of an optically allowed transition and very similar to that for the first negative bands.

3.5. *Dissociative ionization of* N_2

3.5.1. *The production of* N^+ *ions*. The occurrence of a new process

---

† LANGSTROTH, G. O., *Proc. R. Soc.* A146 (1934) 166.
‡ HUTCHISSON, E., *Phys. Rev.* 36 (1930) 410.
§ BATES, D. R., ibid. A196 (1949) 217.
‖ STEWART, D. T., *Proc. phys. Soc.* A69 (1956) 437.
†† SHERIDAN, W. F., OLDENBERG, O., and CARLETON, N. P., *2nd Int. Conf. Phys. Electron. Atom. Collisions*, p. 159 (Berg, New York, 1961).
‡‡ HAYAKAWA, S. and NISHIMURA, N. J., *J. Geomag. Geoelect., Kyoto* 16 (1964) 72.
§§ KISHKO, S. M., *Optiken Spectrosc.* 8 (1960) 160; *Optics Spectrosc.* 8 (1960) 84.
‖‖ STEWART, D. T., *Proc. phys. Soc.* A68 (1955) 404.

in the ionization of nitrogen at an onset potential of $24 \cdot 27 \pm 0 \cdot 1$ eV was first observed by Lozier† by the method described in Chapter 12, § 7.3.1, but he was unable to identify the ions produced. This was first done by Hagstrum and Tate,‡ who were able to show that they were N^+ ions

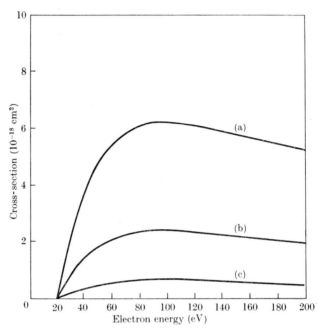

Fig. 13.61. Absolute excitation functions for direct excitation of N_2^+ bands, observed by Stewart. (a) (0, 0) band. (b) (0, 1) band. (c) (0, 2) band.

produced at the onset potential ($24 \cdot 3 \pm 0 \cdot 2$ eV) with negligible kinetic energy. They must be formed by the reaction

$$N_2 + e \rightarrow N^+ + N + 2e.$$

Since the ionization potential of N is $14 \cdot 54$ V, application of (2) of Chapter 12 gives

$$D(N_2) + E(N) + E(N^+) = 9 \cdot 76 \text{ eV},$$

$E(N)$ and $E(N^+)$ being the excitation energies of the N and N^+ produced. If both $E(N)$ and $E(N^+)$ are zero, $D(N_2)$ must be equal to $9 \cdot 76$ eV. This is very close to the now generally accepted value though at the time a lower value of $7 \cdot 38$ eV was also strongly favoured. It so happened that, if the nitrogen atom were formed in the lowest excited state (^2D) for

† Lozier, W. W., *Phys. Rev.* **44** (1933) 575.
‡ Hagstrum, H. D. and Tate, J. T., ibid. **59** (1941) 354.

which $E(N) = 2.37$ eV, the value obtained for $D(N_2)$ agreed closely with the lower value!

Later observations by Burns,[†] Thorburn and Craggs,[‡] and Hagstrum[§] confirmed the earlier work. A further step was made by Clarke,[||] using the electron velocity selector described in Chapter 3, § 2.4.1. He obtained the yield-energy curve for N^+ shown in Fig. 13.63, providing evidence, on analysis into linear portions, of the onset of additional processes, one

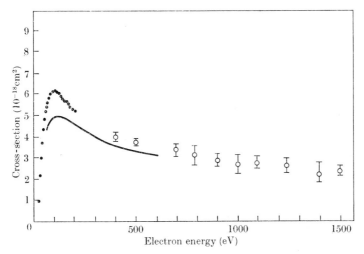

FIG. 13.62. Comparison of observed excitation functions for the (0, 0) first negative band of N_2^+. ● Stewart. —— Sheridan, Oldenberg, and Carleton. ⊈ Hayakawa and Nishimura.

at 2.36 ± 0.57 eV above the threshold for N^+ production and the second 1.4 ± 0.1 eV higher still. The first of these agrees well with that expected if the atoms are formed in the ^2D state, thereby supporting strongly the assumption that the atoms formed at 24·3 eV are indeed in the ground state, and that $D(N_2)$ is 9·76 eV.

The second excited state involved can be identified with the ^2P state of N, which is 1·19 eV above the ^2D state according to optical data.

Rapp et al.[††] have applied the technique described in Chapter 12, § 7.3.4 to measure the cross-section for production of ions in nitrogen with kinetic energy greater than 0·25 eV. Their results are illustrated in

† BURNS, J. F., J. chem. Phys. 23 (1955) 1347.
‡ THORBURN, R. and CRAGGS, J. D., Proc. phys. Soc. B69 (1956) 682.
§ HAGSTRUM, H. D., J. chem. Phys. 23 (1955) 1178.
|| CLARKE, E. M., Can. J. Phys. 32 (1954) 764.
†† RAPP, D., ENGLANDER-GOLDEN, P., and BRIGLIA, D. D., J. chem. Phys. 42 (1965) 4081.

Fig. 13.64. Since some N^+ ions are formed with zero kinetic energy the full dissociative ionization cross-section must be an even larger fraction. It will be noticed that, for incident electron energies above 100 eV, the percentage of total ionization is over 20 per cent.

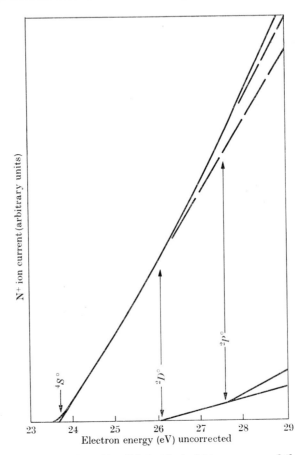

Fig. 13.63. Variation with incident electron energy of the cross-section for production of N^+ by electron impact in N_2, observed by Clarke.

3.5.2. *Direct excitation of NI lines.* Both Stewart[†] and Sheridan, *et al.*[‡] have observed emission of the NI line at $\lambda8210$ arising from the $3p\,^4P^0–3s\,^4P$ transition in atomic nitrogen. The usual linearity checks showed that the excited nitrogen atoms were produced in single impacts

† Stewart, D. T., *Proc. phys. Soc.* A69 (1956) 437.
‡ Sheridan, W. F., Oldenberg, O., and Carleton, N. P., *2nd Int. Conf. Phys. Electron. Atom. Collisions*, p. 139 (Berg, New York, 1961).

with normal molecules. Stewart found the appearance potential to be below 24 V. The absolute excitation function for the line as measured by Sheridan *et al.*† is shown in Fig. 13.65.

3.6. *The formation of* N_2^{++} *and* N^{++} *ions*

Lozier‡ observed the onset of a new ionization process at 47 ± 1 eV and this was confirmed by Hagstrum and Tate§ who identified the ions

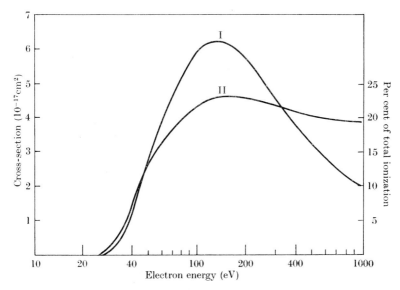

Fig. 13.64. Observed cross-section for production of ions with kinetic energy > 0.25 eV by electron impact in N_2. I, absolute cross-section. II, percentage of total ionization cross-section.

as N_2^{++}. The latter investigators also observed the direct production of N^{++} ions with an appearance potential of 64 ± 2 V.

Dorman and Morrison‖ have studied the variation of the intensity of production of N_2^{++} ions near the threshold, using their method of analysis described in Chapter 3, § 2.5.3. Fig. 13.66 shows the first derivative with respect to electron energy of their observed yield of N_2^{++}. The threshold is at 43.5 ± 0.03 eV and the characteristic quadratic rise of the cross-section for double ionization with energy above the threshold (see Chap. 9, § 10.8) is clearly seen.

† SHERIDAN, W. F., OLDENBERG, O., and CARLETON, N. P., *2nd Int. Conf. Phys. Electron. Atom. Collisions*, p. 139 (Berg, New York, 1961).
‡ LOZIER, W. W., *Phys. Rev.* **44** (1933) 575.
§ HAGSTRUM, H. O. and TATE, J. T., ibid. **59** (1941) 354.
‖ DORMAN, F. H. and MORRISON, J. D., *J. chem. Phys.* **35** (1961) 575.

3.7. *Electronic states of* N_2^-

Atomic nitrogen forms no stable negative ion though there is some evidence, both theoretical and experimental (see Chap. 15, § 5.5), that an ion in the $(1s)^2(2s)^2(2p)^4\,^1D$ state is metastable. This is because the energy of this ion falls below that of the ^2D term of the ground configuration of the neutral atom so that it can only suffer autodetachment to the ^4S ground term. Such a transition violates the spin selection rule and must be very improbable.

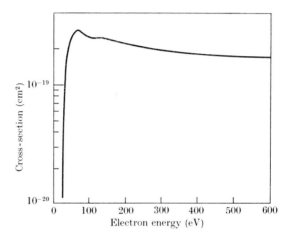

FIG. 13.65. Observed excitation function for the NI
λ 8210 line by electron impact in N_2.

We have described in Chapter 10, § 3.5.2 and Chapter 11, § 3 the remarkable and quite sharp resonance effects observed both in the elastic and vibrational excitation cross-sections of N_2 for electrons with energies in the range 2·3 to 5 eV. These must be interpreted in terms of the capture of the electrons, in this energy range, to form an N_2^- complex. Applying the analysis of Chapter 12, § 6 it is found that the observed resonance effects can be understood if the lifetime of the complex corresponds to a level width of 0·15 eV.† That this is about the correct magnitude is easily seen by noting the width of the resonance peaks in Fig. 11.5 and Fig. 10.32. This lifetime, 4×10^{-15} s, is considerably larger than would be expected for a state of the negative molecular ion of Type II (see Chap. 12, § 6). Thus resonances of Type II arising in collisions with atoms are considerably broader, usually greater than 1 eV.

† CHEN, J. C. Y., *Atomic collision processes*, ed. McDOWELL, M. R. C., p. 428 (North Holland, Amsterdam, 1964).

On the other hand, it is difficult to see what Type I state could exist within the energy range involved.

Thus the $N^-(^3P)$ state arising from addition of an electron to a normal $N(^4S^0)$ is of Type II, as the electron affinity of N is negative. This provides no opportunity for the presence of an N_2^- state above the ground state but still tending to $N^-(^3P)+N(^4S^0)$ in the separated limit

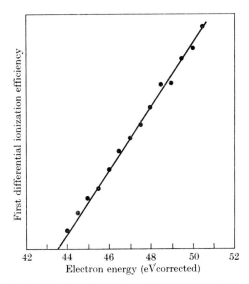

FIG. 13.66. First derivative with respect to electron energy of the observed rate of production of N_2^{++} ions by electron impact in N_2.

and lying only 2 eV or so above the ground state of N_2 near the equilibrium separation. It has been suggested that the N_2^- state responsible for its resonances arises from attachment of an electron to N_2 in the $A^3\Sigma_u^+$ state, but this would require the electron affinity of this latter state to be as high as 4 eV, which seems inconceivable.

The identification of the responsible N_2 state remains a mystery at the time of writing.

3.8. *Analysis of swarm experiments in N_2 in which electronic excitation is important*

In Chapter 11, § 7.2 we discussed the analysis of transport coefficient data for electrons drifting in nitrogen for such values of F/p, the ratio of applied electric field to gas pressure, that the characteristic energy ϵ_k is < 1.3 eV. Within this range electron energy loss through electronic excitation and ionization is unimportant. We now discuss the extension of this analysis to higher values of ϵ_k taking advantage not only of measured values of drift velocity and ϵ_k but also of ionization and photon production coefficients α_i and α_{ph} respectively.

Fig. 13.67 shows the cross-sections assumed by Engelhardt, Phelps, and Risk† in their final analysis. The excitation cross-sections were based on the observations of Schulz, made by the trapped-electron method (see Fig. 13.47 (a)). Of these, three correspond to excitation of $A^3\Sigma_u^+$ with threshold at 6·4 eV, $a^1\Pi_g$ at 8·4 eV, and $C^3\Pi_u$ at 11·2 eV.

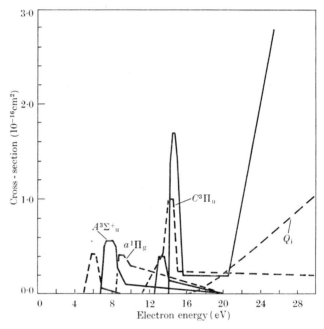

FIG. 13.67. Inelastic cross-sections in N_2 (including the ionization cross-section Q_i) assumed by Engelhardt, Phelps, and Risk, in their analysis of electron swarm data.

Three others are included, one with a threshold at 5·0 eV that is mainly due to vibrational excitation, one at 12·5 eV, and one at 14·0 eV. The latter cross-section was required to be large in order to fit the data on α_{ph}, while a further large energy loss, not accounted for by ionization, had to be allowed for, which rose rapidly at energies above 20 eV. As usual the ionization cross-section was taken to be that obtained by Tate and Smith (see § 3.4.1 and Fig. 13.57).

Fig. 13.68 shows the satisfactory agreement obtained with the observed transport coefficients u, ϵ_k, α_i/p, and α_{ph}/p in the range

† ENGELHARDT, A. G., PHELPS, A. V., and RISK, C. G., *Phys. Rev.* **135** (1964) A1566.

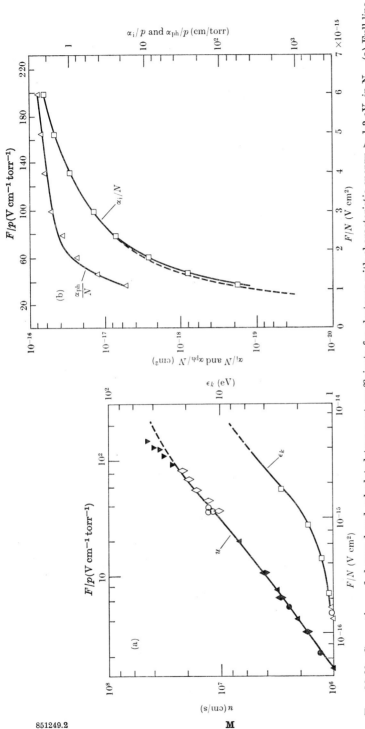

FIG. 13.68. Comparison of observed and calculated transport coefficients for electrons with characteristic energy > 1·3 eV, in N₂. (a) Full line curves give calculated results. u Observed by ▲ Lowke (77·6 °K), ◆ Bradbury and Nielson (293 °K), ○ Riemann, ◇ Frommhold, ▼ Wagner and Raether. ϵ_k Observed by □ Townsend and Bailey (298 °K), △ Crompton and Elford (293 °K), ○ Cochran and Forester (298 °K). (b) α_{ph}/N —— observed by Legler, △ Townsend and Bailey (298 °K), △ Crompton and Elford (293 °K), ○ Cochran and Forester (298 °K). α_i/N —— observed by Frommhold, ––– observed by Heylen, □ calculated.

References: LOWKE, J. J., *Aust. J. Phys.* **16** (1963) 115; BRADBURY, N. E. and NIELSEN, R. A., *Phys. Rev.* **49** (1936) 388; RIEMANN, W., *Z. Phys.* **122** (1944) 216; FROMMHOLD, L., ibid. **160** (1960) 554; WAGNER, K. H. and RAETHER, H., ibid. **170** (1962) 540; TOWNSEND, J. S. and BAILEY, V. A., *Phil. Mag.* **42** (1921) 873; CROMPTON, R. W. and ELFORD, M. T., *Proc. 6th Int. Conf. Ioniz. Phenom. Gases*, Paris 1963; COCHRAN, L. W. and FORESTER, D. W., *Phys. Rev.* **126** (1962) 1785; LEGLER, W., *Z. Phys.* **173** (1963) 169; HEYLEN, A. E. D., *Nature, Lond.* **183** (1959) 1545.

$\epsilon_k > 1{\cdot}3$ eV. The origin of the large energy loss with an apparent threshold near 20 eV is not clear. It may be due to excitation of the optically allowed transition to the $b^1\Pi_u$ state, which certainly dominates at high electron energies. Also, referring to Fig. 13.49 it will be noted that, whereas excitation of this state by 15·7-eV electrons is very weak, it is already strongly dominant for 35-eV electrons. According to the Born approximation, based on generalized oscillator strengths derived from the observations of Lassettre and Krasnow (see § 3.2 and Fig. 13.54) the maximum cross-section for excitation of the $b^1\Pi_u$ state is $\simeq 1{\cdot}7 \times 10^{-16}$ cm^2. This is almost certainly an over-estimate (see Chap. 7, § 5.6) but is not inconsistent with the cross-section assumed by Engelhardt *et al.*

4. Oxygen

Special interest attaches to the study of electron impact phenomena in oxygen because the gas is a major atmospheric constituent. This is also true for nitrogen but there is the additional factor in that oxygen forms negative ions that play an important role in many phenomena. A great deal of attention has therefore been devoted to the experimental investigation of electron reactions in oxygen even though the gas is difficult to work with because of its poisoning effect on oxide-coated filaments and high reactivity with hot tungsten.

4.1. *The electronic states of* O_2 *and* O_2^+

Fig. 13.69 shows the potential energy curves for the low-lying states of O_2 and O_2^+.

The ground state is a triplet, $X^3\Sigma_g^-$, and a number of metastable states lie above, including particularly $a^1\Delta_g$, $b^1\Sigma_g^+$, and $A^3\Sigma_u^+$, all dissociating into ground state atoms. Optical transitions between the ground state and the $A^3\Sigma_u^+$ state have been observed in the laboratory in absorption and are known as the Herzberg bands. They are present in emission in the spectrum of the night airglow which occurs at such a high altitude in the atmosphere that long-lived excited states are rarely deactivated by collisions and eventually radiate. It will be seen from Fig. 13.69 that some vertical transitions from the ground state lead to dissociation into normal atoms. This dissociation process is considered to be basic in the production of the ozone layer in the atmosphere.

Emission bands arising from $b^1\Sigma_g^+ - X^3\Sigma_g^-$ were first observed by Kaplan in the oxygen afterglow and the (0, 1) band is present in the

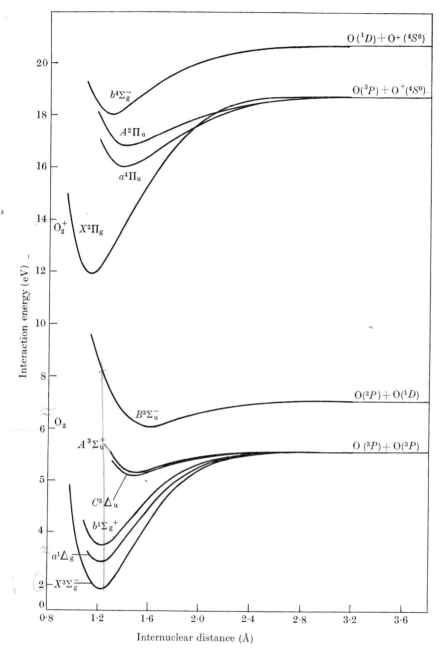

FIG. 13.69. Potential energy curves for some states of O_2 and O_2^+.

night airglow. There is some evidence that bands arising from $a^1\Delta_g$–$X^3\Sigma_g^-$ are also present.

The $B^3\Sigma_u^-$ state is of special interest as it is the upper level of the Schumann–Runge band system, which is such a prominent feature of the oxygen spectrum. It will be seen that vertical transitions from the ground state lead to upper levels above the dissociation limit. The discrete bands are therefore very weak as compared with a strong continuum. Photodissociation through ultra-violet absorption in this continuum is considered to be the basic process that atomizes oxygen in the upper atmosphere so that, above 100 km, oxygen is mainly in the atomic form.

The ground state of O_2^+ is $X^2\Pi_g$. Potential energy curves are shown for the upper states of two prominent band systems, the first negative ($b^4\Sigma_u^-$) and second negative ($A^2\Pi_u$).

Fig. 13.70 illustrates an excitation spectrum for electrons in oxygen obtained by Schulz and Dowell,[†] using the trapped-electron method (Chap. 5, § 2.3) with a well depth of 0·16 eV. A strong peak near 8·0 eV may be identified with excitation of the Schumann–Runge continuum. The peak at smaller electron energy is probably due to excitation of the $A^3\Sigma_u^+$ state, as judged by the appearance potential. The other prominent peak near 9·8 eV is also apparent in the optical absorption spectrum but its interpretation, as well as that of the smaller peaks, is not clear.

As expected, much of the detail disappears at high impact energies, as may be seen by reference to Fig. 13.71, taken by Lassettre, Silverman, and Krasnow.[‡] The only features clearly present below the ionization limit at 12·05 eV are the peaks at 8·4 and 9·76 eV, both of which are the most prominent in the trapped-electron spectra, the former arising from excitation of the Schumann–Runge continuum.

Dissociation of oxygen molecules by homogeneous electron beams with energy insufficient to produce ionization has been studied by Dalton,[§] Henry,[‖] and Glockler and Wilson.[††] In all cases dissociation was observed by a pressure drop in the experimental chamber as, for example, in the corresponding experiments of Glockler, Baxter, and Dalton in hydrogen (§ 1.2). In Glockler and Wilson's experiments a mercury surface and the glass wall of the containing vessel were used to condense any active form of oxygen produced by electron bombard-

† SCHULZ, G. J. and DOWELL, J. T., *Phys. Rev.* **128** (1962) 174.

‡ LASSETTRE, E. N., SILVERMAN, S. M., and KRASNOW, M. E., *J. chem. Phys.* **40** (1964) 1261.

§ DALTON, R., *J. Am. chem. Soc.* **51** (1929) 1366.

‖ HENRY, P., *Bull. Soc. chim. Belg.* **40** (1931) 339.

†† GLOCKLER, G. and WILSON, J. L., *J. Am. chem. Soc.* **54** (1932) 4544.

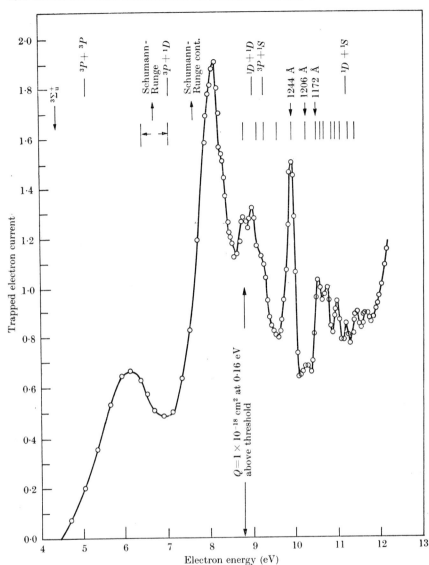

FIG. 13.70. Excitation spectra of oxygen obtained by Schulz and Dowell using the trapped-electron method with a well depth of 0·16 eV. The locations of peaks observed in optical absorption are indicated.

ment. They found a very small pressure change when the electron energy exceeded 3 eV and an enhanced effect at 8 eV. The latter was almost certainly due to excitation of the upper state of the Schumann–Runge bands. The effect observed at 3 eV is difficult to explain but may be due to some secondary reactions involving $b^1\Sigma_g^+$ molecules.

4.2. *Excitation functions for bound states of* O_2

4.2.1. *The upper* $(B^3\Sigma_u^-)$ *state of the Schumann–Runge bands.* Lassettre *et al.*† have measured differential cross-sections and generalized oscillator strengths for excitation of the Schumann–Runge continuum by the method described in Chapter 5, § 5. They find that the peak in the electron energy loss spectrum occurs at 8·44 eV, which does not agree within experimental error (about $\pm0\cdot04$ eV) with the corresponding

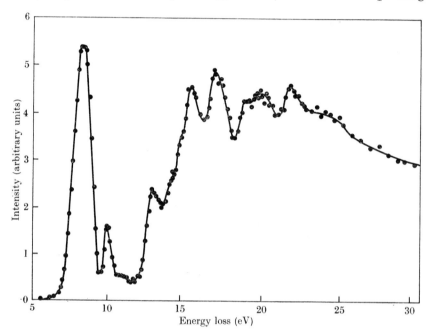

FIG. 13.71. Energy-loss spectrum of electrons of incident energy 518 eV scattered through 5·1° in O_2 as observed by Lassettre, Silverman, and Krasnow.

maximum in the ultra-violet absorption spectrum. Observed values of 8·57,† 8·73,‡ and 8·57 eV§ have been found (see Table 13.13). No reason for this discrepancy has yet been given.

The integrated generalized oscillator strengths found are given in Table 13.14 as a function of K^2, where K is the magnitude of the change of the wave vector of the incident electron. Extrapolation to zero K^2 gives the optical oscillator strength (see Chap. 7, § 5.6.7) as 0·230. This

† LASSETTRE, E. N., SILVERMAN, S., and KRASNOW, M. E., *J. chem. Phys.* **40** (1964) 1261.

‡ LADENBURG, R. and VAN VOORHIS, C. C., *Phys. Rev.* **43** (1933) 315.

§ WATANABE, K., INN, E. C. Y., and ZELIKOFF, M. J., *Chem. Phys.* **20** (1952) 1969 **21** (1953) 1026.

|| DITCHBURN, R. W. and HEDDLE, D. W. O., *Proc. R. Soc.* A**220** (1953) 61.

is compared in Table 13.13 with values observed optically. In addition to integrated oscillator strengths comparison is also made between differential oscillator strengths at the peak. Agreement is seen to be good except with the measurements of Watanabe, Inn, and Zelikoff,[†] who used photoelectric as distinct from photographic methods of detection.

TABLE 13.13

Comparison of excitation potentials and of optical oscillator strengths for the upper states of the Schumann–Runge bands

	From electron[‡] impact	From optical absorption		
Energy at maximum (eV)	8·44	8·57§	8·73†	8·57‖
Differential oscillator strength ($/V$) at maximum	0·170	0·164	0·129	0·166
Integrated oscillator strength	0·230	0·193	0·161	0·215

TABLE 13.14

Integrated generalized oscillator strengths f for electron impact excitation of discrete electronic states of O_2

	$K^2a_0^2$	0	0·1	0·2	0·3	0·4	0·5	0·6	0·7	0·8	0·9	1·0
Energy loss	8·4 eV ($B^3\Sigma_u^-$)	0·230	0·192	0·158	0·130	0·110	0·092	0·080	0·069	0·061	0·054	0·049
	9·9 eV	0·025	0·024	0·022	0·021	0·020	0·019	0·018	0·017	0·015	0·014	0·013

Using the generalized oscillator strengths the total cross-section for excitation of the $B^3\Sigma_u^-$ state may be obtained by the procedure described in Chapter 7, § 5.6.7. The results are illustrated in Fig. 13.72. As usual, the broken part of the curve is not reliable because of the failure of the first Born approximation.

Little information is available about the cross-section at low energies. The measurements by Schulz and Dowell[††] using the trapped-electron method, give a cross-section (see Fig. 13.70) of about $1·0 \times 10^{-18}$ cm² at 0·16 eV above the threshold but as the maximum is likely to occur at an energy two or three times the threshold value it is not possible to derive much from this observation.

† loc. cit., p. 990.
‡ LASSETTRE, E. N., SILVERMAN, S., and KRASNOW, M. E., loc. cit., p. 990.
§ LADENBURG, R. and VAN VOORHIS, C. C., loc. cit., p. 990.
‖ DITCHBURN, R. W. and HEDDLE, D. W. O., loc. cit., p. 990.
†† loc. cit., p. 988.

4.2.2. *The upper state of the 9·94 eV energy loss.* Integrated generalized oscillator strengths for the excitation processes that lead to the 9·94-eV energy loss have been measured by Silverman and Lassettre† and are given in Table 13.14.

4.2.3. *The metastable states.* Only a little information is available about cross-sections for excitation of the metastable states. Fig. 13.70 shows that at 0·16 eV above the threshold the cross-section for excitation of the $A^3\Sigma_u^+$ state is about 7×10^{-19} cm². Schulz and Dowell‡ were

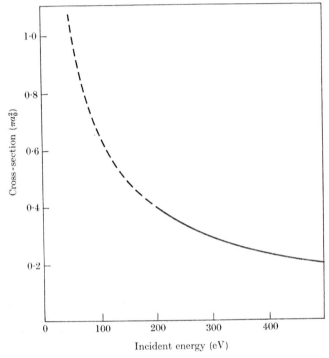

Fig. 13.72. Total cross-section for excitation of the $B^3\Sigma_u^-$ state of O_2 calculated, by Born's approximation, using generalized oscillator strengths derived by Lassettre, Silverman, and Krasnow from their energy-loss observations.

also able to obtain corresponding results for the $a^1\Delta_g$ and $b^1\Sigma_g^+$ states by applying the trapped-electron method with incident electrons with energies less than 2 eV. It is necessary to disentangle contributions from vibrational excitation (see Chap. 11, § 3 and Fig. 11.11) to obtain these values and as this cannot be done with certainty the values are only provisional. They are, however, certainly very small, being 3 and

† SILVERMAN, S. and LASSETTRE, E. N., *J. chem. Phys.* **40** (1964) 2922.
‡ loc. cit., p. 988.

0.8×10^{-20} cm^2 for the $a^1\Delta_g$ and $b^1\Sigma_g^+$ states respectively. These are two orders of magnitude lower than for the metastable states of helium and even for the optically allowed excitation of the $B^3\Sigma_g^-$ state. Indirect evidence in support of low cross-sections for excitation of metastable states of O_2, at least for electrons with energy near 14 eV, has been derived by Thompson[†] from a study of electron velocity distributions in the positive column of an electric discharge in oxygen.

4.3. Ion production in O_2

A great deal of work has been carried out, using both mass spectrometers and the Lozier technique, on the processes of ion production by electron beams in oxygen. Special attention was devoted to the study of appearance potentials for negative ions as it was hoped to obtain an accurate value for the electron affinity of atomic oxygen. The inherent difficulties in this technique, which were not clearly realized, led to much confusion in interpretation. Nevertheless, much of value remains.

The processes which have been investigated using electron beams are as follows:

$$O_2 + e \rightarrow O_2^+ + 2e \qquad (43\,a)$$

$$\rightarrow O^+ + O + 2e \qquad (43\,b)$$

$$\rightarrow O_2^{++} + 3e \qquad (43\,c)$$

$$\rightarrow O^{++} + O + 3e \qquad (43\,d)$$

$$\rightarrow O + O^- \qquad (43\,e)$$

$$\rightarrow O^+ + O^- + e. \qquad (43\,f)$$

We shall discuss first the processes in which positive ions alone are produced. It is convenient to discuss negative ion formation in conjunction with results from swarm experiments and in relation to the value of the electron affinity of atomic oxygen derived by other means.

4.3.1. Total ionization cross-section. Observations of the total ionization cross-section of O_2 at different electron energies are shown in Fig. 13.73. On the whole the agreement between results obtained by different investigators is satisfactory, considerably better than for N_2.

4.3.2. Production of O_2^+ ions near the threshold. Just as for the production of H_2^+ near the threshold, there is some confusion between the results obtained in different experiments. McGowan, Clarke, Hanson, and Stebbings,[‡] using the molecular beam technique described in

---

[†] THOMPSON, J. B., *Proc. R. Soc.* A**262** (1961) 503.
[‡] McGOWAN, J. W., CLARKE, E. M., HANSON, H. P., and STEBBINGS, R. F., *Phys. Rev. Lett.* **13** (1964) 620.

Chapter 3, § 2.4.5, which was applied by McGowan *et al.*† to H_2 (§ 1.4.2), found that the derivative of the yield curve of O_2^+ ions with respect to electron energy agreed quite well with the variation of the photo-ionization cross-section as shown in Fig. 13.74. The peaks in the latter can be associated with auto-ionization (see Chap. 14, § 8.4).

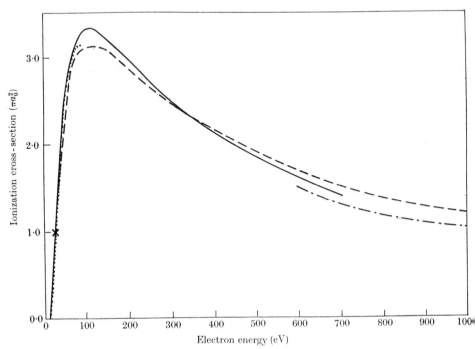

Fig. 13.73. Observed total ionization cross-sections for O_2. ———— Tate and Smith (loc. cit., p. 972). ✕ Schulz, G. J., *Phys. Rev.* **128** (1962) 178. •••• Asundi, R. K., Craggs, J. D., and Kurepa, M. V., *Proc. phys. Soc.* **82** (1963) 967. —·—·— Schram *et al.* (loc. cit., p. 972) ———— Rapp and Englander-Golden (loc. cit., p. 972).

The ionization potential observed by McGowan *et al.* is $12·04\pm0·02$ V, which agrees well with that ($12·065\pm0·003$ V) measured by Nicholson‡ from photo-ionization experiments (Chap. 14, § 8.4). Earlier electron impact data gave values near 12·2 eV. Thus Frost and McDowell,§ using the retarding potential difference method (Chap. 3, § 2.4.2) in conjunction with a mass spectrometer, found $12·21\pm0·04$ V after calibration of their energy scale from the ionization potential of krypton. As for the corresponding curve for N_2^+ production (Fig. 13.58), a number of

† McGowan, J. W., Fineman, M. A., Clarke, E. M., and Hanson, H. P., *Phys. Rev.* **167** (1968) 52.
‡ Nicholson, A. J. C., *J. chem. Phys.* **39** (1963) 954.
§ Frost, D. C. and McDowell, C. A., *J. Am. chem. Soc.* **80** (1958) 6183.

changes of shape occur that presumably arise from the onset of new processes. The thresholds for these processes occur at $16\cdot30\pm0\cdot03$, $17\cdot8\pm0\cdot02$, $18\cdot42\pm0\cdot62$, and $21\cdot34\pm0\cdot02$ eV. The first three of these may be identified as due to formation of O_2^+ ions in the $a^4\Pi_u$, $A^2\Pi_u$, and $b^4\Sigma_g^-$ states for which the respective threshold energies obtained spectroscopically are $16\cdot11$, $16\cdot97$, and $18\cdot16$ eV. These are all a little smaller than the electron impact values just as for ionization to the ground state of O_2^+.

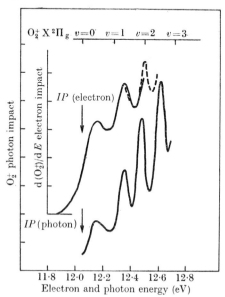

FIG. 13.74. Comparison of the yield curve for O_2^+ ions from O_2 by photo-ionization with the derivative of the yield curve for electron impact ionization.

4.3.3. *Energy loss spectra beyond the ionization threshold.* Silverman and Lassettre† have made an experimental study of the probability of energy losses beyond the ionization limit by electrons in oxygen comparable to their thorough analysis of these losses in nitrogen. Fig. 13.71 illustrates the structure observed in the energy-loss spectrum. Differential generalized oscillator strengths obtained for different energy losses above the ionization limit are shown in Fig. 13.75 as a function of K^2, where K is the magnitude of the change of the wave vector of the incident electron. Carrying out the usual extrapolation to zero to obtain the differential optical oscillator strength, integrating over the ionization continuum, and including the contributions from the excitation of the

† loc. cit., p. 992.

discrete states, the integrated optical oscillator strength is found to be 11·81. According to theory it should be between 12 and 16.

Comparison with optical oscillator strengths in the continuum is difficult because the absorption spectrum is a superposition of discrete and continuous spectra. However, it is possible to isolate the continuum by assuming the variation with wavelength to be gradual except for the absorption lines, so that the latter can be eliminated by drawing a smooth curve through the minima in a graph of absorption coefficients against

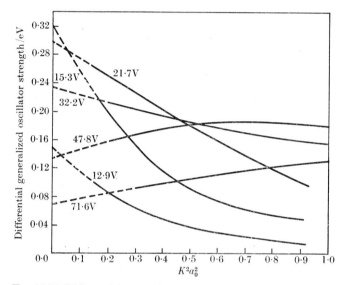

Fig. 13.75. Differential generalized oscillator strengths for transitions excited by electron impact in O_2 beyond the ionization limit, as functions of $K^2a_0^2$.

wavelength. Comparison of continuum oscillator strengths obtained in this way with those obtained by Silverman and Lassettre,† over a wavelength range from 207·2 to 515·6 Å, shows that the latter are somewhat higher, as they include contributions from unresolved discrete transitions. At the highest frequencies, where the contribution from such transitions is small, the agreement is good.

The value of $\mu-1$, where μ is the refractive index, calculated from the impact-derived interpolated oscillator strengths, comes out to be about 22 per cent too high. A somewhat similar discrepancy has been found for nitrogen (see § 3.4.3) but not for hydrogen (see § 1.4.3).

4.3.4. *Direct excitation of excited states of* O_2^+. Stewart and Gaba-

† loc. cit., p. 992.

thuler† have measured optical excitation functions for direct excitation of the first and second negative bands of O_2^+. It was necessary to work at pressures below 8×10^{-3} torr to eliminate secondary effects, while complications were introduced due to the complexity of the first negative bands and the doublet structure of the second negative.

The relative excitation functions of the two systems are similar and of typical optically allowed type. They rise sharply from threshold to

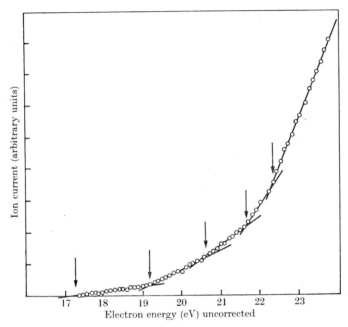

FIG. 13.76. Variation with incident electron energy of the cross-section for production of O^+ by electron impact in O_2, observed by Frost and McDowell.

a broad maximum at 85 eV for the first negative and 100 eV for the second negative and then fall off gradually as the electron energy increases.

4.3.5. *Production of O^+ ions.* Fig. 13.76 illustrates the variation of the cross-section for the direct production of O^+ ions from O_2 as a function of electron energy near the threshold, as observed by Frost and McDowell‡ with the same equipment as that which yielded the O_2^+ results discussed in § 4.3.2. The threshold is at $17 \cdot 30 \pm 0 \cdot 10$ eV. Just as for O_2^+ the onset of further processes can be detected from changes in shape in the cross-section curve. These begin at $18 \cdot 99 \pm 0 \cdot 05$,

† STEWART, D. T. and GABATHULER, E., *Proc. phys. Soc.* **72** (1958) 287.
‡ loc. cit., p. 994.

20.42 ± 0.04, 21.30 ± 0.03, and 22.03 ± 0.03 eV. All of these threshold values agree generally with results obtained by earlier investigators though discrepancies as great as 0.5 eV are found in some cases. In view of the inherent difficulties in determining appearance potentials for processes in which ions are formed with appreciable kinetic energy such discrepancies are not surprising.

The appearance potential for the process (43 b) in which both atom and ion are formed in their ground states with zero kinetic energy is given by
$$D(O_2)+I(O),$$
where $D(O_2)$ is the dissociation energy of $O_2 = 5.09$ eV and $I(O)$ the ionization energy of O $= 13.54$ eV. This value, 18.63 eV, is not far from the second threshold (18.99 eV according to Frost and MacDowell) and it is likely that this may be interpreted as due to
$$O_2+e \to O(^3S)+O^+(^4S), \tag{44}$$
in which the ions are formed with little kinetic energy.

If this is so the first threshold must be ascribed to
$$O_2+e \to O^+(^4S)+O^-(^2P)+e, \tag{45}$$
which is of the type discussed in Chapter 12, § 3.6.3. In this reaction, producing ions of zero kinetic energy, the threshold will occur at an energy less than that for (44) by $A(O)$, the electron affinity of O. Taking this to have the now generally accepted value 1.45 eV (see § 4.4 below), we find a threshold value of 17.18 eV as compared with 17.30 eV found by Frost and MacDowell.

Confirmatory evidence for this interpretation is obtained from the existence of a threshold peak for O^- production at the same value within experimental error.

Identification of the processes that occur at the higher thresholds is more difficult. Thus a second threshold for associated negative ion production certainly exists but, whereas Frost and MacDowell found this at 21.22 ± 0.05 eV, Thorburn,† using the Lozier technique, obtained 20.2 ± 0.2. Part of this discrepancy is probably due to inherent instrumental difficulties but it prevents definite assignment of reactions to the higher thresholds.

Fig. 13.77 shows the total cross-section of production of ions (presumably O^+) in O_2 with kinetic energy 0.25 eV measured by Rapp et al.‡

† THORBURN, R., Applied mass spectrometry, p. 185 (London, Institute of Petroleum, 1954).

‡ RAPP, D., ENGLANDER-GOLDEN, P., and BRIGLIA, D. D., J. chem. Phys. **42** (1965) 4081.

using the technique described in Chapter 12, § 7.3.4. The results are generally comparable with those for N_2 (Fig. 13.64) but the maximum cross-section is somewhat higher and for electrons of 200-eV incident energy as many as 30 per cent of all ionizing collisions lead to ions with appreciable kinetic energy.

4.3.6. *Production of O_2^{++} and O^{++} ions.* The threshold for production of O_2^{++} ions has been measured by Hagstrum and Tate† as $50 \cdot 0 \pm 0 \cdot 05$ eV, but the process (43 d) leading to direct O^{++} production is so weak that no measurements of its threshold energy have been made.

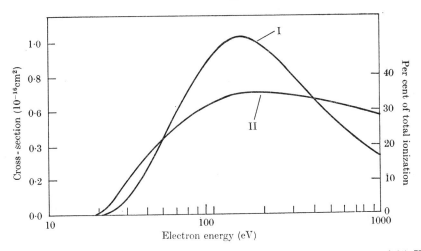

FIG. 13.77. Observed cross-sections for production of ions with kinetic energy > 0·25 eV by electron impact in O_2. I, absolute cross-section. II, percentage of total ionization cross-section.

4.4. Negative ion formation in O_2

4.4.1. *Beam experiments—historical account.* The history of the investigation of processes leading to production of negative ions in oxygen by impact of electron beams with energies ranging from 1 to 2 eV upwards is remarkably interesting and complex.

In 1934 Lozier,‡ using the method described in Chapter 12, § 7.3.1, observed the variation of negative ion current with electron energy and found it to be of the general form shown in Fig. 13.78. The low-energy production exhibited the resonance shape expected for a dissociative attachment process of the type

$$O_2 + e \to O + O^-, \tag{46}$$

---

† HAGSTRUM, H. D. and TATE, J. T., *Phys. Rev.* **59** (1941) 354.
‡ LOZIER, W. W., ibid. **46** (1934) 268.

as discussed in Chapter 12, § 3.6.1. The process operative at higher energies has already been discussed above and is definitely due to

$$O_2 + e \rightarrow O^+ + O^-. \tag{47}$$

Confirmation that the negative ions produced at all electron energies studied were O^- was provided by the observations of Hagstrum and Tate,[†] using the mass spectrograph.

A retarding potential analysis showed that the O^- ions produced by the capture process (46) possessed considerable kinetic energy (1–2 eV). The appearance potential for ions of zero kinetic energy was determined by the linear extrapolation method as $2 \cdot 8 \pm 0 \cdot 2$ eV. Since, as shown in Chapter 12, § 3.6.1, this appearance potential is given by $D(O_2) - A(O)$, where $D(O_2)$, the dissociation energy of O_2, is $5 \cdot 09$ eV, the electron affinity $A(O)$ of O was determined as $2 \cdot 5 \pm 0 \cdot 2$ eV. This value was reduced somewhat in later observations but after Hagstrum's work[‡] in 1951 with the mass spectrograph the electron affinity was accepted as a little greater than $2 \cdot 0$ eV.

Confidence in this value came from the measurements made of the equilibrium constant for reactions leading to O^- formation in the neighbourhood of a hot filament. The first application of this method to oxygen, by Vier and Mayer[§] in 1944, using O_2 as the carrier gas, gave $A(O)$ as $3 \cdot 1$ eV. However, Metlay and Kimball[||] in 1948 pointed out that O_2 was an unsatisfactory carrier as it was only partially dissociated at the filament. They used N_2O instead and obtained $2 \cdot 33 \pm 0 \cdot 03$ eV for $A(O)$ in good agreement with the impact value!

Further evidence about $A(O)$ could have been derived at the time from impact studies in CO and NO but was confused by the uncertainty in the dissociation energies of these molecules. Two values, $9 \cdot 6$ and $11 \cdot 11$ eV, were extant for CO and two, $5 \cdot 29$ or $6 \cdot 48$ eV, for NO. Assuming the lower values in each case the O^- ions formed in these gases appeared to possess a net energy only slightly below that of normal O atoms. This was taken as evidence that these O^- ions were formed in an excited state with excitation energy not much less than the electron affinity.

The existence of a stable excited state for O^- was shown by Bates and Massey[††] to be very unlikely on theoretical grounds, while Hasted[‡‡] found experimentally that the O^- ions formed from O_2, CO, and NO

[†] HAGSTRUM, H. D. and TATE, J. T., *Phys. Rev.* **59** (1941) 354.
[‡] HAGSTRUM, H. D., *Rev. mod. Phys.* **23** (1951) 185.
[§] VIER, D. T. and MAYER, J. E., *J. chem. Phys.* **12** (1944) 28.
[||] METLAY, M. and KIMBALL, G. E., ibid. **16** (1948) 774.
[††] BATES, D. R. and MASSEY, H. S. W., *Phil. Trans. R. Soc.* A**239** (1943) 269.
[‡‡] HASTED, J. B., *Proc. R. Soc.* A**222** (1954) 74.

all behaved in the same way in charge transfer collisions with other gas atoms or molecules. This would be very unlikely if they differed by 2·0 eV in the energy required to remove an electron.

In 1955 Smith and Branscomb† measured $A(O)$ directly by observing the threshold for photodetachment of electrons from O atoms, using the crossed beam method described in Chapter 15, § 2. They found $A(O) = 1·465$ eV and further experiments on these lines have confirmed these values (see Chap. 15, § 5.2). Reconciliation of this result with the impact data was assisted by the strength of the evidence which became available from spectroscopic and thermochemical sources about $D(CO)$ and $D(NO)$. It became clear that the larger and not the smaller values were the more plausible in each case. This being so, the energy of the O⁻ ions produced from CO was reduced by 1·5 eV and from NO by 1·2 eV. The energies of these ions thus became compatible with ground state O⁻ ions and an electron affinity as found by Smith and Branscomb. Furthermore, in 1961, Page‡ re-examined in detail the technique of measurement of electron affinities by the hot-wire equilibrium method and, using both N_2O and O_2 as carrier gases, obtained $A(O) = 1·45$ eV.

There remained the question of the appearance potential for the O⁻ produced from O_2 by the process (46). Every effort was made to refine the technique with the aim of finding a lower appearance potential compatible with the lower electron affinity. Thorburn,§ using the Lozier technique, measured the appearance potential for ions of zero kinetic energy as $3·2\pm1$ eV giving $A(O) = 1·89$ eV, but no lower values were reported. Finally, Schulz‖ carried out a careful experiment using the retarding potential difference method and a modified Lozier technique designed to collect all the ions produced. Using the linear extrapolation technique he found $A(O) = 2·0$ eV.

The final step towards resolving the difficulty seems to have been that of Chantry and Schulz†† who realized that the thermal motion of the target molecules leads to serious broadening of the energy distribution of ions with considerable kinetic energy (Chap. 12, § 7.1). When allowance is made for this effect it appears that the retarding potential data for the process (46) are consistent with $A(O) = 1·45$ eV after all. The same problem did not arise for O⁻ produced from CO, because a

† SMITH, S. J. and BRANSCOMB, L. M., *J. Res. natn. Bur. Stand.* **55** (1955) 165.
‡ PAGE, F. M., *Trans. Faraday Soc.* **57** (1961) 359.
§ THORBURN, R., *Applied mass spectrometry*, p. 185 (London, Institute of Petroleum, 1954).
‖ SCHULZ, G. J., *Phys. Rev.* **128** (1962) 178.
†† CHANTRY, P. J. and SCHULZ, G. J., *Phys. Rev. Lett.* **12** (1964) 449.

considerable fraction are produced with zero kinetic energy. This also seems to be true for the ions produced by the ionic breakup process (47) and to a lesser degree to the ions formed by capture and by break-up in NO.

Because of the serious effect of temperature motion it is not possible to obtain accurate appearance potentials for reactions in which ions are produced with considerable kinetic energy and little weight can be given to data derived by the linear extrapolation method in such cases.

4.4.2. *Beam experiments—results obtained.* Unlike the appearance potential the observed variation of the negative-ion production rate with electron energy is not subject to error due to the thermal motion of the target molecules. It is necessary, however, that all the ions be collected. Otherwise the effects of an anisotropic angular distribution of ionic velocities could be serious. This is obviously so if the absolute cross-section for production of negative ions is to be measured but it may not be so serious if the relative cross-sections at different electron energies alone is required.

The general form of the variation of the cross-section with electron energy has been revised in detail but is essentially the same as that first observed by Lozier[†] and by Hagstrum and Tate.[‡] The first attempt to measure the absolute cross-section was made by Craggs, Thorburn, and Tozer[§] using equipment of the Lozier type which, for reasons now well realized, is not suitable for this purpose. Methods which ensure full collection of the ions produced have been described in Chapter 12, § 7.4.

Fig. 13.78 illustrates the absolute cross-section for negative ion production observed by Rapp and Briglia.[∥] The sharp peak is due to dissociative attachment, the peak cross-section being much greater than for the corresponding process in H_2 (see § 1.6 and Fig. 13.35). There is good agreement in the height, shape, and location of the peak for oxygen (see Table 12.7) as measured by different observers.[††]

A very remarkable effect was observed by Fite and Brackmann.[‡‡] Using a crossed-beam apparatus, similar to that with which they measured the ionization cross-section of atomic hydrogen (Chap. 3, § 2.3)

† loc. cit., p. 999.   ‡ loc. cit., p. 999.
§ CRAGGS, J. D., THORBURN, R., and TOZER, B. A., *Proc. R. Soc.* A240 (1957) 473.
∥ RAPP, D. and BRIGLIA, D. D., *J. chem. Phys.* 43 (1965) 1480.
†† BUCKEL'NIKOVA, I. S., *Zh. eksp. teor. Fiz.* 35 (1958) 1119; *Soviet Phys. JETP* 35 (8) (1959) 783; SCHULZ, G. J., *Phys. Rev.* 128 (1962) 178; ASUNDI, R. K., CRAGGS, J. D., and KUREPA, M. V., *Proc. phys. Soc.* 82 (1963) 967.
‡‡ FITE, W. L. and BRACKMANN, R. T., *Proc. 6th Int. Conf. Ioniz. Phenom. Gases* (1963), p. 21.

they studied the production of negative ions from a beam of molecular oxygen as a function of the energy of the electrons in the intersecting beam. When the O_2 beam issued from an iridium oven at room temperature the production cross-section for O^- ions varied as in Fig. 13.78 but when the oven was heated to 2100 °K the detachment peak was shifted by as much as 2 eV to lower energies (see Fig. 13.79 (a)). Since the equipment operated by phase-sensitive detection it was possible to check that

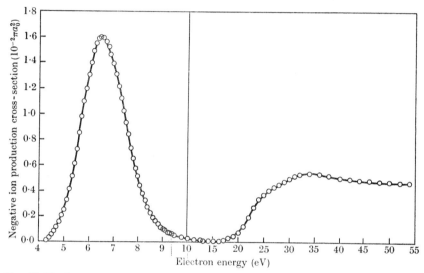

FIG. 13.78. Cross-sections for production of negative ions by electron impact in O_2, as a function of electron energy, observed by Rapp and Briglia.

the phases of the O^- signals were not consistent with their production by attachment to metallic oxides produced in the furnace. Mass analysis of the O_2 beam also failed to reveal the presence of any species likely to contribute appreciably to the O^- production.

O'Malley[†] has shown how these remarkable results may be understood in terms of the formula (19) for the attachment cross-section which includes the factor $\exp(-\overline{\Gamma}\tau/\hbar)$, τ being the time taken for the O atom and O^- ion to separate from the initial distance R_0 to the distance R_s beyond which autodetachment is no longer possible. $\overline{\Gamma}$ is as defined on p. 935. R_0, and hence τ, change appreciably with the initial vibrational state of the molecule and this effect is magnified by the exponential factor. Thus, even though the small degree of thermal excitation of the molecules can do little more than populate the first vibrational state,

[†] O'MALLEY, T. F., *Phys. Rev.* **155** (1967) 59. See also DEMKOV, Y. N., *Phys. Lett.* **15** (1965) 235.

O'Malley finds that, with a quite realistic form for the upper potential energy curve, the results of Fite and Brackmann can be reproduced. It was also found that, if the source was subjected to a discharge, the O^-

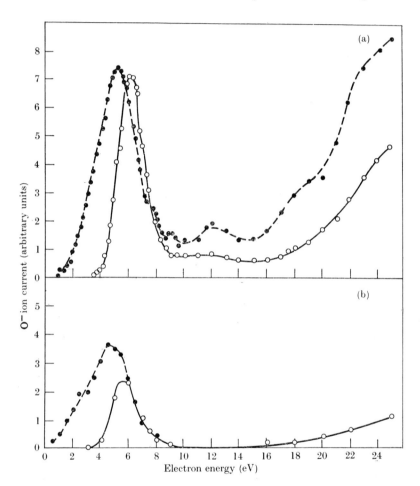

FIG. 13.79. Variation with electron energy of the yield of O^- ions by electron impact with a beam of O_2 under different conditions in the oven source. (a) O—— source at 300° K. ●— — — source at 2100° K. (b) O—— source at 300 °K and no exciting discharge. ◉— — — source at 300 °K but excited by r.f. discharge.

production, again found to arise from O_2, was profoundly affected (see Fig. 13.79 (b)), presumably through the presence of metastable molecules under these conditions.

   4.4.3. *Swarm experiments—introduction.* The history of the investigation of electron attachment in oxygen at the low electron energies that require the use of swarm techniques is almost as remarkable as that of the beam investigations at somewhat higher energies.

Early investigations were carried out both by the electron filter† and diffusion‡ methods (see Chap. 12, §§ 7.5.1, 7.5.3) with results as illustrated in Fig. 13.80. The two methods agreed in giving much the same variation of the attachment probability per collision, h, with mean electron energy and differed only in a factor of about 2 in absolute magnitude. The most distinctive feature is the rapid rise as the mean electron energy decreases below 0·5 eV.

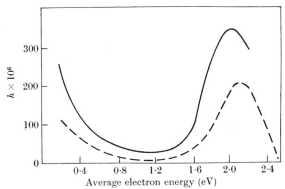

FIG. 13.80. General form of the observed variation of attachment probability per collision (h) for electrons in O_2, in early experiments. — — — measured by filter method. ——— measured by diffusion method.

From the shapes of the curves shown in Fig. 13.80 it appeared that there are two attachment processes involved, one which we shall refer to as process A, effective at very low electron energies, and the other, process B, at energies in the neighbourhood of 2·0 eV.

In 1935 Bloch and Bradbury§ suggested that the very low energy attachment process A is of the type in which a vibrationally excited molecular ion O_2^- is first formed (see Chap. 12, § 3.6.1 and Fig. 12.6 (c)) and is subsequently stabilized by getting rid of its excess energy in a collision. The early experiments showed no definite evidence of any dependence of attachment probability on pressure. This required a very high rate of collision deactivation (see Chap. 12, p. 825) involving an effective collision cross-section as high as 10^{-12} cm². This seemed

† BRADBURY, N. E., *Phys. Rev.* **44** (1933) 883; *J. chem. Phys.* **2** (1934) 827.
‡ HEALEY, R. H. and KIRKPATRICK, C. B., as quoted in HEALEY, R. H. and REED, J. W., *The behaviour of slow electrons in gases*, p. 19 (Amalgamated Wireless, Sydney, 1941).
§ BLOCH, F. and BRADBURY, N. E., *Phys. Rev.* **48** (1935) 689.

most unlikely because of the difficulty of vibrational deactivation on collision (Chap. 17, § 3). Apart from this, detailed analysis showed the explanation to be reasonable, provided the energy excess of the ground state of O_2 above that of the state in which O_2^- is first formed is between 0·17 and 0·07 eV.

The second process, B, occurring at mean energies of 1 eV or so could be explained in a similar way, a transition to a different potential energy curve of O_2^- being involved.

In 1951 Biondi[†] reported the first results obtained on the decay of electron concentration in an oxygen afterglow. He found an attachment-like rate of loss (see Chap. 12, § 7.5.5) at thermal energies and certain conditions of pressure and electron concentration, but at a very low rate, about 10^{-3} times that to be expected by extrapolation of the swarm data using the Bloch–Bradbury interpretation. These results were confirmed in later measurements by other investigators.[‡]

At this stage the subject seemed very confused indeed but, in 1959, two independent investigations led to major clarification. Chanin, Phelps, and Biondi[§] applied the drift-tube method described in Chapter 12, § 7.5.2, to study attachment in oxygen. They found that, for ratios F/p of electric field to gas pressure < 3 V cm^{-1} torr^{-1}, the ratio α_a/p of attachment coefficient to gas pressure depends on the pressure. This is clearly shown by their results, which may be seen in Fig. 13.81. Comparison with results obtained in earlier investigations is made in Fig. 13.82. At much the same time Hurst and Bortner,[||] using the pulse method described in Chapter 12, § 7.5.2, also found, from a study of attachment rates in nitrogen and oxygen mixtures, that α_a/p should, for pure oxygen, be pressure dependent over much the same range of F/p.

Since α_a/p, for fixed F/p, is proportional to the mean attachment cross-section Q_a it follows that, if $F/p < 3$ V cm^{-1} torr^{-1} Q_a is proportional to p, whereas for $F/p > 3$ V cm^{-1} torr^{-1} it is independent of p. The measured characteristic energy for electrons in O_2 for $F/p = 3$ V cm^{-1} torr^{-1} is 1 eV so that the data shown in Figs. 13.81 and 13.82 indicate that, at lower mean energies, the reaction is predominantly a three-body one and at higher a two-body. We may identify these with processes A and B respectively.

† BIONDI, M. A., *Phys. Rev.* **84** (1951) 1072A.

‡ HOLT, E. H., *Bull. Am. phys. Soc.* Ser. II 4 (1959) 112; SAXTON, M. C., MULCAHY, M. J., and LENNON, J. J., *Proc. Int. Conf. Ioniz. Phenom. Gases*, p. 94 (North Holland, Amsterdam, 1960).

§ CHANIN, L. M., PHELPS, A. V., and BIONDI, M. A., *Phys. Rev. Lett.* **2** (1959) 344; *Phys. Rev.* **128** (1962) 219.

|| HURST, G. S. and BORTNER, T. E., ibid. **114** (1959) 116.

The discovery of the pressure dependence removed or, at least alleviated, the first difficulty of the Bloch–Bradbury interpretation—the cross-section for vibrational deactivation required is reduced by two orders of magnitude. There remained the discrepancy with the afterglow

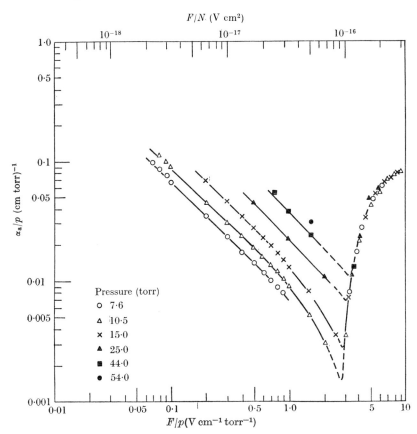

FIG. 13.81. Variation of α_a/p with F/p at different pressures in pure O_2 at 300 °K as observed by Chanin, Phelps, and Biondi.

results. It was suggested by several authors that the small effective attachment rate may be due to replenishment of free electrons through interaction of negative ions with excited species present in the afterglow. This would not be at all surprising (see Chap. 19, § 4) but it is difficult to verify. Evidence in support of this interpretation has come from microwave probing studies† of electron loss in afterglows which follow

† VAN LINT, V. A. J., WIKNER, E. G., and TRUEBLOOD, D. L., Bull. Am. phys. Soc. Ser. II 5 (1960) 122; CRAIN, C. M., J. appl. Phys. 29 (1958) 1605 and J. geophys. Res. 66 (1961) 1117.

the production of low density plasmas by bombardment of the gas with pulses of electrons of some MeV energy. In such an afterglow the distribution of excited species is likely to be very different from that in one that follows a discharge produced by microwave breakdown. Thus it would be

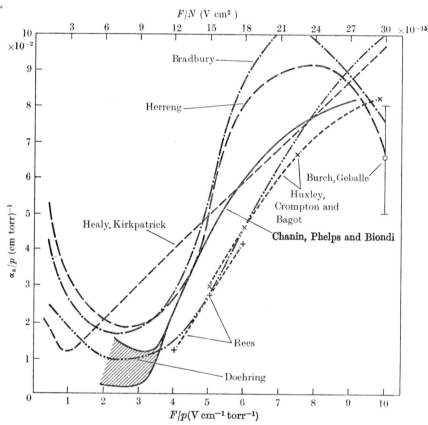

Fig. 13.82. Comparison of variations of α_a/p with F/p for O_2, observed by different experimenters. The shaded region for $F/p < 3$ V cm^{-1} torr^{-1} corresponds to the region in which α_a/p varies with pressure.

References: BRADBURY, N. E., *Phys. Rev.* **44** (1933) 883; BURCH, D. S. and GEBALLE, R., ibid. **106** (1957) 188; HEALEY, R. H. and KIRKPATRICK, C. B., from HEALEY, R. H. and REED, J. W., *The behaviour of slow electrons in gases*, p. 94 (Amal. Wireless, Sydney, 1941); DOEHRING, A., *Z. Naturf.* **7a** (1952) 253; HERRENG, P., *Cah. Phys.* **38** (1952) 7; CHANIN, L. M., PHELPS, A. V. and BIONDI, M. A., *Phys. Rev.* **128** (1962) 219.

expected that there would be a smaller relative concentration in low-lying excited states that are likely to be the most effective in causing detachment of electrons from negative ions. The observed attachment rates in these afterglows agree quite closely with those measured by Chanin *et al.* and exhibit the same pressure dependence.

Finally, the origin of process B has definitely been traced (see § 4.4.5) to the dissociative capture reaction (46) that has been so extensively studied by beam methods (§ 4.4.2).

4.4.4. *Analysis of Process A.* According to the Bloch–Bradbury interpretation, in pure O_2 process A proceeds by two stages

$$O_2 + e \rightarrow O_2^- {}^*,$$

where the * denotes the presence of vibrational excitation, followed by

$$O_2^- {}^* + O_2 \rightarrow O_2^- + O_2 + \text{energy}.$$

An attractive possibility is to consider the first stage as the formation of a vibrationally excited O_4^- molecule in a three-body collision

$$O_2 + O_2 + e \rightarrow O_4^- {}^*$$

followed by break-up of O_4^- into O_2 and O_2^-. In either case the overall result can be represented as a three-body reaction

$$O_2 + O_2 + e \rightarrow O_2^- + O_2 + \text{energy}. \tag{48}$$

We can imagine, in a mixture of O_2 with some non-attaching gas molecules X, that these molecules will also act as third-bodies in the reaction

$$O_2 + X + e \rightarrow O_2^- + X + \text{energy}. \tag{49}$$

Hence in such a mixture the rate of decrease of electron concentration

$$\frac{dn_e}{dt} = -k_0 \{n(O_2)\}^2 n_e - k_1 n(O_2) n(X) n_e, \tag{50}$$

where n_e, $n(X)$, $n(O_2)$ are the respective concentrations of electrons, X molecules, and O_2 molecules. The three-body coefficients k_0, k_1 will depend on the gas temperature and the mean electron energy. If α_a is the attachment coefficient and u the drift velocity of the electrons then

$$\frac{dn_e}{dt} = -\alpha_a u n_e, \tag{51}$$

so that $\qquad v_a = \alpha_a u = k_0 \{n(O_2)\}^2 + k_1 n(O_2) n(X), \tag{52}$

where v_a is the attachment frequency.

The three-body coefficient k_0 may be obtained either from experiments in pure oxygen or in mixtures with non-attaching gases. Thus v_a may be measured as a function of the fractional concentration of X. Extrapolation to the limit $n(X) = 0$ then gives k_0.

Fig. 13.83 illustrates the variation of k_0 with characteristic electron energy for different gas temperatures. The bulk of the observations were made by Chanin et al.† using the method described in Chapter 12,

† loc. cit., p. 1006.

§ 7.5.2 applied either to pure O_2 or to O_2–He mixtures. Some values, derived from the observations of Hurst and Bortner† for O_2–N_2 mixtures using their method described in Chapter 12, § 7.5.2, are also included, while the remaining data are those of Pack and Phelps.‡

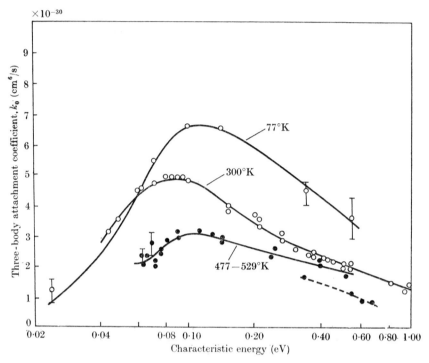

FIG. 13.83. Observed variation of the three-body attachment coefficient k_0 for O_2 with the characteristic electron energy ϵ_k for different gas temperatures. ○——— Chanin, Phelps, and Biondi (loc. cit., p. 1006). ●— — — Hurst and Bortner, 293 °K (loc. cit., p. 1006). ●——— Pack and Phelps.‡

Fig. 13.84 illustrates results obtained for O_2–CO_2 mixtures at 300 °K for values of $F/n(CO_2)$, which are such that the mean electron energy is close to thermal. In the figure $\nu_a/n(O_2)$ is plotted against $n(CO_2)$ under conditions in which the fractional concentration of O_2 is kept very small. According to (52) the resulting plot should be linear, passing through the origin and of slope k_1 for CO_2 as the third body. The observations of Pack and Phelps fall on such a plot quite accurately and provide strong evidence that the effect of the CO_2 is simply to provide a third body as in (49). This being so it is convenient to evaluate both k_0 and k_1

† loc. cit., p. 1006.
‡ PACK, J. L. and PHELPS, A. V., J. chem. Phys. 44 (1966) 1870.

by plotting $v_a/n(O_2)n(CO_2)$ as a function of $n(O_2)/n(CO_2)$ as in Fig. 13.85. The intercept of the linear plot gives $k_1 = 3\cdot1\pm0\cdot3\times10^{-3}$ cm^6/s and the slope gives $k_0 = 2\cdot0\pm0\cdot2\times10^{-30}$ cm^6/s.

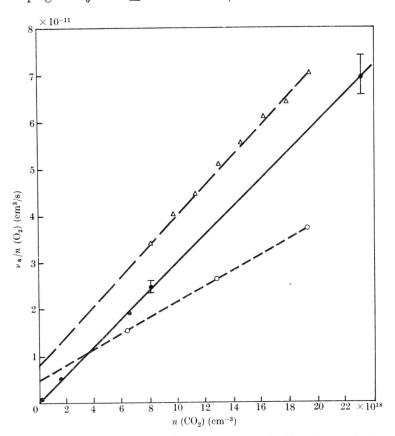

FIG. 13.84. Variation of $v_a/n(O_2)$ with $n(CO_2)$ for O_2–CO_2 mixtures in the limit of low O_2 concentrations. Observed by ● Pack and Phelps,[†] ○ Smith and Conway,[‡] △ O'Kelly, Hurst, and Bortner.[§]

Experiments in O_2–H_2O mixtures are of special interest in view of the long-known importance of water vapour in facilitating electron attachment. It was believed until quite recently that attachment occurred directly to water molecules but it is now known that this does not occur for low energy electrons (see § 7.2). Fig. 13.86 shows a plot of

$$v_a/n(O_2)n(H_2O)$$

[†] loc. cit., p. 1010.
[‡] SMITH, C. F. and CONWAY, D. C., Rev. scient. Instrum. 33 (1962) 726.
[§] O'KELLY, L. B., HURST, G. S., and BORTNER, T. E., Oak Ridge Nat. Lab. Rep. 2887 (1960).

against the ratio $n(O_2)/n(H_2O)$ under nearly thermal electron energy conditions, for two vapour temperatures, using observations made by Pack and Phelps. In obtaining these results special attention had to be paid to the preparation of the mixture. O_2–CO_2 mixtures were prepared in a mixing chamber of 20-cm diameter and 13-cm height and allowed to stand for about 1 h at room temperature before expansion into the main system. This free-mixing technique proved unsatisfactory for O_2–H_2O mixtures at 300 °K, due apparently to marked absorption of water on the walls of the system. An alternative 'freezing'

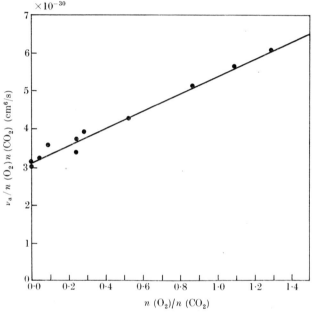

Fig. 13.85. Variation of $\nu_a/n(O_2)n(CO_2)$ with $n(O_2)/n(CO_2)$ for O_2–CO_2 mixtures for 1.1×10^{17} cm$^{-3} < n(CO_2) < 2.3 \times 10^{19}$ cm^{-3} and $F/n(CO_2) \leqslant 1.6 \times 10^{-17}$ cm^2 at 300 °K as observed by Pack and Phelps.

procedure was adopted in which the water vapour was first admitted to the system and allowed to stand for about 1 h during which an equilibrium pressure was reached. This pressure was read and the water vapour frozen out in a trap cooled with liquid nitrogen. Oxygen was then admitted at a pressure a little below the desired value and the trap then allowed to warm up to room temperature. As many as 4 h were necessary for the water vapour to diffuse throughout the system at the higher total gas densities used. This long time limited use of the method to temperatures below 450 °K—otherwise the risk of contamination was considerable. Results obtained using both methods of mixture preparation are shown in Fig. 13.86. The two points taken at 395 °K after the mixture had been aged for 60 h show that the ageing effects were small under the conditions involved.

Not only are the plots reasonably linear, particularly when mixture preparation by freezing is used, but the values of k_0 derived from the

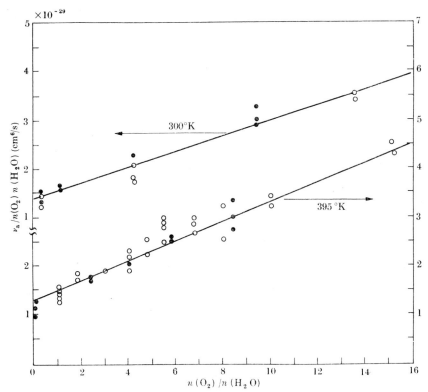

FIG. 13.86. Variation of $\nu_a/n(O_2)n(H_2O)$ with $n(O_2)/n(H_2O)$ for O_2–H_2O mixtures for $1 \cdot 1 \times 10^{17}$ cm^{-3} $< n(H_2O) < 3 \cdot 9 \times 10^{17}$ cm^{-3} and $F/n(H_2O) \leqslant 1 \times 10^{-16}$ V cm^2, at 300 and 395 °K, as observed by Pack and Phelps. ● observed using 'freezing' mixture technique. ○ observed using premixing.

slopes agree quite well with those obtained from the corresponding plots for O_2–CO_2 mixtures and from observations in pure O_2. Thus for 300 °K k_0 from the results obtained from O_2–H_2O mixtures prepared by the freezing technique is $1 \cdot 8 \pm 0 \cdot 2 \times 10^{-30}$ cm^6/s as compared with $2 \cdot 0 \pm 0 \cdot 2 \times 10^{-30}$ cm^6/s from O_2–CO_2 mixtures.

It seems likely that the interpretation of the role of the water vapour under the experimental conditions is only that of a third body as in (49). The striking result is that the values of k_1 derived from the intercepts

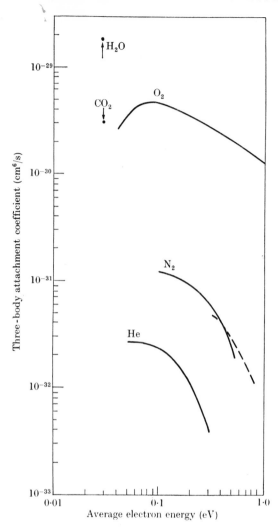

Fig. 13.87. Comparison of three-body attachment coefficients for different 'third' bodies at a gas temperature of 300 °K. —— Chanin, Phelps, and Biondi. — — — Hurst and Bortner.

of the linear plots in Fig. 13.86 on the vertical axis are nearly ten times larger than k_0, being $1\cdot4\pm0\cdot15\times10^{-29}$ cm^6/s both at 300 and 395 °K. It is this that probably explains the importance of eliminating water vapour if attachment of low energy (nearly thermal) electrons is to be avoided. Under most conditions oxygen is present as an impurity and the additional presence of water vapour markedly increases the ability of the oxygen molecules to attach electrons.

Fig. 13.87 gives comparative data on the effectiveness of different thermal bodies. Helium is much less effective than the molecular gases, as would be expected from the absence of internal degrees of freedom apart from electronic motion. N_2 is not very effective, presumably because of its comparative inertness. CO_2 is a little more effective than O_2 but H_2O is much more effective still.

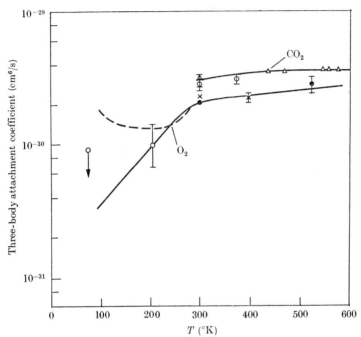

FIG. 13.88. Variation of the three-body attachment coefficients for thermal electrons with the gas temperature for O_2 and CO_2 as third bodies. Observed for O_2, ○ by Chanin, Phelps, and Biondi[†] from experiments in O_2. ● by Pack and Phelps[‡] from experiments in O_2–CO_2 mixtures. ✕ by Pack and Phelps[‡] from experiments in O_2–H_2O mixtures. — — — by van Lint, Wikner, and Trueblood[§] by microwave probing of an afterglow resulting from high-energy particle excitation. Observed for CO_2, △ by Pack and Phelps.[‡]

Finally, in Fig. 13.88 we give data on the variation of the three-body coefficients with gas temperature for thermal electrons, for O_2 and for CO_2. The data at the lowest temperatures are less reliable.

To carry out a more detailed analysis of the Bloch–Bradbury process it is necessary to use available information on the electron affinity $A(O_2)$ of O_2. This is still far from well known. Bates and Massey[||] estimated about 1 eV by empirical arguments depending on the bonding powers

† loc. cit., p. 1006.   ‡ loc. cit., p. 1010.   § loc. cit., p. 1007.
|| BATES, D. R. and MASSEY, H. S. W., Phil. Trans. R. Soc. A239 (1943) 269.

of different molecular orbitals. From analysis of the lattice energies of the polar crystals, $K^+O_2^-$, $Rb^+O_2^-$, and $Cs^+O_2^-$ Kazarnovski[†] found 1.0 ± 0.5 eV and Evans and Uri[‡] 0.75 eV. Other thermochemical studies[§] gave values between 0.15 and 0.4 eV. Photodetachment studies (Chap. 15, § 6.3) proved inconclusive. Curran[||] (see § 10) determined the appearance potential for O_2^- ions resulting from the reaction

$$O_3 + e \rightarrow O_2^- + O,$$

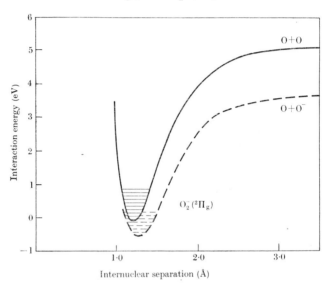

FIG. 13.89. Possible forms of the lowest potential energy curves of O_2^- in relation to O_2.

from which he deduced $A(O_2) \geqslant 0.58$ eV. Finally, Phelps and Pack,[††] from a study of the attachment–detachment equilibrium in a diffusing electron swarm by a technique described in Chapter 19, § 4, obtained $A(O_2) = 0.44$ eV.

Assuming for the moment that $A(O_2) = 0.44$ eV then we may tentatively relate the potential energy curve for O_2^- to that for O_2 as in Fig. 13.89.

If the lowest electronic ($^2\Pi_g$) state of O_2^- is the one involved in the attachment process, the initial capture of the electron must produce the ion with three vibrational quanta excited. This must be taken into

† KAZARNOVSKI, I. A., *Dokl. Akad. Nauk SSSR* **29** (1948) 67.
‡ EVANS, M. G. and URI, N., *Trans. Faraday Soc.* **45** (1949) 224.
§ PRITCHARD, H. O., *Chem. Rev.* **52** (1953) 529.
|| CURRAN, R. K., *J. chem. Phys.* **35** (1961) 1849.
†† PHELPS, A. V. and PACK, J. L., *Phys. Rev. Lett.* **6** (1961) 111.

account in discussing the rate of deactivation. It may, of course, be an upper electronic state of O_2^- such as $^4\Sigma_u^-$, which is concerned in the attachment process, but in any case the state must be either a stable one or of Type I in the sense of Chapter 12, § 6. The further alternative that the process occurs through formation of a complex such as O_4^- in a three-body collision cannot be ruled out.

Discussion of detachment processes and the composition of the ions under different conditions in pure O_2 and in mixtures of O_2 with other molecules will be deferred to Chapter 19, § 4.

4.4.5. *Analysis of swarm data for characteristic energies* $> 1 \cdot 0$ *eV* (*including attachment process B*). In Chapter 11, § 7.4 we discussed the analysis of swarm data in O_2 for electron characteristic energies $\epsilon_k < 1 \cdot 0$ eV ($F/p < 10$ V cm^{-1} torr^{-1}, $F/N < 3 \times 10^{-16}$ V cm^2). For this purpose the only inelastic energy losses are due to excitation of molecular rotation and vibration. At higher ϵ_k it is necessary to take account of the energy losses to electronic excitation and ionization.

Fig. 13.90 illustrates the cross-sections for these loss processes as well as the momentum-transfer cross-section assumed by Hake and Phelps† in their extension of the analysis to these higher values of ϵ_k. The excitation cross-section is based on the trapped-electron data of Schulz and Dowell‡ (see § 4.1 and Fig. 13.70) showing peaks in the energy loss spectrum with thresholds at 3·1, 7·0, and 10·5 eV. The relative magnitudes and energy dependence of the cross-sections corresponding to the first two of these processes were kept fixed at the values suggested by these observations near the thresholds, while the absolute values were determined to fit primarily the observed data on the energy-loss collision frequency (see Chap. 11 (76) for definition) for $0 \cdot 8 \leqslant \epsilon_k < 2$ eV. The shape and magnitude of the 10·5 eV loss were adjusted to provide a good fit of the observed values of the attachment frequency ν_a for $\epsilon_k > 2$ eV and of the net ionization coefficient $\alpha_i - \alpha_a$.

Fig. 13.91 shows the extent of the agreement achieved with the swarm data. Observed results for drift velocities u and characteristic energies ϵ_k at values of F/N beyond the range covered by the analysis are also included. The agreement for u is quite satisfactory. For ϵ_k there is considerable disagreement between the earlier and the most recent data with which the analysis agrees well.

The attachment coefficient α_a was calculated using the attachment cross-section measured by Schulz.§ This differs little from that observed

† HAKE, R. D. and PHELPS, A. V., *Westinghouse Research Report* 66–1E2–P1 (1966).
‡ loc. cit., p. 988.　　　§ loc. cit., p. 861.

by Rapp and Briglia,† which is illustrated in Fig. 13.78. It will be seen
that the calculated attachment coefficient agrees quite well with the
observed values obtained for F/N between 10^{-16} and 6×10^{-16} V/cm².
In this range the results of different experiments are not in too marked

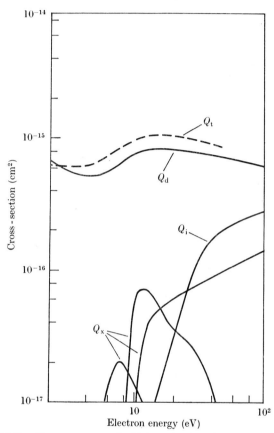

FIG. 13.90. Momentum transfer (Q_d), excitation (Q_x), and ionization
(Q_i) cross-sections assumed by Hake and Phelps in their analysis of
swarm data in O_2 at characteristic energies between 1 and 3 eV.
— — — observed total cross-section (Q_t).

disagreement. The observations at higher F/N do not agree well among
themselves but the calculated coefficient agrees closely with the observations of Huxley, Crompton, and Bagot.‡ For $(\alpha_i - \alpha_a)/N$ the agreement
between the calculated and observed values is good but does not cover
a wide range of F/N.

  Although there are many discrepancies to clear up before the swarm

† loc. cit., p. 1002.
‡ HUXLEY, L. G. H., CROMPTON, R. W., and BAGOT, C. H., *Aust. J. Phys.* **12** (1959) 303.

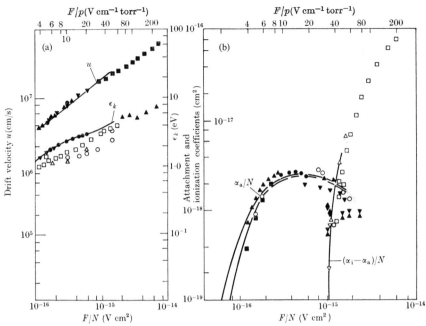

FIG. 13.91. Comparison of observed and calculated transport coefficients for electrons with characteristic energy between 1 and 3 eV in O_2. Full line curves give calculated results. (a) u, observed by ▲ Nielsen and Bradbury; ▼ Doehring; ● Pack and Phelps; ■ Frommhold. ϵ_k, observed by ● Huxley, Crompton, and Bagot; ▼ Rees; □ Brose; ○ Healey and Kirkpatrick; △ Townsend and Bailey; ▲ Schlumbohm; (b) $(\alpha_i - \alpha_a)/N$ observed by ▽ Dutton, Llewellyn-Jones, and Morgan; △ Freely and Fisher; □ Frommhold. α_a/N observed by ● Huxley, Crompton, and Bagot; ▲ Chanin, Phelps, and Biondi; ■ Rees; ▼ Freely and Fisher; □ Prasad and Craggs; ⅄ Frommhold; ○ Harrison and Geballe; ◆ Dutton, Llewellyn-Jones, and Morgan.

References: NIELSEN, R. A. and BRADBURY, N. E., *Phys. Rev.* **51** (1937) 69; DOEHRING, A., *Z. Naturf.* **7a** (1952) 253; PACK, J. L. and PHELPS, A. V., *J. chem. Phys.* **44** (1966) 1870; FROMMHOLD, L., *Fortschr. Phys.* **12** (1964) 597; HUXLEY, L. G. H., CROMPTON, R. W., and BAGOT, C. H., *Aust. J. Phys.* **12** (1959) 303; REES, J. A., ibid. **18** (1965) 41; BROSE, H. L., *Phil. Mag.* **50** (1925) 536; HEALEY, R. H. and KIRKPATRICK, C. B., from HEALEY, R. H. and REED, J. W., *The behaviour of slow electrons in gases* (Amal. Wireless, Sydney, p. 94, 1941); TOWNSEND, J. S. and BAILEY, V. A., loc. cit. (Fig. 13.68); SCHLUMBOHM, H., *Z. Phys.* **184** (1965) 492; DUTTON, J., LLEWELLYN-JONES, F., and MORGAN, G. B., *Nature, Lond.* **198** (1963) 680; FREELY, J. B. and FISHER, L. H., *Phys. Rev.* **133** (1964) A304; CHANIN, L. M., PHELPS, A. V., and BIONDI, M. A., ibid. **128** (1962) 219; PRASAD, A. N. and CRAGGS, J. D., *Proc. phys. Soc.* **77** (1961) 385; HARRISON, M. A. and GEBALLE, R., *Phys. Rev.* **91** (1953) 1.

data for O_2 and their analysis can be regarded as satisfactory for $F/p > 10$ V cm^{-1} torr^{-1} it seems quite clear that the attachment process observed in swarm experiments, which has a maximum for F/p close to 15 V cm^{-1} torr^{-1} and which we have referred to in § 4.4.3 as process B, is the two-body dissociative attachment

$$O_2 + e \rightarrow O + O^-$$

observed in the beam experiments (§ 4.4.2).

4.4.6. *Transport coefficients in dry air.* Using the cross-sections that were found to provide good results for analysis of swarm data in N_2 (see § 7.2 of Chap. 11 and § 3.8 of this chapter) and in O_2 (see § 7.4 of Chap. 11) it is possible to calculate the effective momentum transfer and energy-loss collision frequencies ν_m and ν_u for dry air containing 21 per cent O_2 and 78 per cent N_2. In Fig. 13.92 comparison is made

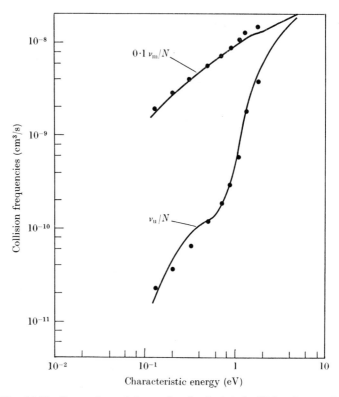

FIG. 13.92. Comparison of observed and calculated collision frequencies ν_m and ν_u in dry air. —— observed. ● calculated.

between values of ν_m/N and ν_u/N calculated in this way,[†] and derived from observations of the drift velocity made by Nielsen and Bradbury[‡] (using the filter method described in Chapter 2, § 3.1) and of ϵ_k made by Crompton, Huxley, and Sutton[§] and by Rees and Jory[||] (using the diffusion method described in Chap. 2, § 2). The agreement is quite

† HAKE, R. D. and PHELPS, A. V., loc. cit., p. 1017.
‡ NIELSEN, R. A. and BRADBURY, N. E., *Phys. Rev.* **51** (1957) 69.
§ CROMPTON, R. W., HUXLEY, L. G. H., and SUTTON, D. J., *Proc. R. Soc.* **A218** (1953) 507.
|| REES, J. A. and JORY, R. L., *Aust. J. Phys.* **17** (1964) 307.

good. It is of interest to note that vibrational excitation of O_2 is the dominant loss process in dry air for characteristic energies between 0·13 and 0·8 eV.

5. Carbon monoxide

Although little attention has been paid to the study of the excitation of discrete states of carbon monoxide by optical methods a considerable

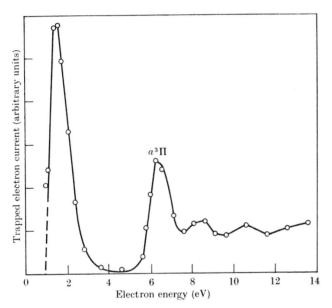

FIG. 13.93. Excitation spectrum of carbon monoxide obtained by Schulz using the trapped-electron method with a well depth of 0·7 V.

amount of work has been done using electrical methods to investigate excitation and ionization processes in the gas. Associated with this work was the hope of throwing light on the dissociation energy of CO and the electron affinity of O. A certain astrophysical interest attaches to the study of processes leading to excited CO^+ as the spectra of comets include quite strong bands arising from this ion.

5.1. Excitation of discrete states

Carbon monoxide is isoelectronic with nitrogen so the low-lying electronic states are of the same character and distribution except for the absence of splitting due to the equivalence of the force fields of the two nitrogen nuclei.

Fig. 13.93 shows an excitation spectrum for low energy electrons in CO

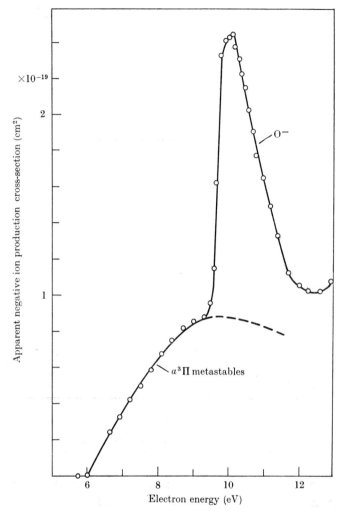

Fig. 13.94. Effective cross-section for production of negative ions from CO by electron impact as observed by Schulz. The anomalous contribution due to collection of secondary electrons ejected from the grid by $a^3\Pi$ metastable CO molecules is indicated.

obtained by Schulz,† using the trapped-electron method (Chap. 5, § 2.3) with a well depth of 0·7 V. The most prominent peak at 6·4 eV is almost certainly due to excitation of the metastable $a^3\Pi$ state for which the spectroscopic threshold is 6·0 eV. Further information about the excitation function for this state is forthcoming from the experiments of Schulz‡ on the production of O⁻ ions by electron beams in CO. With

† Schulz, G. J., Phys. Rev. 116 (1959) 1141.   ‡ Ibid. 128 (1962) 178.

the particular apparatus used (see Chap. 12, Fig. 12.19) some of the metastable atoms produced by the electron beam ejected secondary electrons from the grid that were collected by the plates together with

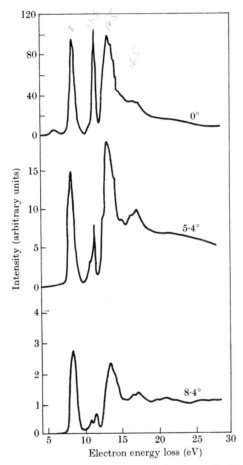

FIG. 13.95. Energy loss spectra of electrons of incident energy 508 eV scattered through various angles in CO as observed by Lassettre and Silverman.

any negative ions. Reference to Fig. 13.94, which shows the total yield of negative current, will indicate how the two contributions may be disentangled. Fig. 13.95 shows typical energy loss spectra obtained for electrons of a few hundred eV kinetic energy by Lassettre and Silverman.† Under these conditions the excitation of the $a^3\Pi$ state is inappreciable. The peak at 8.35 ± 0.06 eV is due to excitation of the $A^1\Pi$ state with a

† LASSETTRE, F. N. and SILVERMAN, S. M., *J. chem. Phys.* **40** (1964) 1256.

spectroscopic threshold at 8·02 eV. As the nuclear separations of the initial $X^1\Sigma^+$ and final $A^1\Pi$ states differ considerably (1·128 and 1·235 Å respectively) the transition is a suitable one for testing the application of the Franck–Condon principle. This was done by Lassettre and Silverman[†] who investigated the variation of the cross-section for

$$X^1\Sigma^+(v=0) \to A^1\Pi(v')$$

with the vibrational quantum number v' of the final state.

As discussed in Chapter 12, § 3.2, the shape of the energy-loss peak is independent of scattering angle so average values were taken at six scattering angles for electrons of 508-eV energy. The observed peak is not resolved as far as vibrational levels are concerned and correction

TABLE 13.15

Comparison of observed and calculated variation of peak height of $X^1\Sigma^+(v=0) \to A^1\Pi(v')$ transitions in CO, with v'

v'	Energy loss (eV)	Ratio of peak heights	
		Observed	Calculated
0	8·02	0·59	0·57
1	8·21	0·90	0·89
2	8·39	1·00	1·00
3	8·56	0·90	0·90
4	8·73	0·67	0·69
5	8·90	0·45	0·45
6	9·06	0·28	0·27
7	9·22	0·15	0·14

was made for the energy distribution of the incident electrons. This distribution was determined from observation of the peak shape for the 21·21-eV energy loss in helium (Chap. 5, § 4.2). Table 13.15 gives the comparison between the observed and calculated relative heights of the peak at the various energies corresponding to the different values of v'. A small energy displacement of 0·04 eV was introduced in the experimental scale to bring the peak maxima into the same position. The calculations[‡] simply involved the evaluation of the overlap integrals for the vibrational wave functions (see Chap. 12, § 4.1 and §§ 1.3.1, 1.4.2, and 3.2 of this chapter), which were represented by the Morse approximation.

The agreement is remarkably good even out to quite large values of v'

[†] LASSETTRE, F. N. and SILVERMAN, S. M., loc. cit. p. 1023.
[‡] JARMAIN, W. R., EBISUZAKI, R., and NICHOLLS, R. W., *Can. J. Phys.* **38** (1960) 510.

(cf. comparisons of a similar kind in Chapter 14, Tables 14.14, 14.19, 14.22, 14.24, and 14.26 as well as Tables 13.5 and 13.12 of this chapter).

As for H_2 (§ 1.3.1), N_2 (§ 3.2), and O_2 (§ 4.2), Lassettre and Silverman† determined the differential and integral generalized oscillator strengths for this transition and found by extrapolation 0·27 for the optical oscillator strength.

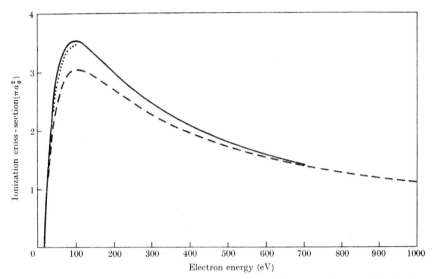

FIG. 13.96. Observed total ionization cross-sections for CO. —— Tate and Smith (loc. cit., p. 972). — — — Rapp and Englander-Golden (loc. cit., p. 972). • • • • Asundi, Craggs, and Kurepa (loc. cit., p. 1002).

The energy loss peaks at 10·78 and 11·41 eV may be identified as arising from excitation of the $B^1\Sigma^+$ and $C^1\Sigma^+$ levels with spectroscopic thresholds at 10·78 and 11·41 eV respectively. From extrapolation of the data obtained on the generalized oscillator strengths the optical oscillator strengths for the two transitions were derived as 0·034 and 0·28 respectively.

Lassettre and Silverman also obtained differential generalized oscillator strengths for the 13·55 eV transition and for a number of energies in the ionization continuum (17·09, 20, 24, 28, and 32 eV).

5.2. Ion production in CO

Fig. 13.96 illustrates the observed ionization cross-section as measured by different observers. The agreement between these results is not as satisfactory as for O_2 but is comparable with that for N_2.

† loc. cit., p. 1023.

Fox and Hickam,† using the retarding potential difference method, have investigated the variation of positive-ion production with electron energy just above the threshold. Fig. 13.97 illustrates the results they obtained. The first ionization potential is found to be $13 \cdot 98 \pm 0 \cdot 02$ eV, which agrees closely with the spectroscopic value $14 \cdot 013$ eV. Evidence of two changes of slope at $2 \cdot 4 \pm 0 \cdot 1$ and $5 \cdot 6 \pm 0 \cdot 1$ eV above the onset energy may be seen. These are interpreted as thresholds for production

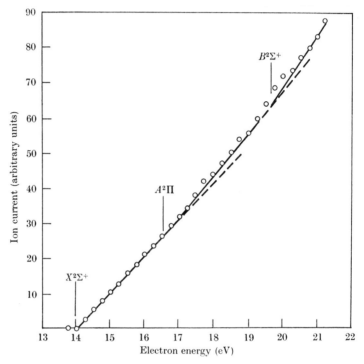

FIG. 13.97. Variation of the ionization cross-section of CO near the threshold as observed by Fox and Hickam.

of CO^+ ions in the $A^2\Pi$ and $B^2\Sigma^+$ states respectively which, according to spectroscopic data, lie at $2 \cdot 52$ and $5 \cdot 66$ eV above the ground state. The optical excitation function of the comet-tail bands of CO^+ ($A^2\Pi - X^2\Sigma^+$) has been observed by Bernard.‡ It is very similar to that for the first negative bands of N_2^+ (see Fig. 13.61).

The study of the various processes which yield positive and/or negative ions is of special interest because it seems that O^- ions are formed with

† Fox, R. E. and Hickam, W. M., *J. chem. Phys.* **22** (1954) 2059.
‡ Bernard, R., *C.r. hebd. Séanc. Acad. Sci. Paris,* **205** (1937) 793.

very little kinetic energy. Appearance potentials for their production should therefore be more reliable than for dissociative capture in O_2 which yields O^- with an energy of nearly 2 eV. Furthermore, processes have been observed which yield C^- ions and some evidence may be obtained about $A(C)$, the electron affinity of carbon.

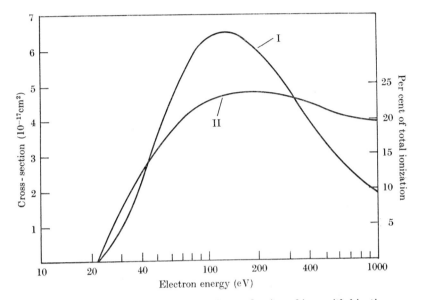

FIG. 13.98. Observed cross-section for production of ions with kinetic energy > 0.25 eV by electron impact in CO. I, absolute cross-section. II, percentage of total ionization cross-section.

Fig. 13.98 illustrates the total cross-section for production of positive ions with kinetic energy greater than 0.25 eV as observed by Rapp *et al.*,† using the method described in Chapter 12, § 7.3.4. Their results are very closely similar to those for N_2 (Fig. 13.64).

Table 13.16 summarizes the results obtained for appearance potentials for O^+, C^+, O^-, and C^- ions using mass-spectrograph and Lozier techniques. The proposed identification of the process involved is given, together with the predicted appearance potentials, on the assumption that the ions are formed with zero kinetic energy and that the following are the values of the atomic and molecular energies concerned:

$$D(CO) = 11.11 \text{ eV}, \qquad A(O) = 1.45 \text{ eV},$$
$$A(C) = 1.23 \text{ eV}, \qquad E(C^1D)‡ = 1.25 \text{ eV}.$$

† loc. cit., p. 979.
‡ $E(C^1D)$ denotes the excitation energy of the ^1D term of the ground configuration of carbon.

TABLE 13.16

Observed appearance potentials for positive and negative ions in carbon monoxide compared with theoretical expectation

Process	Ion	Observed (eV) a	b	c	d	e	f	g	h	Theoretical (eV)
$C(^3P)+O^-(^2P^0)$			$9·5\mp1·0$	$9·5\mp0·1$	$9·5\mp0·2$	$9·6\mp0·2$	$9·5\mp0·2$	$9·35\mp0·1$	$9·39\mp0·5$	$9·64$
$C^*(^1D)+O^-(^2P^0)$							$10·94\mp0·15$			$10·92$
$C^-(^4S^0)+O(^3P)$							$10·07\mp0·10$			$9·88$
$C^+(^2P_0)+O^-(^2P_0)$	C⁺			$20·9\mp0·1$	$20·9\mp0·2$	$20·9\mp0·2\ /\ 21·1\mp0·2$	$20·95\mp0·05$	$20·9\mp0·2$	$20·89\mp0·09\ /\ 20·92\mp0·05$	$20·92$
$C^+(^2P^0)+O(^3P)$	O⁻	$22·8\mp0·5$	$22·5\mp0·2$	$22·8\mp0·1$	$22·8\mp0·2$	$22·8\mp0·2$	$22·69\mp0·05$		$22·57\mp0·20$	$22·37$
$C^-(^4S^0)+O^+(^4S^0)$	C⁻						$23·65\mp0·01\ /\ 23·7\mp0·2$		$22·97\mp0·05$	$23·63$
$C(^3P)+O^+(^4S^0)$	O⁺	$24·8\mp0·5$							$23·41\mp0·17$	$24·72$
$C^*(^1D)+O^+(^4S^0)$							$25·8\mp0·2$		$24·78\mp0·17$	$25·99$

(a) Hogness, T. R. and Harkness, R. W., Phys. Rev. 32 (1928) 936 (M.S.).

(b) Vaughan, A. L., ibid. 38 (1931) 1687 (M.S.).

(c) Lozier, W. W., ibid. 46 (1934) 268 (L.).

(d) Hagstrum, H. D. and Tate, J. T., ibid. 59 (1941) 354 (M.S.).

(e) Hagstrum, H. D., Rev. mod. Phys. 23 (1951) 185 (M.S.).

(f) Lagergren, C. R., Dissertation, Univ. of Minnesota (1955) (M.S.).

(g) Craggs, J. D. and Tozer, B. A., Proc. R. Soc. A247 (1958) 337 (L.).

(h) Fineman, M. A. and Petrocelli, A. W., J. chem. Phys. 36 (1962) 25 (L.).

M.S. denotes mass spectrograph; L. Lozier method.

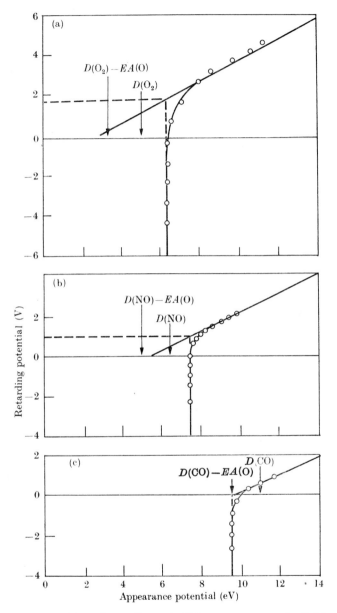

FIG. 13.99. Plots of retarding potential v. appearance potential for O^- ions arising from dissociative attachment in CO, NO, and O_2, as observed with a Lozier type apparatus. (a) O^- from O_2. (b) O^- from NO. (c) O^- from CO.

The assumed electron affinities are those obtained by Branscomb and his collaborators using the photodetachment method (see Chap. 15, §§ 5.2, 5.3).

It will be seen that the agreement both between the observed values and of these with the predicted is surprisingly good and suggests that in most cases the ions are formed with small kinetic energy only. This

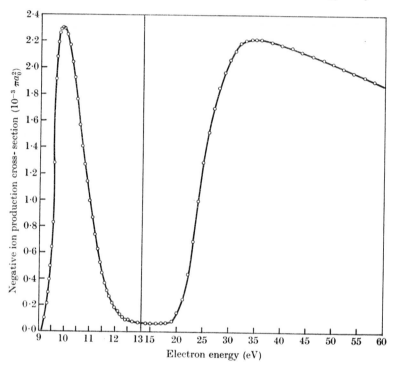

FIG. 13.100. Cross-section for production of negative ions in CO, as a function of electron energy, as observed by Rapp and Briglia.

is certainly true for the O⁻ formed by electron capture as may be seen by reference to a typical retarding potential analysis of the ion energies in this case, shown in Fig. 13.99 (c). The contrast with the corresponding results for O⁻ from capture in O_2 shown in Fig. 13.99 (a) is apparent.

Fig. 13.100 shows the total cross-section for production of negative ions from CO by electron impact, observed by Rapp and Briglia.† Although not so good as for O_2 the agreement with the results of other observers‡ is not unsatisfactory (see Chap. 12, Table 12.7).

† RAPP, D. and BRIGLIA, D. D., *J. chem. Phys.* **43** (1965) 1480.
‡ SCHULZ, G. J., *Phys. Rev.* **128** (1962) 178; ASUNDI, R. K., CRAGGS, J. D., and KUREPA, M. V., *Proc. phys. Soc.* **82** (1963) 967.

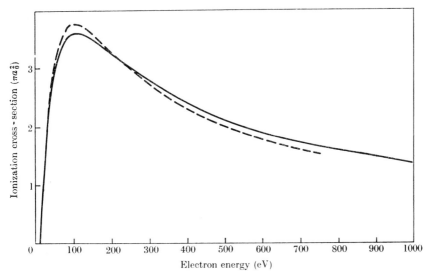

Fɪɢ. 13.101. Observed total ionization cross-sections for NO. ——— Tate and Smith (loc. cit., p. 972). ——— Rapp and Englander-Golden (loc. cit., p. 972).

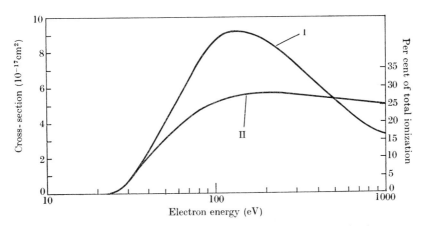

Fɪɢ. 13.102. Observed cross-sections for production of ions with kinetic energy > 0·25 eV by electron impact in NO. I, absolute cross-section. II, percentage of total ionization cross-section.

The shape of the cross-section is very similar to that for O_2 (Fig. 13.78) but it must be remembered in this case that the results include a contribution from C^- as well as O^- production in both the dissociative attachment and polar dissociation regions.

6. Nitric oxide

Fig. 13.101 illustrates the total ionization cross-section measured by

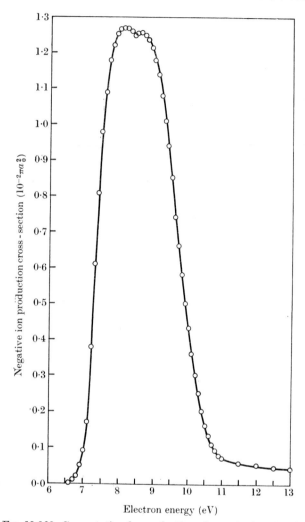

Fig. 13.103. Cross-section for production of negative ions in NO, as a function of energy, as observed by Rapp and Briglia.

Tate and Smith† and by Rapp and Englander-Golden.‡ The agreement between these observations is not unsatisfactory.

The cross-section for production of ions with kinetic energy greater than 0·25 eV§ is illustrated in Fig. 13.102 and that for production of negative ions|| in Fig. 13.103.

Appearance potentials for processes leading to production of positive

† loc. cit., p. 972.   ‡ loc. cit., p. 972.
§ Rapp, D., Englander-Golden, P., and Briglia, D. D., loc. cit., p. 979.
|| Rapp, D. and Briglia, D. D., loc. cit., p. 1002.

and negative ions in nitric oxide have been measured by many investigators using both Lozier and mass-spectrometric methods. Results obtained in some of the later work, that of Hagstrum[†] using the spectrograph, are given in Table 13.17. The interpretation in terms of particular reactions is also included together with the predicted appearance potential, on the assumption that ions are formed with zero initial kinetic energy and that the atomic and molecular energies are as shown in the table.

TABLE 13.17

Observed and predicted appearance potentials for positive and negative ions produced by electron impact on nitric oxide

Ion	Appearance potl. (V) for ions of zero kinetic energy (obs.)	Reaction	Appearance potl. (V) (calc.)
NO^+	9.4 ± 0.2	$\rightarrow NO^+ + 2e$	
O^-	5.3 ± 0.4	$N(^4S) + O^-(^2P)$	4.92_5
O^-	19.8 ± 0.2	$\left.\begin{array}{l} \\ \end{array}\right\} N^+(^3P) + O^-(^2P) + e$	19.57_5
N^+	19.9 ± 0.2		
N^+	21.7 ± 0.2	$N^+(^3P) + O(^3P) + 2e$	21.03
O^+	20.6 ± 0.2	$N(^4S) + O^+(^4S) + 2e$	20.10

$D(NO) = 6.49$ eV, $A(O) = 1.46_5$ eV, $I(N) = 14.55$ eV.

The agreement between observed and calculated appearance potentials is not very satisfactory even for the processes that give rise to positive ions only. Some discrepancy for the electron-capture process would not be surprising, for the ions have appreciable kinetic energy as may be seen by reference to a typical retarding potential curve shown in Fig. 13.99 (b).

7. Water vapour and hydrogen peroxide

A considerable amount of attention has been devoted to the experimental study of electron impact phenomena in water vapour.

7.1. *Excitation and ionization of* H_2O *by electrons of a few hundred eV energy*

The technique described in Chapter 5, § 4 has been applied by Lassettre and Francis,[‡] White,[§] Skerbele and Lassettre[||] and, using an apparatus of increased energy resolution, by Skerbele, Meyer, and Lassettre.[††]

[†] HAGSTRUM, H. D., *J. chem. Phys.* 23 (1955) 1178.
[‡] LASSETTRE, E. N. and FRANCIS, S. A., *J. chem. Phys.* 40 (1964) 1208.
[§] WHITE, E. N., Ph.D. thesis, Ohio State University (1956).
[||] SKERBELE, A. M. and LASSETTRE, E. N., *J. chem. Phys.* 42 (1965) 395.
[††] SKERBELE, A. M., MEYER, V. D., and LASSETTRE, E. N., ibid. 43 (1965) 817.

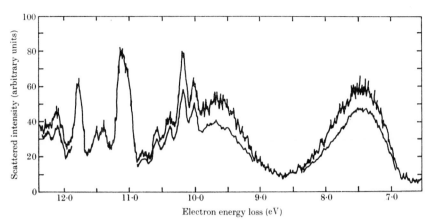

FIG. 13.104. Energy loss spectra of electrons of incident energy 200 eV scattered in the forward direction in H_2O as observed by Skerbele, Meyer, and Lassettre.

TABLE 13.18

Observed excitation potentials (V) of states in the water molecule

State	Electron impact	Ultra-violet spectra	
		(a)	(b)
Continuum	$7\cdot42\pm0\cdot02$	$7\cdot49$	
Continuum	$9\cdot67\pm0\cdot05$	$9\cdot75$	$9\cdot52$
$A_1 v = 0$	$10\cdot00\pm0\cdot00$	$10\cdot00$	$10\cdot00$
$B_1 v = 0$	$10\cdot17\pm0\cdot01$	$10\cdot16$	$10\cdot16$
$A_1 v = 1$	$10\cdot38\pm0\cdot01$	$10\cdot39$	$10\cdot39$
$B_1 v = 1$	$10\cdot57\pm0\cdot01$	$10\cdot56$	$10\cdot56$
$A_1 v = 2$	$10\cdot76\pm0\cdot02$	$10\cdot76$	
C_1	$11\cdot01\pm0\cdot01$	$11\cdot00$	$11\cdot00$
D_1	$11\cdot11\pm0\cdot001$	$11\cdot13$	$11\cdot11$
A_2	$11\cdot37\pm0\cdot01$	$11\cdot37$	$11\cdot36$
B_2	$11\cdot49\pm0\cdot02$	$11\cdot52$	$11\cdot50$
C_2, D_2	$11\cdot75\pm0\cdot00$		$11\cdot75$
C_3, D_3	$12\cdot07\pm0\cdot01$		$12\cdot06$
C_4, D_4	$12\cdot23\pm0\cdot01$		$12\cdot23$

(a) WATANABE, K. and ZELIKOFF, M., *J. opt. Soc. Am.* **43** (1953) 753.
(b) PRICE, W. C., *J. chem. Phys.* **4** (1936) 147.

Fig. 13.104 illustrates typical energy-loss spectra obtained in the last of these investigations. Table 13.18 lists the excitation potentials corresponding to the observed peaks, the energy scale being calibrated so that two of the peaks occur at the spectroscopically determined excitation energies, 10·00 and 11·75 eV, of two states that are the first members of Rydberg series. The first ionization potential is at 12·61 V.

Comparison of the excitation potentials observed by electron impact and by ultra-violet absorption spectroscopy reveals very good agreement. The identification of the excited states in terms of electronic states which are members of Rydberg series and the associated vibrational levels is also given in Table 13.18. Less precise observations are available about the magnitude of differential cross-sections and oscillator strengths. However, Skerbele *et al.* have determined oscillator strengths, using the extrapolation technique described in earlier sections, of the different transitions relative to that of the peak at 10·17 eV. Their results are compared with relative ultra-violet absorption coefficients in Table 13.19 and again good agreement is found.

TABLE 13.19

Comparison of relative optical oscillator strengths, for excitation of different states in the water molecule, by electron impact and by ultra-violet absorption spectroscopy

Energy loss (eV)	Relative oscillator strength	
	Electron impact	Absorption spectroscopy
7·42	0·29	0·26
9·67	0·60	0·43
10·00	0·79	0·72
10·17	1·00	1·00
10·38	0·60	0·63
10·57	0·49	0·47
10·76	0·27	0·25

7.2. *Excitation and ion production by low energy electrons*

Schulz† has used the trapped-electron method (Chap. 5, § 2.3) to investigate both excitation and negative ion formation by low energy electrons in H_2O. The procedure adopted to separate the two processes was as follows. With the potential on the collector F (Fig. 5.3) such as to produce a well depth of order of a fraction of 1 V, the current received at the collector will include negative ions as well as the trapped electrons. If the potential difference between F and the grid is reduced to zero no electrons will be trapped and the only current received will be due to the negative ions. By subtraction of this current from that obtained when the well depth has a finite depth W V the trapped electron current is obtained. In carrying out this subtraction the electron energy scale must be displaced by W eV to allow for the fact that the electron beam energy is greater by W eV when the well is present.

† SCHULZ, G. J., *J. chem. Phys.* **33** (1960) 1661.

Fig. 13.105 illustrates the excitation spectrum up to electron energies of 15 eV obtained in this way for a well depth of 0·3 eV. Peaks are found at 7·3, 9·2, 10·1, and 12·4 eV. It is probable that the first of these arises from the same excitation as that observed by Skerbele, Meyer, and Lassettre† with 200-eV electrons.

The variation of negative ion production with electron energy obtained by Schulz‡ is illustrated in Fig. 13.106 (a). Peaks occur at 6·5±0·1 and 8·8 eV in agreement with the earliest observations by Lozier§ and those of Buchel'nikova, using the equipment described in Chapter 12, § 7.4,

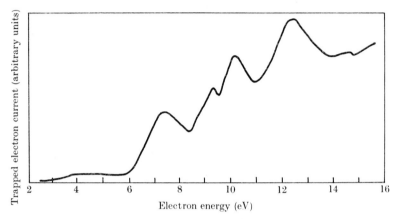

FIG. 13.105. Excitation spectrum of H_2O obtained by Schulz using the trapped-electron method and a well depth of 0·3 eV.

who found the peaks at 6·4 and 8·6 eV respectively. Identification of the ions requires the use of a mass spectrograph. Mann, Hustrulid, and Tate‖ found peaks of H^- production at 7·1±0·5 and 8·9±0·5 eV, which almost certainly coincide with the first two peaks in the curve shown in Fig. 13.106.

Compton and Christophorou†† have measured the variation with electron energy of both the H^- and O^- production. They paid special attention to the importance of avoiding loss of H^- ions, which possess considerable initial kinetic energy, before extraction into the mass analysis section. To do this they pulsed the electron current and applied the ion collecting potential as a pulse within 1 μs after the electron gate pulse. The O^- production is less than 3 per cent of the H^-, so that they were able to check that the H^- current they observed varied with

† loc. cit. p. 1033.   ‡ loc. cit. p. 1030.
§ LOZIER, W. W., *Phys. Rev.* **36** (1930) 1417.
‖ MANN, M. M., HUSTRULID, A., and TATE, J. T., ibid. **58** (1940) 340.
†† COMPTON, R. N. and CHRISTOPHOROU, L. G., ibid. **154** (1967) 110.

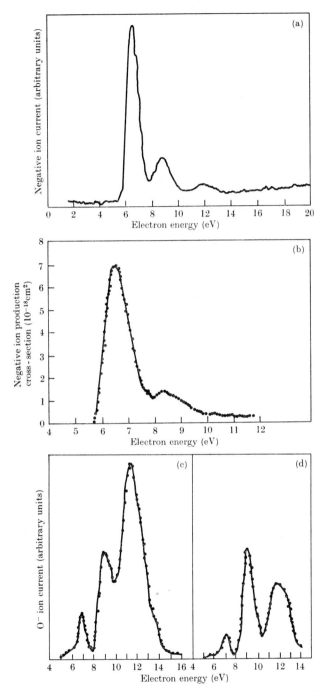

Fig. 13.106. Cross-section for production of negative ions in H_2O and D_2O as a function of electron energy. (a) Variation of total negative ion production from H_2O with electron energy as observed by Schulz. (b) Cross-sections for production of H^- ions as observed by Compton and Christophorou. (c) Variation of the production of O^- from H_2O with electron energy as observed by Compton and Christophorou. (d) Variation of the production of O^- from D_2O with electron energy as observed by Compton and Christophorou.

electron energy in very nearly the same way as the total ion current observed by Schulz (compare Fig. 13.106 (a) and (b)). It will be seen from Fig. 13.106 (b) that the H^- production is a maximum at 6·5 and 8·6 eV as compared with the earlier observations of Mann, Hustrulid, and Tate at $7·1 \pm 0·5$ and $8·9 \pm 0·5$ eV. There is good agreement also with the recent measurements of Dorman† who finds the peaks at $6·7 \pm 0·2$ and $8·8 \pm 0·2$ eV.

Compton and Christophorou‡ also observed D^- production from D_2O and found the width of the first peak to be 0·3 eV narrower than for H^- from H_2O, a result which can be accounted for by the narrower spread of the ground state vibrational wave function in D_2O. For O^- production a marked isotope effect was found as may be seen from Fig. 13.106 (c) and (d). This can be interpreted in much the same way as for H_2 and D_2 if the mean lifetime of the intermediate complex, H_2O^- or D_2O^-, is about 2×10^{-14} s. The shape of the O^- production curve from H_2O agrees well with observations by Dorman.

Buchel'nikova§ has measured the cross-sections for production of H^- ions as in her work on O^- from O_2 and CO. She finds peak cross-sections of 4·8 and $1·3 \times 10^{-18}$ cm² for processes occurring at 6·4 and 8·8 eV.

No evidence was found of the production of OH^- ions in single impacts although these were found in impact studies in H_2O_2 (see § 7.3). OH^- ions are, however, produced quite rapidly in ion-molecule reactions.

For the production of H^- from H_2O the effect of the temperature of the target molecules in invalidating the linear extrapolation method for determining the appearance potential for zero energy ions is quite small. Fig. 13.107 shows the linear extrapolation method applied to data obtained by Schulz for the two H^- production processes. The remarkable thing is that the same straight line, of the theoretical slope (17/18), passes through the points for both processes, giving the same appearance potential 4·4 eV for ions of zero kinetic energy. This would be equal to $D(HO-H) - A(H)$, where $D(HO-H)$ is the energy required to dissociate H_2O into HO and H and A is the electron affinity of H. Taking D as having the thermochemical value 5·11 eV and $A(H)$ as 0·74 eV, as derived from theory and photodetachment threshold observation, we obtain $D - A = 4·37$ eV, which agrees very well with the extrapolated value.

It seems that both the H^- ions and OH radicals are produced by each

† DORMAN, F. H., *J. chem. Phys.* **44** (1966) 3856.
‡ COMPTON R. N. and CHRISTOPHOROU, L. G., *Phys. Rev.* **154** (1967) 110.
§ BUCHEL'NIKOVA, I. S., loc. cit., p. 1002.

electron capture process in the same states. The processes must differ then in the nature of the unstable intermediate H_2O^- ion.

Attachment of electrons drifting in water vapour has been studied by a number of investigators. The diffusion method described in Chapter 12, § 7.5.3 was used in its original form by Bailey and Duncanson† in 1930 and in the form incorporating the modifications due to Huxley, by

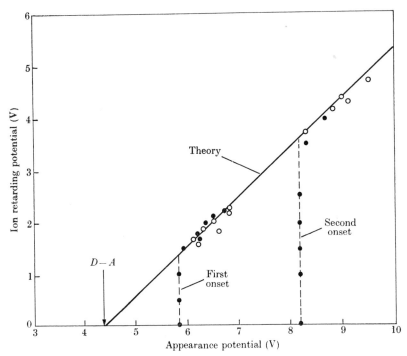

FIG. 13.107. Plot of retarding potential v. appearance potential for H^- ions from H_2O as observed by Schulz.

Crompton, Rees, and Jory‡ thirty-five years later. In contrast to most other substances investigated the two sets of observations of α_a/p, where α_a is the attachment coefficient and p the pressure of water vapour, do not agree very well as may be seen by reference to Fig. 13.108. We include also in Fig. 13.108 results obtained by Kuffel,§ using the Bradbury electron-filter method, and by Prasad and Craggs,‖ using the method described in Chapter 12, § 7.5.4, which depends on the measurements of the current passing in a pre-breakdown discharge through the vapour.

† BAILEY, V. A. and DUNCANSON, W. E., *Phil. Mag.* **10** (1930) 145.
‡ CROMPTON, R. W., REES, J. A., and JORY, R. L., *Aust. J. Phys.* **18** (1965) 541.
§ KUFFEL, E., *Proc. phys. Soc.* **74** (1959) 297.
‖ PRASAD, A. N. and CRAGGS, J. D., ibid. **76** (1960) 223.

Attempts to correlate the swarm measurements of α_a with the attachment cross-sections measured by Buchel'nikova were based first on the only available data about the characteristic energy of electrons in water vapour—those of Bailey and Duncanson.† Even when full allowance was made for the crudity of the comparison, based as it was on the assumption that the electron energy distribution in the drifting swarm

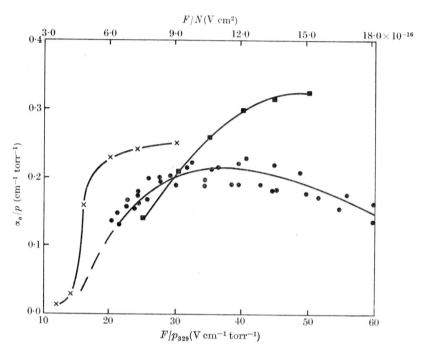

FIG. 13.108. Observed attachment coefficients α_a/p for electrons in water vapour. Observed by ——● Crompton, Rees, and Jory; ——× Bailey and Duncanson; —— — — Kuffel; ——■ Prasad and Craggs.

was Maxwellian about a mean energy $\frac{3}{2}\epsilon_k$, the two sets of data were clearly inconsistent. However, measurement by Crompton, Rees, and Jory‡ gave grossly different results (see Fig. 13.109). If these are used in place of the earlier data no inconsistency appears between the beam and the swarm data.

Hurst, O'Kelly, and Bortner§ applied their pulse-height method (Chap. 12, § 7.5.2) to measure α_a for argon containing varying fractional concentrations of water vapour. By extrapolation to zero concentration

† loc. cit., p. 1039.   ‡ loc. cit., p. 1039.
§ HURST, G. S., O'KELLY, L. B., and BORTNER, T. E., *Phys. Rev.* **123** (1961) 1715.

they obtained the attachment cross-section for H_2O averaged over the energy distribution appropriate to pure argon at each value of F/p where p is the argon pressure. They also found their results to be consistent with the beam observations of Buchel'nikova.† It appears that the observed attachment in a drifting swarm can all be assigned to dissociative attachment as observed in the beam experiments and no new process of attachment occurs at very low electron energies.

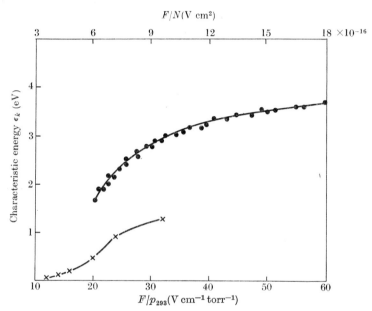

FIG. 13.109. Observed characteristic energy ϵ_k for electrons in water vapour. Observed by ———● Crompton, Rees, and Jory ; ———✕ Bailey and Duncanson.

The appearance potentials observed by Mann, Hustrulid, and Tate‡ for the positive ions produced in water vapour are given in Table 13.20, together with the suggested process responsible. We also include in the table the appearance potentials for O^- production by polar dissociation observed by the same authors. In all cases the appearance potentials are directly observed and do not necessarily equal the apparent appearance potential for production of ions of zero kinetic energy obtained by linear extrapolation. Because of this and lack of adequate knowledge of the kinetic energies of the ions, the identifications are very uncertain.

† See also COMPTON, R. N. and CHRISTOPHOROU, L. G., *Phys. Rev.* **154** (1967) 110.
‡ loc. cit. p. 1036.

7.3. *Negative ion production in* H_2O_2

Curran[†] has used a mass spectrometer in conjunction with an ion source equipped for application of the retarding-potential difference method to study the formation of negative ions by electron impact in hydrogen peroxide. In contrast to H_2O, OH^- ions were observed in addition to O^-. The variation of OH^- production with electron energy is shown in Fig. 13.110.

TABLE 13.20

Appearance potentials and probable reactions responsible for production of positive ions by electron impact in water vapour

Ion	Appearance potential (V)	Probable process
OH^+	$18\cdot7\pm0\cdot02$	$\rightarrow$ $H+OH^+$
O^+	$18\cdot8\pm0\cdot5$	$\rightarrow$ H_2+O^+
	$28\cdot1\pm1\cdot0$	$\rightarrow$ $2H+O^+$
H^+	$19\cdot5\pm0\cdot2$	$\rightarrow$ $OH+H^+$
H_2^+	$23\cdot0\pm2\cdot0$	$\rightarrow$ $O+H_2^+$
O^-	$23\cdot7\pm0\cdot5$	$\rightarrow$ $H+H^++O^-$
	$36\cdot0\pm3$	$\rightarrow$ $H^++H^++O^-$

The appearance potential is between 0 and 0·05 eV and the kinetic energy of the OH^- at the peak is less than 0·05 eV. Assuming the production process to be

$$H_2O_2+e \rightarrow OH+OH^-$$

then we must have $D(OH–OH)-A(OH) \simeq 0$. The dissociation energy $D(OH–OH)$ is known from thermochemical studies to be $2\cdot12\pm0\cdot05$ eV,[‡] which must therefore be close to the electron affinity of OH. It is somewhat larger than that, 1·83 eV, determined from the observed photodetachment threshold by Branscomb (Chap. 15, § 6.1).

8. Carbon dioxide

Considerable interest attaches to the study of electron impact phenomena in CO_2. The gas is an important minor constituent of the earth's atmosphere and is probably of even greater importance in the atmospheres of Venus and Mars. Furthermore, the main optical absorption by the gas in the ultra-violet occurs in the experimentally difficult wavelength range 720–1126 Å. This, however, corresponding to energy losses between 11 and 17 eV, offers no special difficulty for electron

[†] CURRAN, R. K., *Westinghouse Research Report* 908–3901–P4 (1961).
[‡] COTTRELL, T. L., *The strength of chemical bonds* (Butterworth, London, 1956).

impact studies which provide much useful information about the optical behaviour. In addition the molecule CO_2 exhibits in a comparatively simple form a number of the features characteristic of transitions in polyatomic molecules.

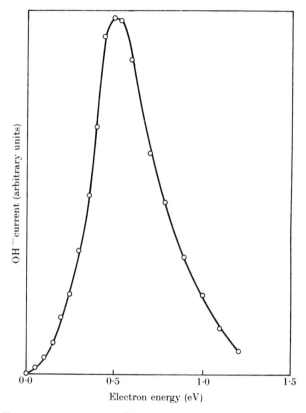

FIG. 13.110. Variation with electron energy of the production of OH^- ions by electron impact in H_2O_2 as observed by Curran.

Lassettre and Shiloff† have made a detailed study of inelastic processes in CO_2 produced by the impact of 500-eV electrons. The technique was the same as described in Chapter 5, § 4 and applied to various atomic and molecular gases (see Chap. 7, § 5.6.7 and §§ 1.3.1, 3.2, 4.2, 5.1, and 7.1 of this chapter). Fig. 13.111 illustrates a typical energy-loss spectrum. The ionization potential is 13·79 eV. Below this limit a number of peaks due to discrete excitation occur, namely at 8·61±0·06, 9·16±0·06, 11·09±0·03, 11·51±0·06, and 12·39±0·03 eV.

† LASSETTRE, E. N. and SHILOFF, J. C., J. chem. Phys. 43 (1965) 560.

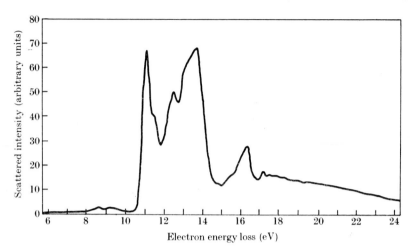

FIG. 13.111. Energy loss spectrum of electrons of incident energy 512 eV scattered in the forward excitation in CO_2, observed by Lassettre and Shiloff.

TABLE 13.21

Comparison of excitation potentials in CO_2 as observed by electron impact and by optical absorption methods

Excitation potentials (V)				
Electron impact		Optical absorption		
a†	b‡	a§	b‖	c††
$8{\cdot}41\pm0{\cdot}22$	$8{\cdot}61\pm0{\cdot}06$	$8{\cdot}43$	$8{\cdot}29$	$8{\cdot}40$
	$9{\cdot}16\pm0{\cdot}06$	$9{\cdot}28$		$9{\cdot}30$
		$10{\cdot}96$		
$11{\cdot}22\pm0{\cdot}25$	$11{\cdot}09\pm0{\cdot}03$	$11{\cdot}06$		$11{\cdot}05$
	$11{\cdot}51\pm0{\cdot}06$			$11{\cdot}40$
	$12{\cdot}39\pm0{\cdot}03$	$12{\cdot}45$		

Comparison with data obtained in optical absorption measurements is given in Table 13.21. The agreement is not unsatisfactory.

Lassettre and Shiloff derived the generalized differential oscillator strengths f′ in the usual way (see Chap. 7, § 5.6.7) not only for the discrete transitions we have been discussing but also for transitions in the ionization continuum out to energy losses of 75 eV.

† RUDBERG, E., *Proc. R. Soc.* A130 (1930) 182.
‡ LASSETTRE, E. N. and SHILOFF, J. C., loc. cit., p. 1043.
§ PRICE, W. C. and SIMPSON, D. M., *Proc. R. Soc.* A169 (1939) 501.
‖ WILKINSON, P. G. and JOHNSTON, H. L., *J. chem. Phys.* 18 (1950) 190.
†† INN, E. C. Y., WATANABE, K., and ZELIKOFF, M., ibid. 21 (1953) 1648.

Fig. 13.112 shows f' as a function of the square of the wave vector change, K^2, of the incident electron for the peaks of the 8·61- and 9·16-eV transitions. In both cases the transitions are extremely weak but it is clear that f' does not extrapolate to zero at vanishing momentum change. The transitions must therefore be optically allowed (see Chap. 7, § 5.6.7).

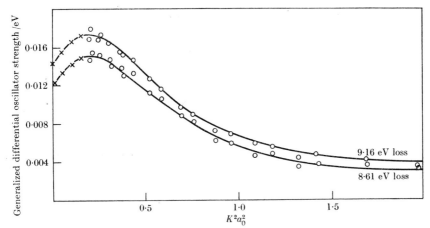

FIG. 13.112. Generalized differential oscillator strengths f' for the processes giving rise to the 8·61- and 9·16-eV energy losses for electrons in CO_2, as functions of $K^2 a_0^2$. ○ from observations of Lassettre and Shiloff. ✕ extrapolated using Newton's formula.

Lassettre and Shiloff produced arguments which suggest that, although the initial and final electronic states have different symmetries, the transition is not forbidden because vibrational distortion disturbs the symmetries.

Fig. 13.113 shows the corresponding results for the discrete transitions at 11·09 and 12·39 eV and for a number of transitions in the continuum.

Fig. 13.114 illustrates the total ionization cross-section of CO_2 as observed by Rapp and Englander-Golden† and by Asundi, Craggs, and Kurepa.‡ The agreement is not very close but not much worse than for O_2 (Fig. 13.73).

The cross-section for negative ion production as a function of electron energy observed by Rapp and Briglia§ (see Chap. 12, § 7.4) is illustrated in Fig. 13.115. These results agree quite well with earlier measurements‖ (see Chap. 12, Table 12.7).

The kinetic energy of the ions formed in the first process appears to

† RAPP, D. and ENGLANDER-GOLDEN, P., loc. cit., p. 972.
‡ ASUNDI, R. K., CRAGGS, J. D., and KUREPA, M. V., *Proc. phys. Soc.* **82** (1963) 967.
§ RAPP, D. and BRIGLIA, D. D., loc. cit., p. 1002.
‖ SCHULZ, G. J., *Phys. Rev.* **128** (1962) 178.

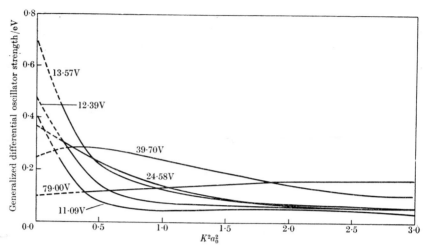

FIG. 13.113. Generalized oscillator strengths f' for processes giving rise to various energy losses for electrons in CO_2, including transitions to the continuum, as functions of $K^2a_0^2$.

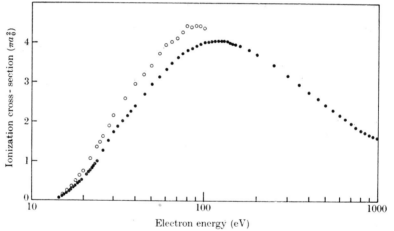

FIG. 13.114. Total ionization cross-sections for CO_2. Observed ● by Rapp and Englander-Golden; ○ by Asundi, Craggs, and Kurepa.

be less than 0·5 eV. For ions of zero kinetic energy the onset energy should be given by $D(CO–O)—A(O)$. Taking the thermochemical value 5·6 eV for $D(CO–O)$ and the photodetachment value 1·465 eV for $A(O)$ this gives 4·1 eV. This is to be compared with 3·83 eV as the actual onset for ions produced by the first process and 3·65 eV obtained by linear extrapolation of appearance potential curves for both processes. These both fit on the same straight line which lent additional but unjustified evidence to the reliability of the extrapolated value.

In Chapter 11, § 7.5 we discussed the analysis carried out by Hake and Phelps[†] of swarm data in CO_2 for electrons with characteristic energy $\epsilon_k < 1 \cdot 0$ eV. To extend this analysis to higher values of ϵ_k allowance must be made for electronic excitation and ionization. Hake and Phelps based their assumed excitation cross-section on the observations made by

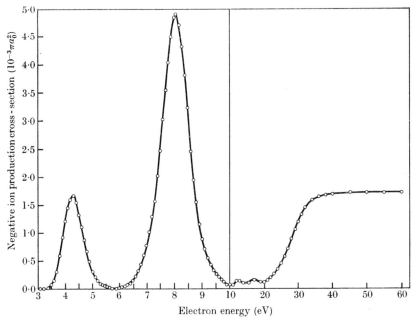

FIG. 13.115. Cross-sections for production of negative ions from CO_2 by electron impact, as a function of electron energy, observed by Rapp and Briglia.

Schulz[‡] using the trapped-electron method. They are shown in Fig. 13.116 together with the assumed ionization cross-section, which agrees quite well with that observed by Rapp and Englander-Golden[§] (see Fig. 13.114) and the assumed momentum-transfer cross-section, which is compared with the total cross-section measurements of Brüche.[||] It was also necessary to extend the vibrational excitation cross-sections to higher energies than required for analysis of data for $\epsilon_k < 1 \cdot 0$ eV (cf. Fig. 11.45).

Using these cross-sections Hake and Phelps extended their analysis to the calculation of the drift velocity u and characteristic energy ϵ_k for $F/N > 3 \times 10^{-16}$ V cm² with results shown in Fig. 13.117 (a). The good agreement obtained here persists when comparison is made with observed

[†] HAKE, R. D. and PHELPS, A. V., loc. cit., p. 1017.
[‡] SCHULZ, G. J., unpublished.   [§] loc. cit., p. 972.
[||] BRÜCHE, E., *Annln Phys.* **83** (1927) 1065.

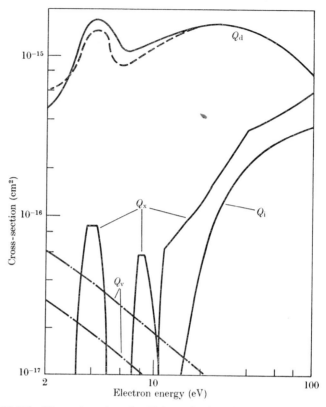

FIG. 13.116. Momentum transfer (Q_d), excitation (Q_x), and ionization (Q_i) cross-sections assumed by Hake and Phelps in their analysis of swarm data in CO_2 at characteristic energies > 1 eV. ——— observed total cross-section. —·—·— assumed cross-sections for vibrational excitation. The upper and lower curves refer respectively to the excitation of the vibrational levels at 0·300 and 0·083 eV.

and calculated values of the net ionization coefficient $\alpha_i - \alpha_a$ but is not quite so satisfactory for α_a alone (see Fig. 13.117 (*b*)).

9. Ion production by electron impact in NO_2 and N_2O

Apart from the general interest of these molecules as minor atmospheric constituents attention has been drawn to the frequent observation of NO_2^- in experiments involving nitrogen and oxygen mixtures and its observation in the upper atmosphere by rocket-borne mass-spectrometers.

Fox† has studied negative-ion formation by impact of electron beams in NO_2, using a mass spectrometer of 90° sector magnetic field type in

† Fox, R. E., *J. chem. Phys.* **32** (1960) 285.

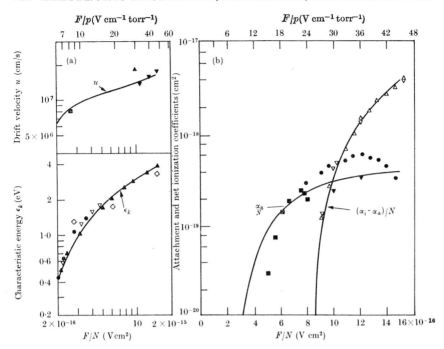

Fig. 13.117. Comparison of observed and calculated transport coefficients for electrons with characteristic energy > 1 eV in CO_2. (a) Full line curves give calculated results. u observed by ■ Errett; ▼ Frommhold; ▲ Pack, Voshall, and Phelps. ϵ_k observed by ▲ Rees; ◇ Skinker; ▽ Rudd; ● Warren and Parker. (b) Full line curves give calculated results. $(\alpha_i - \alpha_a)/N$ observed by △ Bhalla and Craggs; ▽ Schlumbohm. α_a/N observed by ● Bhalla and Craggs; ■ Chatterton and Craggs; ▼ Schlumbohm.

References: ERRETT, D., Ph.D. Thesis, Purdue University (1951); FROMMHOLD, L., loc. cit. (Fig. 13.68); PACK, J. L., VOSHALL, R. E., and PHELPS, A. V., *Phys. Rev.* **127** (1962) 2084; REES, J. A., *Aust. J. Phys.* **18** (1965) 41; SKINKER, M. F., *Phil. Mag.* **44** (1922) 994; RUDD, J. B., from HEALEY, R. H. and REED, J. W., *The behaviour of slow electrons in gases* (Amal. Wireless, Sydney, 1941); WARREN, R. W. and PARKER, J. H., *Phys. Rev.* **128** (1962) 2661; BHALLA, M. S. and CRAGGS, J. D., *Proc. phys. Soc.* **76** (1960) 369; SCHLUMBOHM, H., *Z. Phys.* **166** (1962) 192; CHATTERTON, P. A. and CRAGGS, J. D., *Proc. phys. Soc.* **85** (1965) 355.

conjunction with the retarding potential difference method. The only negative ion observed in appreciable quantity was O^-. A very weak NO_2^- peak was seen but may well have arisen in some secondary process (see Chap. 19, § 4).

Schulz† has studied excitation and negative ion formation in N_2O by the same methods as those he used for H_2O (§ 7.2). The only negative ions formed are O^- as identified in mass-spectrometer studies by Curran and Fox.‡ They found peaks of O^- production at $0·5\pm0·2$ and $2·0\pm0·2$ eV respectively.

† SCHULZ, G. J., *J. chem. Phys.* **34** (1961) 1778.
‡ CURRAN, R. K. and FOX, R. E., ibid. **34** (1961) 1590.

O

Cross-sections for total ionization,† for production of ions with kinetic energy > 0.25 eV‡ and for negative ion formation§ have been measured for N_2O and are shown in Fig. 13.118.

10. Negative ion production in ozone

The principal interest of this work is that it provides some further evidence about the electron affinity of O_2. Curran‖ studied the negative ion production using a mass spectrometer in conjunction with the retarding potential difference method. Ozone, purified from oxygen by liquefaction at liquid nitrogen temperature, was distilled into the mass spectrometer and again liquefied. Ozone evaporating from the liquid surface was admitted to the ion source through a glass tube about 1 m long. A filament of thoriated iridium ribbon was used in the ion source.

The ionization potential was measured as 12.89 ± 0.10 eV, which agrees well with that obtained spectroscopically (12.80 ± 0.05 eV). O^- and O_2^- ions were both observed. In view of the possible presence of impurities check was made of the linear dependence of the negative ion current on pressure. This was done by using the measured O_2^+ current as a measure of ozone pressure, O_2^+ being the dominant positive-ion fragment produced.

The appearance potentials of O^- and O_2^- were found to be in the range 0 to 0.05 eV and 0.42 ± 0.03 eV respectively. By retarding potential methods the kinetic energies of the ions were found to be 0.3 ± 0.1 eV and in the range 0 to 0.05 eV respectively.

Since the O_2^- ions appear to be formed with nearly zero kinetic energy we must have

$$D(O_2\text{–}O) - A(O_2) - Ex(O_2^-) \simeq 0.42 \text{ eV}.$$

The thermochemical value of the dissociation energy $D(O_2\text{–}O)$ of O_3 into $O_2 + O$ is 1.0 eV so that

$$A(O_2) = 0.58 \text{ eV} + Ex(O_2^-).$$

Here $Ex(O_2^-)$ is the vibrational energy content of the O_2^- ion, which may or may not be zero. Other evidence about $A(O_2)$, the electron affinity of O_2, is discussed in § 4.4.4 and in Chapter 19, § 4.

The appearance of O^- ions at zero electron energy is consistent with

† RAPP, D. and ENGLANDER-GOLDEN, P., loc. cit., p. 972.
‡ RAPP, D., ENGLANDER-GOLDEN, P., and BRIGLIA, D. D., loc. cit., p. 998.
§ RAPP, D. and BRIGLIA, D. D., loc. cit., p. 1002.
‖ CURRAN, R. K., J. chem. Phys. 35 (1961) 1849.

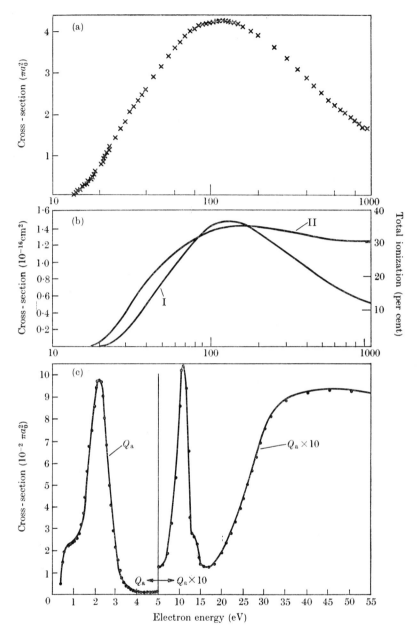

Fig. 13.118. Observed cross-sections (a) for total ionization, (b) for production of ions with kinetic energy > 0.25 eV, and (c) for negative ion formation, by electron impact in N_2O.

the fact that $D(O_2\text{--}O) - A(O) < 0$, $A(O)$, the electron affinity of O, being taken as 1·465 eV (Chap. 15, § 5.2).

11. Negative ion production from halogen-containing molecules

The halogens are the most electronegative of the elements and their atoms have the highest electron affinities. It is therefore of considerable interest to study negative ion production by electron impact with molecules containing halogen atoms. We shall begin by discussing the case of sulphur hexafluoride SF_6 because it shows some special features, and turns out to be of considerable importance in practice.

11.1. Sulphur hexafluoride

Considerable interest attaches to the study of the formation of negative ions by electron impact in sulphur hexafluoride. In the first instance the interest was largely a technical one because of the high dielectric strength of SF_6, which has led to its use in transformer insulation, etc. However, it turned out that metastable SF_6^- ions are formed by resonance capture of electrons with almost zero energy. The capture cross-section is large and the lifetime of the unstable ion is quite high enough for it to be observed readily in mass spectrometers with normal path length. Because of the long lifetime the resonance is very sharp. This means that the observed variation of the cross-section for SF_6^- production with electron energy is determined by the energy distribution of the electrons in the incident beam. A very convenient means is therefore provided for obtaining this distribution in mass-spectrograph studies of negative ion formation. Furthermore, it appears that the resonance occurs very close indeed to zero electron energy so that the appearance potential for SF_6^- production can be used to calibrate the voltage when studying negative ion formation from any substance. SF_6 is not very reactive so it can usually be mixed at low pressure with the gas or vapour under investigation.

The evidence for the metastability of SF_6^- resulting from impact of very low energy electrons is twofold. It is found that the production rate is independent of pressure and electron current under the experimental conditions so the ions are formed in single collisions. Under these circumstances the energy of the ion must exceed that of the neutral molecule as the only way of disposing of the energy introduced with the captured electron, except by dissociation or autodetachment, is by radiation, a most unlikely process. Moreover, non-integral peaks in the mass spectra of negative ions from SF_6 have been observed by Ahearn and Hannay† and by Marriott and Craggs.‡ These probably arise through dissociation of metastable ions during flight.

The lifetime of a complex ion such as SF_6^- before break-up is likely to be quite long. Once the electron is captured the surplus energy is rapidly distributed among a number of different modes of motion and, on the average, a considerable time will elapse before it concentrates on a mode which leads to dissociation or autodetachment.

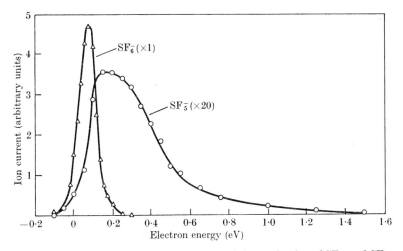

FIG. 13.119. Variation with electron energy of the production of SF_6^- and SF_5^- ions by electron impact in SF_6, as observed by Hickam and Fox.

In their investigation of electron capture in SF_6^-, which first established the main features of SF_6^- and SF_5^- production by slow electrons (< 1.5 eV), Hickam and Fox§ used two instruments. One was a 90° sectored field mass-spectrometer and the other an apparatus which measured total ion production. In both cases the electrode system was designed for use of the retarded potential difference method so that the electron energy distribution could be confined to a width of order 0.1 eV. The method was used with pulsed fields (see Chap. 3, § 2.4.2) with the mass spectrometer but steady field with the total ion collection apparatus.

Fig. 13.119 illustrates the variation with electron energy of the rate of production of SF_6^- and SF_5^- ions observed with the mass spectrometer. Good agreement was found between the variation with electron energy of the total negative ion current, observed with the total ion collection apparatus, and that obtained by summing the SF_6^- and SF_5^- currents

† AHEARN, A. J. and HANNAY, N. B., *J. chem. Phys.* **21** (1953) 119.
‡ MARRIOTT, J. and CRAGGS, J. D., *Brit. J. Electron.* ser. iv (1956) 405.
§ HICKAM, W. M. and FOX, R. E., *J. chem. Phys.* **25** (1956) 642.

observed in the mass spectrometer. In Fig. 13.120 the shape of the
SF_6^- peak is compared with the electron energy distribution obtained
by a retarding potential analysis at the electron collector electrode. It
will be seen that the maximum rate of SF_6^- production occurs 0·03 eV
above the maximum of the electron energy distribution, the shape of
which is very closely similar to that of the SF_6^- peak. In fact, within
experimental error, it can be said that the SF_6^- production occurs,

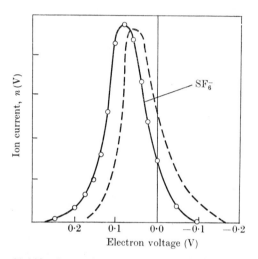

FIG. 13.120. Comparison of the electron energy distribution (— — —) as determined by a retarding potential
analysis, with the shape of the SF_6^- peak (——).

within an energy range of less than 0·05 eV, about a mean energy of
less than 0·1 eV. If this is so the cross-section for SF_6^- production,
derived from the negative ion current observed in the total ion collection
apparatus, corrected for the proportion of SF_5^- present, comes out
to be greater than 10^{-15} cm². From the observation of SF_6^- in these
experiments it follows that the mean lifetime of the metastable ion must
be greater than 10^{-6} s.

The production of SF_6^- ions has also been studied by Buchel'nikova[†]
and by Rapp and Briglia[‡] with results very similar to those of Hickam
and Fox. They obtain as peak cross-sections 2·4 and $5·7\pi a_0^2$ respectively,
but the peak is so sharp that the observed maximum depends very much
on the electron energy distribution.

† BUCHEL'NIKOVA, I. S., *Zh. éksp. teor. Fiz.* **35** (1958) 1119; *Soviet Phys. JETP* **35**(8)
(1959) 783.

‡ RAPP, D. and BRIGLIA, D. D., loc. cit., p. 1002.

The total ionization cross-section of SF_6 has been measured by Rapp and Englander-Golden† and is as shown in Fig. 13.121.

Negative and positive ion production by more energetic electrons have been studied by several investigators. The ratio of the abundances of

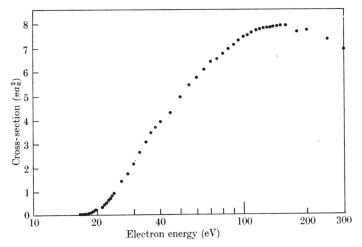

FIG. 13.121. Total ionization cross-section of SF_6 as measured by Rapp and Englander-Golden.

different positive ions produced by impact of 70-eV electrons was found to be as follows:

$$SF_5^+ \ (100), \ SF_4^+ \ (10, \ 9\cdot6), \ SF_3^+ \ (22, \ 29\cdot6), \ SF_2^+ \ (3\cdot2, \ 5\cdot6), \ SF^+ \ (4\cdot1, \ 9\cdot8),$$

$$F^+ \ (0\cdot7, \ 5\cdot0), \ S^+ \ (-, \ 8\cdot8), \ SF_5^{++} \ (2\cdot4, \ -), \ SF_4^{++} \ (5\cdot8, \ 9\cdot8), \ SF_3^{++} \ (1\cdot3, \ 1\cdot1),$$

$$SF_2^{++} \ (2\cdot4, \ 6\cdot0), \ SF^{++} \ (-, \ 0\cdot7).$$

In each case the first numbers are those found by Dibeler and Mohler,‡ the second by Marriott.§ They cannot be regarded as providing more than a rough guide to the true results because of uncertain discrimination in the collection of ions by the mass spectrograph.

A noteworthy feature is the negligibly small production rate of SF_6^+ ions at these electron energies. There are other cases in which the positive ion resulting from removal of a single electron is formed with very low probability by electrons of energy of the order 100 eV (e.g. CCl_4,‖ CCl_3F,‖ CCl_2F_2,‖ $CClF_3$,‖ CF_4,‖ SiF_4‡) and in others it is not

† loc. cit., p. 972.
‡ DIBELER, V. H. and MOHLER, F. L., *J. Res. natn. Bur. Stand.* **40** (1948) 25.
§ MARRIOTT, J., Thesis, Liverpool (1954).
‖ MARRIOTT, J. and CRAGGS, J. D., *E.R.A. Report No. L/T* 301 (1953); CRAGGS, J. D. and MCDOWELL, C. A., *Rep. Prog. Phys.* **18** (1955) 374.

the most abundant ion (e.g. CF_3Br,[†] $SiCl_4$,[‡] $GeCl_4$,[‡] $TiCl_4$,[§] PCl_3,[||] $AsCl_3$, $SbCl_3$).

Buchel'nikova[††] has measured cross-sections for production of negative ions by slow electrons in a number of halogen-containing compounds and in most cases observes high maximum cross-sections at very

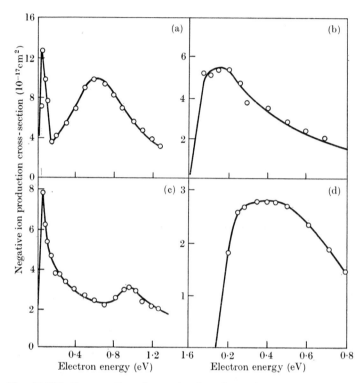

FIG. 13.122. Cross-sections for production of negative ions by impact of slow electrons in (a) CCl_4, (b) CCl_2F_2, (c) CF_3I, and (d) BCl_3, observed by Buchel'nikova.

low energies. Fig. 13.122 illustrates her results for CCl_4, CCl_2F_2, CF_3I, and BCl_3. Fox and Curran[‡‡] have shown by mass analysis that, in contrast to SF_6, the dominant ion produced by very low energy electrons in CCl_4 is not CCl_4^- but Cl^-.

† WARREN, J. W., Thesis, London (1950).
‡ VOUGHT, R. H., *Phys. Rev.* **71** (1947) 93; MARRIOTT, J., Thesis, Liverpool (1954).
§ MARRIOTT, J., THORBURN, R., and CRAGGS, J. D., *Proc. phys. Soc.* B**67** (1954) 437.
|| KUSCH, P., HUSTRULID, A., and TATE, J. T., *Phys. Rev.* **52** (1937) 840.
†† loc. cit., p. 1054.
‡‡ FOX, R. E. and CURRAN, R. K., *J. chem. Phys.* **34** (1961) 1595.

11.2. *Iodine*

Experimental investigation of electron attachment processes in the halogens is specially difficult because of the high reactivity of these substances. Most attention has been paid to iodine because it is somewhat easier to handle. Even then there are difficulties due to production of hydrogen iodide while the experiment is in progress. Furthermore, introduction of iodine vapour into the collision chamber produces serious changes in contact potentials. It is not surprising that the earlier experiments failed to produce consistent results. The earliest observations of Mohler[†] and of Hey and Leipunsky[‡] agreed only in indicating that the cross-section for negative ion formation is high at low electron energies—they observed different variations of production rates with electron energy and different absolute cross-sections. Hogness and Harkness[§] using a mass spectrometer showed that I^-, I_2^-, and I_3^- are all formed and their results suggested that the most abundant ion, I^-, is formed by capture of electrons of zero kinetic energy.

Somewhat later Buchdahl[||] investigated negative ion formation using a Lozier type apparatus and obtained a negative ion production curve with maxima at energies of 0·4, 1·75, and 2·5 eV. The peak cross-section, at 0·4 eV, was determined as about 4×10^{-17} cm².

Swarm experiments carried out by Healey[††] gave a single broad maximum of about 10^{-17} cm² at a mean electron energy of 2 eV.

In view of this unsatisfactory position, Biondi and Fox undertook a thorough investigation using microwave,[‡‡] mass spectrograph,[§§] and total ion collection[§§] techniques.

The microwave technique used was the same as that described in Chapter 12, § 7.5.5. Because of the large mass of the iodine molecule the time for electrons in the afterglow to come to thermal equilibrium with the ions and neutral molecules is relatively long, comparable with the time interval between successive probings. To remove this difficulty a partial pressure of 3 torr of helium was added to the iodine vapour in the quartz cavity. In this way the time for electrons to come to thermal equilibrium was reduced to below 50 μs. It has the additional advantage of reducing electron loss to the walls by diffusion.

† MOHLER, F. L., *Phys. Rev.* **26** (1925) 614.
‡ HEY, W. and LEIPUNSKY, A., *Z. Phys.* **66** (1930) 669.
§ HOGNESS, T. R. and HARKNESS, R. W., *Phys. Rev.* **32** (1928) 784.
|| BUCHDAHL, R., *J. chem. Phys.* **9** (1941) 146.
†† HEALEY, R. H., *Phil Mag.* **26** (1938) 940.
‡‡ BIONDI, M. A., *Phys. Rev.* **109** (1958) 2005.
§§ FOX, R. E., ibid. 2008.

Before experiments the quartz bottle was first evacuated and baked out at 450 °C on an ultra-high vacuum system prior to distilling in chemically pure iodine and the working pressure of helium. It was then sealed off, leaving a projecting tip which extended into an oven. By varying the oven temperature between 160 and 230 °K the iodine vapour pressure could be varied between 10^{-10} and 10^{-4} torr.

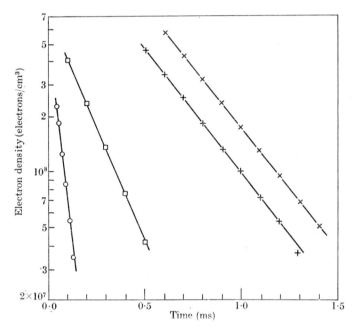

FIG. 13.123. Observed decay of electron concentration in an afterglow in an iodine–helium mixture at 300 °K. Helium pressure 3 torr. Iodine pressure ◯ 1.5×10^{-4}, ▢ 3.1×10^{-5}, + 1×10^{-7}, ✕ 3×10^{-10} torr.

To avoid appreciable negative ion formation during the discharge a short exciting pulse length ($< 10^{-5}$ s) was used. In order that the conditions in the afterglow were such that electron loss by attachment dominated it was necessary to work with a negative ion concentration less than 10–100 times that of the electrons.

Fig. 13.123 shows some typical logarithmic plots of observed electron density against time from the cessation of power input. The curves taken over a wide range of pressures are accurately linear, verifying that the exponential decay rate is closely followed. When, however, the attachment rate ν_a', derived from these curves, is plotted as a function of iodine pressure the curve shown in Fig. 13.124 is obtained. It will be seen that, as the iodine pressure falls, the apparent attachment rate tends to a steady value which is far above that due to loss by ambipolar diffusion. As the mass spectrograph investigations carried out by Fox, which will shortly be described, showed that the HI concentration built up quite rapidly in his equipment, it was assumed that the steady background value is due to a constant pressure of HI, presumably formed during seal-off. This value was then subtracted to give the decay rate shown as in Fig. 13.124. This satisfies the requirement of being a straight line on the log–log plot, with unit slope.

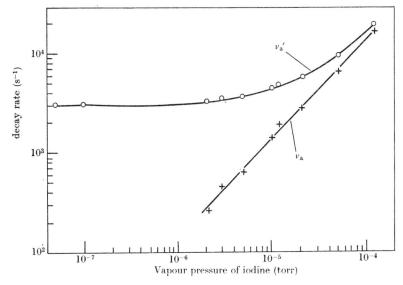

FIG. 13.124. The measured electron decay rate ν_a' at 300 °K as a function of the vapour pressure of iodine. ν_a is the derived attachment rate when the constant background rate is subtracted.

The derived value of the mean attachment cross-section for thermal electrons is as high as $3\cdot9 \times 10^{-16}$ cm^2, which is nearly ten times greater than the maximum observed by Buchdahl at an energy of $0\cdot4$ eV.

In the experiments of Fox, using a 90° sector mass spectrometer and a total ion collection tube as in the work of Fox and Hickam on SF$_6$ (§ 11.1), the apparatus in each case was baked out at 400 °C before introducing the iodine, thereby reducing the background pressure to 10^{-8} torr. The mass spectrum of positive ions just after bake-out showed only traces of H$_2$O$^+$ and CO$^+$. When iodine was admitted at a pressure of about 10^{-5} torr no HI$^+$ ions were found initially. However, a peak due to these ions began to appear after 10 or 15 min and increased in intensity over 30–40 min to about 10 per cent of the I^+ peak. No evidence was obtained of the production of either I_2^- or I_3^- ions.

Using the total ionization tube and working under conditions in which the presence of HI is likely to be small, Fox obtained the results shown in Fig. 13.125 for the variation of total negative ion current with electron energy. To calibrate the energy scale and determine the electron energy distribution, the total negative ion current produced when SF_6 was

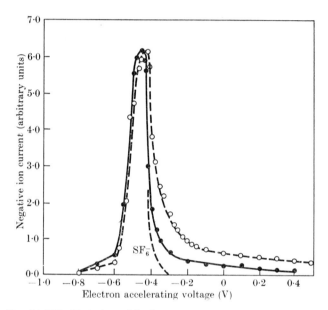

Fig. 13.125. Variation with electron energy of the total negative ion current arising from electron impact in I_2 compared with that in SF_6. ○ I^- from I_2; ● $SF_6^- + SF_5^-$ from SF_6. ——— SF_6^- derived from relative abundance measurements.

substituted for the iodine was measured as a function of the electron accelerating voltage. Care had to be taken in carrying out the comparison because, initially, introduction of iodine changed the energy scale as determined by the accelerating voltage. To eliminate this, I_2 and SF_6 were alternatively admitted to the tube several times until the various electrode surfaces were brought to a stabilized condition.

Fig. 13.125 illustrates the observed negative ion current when the tube contained SF_6, after stable conditions were reached. The known proportion of SF_5^- (see § 11.1) was subtracted off to give the SF_6^- production shown. The I^- and SF_6^- production curves are very nearly the same in shape and position in the energy scale on the low energy side.

This shows that I⁻ ions are produced from I_2 with a maximum probability by electrons of very nearly zero energy just as are SF_6^- from SF_6. At electron energies on the high energy side of the maximum the I⁻ production falls off much more slowly than the SF_6^- showing that there is a considerable probability of I⁻ formation by impact of electrons with an energy of 0·4 eV or higher.

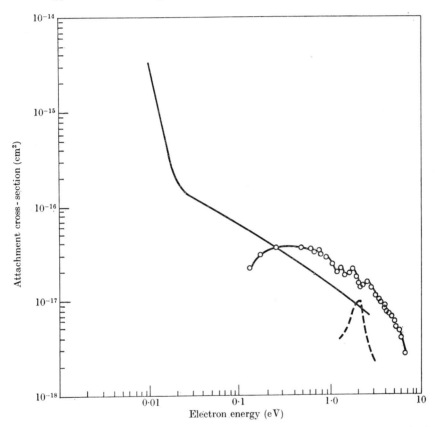

FIG. 13.126. Cross-section for dissociative attachment of electrons to I_2. —— derived by Biondi and Fox. –O–O–O– observed by Buchdahl. — — — observed by Healey.

If $\bar{Q}_a(\bar{E})$ is the mean attachment cross-section when the mean electron energy is $\bar{E}$, then

$$\bar{Q}_a(\bar{E}) = \int Q_a(E) f(E, \bar{E})\, dE, \tag{53}$$

where $Q_a(E)$ is the attachment cross-section for electrons of energy E and $f(E, \bar{E})\, dE$ is the fraction of electrons with energy between E and $E+dE$ when the mean energy is $\bar{E}$. Knowing $f(E, \bar{E})$ from the shape of the SF_6^- production curve and $Q_a(E)$ from that of I⁻ we may invert

(53) by a numerical procedure to obtain $Q(E)$ as a function of E. This was done by Biondi and Fox† with the result shown in Fig. 13.126.

To normalize these cross-sections use may now be made of Biondi's microwave result for thermal electrons, $\bar{E}$ being now 0·039 eV and $f(E, \bar{E})$ the Maxwellian distribution function for the corresponding temperature. In this way the absolute scale shown in Fig. 13.126 is obtained. On the same diagram results obtained by Buchdahl‡ and by Healey§ are shown. The agreement is poor at best.

TABLE 13.22

Observed and calculated appearance potentials for production of atomic negative ions through capture of electrons by hydrogen halides

Halide	Dissociation energy (eV)	Electron affinity of halogen atom (eV)	Appearance potential for ions of zero kinetic energy	
			Calculated	Observed
HF	5·83	3·63	2·00	1·88±0·06‖
HCl	4·43	3·78	0·65	0·8±0·3,†† 0·4,‡‡ 0·66±0·02§§ 0·46±0·02,‖‖ 0·62±0·05‖
HBr	3·75	3·65	0·10	0·6±0·3,‡‡ 0·43±0·001,‖‖ 0·10±0·05‖
HI	3·06	3·24	− 0·18	0·03±0·03‖

The mass-spectrometer observations made by Fox, while not contributing any new information, are of interest in that a technique was employed to overcome difficulties due to formation of the negative ions with considerable kinetic energy. As the dissociation energy of I_2 is considerably smaller (1·5 eV) than the electron affinity of I (3·0 eV), I^- ions formed by electron capture possess a kinetic energy greater than 1·5 eV. Such fast-moving ions would rapidly be lost to the walls of the ion sources before being drawn out into the mass spectrometer. Fox therefore applied a d.c. potential across the ionization chamber in a sense so as to reduce this loss. Inevitably this broadened the electron energy distribution as well as changing the energy scale. Use of the

† BIONDI, M. A. and FOX, R. E., *Phys. Rev.* **109** (1958) 2012.
‡ loc. cit., p. 1057.   § loc. cit., p. 1057.
‖ FROST, D. C. and McDOWELL, C. A., *J. chem. Phys.* **29** (1958) 503.
†† GUTBIER, H. and NEUERT, H., *Z. Naturf.* **9a** (1954) 335.
‡‡ REESE, R. M., DIBELER, V. H., and MOHLER, F. L., *J. Res. natn. Bur. Stand.* **57** (1956) 367.
§§ FOX, R. E., *J. chem. Phys.* **26** (1957) 1281.
‖‖ BUCHEL'NIKOVA, I. S., loc. cit., p. 1054.

retarding potential difference method eliminates most of the effects due to the thermal distribution of the electrons and to the contact potential between the filament and ionization chamber. The procedure adopted was to fix the d.c. ion repeller potential across the ionization chamber and then measure the I^- and electron currents as functions of the applied

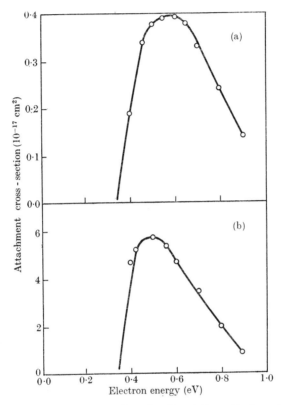

FIG. 13.127. Cross-sections for dissociative attachment of electrons: (a) to HCl; (b) to HBr, observed by Buchel'-nikova.

electron accelerating voltage. Differentiation of the electron current-voltage characteristics then yielded an electron energy distribution curve determined by the change of scale and spread due to the applied ion-repeller voltage of the current. This was repeated with other ion-repeller voltages. The onset of the ion current and the electron energy distribution were then plotted against this voltage giving in each case a linear relation which extrapolated to zero at zero voltage on the ion repeller. Similarly, when the location of the maxima were plotted in the same way the extrapolation again gave zero volts as the position of the maximum at zero ion repeller voltage.

These results are in agreement with those obtained using the total ion-collection tube in that the ion-production cross-section is a maximum for zero energy electrons. Any apparent spread is due to the effect of the ion repeller voltage.

Frost and McDowell[†] also observed the production of I^- ions from I_2 using the retarded potential difference method in conjunction with a mass spectrograph, but did not investigate the effect of the high ion kinetic energy. This may be the reason why their results show a maximum cross-section for production of I^- ions at a finite rather than zero electron energy.

11.3. *Hydrogen halides*

Table 13.22 gives the dissociation energies of the hydrogen halides, the electron affinities of the appropriate halogen atom, and the derived appearance potential for halogen atomic negative ions. These are compared with observed appearance potentials for ions of zero kinetic energy obtained in different electron impact experiments. The spread of the observed data is not surprising in view of the inherent difficulties in determining these appearance potentials. Quite good agreement exists with the observations of Frost and McDowell,[‡] who used the production of SF_6^- to calibrate their energy scale and allow for residual electron energy spread (they used the retarding potential difference method to reduce this spread).

Buchel'nikova has measured the absolute cross-sections for production of negative ions in HCl and HBr. Her results are shown in Fig. 13.127. As with all halogen-containing molecules to be studied the peak cross-section is relatively large.

† FROST, D. C. and McDOWELL, C. A., *J. chem. Phys.* **29** (1958) 964.
‡ FROST, D. C. and McDOWELL, C. A., ibid. 503.

COLLISIONS INVOLVING ELECTRONS AND PHOTONS—RADIATIVE RECOMBINATION, PHOTO-IONIZATION, AND BREMSSTRAHLUNG

1. Introduction

So far we have limited ourselves to the discussion of collisions between electrons and atoms or molecules, charged or uncharged, in which no emission of radiation occurs. In this and the following chapter we consider collisions in which quanta are emitted. These may occur through the incident electron undergoing a free–bound transition so that it is captured by the target atom or molecule, or a free–free transition in which, while losing energy, it remains free. Although we are not primarily concerned in this book with effects due to collisions of quanta with neutral or charged atoms and molecules it would be absurd in these two chapters to limit ourselves to the discussion of radiative electron capture while ignoring the inverse process of photo-ionization which is, in general, much easier to study experimentally. Through the principle of detailed balancing the cross-section for radiative capture into the ground state of an atom or an ion can be derived from the observed cross-section for absorption of radiation by photo-ionization. The importance of the latter process in many applications and the close relation between it and ionization by electron impact are further reasons why it is natural to include a thorough discussion of photo-ionization in this chapter as we shall do from the beginning.

We shall distinguish the selection of material between this and the succeeding chapter in that, in this chapter, we shall deal with photo-ionization from neutral or positively ionized atoms or molecules and the inverse process of radiative recombination of an electron to a positive ion. An account will also be given of free–free transitions, the emission of bremsstrahlung, by an electron moving in the field of a positive ion. Chapter 15, on the other hand, will be concerned with photodetachment of electrons from negative ions, the inverse process of radiative attachment, and the emission of radiation in free–free transitions within the field of a neutral atom or molecule.

In general we shall discuss the ratio of the various processes in terms of cross-sections Q^c for radiative capture and Q^a for photo-ionization respectively, but for many purposes it is better to work in terms of the radiative recombination coefficient α^r, which is given by

$$\alpha^r = vQ^c \tag{1}$$

where v is the velocity of the electron before capture.

2. Basic formulae for bound–free and free–bound transitions

Most of the basic formulae have already been summarized in Chapter 7, § 5.2.1. Thus, according to (69) of that chapter, the cross-section for a transition from a continuous state of energy E to a discrete state i, in which a single electron is alone involved, is given by

$$Q_E^i = \frac{16\pi^3 \nu^3 e^2}{3hc^3 v} \int |\mathbf{r}_{iE}|^2 \, d\omega, \tag{2}$$

where ν is the frequency radiated and v is the electron velocity $\mathbf{r}_{iE}$ is the matrix element

$$\mathbf{r}_{iE} = \int \psi_i^* \, \mathbf{r} \psi_E \, d\mathbf{r}, \tag{3}$$

where ψ_i is the wave function for the bound state i and ψ_E that for the initial 'free' state, which has the asymptotic form of a Coulomb-modified plane wave (see Chap. 6, § 3.9) of unit amplitude and the corresponding outgoing scattered wave. The angular integration is over the direction of the vector $\mathbf{r}$ with respect to that of the incident electron so that

$$Q_E^i = \frac{64\pi^4 \nu^3 e^2}{3hc^3 v} \{|x_{iE}|^2 + |y_{iE}|^2 + |z_{iE}|^2\}. \tag{4}$$

The cross-section Q_i^E for photo-ionization of the atom or ion in the state of energy i to produce a free electron of energy E is given according to Chapter 7, (68) by

$$Q_E^i = \frac{2h^2\nu^2}{m^2v^2c^2} \, Q_i^E. \tag{5}$$

If a number of electrons is involved the formulae apply but with $\mathbf{r}_{iE}$ replaced by the sums $\sum_s \mathbf{r}_{s,iE}$ of the contributions from separate electrons s. The wave functions ψ_i, ψ_E must be properly antisymmetric in all electrons.

The question of degeneracy of the wave functions frequently arises. If g_i, g_E are the respective statistical weights of the states i, E then

$$Q_E^i = \frac{64\pi^4 \nu^3 e^2}{3hc^3 v} \frac{\sum |\mathbf{r}_{iE}|^2}{g_E}, \tag{6}$$

where $\sum |\mathbf{r}_{iE}|^2$ denotes the sum of $|\mathbf{r}_{iE}|^2$ over all $g_i g_E$ pairs of degenerate

states, and
$$Q_i^E = \frac{m^2 v^2 c^2}{2h^2 \nu^2} \frac{g_E}{g_i} Q_E^i. \tag{7}$$

In most cases we may write $g_E = g_+ g_e$, where g_+ is the statistical weight of the state of the positive ion in the field of which the incident free electron moves and g_e that of a free electron ($= 2$).

If the wave functions are accurate the formulae can be relied upon to give the cross-sections to a very good approximation, provided the wavelength of the emitted (absorbed) radiation is small compared with the dimensions of the atom or molecule in the state i, a condition satisfied in all cases of practical importance. Hence there is no difficulty in obtaining accurate data for radiative capture by bare nuclei. For all other cases in which more than one electron is involved the accuracy is limited by the fact that only approximate wave functions ψ_i, ψ_E are available. In general, such functions will only be good approximations within limited ranges of r. It is often an advantage to use, not the dipole matrix element $\mathbf{r}_{iE}$, but some transformation of it that depends more strongly on regions of r for which the wave functions are good approximations.

Such transformations have already been discussed in Chapter 7, § 5.2.3. It is shown that
$$\mathbf{r}_{iE} = i\hbar(E_i - E)^{-1}\mathbf{v}_{iE} = -\hbar^2(E_i - E)^{-2}\mathbf{a}_{iE}, \tag{8}$$
where $\mathbf{v}$ and $\mathbf{a}$ are the velocity and acceleration operators
$$\mathbf{v} = -(i\hbar/m)\text{grad}, \qquad \mathbf{a} = -m^{-1}\text{grad}\,V, \tag{9}$$
V being the potential energy of the field in which the electron moves. The main contribution to the matrix element comes from smaller values of r in going from $\mathbf{r}_{iE}$ to $\mathbf{v}_{iE}$ to $\mathbf{a}_{iE}$. In general, an approximate wave function determined by a variational method is most nearly correct at close distances r, which contribute most to the determination of the energy of the state. It seems likely that in most cases such distances are more nearly comparable with those important in $\mathbf{v}_{iE}$ than in $\mathbf{r}_{iE}$ or $\mathbf{a}_{iE}$. If, in carrying out a calculation with approximate functions, the approximate values of $\mathbf{r}_{iE}$, $\mathbf{v}_{iE}$, and $\mathbf{a}_{iE}$ obtained are nearly equal then it may be assumed that the approximation is a good one.

For a many-electron problem the velocity and acceleration formulae may be obtained by summing the contributions from the separate electrons just as for the dipole formula.

An important additional check on calculated cross-sections is often provided by the various sum rules that are satisfied by the optical oscillator strengths (see Chap. 7, § 5.2.2).

2.1. *Threshold behaviour*

The wave function ψ_E for the final continuum state in a photo-ionization event is an attractive Coulomb wave distorted at small distances by screening effects due to the inactive electrons. Because of this the cross-section Q_i^E is finite at the threshold. This is to be contrasted with the cross-section for ionization by electron impact (Chap. 9, § 10.8), which rises from zero at the threshold as $(E-E_t)^s$, where E_t is the threshold energy and s is certainly close to and is probably unity. Thus the threshold behaviour of the electron impact cross-section is closely similar to that of the energy integral from the threshold of the corresponding cross-section for photon impact. This is probably true in general. It has already been referred to and illustrated in Chapter 13, §§ 1.4.2 and 4.3.2.

3. Theory of radiative capture by bare nuclei—photo-ionization of atomic hydrogen

As explained in § 2 it is possible to make calculations as accurate as desired of the cross-sections for radiative capture of electrons by bare nuclei and for the inverse processes, including the photo-ionization of atomic hydrogen. In contrast, the measurement of these cross-sections is a matter of great difficulty and the results obtained, by methods described in § 7.5.1, are both of low accuracy and abundance.

There has been a considerable stimulus towards carrying out extensive calculations because of astrophysical applications—H^+ and H are the most abundant constituents of matter in the universe. There are also important applications in the study of high-temperature plasmas. Furthermore, the hydrogen cross-sections for capture with highly excited states provide good approximations to corresponding cross-sections for capture into excited states of complex atoms.

It is convenient to introduce at this stage a different notation for the cross-sections. We take $Q_{nl}^c(Z, k^2)$, $Q_{nl}^a(Z, k^2)$ to be the respective radiative capture and photo-ionization cross-sections for transitions between a bound state, with quantum numbers n, l and a continuum state of wave number k in the field of a bare nucleus of charge Ze. Then

$$Q_{nl}^c(Z, k^2) = (2l+1)(\hbar/mc)^2(k^2+Z^2/n^2a_0^2)^2k^{-2}Q_{nl}^a(Z, k^2). \tag{10}$$

But
$$Q_{nl}^a(Z, k^2) = Z^{-2}Q_{nl}^a(1, k^2/Z^2), \tag{11}$$

so
$$Q_{nl}^c(Z, k^2) = Q_{nl}^c(1, k^2/Z^2). \tag{12}$$

Sometimes it is convenient to express these cross-sections as ratios to the corresponding classical cross-sections, calculated from electromagnetic theory in terms of the Kramers–Gaunt factor $g_2(n, k/Z)$ where

$$\sum_l \left(\frac{2l+1}{n^2}\right) Q_{nl}^{\mathrm{a}}(Z, k^2) = \frac{n}{Z^2} \frac{g_2(n, k/Z)}{(1+n^2k^2a_0^2/Z^2)^3}. \qquad (13)$$

The earliest calculations of these cross-sections were carried out by Stueckelberg and Morse[†] and by Wessel.[‡] These covered only a small range of values of n, l but Burgess[§] has now tabulated Q_{nl}^{a} to five significant figures for all n, l up to $n = 20$, $l = 19$ inclusive. In fact a quantity $\Theta(n, l; ka_0/Z, l')$ is tabulated, where

$$Q_{nl}^{\mathrm{a}}(Z, k^2) = \frac{4\pi\alpha a_0^2}{3}(n^2/Z^2) \sum_{l'=l\pm1} \frac{l_>}{2l+1}\, \Theta(n, l; ka_0/Z, l'), \qquad (14)$$

$l_>$ being the greater of l, l' and $\alpha = e^2/\hbar c$.

In addition Burgess also includes tables of the corresponding recombination coefficients $\alpha_{nl}(k^2)$ averaged over electron energy distributions of Maxwellian form at various temperatures T. In this case functions $\Phi(n, l, l', t)$ are tabulated, in terms of which $\alpha_{nl}(T)$, the averaged recombination coefficient, is given by

$$\alpha_{nl}(T) = (2\pi^{\frac{1}{2}}\alpha^4 a_0^2 c/3)(2y^{\frac{1}{2}}/n^2)Z \sum_{l'=l\pm1} I(n, l, l', t), \qquad (15)$$

where
$$\Phi(n, l, l', t) = \begin{cases} I(n, l, l', t) & (t \leqslant 1), \\ tI(n, l, l', t) & (t \geqslant 1), \end{cases}$$

and
$$\alpha = e^2/\hbar c, \qquad y = Z^2 e^2/2a_0 \kappa T, \qquad t = T/10^4 Z^2.$$

The tables of Φ cover the same range of n, l, l' as those of Θ and $t = 0$, ∞ and 2^n, where n ranges in unit steps from -4 to $+4$.

In Table 14.1 typical values of Q_{nl}^{c} are given as functions of l for $n = 2, 3$ and electrons of near-thermal energy, while in Table 14.2 mean radiative recombination coefficients are given for a number of temperatures T for the same values of n.

Radiative capture cross-sections summed over all values of l for a given n may be computed directly from a formula given by Oppenheimer[||] based on expression of the wave functions in terms of parabolic coordinates. Using this formula Bates, Buckingham, Massey, and Unwin[††] carried out extensive calculations that enabled the total capture cross-section for thermal electrons, summed over all final states, to be

† STUECKELBERG, E. C. G., and MORSE, P. M., *Phys. Rev.* **36** (1930) 16.
‡ WESSEL, W., *Annln Phys.* **5** (1930) 611.
§ BURGESS, A., *Mem. R. astr. Soc.* **69** (1964) 1.
|| OPPENHEIMER, J. R., *Z. Phys.* **55** (1929) 725.
†† BATES, D. R., BUCKINGHAM, R. A., MASSEY, H. S. W., and UNWIN, J. J., *Proc. R. Soc.* **A170** (1939) 322.

TABLE 14.1

Variation of radiative capture cross-section with azimuthal quantum number of the final state

Electron energy (eV)		0·28	0·13	0·069	0·034
Quantum numbers of final states					
n	l	Cross-section in 10^{-21} cm²			
2	0	1·20	2·45	4·80	9·81
	1	3·04	6·48	12·90	26·71
3	0	0·402	0·826	1·62	3·30
	1	1·15	2·46	4·95	10·11
	2	1·15	2·62	5·47	11·47

TABLE 14.2

Variation of mean recombination coefficient $\alpha_{nl}(T)$ with azimuthal quantum number of the final state

Temperature (T °K)		312·5	1250	10^4	8×10^4	$6 \cdot 4 \times 10^5$
Quantum numbers of final states						
n	l	Recombination coefficient in 10^{-14} cm³/s				
2	0	136	68	23	6·7	1·1
	1	372	182	54	9·4	0·8
3	0	46	23	8	2·1	0·3
	1	141	69	20	3·5	0·3
	2	161	75	17	1·9	0·1

obtained. Seaton† has greatly extended this work by using the formula (13) together with the asymptotic expansion‡ of the Kramers–Gaunt factor

$$g_2(n, \epsilon) = 1 + 0 \cdot 1728 n^{-2/3}(u+1)^{-2/3}(u-1) -$$
$$- 0 \cdot 0496 n^{-4/3}(u+1)^{-4/3}(u^2 + \tfrac{4}{3}u + 1), \quad (16)$$

where $u = n^2 \epsilon$. If the first term only of (16) is retained

$$\alpha_n(T) = \mathscr{D} Z x_n^{3/2} S_n^0(\lambda), \quad (17)$$

where

$$\mathscr{D} = \frac{2^6}{3}\left(\frac{\pi}{3}\right)^{\frac{1}{2}}\left(\frac{e^2}{\hbar c}\right)^4 c a_0^2 = 5 \cdot 197 \times 10^{-14} \text{ cm}^3/\text{s}, \quad (18)$$

$$\lambda = 2\pi^2 m e^4 Z^2 / h^2 \kappa T = 157\,890 Z^2 / T, \quad (19)$$

$$x_n = \lambda / n^2, \quad (20)$$

† SEATON, M. J., *Mon. Not. R. astr. Soc.* **119** (1959) 81.
‡ MENZEL, D. H. and PEKERIS, C. L., ibid. **96** (1935) 77; BURGESS, A., ibid. **118** (1958) 477.

and

$$S_n^0(\lambda) = \int\limits_0^\infty \frac{e^{-x_n u}}{1+u}\, du$$

$$= e^{x_n} \mathrm{Ei}(x_n), \qquad (21)$$

where Ei is the exponential integral. Seaton improved the approximation (17) by taking into account the next two terms in (16).

TABLE 14.3

Cross-sections for radiative capture of an electron to various states of a hydrogen atom

Electron energy (eV)	0·28	0·13	0·069	0·034
Total quantum number of atomic state into which electron is captured	Cross-section in 10^{-21} cm^2			
1	8·10	16·63	32·80	66·95
2	4·24	8·93	17·70	36·52
3	2·70	5·91	12·04	24·88
4	1·88	4·24	8·84	18·59
5	1·36	3·20	6·89	14·85
6	0·99	2·49	5·51	12·25
7	0·73	2·00	4·52	10·31
8		1·62	3·81	8·75
9		1·31	3·23	7·55
10		1·04	2·74	6·50
11			2·33	5·72
12			2·00	4·52
13			1·72	4·03
14	0·15		1·47	3·62
15				3·25
16				2·91
17				2·62
18				2·35
19				2·09
20		0·215		
28			0·302	
40				0·432
Sum for all final states	23·0	53·7	119	272

Table 14.3 gives cross-sections for radiative capture of an electron to various states of a hydrogen atom, summed over contributions from states of different azimuthal quantum number, while in Table 14.4 mean radiative recombination coefficients to states of different total quantum number are given as functions of electron temperature T.

For the deepest states the capture cross-section varies nearly as E^{-1} at low electron energies E. The cross-section summed over all final states increases somewhat more rapidly as E decreases, owing to the increased contribution from highly excited states. Even at room temperature energies the cross-section is still quite small, being about 3×10^{-19} cm^2, of which about one-quarter is contributed from capture to the ground state. It is to be expected then that, in a spectrum arising from population of excited states by radiative capture of electrons, a 'recombination spectrum', lines due to transitions from highly excited states should be comparatively strong. This is a general feature of such spectra.†

TABLE 14.4

Radiative recombination coefficients to different final states of a hydrogen atom as functions of electron temperature

Electron temperature (°K)	250	1000	8000	64 000
Total quantum number of state into which electron is captured	Recombination coefficient in 10^{-14} cm^3/s			
1	102	50·7	17·4	5·19
2	56·6	27·9	8·80	1·95
3	39·0	18·8	5·33	0·946
4	29·5	14·0	3·53	0·533
5	23·6	10·8	2·48	0·332
6	19·6	8·7	1·82	0·222
7	16·6	7·16	1·38	0·156
8	14·3	5·99	1·07	0·114
9	12·5	5·08	0·851	0·0866
10	11·1	4·36	0·688	0·067
11	9·88	3·77	0·565	0·053
12	8·87	3·29	0·471	0·043
Total summed over all final states	484	199	48·3	10·0

Finally, in Fig. 14.1 the cross-section Q_{10}^{a} for photo-ionization of atomic hydrogen in the $1s$ state is shown as a function of photon energy.

4. Radiative capture and photo-ionization for complex ions and atoms

4.1. *Further general considerations*

Turning now to cases in which more than one electron is involved we encounter a number of new features. On the one hand, the theoretical analysis is no longer so accurate as the wave functions that were used are no longer exact, but there is the partially compensating feature that direct and accurate observation of cross-sections for photo-ionization from the ground state is readily practicable in many cases. While this

† See, for example, PASCHEN, F., *Sber. preuss. Akad. Wiss.* **16** (1926) 135.

does not provide all the information normally required about radiative recombination it nevertheless goes a considerable way towards meeting these requirements. Thus the most difficult capture cross-section to evaluate theoretically is that for capture to the ground state. Through (7) this may be obtained from the photo-ionization cross-section.

Apart from these considerations a further interesting possibility arises—the existence of resonance effects due to the existence of auto-ionizing states (see Chap. 1, § 6.2, Chap. 3, § 2.5.3, Chap. 5, §§ 4.2.1, 5.3.3,

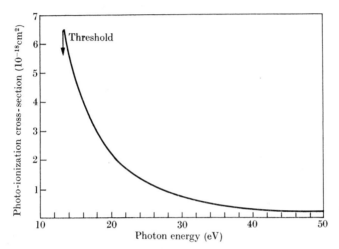

FIG. 14.1. Calculated cross-section for photo-ionization of atomic hydrogen.

Chap. 9, §§ 1–9, Chap. 10, § 3.5, Chap. 11, § 3, and Chap. 13, §§ 1.4.2, 3.4.2, 3.7, and 4.4).

We begin the discussion of the many-electron case by considering what can be done to obtain reasonably reliable calculated cross-sections, ignoring resonance effects, and then proceed to a brief theoretical analysis of what form such effects will take. Consideration of separate cases will then follow after a description of experimental techniques that have been used for the measurement of photo-ionization cross-sections.

4.2. *Calculation of cross-sections—auto-ionization ignored*

The usual procedure for calculating the photo-ionization cross-section of an atom is to use for the bound state wave function that given by the Hartree–Fock self-consistent field method. For the free state a distorted Coulomb wave is used for the free electron, the distortion being calculated in some such way as described in Chapter 8, § 5. Improved approximations may be used for simpler atoms such as helium and lithium as described respectively in §§ 7.1 and 7.4.1. In all cases it is usual to calculate with at least the length and velocity formulae. As mentioned earlier it appears that, with the usual approximation for the wave function, the velocity formula often gives the best results. When high accuracy is aimed at, the close approximation of results given by length, velocity, and acceleration formulae is an important indication of reliability. This is rarely attained except in the simplest cases.

Burgess and Seaton† have given a general formula for atomic photo-ionization cross-sections that is semi-empirical and often gives good results. It is essentially the generalization to bound-free transitions of a method introduced by Bates and Damgaard‡ for calculating probabilities for transitions between bound states. The method depends on the fact that the dipole matrix element is determined mainly by the form of the wave function of the active electron at large values of r and is insensitive to the form for small r. Furthermore, the asymptotic form of both the bound and free wave functions may be determined from observed line spectra.

Consider first the wave function for a bound state with quantum numbers n, l and energy ϵ_n in Rydbergs where

$$\epsilon_n = -Z^2/n^{*2}, \tag{22}$$

n^* is the effective total quantum number and the asymptotic form of the potential acting on the electron is $-Ze^2/r$. The quantum defect $\mu(\epsilon_n)$ of the state is given by $n-n^*$ and as in Chapter 6, § 3.10 we consider this derived function to be generalized through interpolation and extrapolation to give the continuous function $\mu(\epsilon)$ extending to positive energies ϵ.

It may be shown that the radial wave function $P_{nl}(r)$ (see Chap. 6, (5)), for the bound state, outside the atomic core has the form§

$$P_{nl}(r) = Z^{\frac{1}{2}}K(n^*, l)W_{n^*, l+\frac{1}{2}}(2Zr/n^*), \tag{23}$$

where $W_{n^*, l+\frac{1}{2}}(x)$ is the Whittaker function that has the asymptotic form

$$W_{n, l+\frac{1}{2}}(x) \sim x^n e^{-\frac{1}{2}x} \sum_{s=0}^{s_0} b_s(n, l)(\tfrac{1}{2}nx)^{-s} + O\{(\tfrac{1}{2}nx)^{-s_0-1}\}, \tag{24}$$

with
$$b_0 = 1, \qquad b_s = \frac{n}{2s}\{l(l+1)-(n-s)(n-s+1)\}b_{s-1} \quad (s \geqslant 1).$$

† BURGESS, A. and SEATON, M. J., *Mon. Not. R. astr. Soc.* **120** (1960) 121.
‡ BATES, D. R. and DAMGAARD, A., *Phil. Trans. R. Soc.* A**242** (1949) 101.
§ HARTREE, D. R., *The calculation of atomic structures* (Wiley, New York, 1957).

The normalizing factor $K(n^*, l)$ is given by

$$K(n^*, l) = \{\zeta(n^*)n^{*2}\Gamma(n^*+l+1)\Gamma(n^*-l)\}^{-\frac{1}{2}}, \tag{25}$$

where
$$\zeta(n^*) = 1 + \frac{2}{n^{*3}}\frac{\partial \mu(\epsilon)}{\partial \epsilon} \tag{26}$$

and is nearly equal to 1 except for small values of n^*.

An approximation to the true bound state function is now obtained by replacing $W_{n^*,l+\frac{1}{2}}$ by its asymptotic form (24) terminated at $s = s_0$, where s_0 is such that, for $s > s_0$, the terms in the expansion begin to increase again.

For a continuum state of energy ϵ (in Rydbergs) the corresponding wave function $G_{kl}(r)$ has the asymptotic form for large r (see Chap. 6, § 3.10)

$$G_{kl}(r) \sim \{G_l^c(\epsilon, r)\cos \sigma_l - H_l^c(\epsilon, r)\sin \sigma_l\}, \tag{27}$$

where G_l^c and H_l^c are the respective regular and irregular solutions of the wave equation for motion in a Coulomb field of charge Ze and positive energy ϵ, which have the asymptotic forms

$$\begin{matrix} G_l^c \\ H_l^c \end{matrix} \sim \begin{matrix} \sin \\ \cos \end{matrix} (kr - \tfrac{1}{2}l\pi - \alpha \ln 2kr + \eta_l). \tag{28}$$

Here

$$k^2 a_0^2 = \epsilon, \qquad \alpha = -Ze^2/\hbar v, \qquad v = k\hbar/m, \qquad \eta_l = \arg \Gamma(l+1+i\alpha).$$

As discussed in Chapter 6, § 3.10, provided ϵ is small,

$$\sigma_l \simeq \pi\mu_l(\epsilon). \tag{29}$$

Burgess and Seaton therefore chose for the full continuum wave function

$$G_{kl}(r) = G_l^c(\epsilon, r)\cos \pi\mu_l(\epsilon) + \{1 - \exp(-\tau_l Zr/a_0)\}^{2l+1}H_l^c(\epsilon, r)\sin \pi\mu_l(\epsilon), \tag{30}$$

the cut-off factor $\{1 - \exp(-\tau_l Zr/a_0)\}$ being introduced to eliminate the irregular behaviour of H_l^c at small r.

It remains to determine τ_l. In terms of $\rho = Zr/a_0$ the functions G_l^c, H_l^c satisfy

$$\left\{\frac{d^2}{d\rho^2} - \frac{l(l+1)}{\rho^2} + \frac{2}{\rho} + \epsilon\right\}\begin{matrix} G_l^c \\ H_l^c \end{matrix} = 0.$$

The first point of inflexion in the solutions occurs at $\rho = \rho_{i,l}$ where

$$\frac{2}{\rho_{i,l}} + \epsilon - \frac{l(l+1)}{\rho_{i,l}^2} = 0.$$

τ_l is then chosen so $\tau_l \rho_{i,l} \simeq 5$ or $\tau_l \simeq 10/l(l+1)$.

With these approximations

$$Q_{nl}^a = \frac{4}{3}\frac{e^2}{\hbar c}\frac{n^{*2}}{Z^2}(1 + \epsilon'n^{*2})\sum_{l'=l\pm 1}C_{l'}\{g(n^*, l; \epsilon', l')\}^2\pi a_0^2, \tag{31}$$

where

$$g(n^*, l; \epsilon', l') = \frac{G(n^*, l; \epsilon', l')}{\zeta^{\frac{1}{2}}(n^*, l)} \cos[\pi\{n^* + \mu_{l'}(\epsilon') + \chi(n^*, l; \epsilon', l')\}], \quad (32)$$

ϵ' being the energy, in Rydbergs, of the ejected electron and

$$G(n^*, l; \epsilon', l') = (-1)^{l+1} G_{ll'}(n^*)\{1 + \epsilon' n^{*2}\}^{-\gamma_{ll'}(n^*)}. \quad (33)$$

$G_{ll'}$, $\gamma_{ll'}$, and $\chi_{ll'}$ were given by Burgess and Seaton† in tabular form. Their tables of $G_{ll'}$ and $\chi_{ll'}$ were later extended and improved in accuracy for high values of n^* by Peach.‡ Her tables cover the values of $(l, l') = (0, 1)$, $(1, 0)$, $(1, 2)$, $(2, 1)$, $(2, 3)$, and $(3, 2)$ for $0.6 \leqslant n^* \leqslant 12$ and $0 \leqslant \epsilon' \leqslant 1.0$.

On substitution of numerical values for the atomic constants

$$Q_{nl}^{a} = 8.56 \times 10^{-19} (n^*/Z^2)(1 + \epsilon' n^{*2}) \sum_{l'=l\pm 1} C_{l'}\{g(n^*, l; \epsilon', l')\}^2 \text{ cm}^2. \quad (34)$$

Results obtained in this way are likely to be good approximations provided no strong cancellation occurs in the calculation of the dipole matrix element. When this does occur the cross-section will in any case be abnormally small. Writing (32) in the form

$$g(n^*, l; \epsilon', l') = \zeta^{-\frac{1}{2}}(n^*, l)G(n^*, l; \epsilon', l')\cos\phi(\epsilon'),$$

where

$$\phi(\epsilon') = \pi\{n^* + \mu_{l'}(\epsilon') + \chi(n^*, l; \epsilon', l')\}, \quad (35)$$

we see that a small value for $\cos\phi$ may be taken as a warning of strong cancellation and hence unreliability of the calculated results.

Applications of these formulae will be discussed below. The evidence indicates that they give quite good results except under conditions in which high sensitivity to the form of the wave functions is expected. In some cases the effect of distortion from the Coulomb form for the continuum wave function is not important so that $\mu_{l'}$ may be taken as zero as in earlier calculations of Bates.§

4.3. *Resonance effects due to excitation of auto-ionizing states*

In general, effects due to excitation of auto-ionizing states occur on the long-wave side of, but close to, the threshold for onset of a new absorption process arising from the possibility of excitation of states in which more than one electron occupies an excited orbital. Transitions to such states, which are unstable towards auto-ionization, constitute a Rydberg series converging to a limit at the onset of the new process.

Excitation of these auto-ionizing states gives rise to absorption lines broadened by the short lifetime of the state. The shape of the line is,

---

† loc. cit., p. 1074.   ‡ PEACH, G., *Mem. R. astr. Soc.* **71** (1967) 13.
§ BATES, D. R., *Mon. Not. R. astr. Soc.* **106** (1946) 423.

however, modified by the interaction of the doubly or multiply excited electron states with the continuum levels in the same energy range.

This situation has already been discussed in Chapter 9. In particular, formula (75) of Chapter 9, § 1.2.1 may be applied to give for the form of the photo-ionization cross-section in the neighbourhood of a 'line' due to excitation of an auto-ionizing state

$$Q^a = Q_1^a \frac{(q+\epsilon)^2}{1+\epsilon^2} + Q_2^a. \tag{36}$$

Here q, the so-called *line profile index*, is a real number that may have any positive or negative value depending on the shape of the final wave functions (see Chap. 9, § 1.2.1). $\epsilon = (E-E_r)/\frac{1}{2}\Gamma$, where E_r is the resonance energy required to excite the auto-ionizing level from the ground state and Γ is the line width, such that the mean lifetime of the level is $\hbar/\Gamma$. Q_2^a is the contribution from the other levels and from transitions to those levels of the continuum which do not interact with the discrete auto-ionizing state.

At a great distance from the resonance ($|\epsilon| \gg 1$)

$$Q^a = Q_1^a + Q_2^a. \tag{37}$$

The effective contribution of the auto-ionizing level is therefore

$$\int_0^\infty (Q^a - Q_1^a - Q_2^a)\, dE = \tfrac{1}{2}\pi\Gamma(q^2-1)Q_1^a. \tag{38}$$

This will be negative when $|q| < 1$ so that the absorption line under these circumstances will appear reversed—at the location of the resonance the absorption will be less than that of the background. Typical examples of the function $Q^a(\epsilon)$ for different values of q are illustrated in Chapter 9, Fig. 9.4.

The form of the variation of Γ and q along a Rydberg series of auto-ionizing levels has been discussed in Chapter 9, § 1.4. Let E_j be the energy towards which the series converges. Then if $E_{j,n}$ is the energy of the nth term of the series

$$E_{j,n} = E_j - E_H/n^{*2}. \tag{39}$$

n^* is the effective quantum number given by

$$n^* = n - \mu(n), \tag{40}$$

where μ is the quantum defect which varies only slowly with n. It was shown that $q_{j,n}$ is practically independent of n while

$$\Gamma_{j,n} \propto \tfrac{1}{2}(E_{j,n+1} - E_{j,n-1}) \tag{41}$$
$$\simeq E_H\, 2n^*/(n^{*2}-1)$$
$$= 2E_H/n^3 \quad \text{for large } n.$$

The results of experimental observation of the cross-section near the convergence limit of the series will depend on the frequency resolving power of the apparatus. In general the average contribution from the higher terms alone will be observable. Since, according to (41), the line width is proportional to the distance between successive lines, the average excess cross-section due to the series near the limit is simply†

$$\tfrac{1}{2}\pi\overline{\Gamma}(\bar{q}^2-1)\overline{Q}_1^{\mathrm{a}}, \tag{42}$$

where $\bar{q}$ is the constant value of q and

$$\overline{\Gamma} = \Gamma_{j,n}/\tfrac{1}{2}(E_{j,n+1}-E_{j,n-1}). \tag{43}$$

Just as for the effect of a single term, (42) is only positive if $|\bar{q}| > 1$. If $|\bar{q}| < 1$ there will not be a jump but a drop in absorption as the photo-ionizing frequency passes through E_j/h.

Formulae for the calculation of q are given in Chapter 9, §§ 1.1.1 and 1.2.1, the general operator T in that section being now the dipole operator $\mathbf{r}$. We shall discuss detailed applications to the analysis of high resolution absorption spectra of the rare gases in §§ 7.1–7.3.

5. Experimental methods for studying photo-ionization of atoms and molecules

5.1. General introduction

The absorption of light with quantum energy greater than the threshold for primary ionization of an atom can only take place through photo-ionization so that the absorption and photo-ionization cross-sections are equal under these conditions. On the other hand, if the absorber is a molecule, processes of absorption may occur even beyond the primary ionization threshold which do not lead to ionization. In particular, the process may be one that dissociates the molecule into two or more neutral fragments. It is then necessary to distinguish experimentally between the different absorption processes. In any actual experiment a complete analysis would be very difficult. The first experiments measure the ratio of the total photo-ionization cross-section $Q_{\mathrm{i}}^{\mathrm{ph}}$ to the total absorbing cross-section Q^{a}, giving the so-called ionizing efficiency. At a later stage the products of ionization may be analysed with a mass spectrograph to complement information obtained from the corresponding electron-impact studies. Observations with high energy resolution of the distribution of the photoelectrons produced by radiation of a definite frequency can yield valuable

† FANO, U. and COOPER, J. W., *Phys. Rev.* **137** (1965) 1364.

information about the nature of the process involved and their energy thresholds.

The first measurements of photo-ionization cross-sections were undertaken using caesium and rubidium vapour as the absorbing media. This apparently peculiar choice was dictated by the fact that these atoms have the lowest ionization potentials and may be ionized by ultra-violet radiation for which quartz optics may be used. For permanent gases it is necessary to work in the vacuum ultra-violet.

The first quantitative measurements of Q_i^{ph} as a function of wavelength were made by Mohler and Boeckner† in 1929. A few years later Ditchburn and his collaborators‡ developed the technique for absolute measurement of the total absorption, as distinct from the photo-ionization, cross-section for metal vapours and this has yielded many results of importance.

Measurements of cross-sections for vacuum ultra-violet photons only began recently, since the last war, but extensive data are now available both about photo-ionization and total absorption cross-sections for many molecules as well as for the rare gas atoms. The data for molecules include in many cases values for both cross-sections Q^a and Q_i^{ph}. In addition a considerable amount of work has already been done on the mass analysis of the products of photo-ionization and on the measurement of the energies of photoelectrons.

A further recent feature of the work has been made possible through the availability of radiation continua extending down to the far ultra-violet, emitted by the circulating electrons from an electron synchrotron. This has made observations possible, with high energy resolution, over a wide continuum so that fine resonance effects due to excitation of auto-ionizing levels have been detectable down to wavelengths as short as 100 Å.

Before describing some typical apparatus used in these investigations there are a few general matters to discuss.

In carrying out absorption cross-section measurements the radiation is passed through an absorbing cell and the intensity I is measured, after transmission, as a function of the pressure p of the absorbing gas. If l is the path length of the radiation in the cell then, if the cell contains n atoms or molecules/cm³ with absorption cross-section Q^a

$$I = I_0 e^{-nQ^a l}, \tag{44}$$

† MOHLER, F. L. and BOECKNER, C., J. Res. natn. Bur. Stand. 3 (1929) 303.

‡ BRADDICK, H. J. J. and DITCHBURN, R. W., Proc. R. Soc. A143 (1934) 472 and 150 (1935) 478; DITCHBURN, R. W. and HARDING, J., ibid. 157 (1936) 66.

where I_0 is the intensity transmitted when no absorber is present. This involves only measurements of relative intensity, no absolute calibration of the detectors, which may be photographic or photoelectric, being required. In this respect absorption, as distinct from photo-ionization, measurements are relatively simple.

Results of measurements of this kind are often given in terms of an absorption coefficient κ cm^{-1} which is the value of nQ^a when n is Loschmidt's number. Thus the reduction of intensity after passing a distance l through a gas at 0 °C and 760 torr pressure is $e^{-\kappa l}$. Numerically

$$Q^a = 3.72 \times 10^{-20} \, \kappa \, \mathrm{cm}^2$$

$$= 4.228 \times 10^{-5} \, \kappa \pi a_0^2. \tag{45}$$

As it is necessary to obtain measurements for nearly monochromatic radiation as a function of wavelength the equipment required must include not only a source and absorbing cell but also a monochromator or spectrograph. The absorbing gas may either be in a cell between source and spectrograph, may fill the spectrograph, or may be in a cell placed just beyond the exit slit of the spectrograph. Of these the last has the advantage that it reduces very much the chance of photochemical reactions modifying the chemical composition of the absorbing gas— the intensity of radiation passing through the absorbing cell is much reduced. It is an essential arrangement if photo-ionization measurements are also being made. On the other hand, the first two possibilities have advantages with photographic-plate detection as a whole absorption spectrum may be photographed at once. When practicable, the arrangement using the gas-filled spectrograph has the advantage, for work in the far ultra-violet, that the source may be placed close to the entrance slit of the spectrograph thereby increasing the intensity of the radiation passing through the absorbing gas.

It also requires neither differential pumping nor a window between the monochromator and the absorption cell. At wavelengths below the lithium fluoride cut-off at 150 Å it is difficult, though not impossible, to find a suitable window material (see pp. 1091 and 1092). Without a window, special arrangements must be made to ensure, not only that contamination does not occur from the source gas, but also that the pressure in the monochromator remains negligible when gas is admitted to the absorption cell, if separate. With the gas-filled monochromator the need for the latter precaution is eliminated. One disadvantage is the somewhat greater difficulty in calculating the effective path length. Also with a separate absorption cell of quite small volume it is easier to maintain gas purity.

In most experiments that have been carried out a single beam method has been used. This means that, during actual absorption measurements, the intensity of the primary beam is not monitored. Variations in the source can only be tested by observations immediately before and after absorption measurements and from the general reproducibility of results.

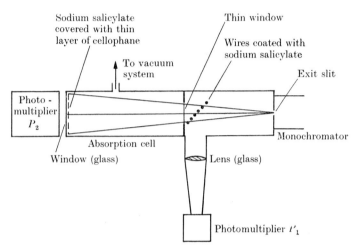

FIG. 14.2. Arrangement of double-beam apparatus for measurements of absorption cross-sections for ultra-violet radiation.

The difficulty about using double-beam methods in which separation is achieved at a semi-reflecting surface is that, in the far ultra-violet, no solid reflector may be used because of its strong absorption. Ditchburn[†] has, however, replaced the reflector by a grid of wires, soaked in sodium salicylate, which fluoresces in the visible when irradiated with ultra-violet light. Fig. 14.2 illustrates the principle of the method. The intensity of the beam before entering the absorption cell is monitored by the photomultiplier P_1, while P_2 monitors that of the beam after absorption.

When photo-ionization as well as absorption measurements are carried out at the same time the absorption cell is now also an ionization chamber measuring the total ion current collected, for example, at one of a pair of plane electrodes parallel to the light beam which passes between them. A potential difference is applied between the plates,

† DITCHBURN, R. W., J. quant. spectrosc. Radiat. Transfer, 2 (1963) 361.

just enough to collect all the photo-ions produced but not so great as to cause secondary ionization.

If I_0 and I are the respective incident and transmitted fluxes through the cell then, as in (44),
$$I = I_0 e^{-nQ^a l}. \tag{46}$$

Suppose first that the absorbing gas is atomic so that all absorption involves ionization. Then, if i_g is the saturated ion current collected,
$$i_g/e = I_0 - I$$
$$= I_0(1 - e^{-nQ^a l}).$$

Having determined Q^a from absorption measurements, the absolute flux of incident radiation can be obtained from
$$I_0 = (i_g/e)/(1 - e^{-nQ^a l}). \tag{47}$$

This in turn makes it possible to calibrate the radiation detector which may, for example, be a platinum plate used photoelectrically. The photoelectric yield γ of the plate for the radiation under study is then obtained from the photoelectric current i_p^0 emitted by the plate when there is no absorbing gas present. Thus
$$\gamma = i_p^0/I_0 e. \tag{48}$$

Having carried out these observations using a rare gas absorber, the method may be applied to determine both Q_a and Q_i^{ph} for molecular gases. Further remarks on application to absolute determination of ultra-violet intensity are made in § 5.3.

Thus
$$\frac{Q_i^{ph}}{Q^a} = \frac{i_g}{eI_0(1 - e^{-nQ^a l})} = \frac{i_g}{\gamma(i_p^0 - i_p)}, \tag{49}$$

where i_p, i_p^0 are the respective photoelectric currents emitted by the detector when the absorbing gas is present to a concentration n and when it is absent.

5.2. Sources of radiation

Apart from the obvious requirements of supplying a constant high intensity of radiation it is an advantage, for absolute measurement of cross-sections, that the source should supply a closely spaced line spectrum. This enables allowance to be made for stray background radiation based on the intensity level observed in the interval between the lines. It also makes it possible, when using a grating, to avoid complications due to the presence of lines of higher order. However, for the study of rapid changes occurring in a very narrow wavelength range, such as the resonance features due to excitation of auto-ionization levels, it is desirable to work with a source of continuous radiation.

At wavelengths above 1000 Å a hydrogen glow discharge source is very convenient. For shorter wavelengths, down to about 150 Å, a low pressure spark through a capillary of high melting-point material, such

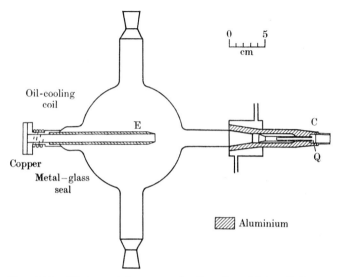

FIG. 14.3. Discharge-tube source for far ultra-violet radiation as used by Ditchburn and his collaborators.

as the ceramic capillary used by Weissler and his collaborators, is often employed. The choice of gas in the capillary depends on the wavelength region under investigation. Fig. 14.3 illustrates diagrammatically a source of this type used by Ditchburn,† the gas being neon. The discharge passes from the high potential electrode E to the earthed aluminium core C through a quartz capillary Q of rectangular cross-section $(5 \times 1 \cdot 5$ mm). This gives a strong glow spreading from the end which is only 25 mm from the slit of the spectrograph. To reduce sputtering the metal near the slit is of aluminium and the electrode E exposed to the gas is aluminized. The working pressure is about 7×10^{-2} torr. During a cycle of the mains supply a breakdown occurs two to three times. In each such phase the potential across the tube builds up to a few kV and suddenly falls as a current pulse passes. Steadiness of emission from the tube is monitored from the steadiness of the CRO trace of the electrical breakdown cycle.

† DITCHBURN, R. W., *Proc. R. Soc.* A229 (1955) 44.

The synchrotron source which has been used so effectively by Madden and Codling (see §§ 7.1–3) depends on the radiation emitted by fast electrons circulating in the 'donut' of a synchrotron.

An electron revolving in a circular path of radius R under the influence of a uniform and constant magnetic field normal to the plane of the path will radiate electromagnetic waves because it is subject to centrifugal acceleration. If the electron energy $E \gg mc^2$ the radiation at any point of the path is emitted very largely in the direction of the tangent. As viewed tangentially, the pulse emitted by the electron as it passes will persist for a time Δt which, because of the Lorentz transformation, will be of order

$$\Delta t \simeq \frac{R}{c} \frac{1}{\gamma^3},$$

where $\gamma = E/mc^2$. The Fourier analysis of this pulse will include angular frequencies ω up to $1/\Delta t$. Taking $R = 1$ m, $\gamma = 400$ (corresponding to 204 MeV electrons) we see that frequencies as high as 3×10^{15} s^{-1} will be radiated, corresponding to a wavelength of 1000 Å, far in the ultraviolet.

The detailed theory of the emission of radiation from a source of this kind has been worked out by Schwinger.† The power radiated in ergs/s/rad/Å is given by

$$P(\phi, \lambda) = \frac{8}{3} \frac{\pi e^2 c^2}{\omega_0 \lambda^4 \gamma^4} (1+x^2)^2 \left\{ K_{\frac{2}{3}}^2(\xi) + \left(\frac{x^2}{1+x^2} \right) K_{\frac{1}{3}}^2(\xi) \right\},$$

where ϕ is the angle the direction of observation from the electron makes with its normal projection on the orbital plane, λ is the wavelength, $x = \gamma\phi$, $\omega_0 = c/R$, and

$$\xi = (2\pi R/3\lambda)\gamma^{-3}(1+x^2)^{3/2}.$$

$K_{\frac{2}{3}}$ and $K_{\frac{1}{3}}$ are Bessel functions in the usual notation.

Fig. 14.4 illustrates the dependence of the power on the angle ϕ for three wavelengths radiated by electrons of 180-MeV energy revolving in a circle of radius 83·4 cm (the dimensions of the NBS electron synchroton). It will be seen that the emission is confined to a very small range of ϕ. In general this range is of order $1/\gamma$.

Because of this small range of γ, the experimentally important quantity is the total power integrated over all angles ϕ. This is given by

$$\overline{P}(\lambda) = \frac{3^{\frac{5}{2}}}{16\pi^2} \frac{e^2 \omega_0^3}{c^2} \gamma^7 G(\lambda/\lambda_c), \tag{50}$$

† Schwinger, J., *Phys. Rev.* **75** (1949) 1912.

where G is the function illustrated in Fig. 14.5. The critical wavelength λ_c is given by

$$\lambda_c = 4\pi R/3\gamma^3.$$

The peak power, for given R and γ occurs where $\lambda = 0\cdot42\lambda_c = \lambda_m^p$, and so the wavelength in which the maximum power is radiated varies as E^{-3}. The power radiated at the peak varies as E^7. Table 14.5 gives values of λ_m^p for different electron energies assuming that $R = 1$ m.

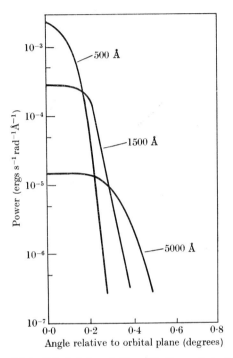

FIG. 14.4. Calculated variation with the angle ϕ of the power emitted per electron at three different wavelengths for electrons of energy 180 MeV circulating in an orbit of 83·4-cm radius.

For application to a source for ultra-violet absorption or photo-ionization the quantity of interest is $N(\lambda)\,d\lambda$, the number of photons emitted per second in the wavelength range λ to $\lambda+d\lambda$. According to (50) this will be given by

$$N(\lambda)\,d\lambda = \frac{3^{\frac{5}{2}}}{16\pi^2}\frac{e^2}{R^3}\frac{\gamma^7}{h}\,\lambda G(\lambda/\lambda_c)\,d\lambda$$

$$= \frac{3^{\frac{3}{2}}}{12\pi}\frac{e^2}{h}\frac{\gamma^4}{R^2}\,F(\lambda/\lambda_c)\,d\lambda,$$

where $F(x) = xG(x)$ and is a maximum when $x = 0\cdot6$ (see Fig. 14.5). Hence $N(\lambda)$ is a maximum when $\lambda = 0\cdot6\lambda_c = \lambda_m$.

Reference to Table 14.5 shows that an electron revolving in an orbit of 1-m radius will radiate strongly in the ultra-violet if it possesses an

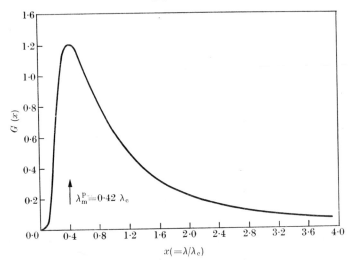

FIG. 14.5. The function $G(x)$.

energy greater than 100 MeV. Electron synchrotrons attain energies of this order in the course of which the electrons involved are circulating for an appreciable part of the duty cycle in orbits with radius of the order of 1 m.

TABLE 14.5

Variation of the wavelength λ_m^p for peak emission of power per unit wavelength range, with electron energy, the radius of the electron orbit being 1 m

Electron energy (MeV)	50	100	150	200	250	300
λ_m (Å)	18 670	2346	695	293	149	86·9

Furthermore, the circulating current in these machines is large enough for the total intensity of the radiation to be adequate as a source for high-resolution spectroscopy in the ultra-violet. Such a source has great advantages in that it provides a continuum, free from overlying emission lines, which has accurately predictable properties. Furthermore, the source exists in high vacuum and is even free from sputtered material.

The first studies of the possibility of using synchrotron radiation as a source for ultra-violet spectroscopy were made by Tomboulian and

Hartman† and by Tomboulian and Bedo.‡ Codling and Madden§ were
the first to use such a source for the observation, with high resolution, of
the far ultra-violet spectra of gases. They used the NBC synchrotron,
which accelerates electrons up to 180 MeV in a 'donut' of radius 83·4 cm.
In the normal mode of operation of this machine the magnetic field varies
sinusoidally at 60 c/s. Electrons are injected as the field goes positive
and only remain in stable orbit within the donut during the first quarter
cycle. To extend the time during which the electrons are at their highest
energy the operating parameters were adjusted so that the stable orbit
is maintained for as much as 1 ms after the peak of positive magnetic
field is reached. Results obtained using this source are described in
§§ 7.1–7.3.

5.3. *Radiation detectors*

A great deal of the observations have naturally been made using
photographic detection. In that case everything depends on accurate
calibration of plate response and various means of doing this have been
employed. As an example Ditchburn and Heddle‖ varied the intensity
of the beam in an exactly calculable way by inserting grids of known
transmission in the path of the beam, in this way covering the range
of intensities to be encountered in the absorption experiments. Rotating
sectors are also frequently used to intercept accurately calculated
portions of a beam.

Photoelectric, including photomultiplier, detectors may be sensitized
for use in the far ultra-violet by coating with a suitable fluorescent
material. Sodium salicylate is particularly useful as the sensitivity of
a fresh coating is constant within ± 10 per cent between 400 and 1250 Å
and within ± 20 per cent between 1250 and 2300 Å. It is also relatively
insensitive to strong light. Photographic plates may also be sensitized
by coating with a thin fluorescent film of lacquer.

Absolute measurements of flux may be made as described in § 5.1 using
an ionization chamber filled with a rare gas. Samson†† has developed
this technique using three designs of ionization chamber. One is the
single chamber used as described in § 5.1 while the others are double
chambers. Fig. 14.6 (a) illustrates the geometry of a double chamber

† TOMBOULIAN, D. H. and HARTMAN, P. L., *Phys. Rev.* **102** (1956) 1423.
‡ TOMBOULIAN, D. H. and BEDO, D. E., *J. appl. Phys.* **29** (1958) 804.
§ CODLING, K. and MADDEN, R. P., ibid. **36** (1965) 380 and references given in §§ 7.1–7.3.
‖ DITCHBURN, R. W. and HEDDLE, D. W. O., *Proc. R. Soc.* A **220** (1953) 61.
†† SAMSON, J. A. R., *J. opt. Soc. Am.* **54** (1964) 6.

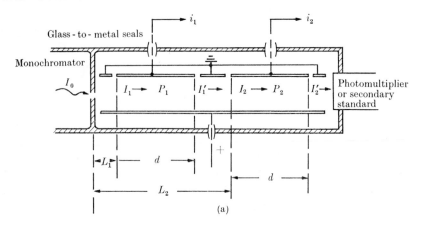

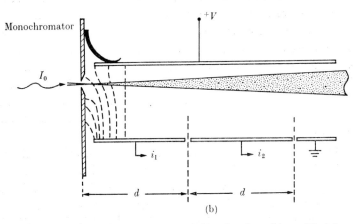

FIG. 14.6. Geometry of (a) double ionization chamber, (b) modified form of double ionization chamber.

of a type first used by Weissler and his associates.† Ion currents i_1 and i_2 are collected at two plates P_1 and P_2. Guard-rings are included between and at the outer edges of the plates so as to provide a uniform field between the plates when all are earthed. If I_0 is the flux passing through the last slit of the monochromator, I_1 and I_2 the fluxes entering and I_1' and I_2' the fluxes leaving the regions of plates P_1 and P_2 respectively, we have

$$I_1 - I_1' = I_0\, e^{-\kappa L_1}(1 - e^{-\kappa d}), \tag{51}$$

where the dimensions L_1, d are indicated in Fig. 14.6 (a). Hence, if

† WAINFAN, N., WALKER, W. C., and WEISSLER, G. L., *J. appl. Phys.* **24** (1953) 1318.

all absorption processes involve ionization,

$$I_0 = \frac{i_1/e}{e^{-\kappa L_1}(1-e^{-\kappa d})},$$ (52)

and similarly

$$I_0 = \frac{i_2/e}{e^{-\kappa L_2}(1-e^{-\kappa d})}.$$ (53)

From these equations the absorption coefficient κ and the flux I_0 can both be determined.

A modified version of the double chamber is to make $L_1 \to 0$, $L_2 \to d$ as in Fig. 14.6 (b). In this case

$$I_0 = \frac{i_1^2/e}{i_1 - i_2}.$$ (54)

When the pressure is high enough to give total absorption this chamber behaves as a single chamber with $I_0 = i_1/e$.

The principal difficulty in this method arises from the requirements of differential pumping. With no suitable windows available at wavelengths below 1000 Å it is necessary to maintain an adequately high vacuum in the monochromator to ensure that the flux of radiation just before entering the ionization chamber does not change when gas at the working pressure is admitted. Pressure gradients within a double ion chamber must also be avoided.

To check the method for self-consistency by verifying the assumption of the equality of absorption and photo-ionization cross-sections for rare gases, the relative photo-ionization yield was determined for different pairs of rare gases with the satisfactory results shown in Fig. 14.7. Deviations from unity in relative values are within ± 0.5 per cent over the wavelength range from 400–900 Å.

The wavelength sensitivity of sodium salicylate referred to above was measured by Samson† using the ionization chamber technique.

5.4. *Sources of error*

Although it is possible in principle to determine the total absorption cross-section from measurements made of the ratio of the transmitted to the incident intensity at one gas pressure only, it is important to verify by observations at a number of pressures that

$$I(n)/I(0) = 1 - e^{-nQ^a l},$$

where I is considered as a function of the concentration n of the absorbing particles. In other words, it should be verified that the absorption

† loc. cit., p. 1087.

coefficient is a constant, independent of pressure over a significant range of pressures from zero upwards.

Care should be taken in experiments with molecular gases that pressure changes or changes of the apparent absorption coefficient do not occur during the observations. Such changes are indicative of photochemical reactions in the gas, although variations of the absorption coefficient could be due to fluctuations of intensity from the source.

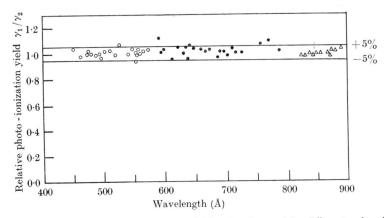

FIG. 14.7. Relative photo-ionization yields γ_1/γ_2 observed for different pairs of rare gases as a function of wavelength △ Kr/Xe, ● Kr/Ar, ○ Ne/Ar.

Scattered light may give rise to spurious signals in both absorption and photo-ionization measurements. Errors may be introduced in the latter by photoelectrons emitted from surfaces struck by reflected or scattered light. A further source of error in ionization chamber measurements may arise from production of metastable radicals which eject electrons from the electrodes.

5.5. *Some typical experiments*

5.5.1. *Total absorption method for permanent gases.* Fig. 14.8 illustrates the arrangement of the apparatus used by Weissler and his associates for measurement of the total absorption cross-sections of permanent gases. It was of the type in which the absorbing gas fills the spectrograph. The absorbing region was not sealed off from the radiating source by any window so that strong differential pumping was necessary to prevent the gas used in the source from contaminating the absorbing region. In this way it was possible to extend the wavelength range below the lithium fluoride cut-off at 1050 Å. Dispersion was achieved with a grating at grazing incidence.

A very similar arrangement has been used by Ditchburn and his associates† down to 350 Å, except that the grating was operated at normal incidence and that a thin window was introduced between the discharge tube and the spectrograph. This was found possible because of the small size required (not more than $6 \times 1 \cdot 5$ mm) and the small pressure difference (less than 3 torr) between the two sides. Films 300 Å thick are quite strong enough. They were successfully prepared from suitable solutions

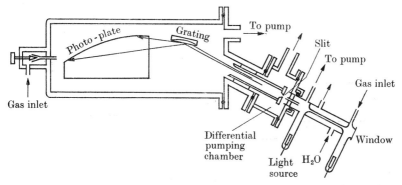

Fig. 14.8. Typical arrangement of apparatus used by Weissler and his associates for measurement of the total absorption cross-sections of permanent gases.

of celluloid in amyl acetate. Special arrangements were made for speedy replacement of damaged windows. To reduce damage by heating from the discharge the window was placed at a distance of $2 \cdot 5$ cm from the 20-μm wide slit of the spectrograph. Allowance for any small change in window transmission during an experiment was made by taking some calibration spectra before and others after the absorption spectra.

Scattered light was reduced to less than 10 per cent in the region 250–400 Å and less than 2 per cent at longer wavelengths. This was achieved by adjusting the grating so that the direct image fell off the plate and was collected on a blackened cylinder 5 mm in diameter and 7 mm long. Scattering from the edges of the diaphragm controlling the beam entering the spectrograph was eliminated by introducing additional stops. Further reductions in scattered light resulted from lining portions of the spectrograph tube with blackened gauze.

Ditchburn estimates that relative values of the absorption coefficient obtained with this apparatus are accurate to better than 10 per cent in the wavelength range 450–1300 Å. Absolute cross-sections are probably correct to about 10 per cent.

† DITCHBURN, R. W. and HEDDLE, D. W. O., *Proc. R. Soc.* A**220** (1953) 61; **226** (1954) 509.

Watanabe and his associates† have carried out an extensive series of measurements using an apparatus designed by Tousey, Johnson, Richardson, and Torau.‡ The geometry is illustrated in Fig. 14.9. The apparatus is of the type in which the absorption cell is placed beyond the exit slit of the spectrograph. Dispersion is produced by a 1-m grating and the wavelength reaching the exit slit is varied by displacing

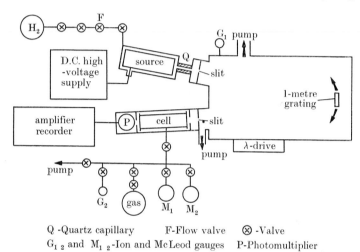

Q -Quartz capillary   F-Flow valve   ⊗ -Valve
$G_{1\,2}$ and $M_{1\,2}$-Ion and McLeod gauges   P-Photomultiplier

FIG. 14.9. Arrangement of apparatus designed by Tousey, Johnson, Richardson, and Torau for measuring total absorption cross-section of permanent gases.

the grating along the Rowland circle. Relative radiation intensity measurements were made using a glass-enclosed photomultiplier, coated externally with sodium salicylate. Once again no windows were introduced either between the source and spectrograph or between the spectrograph and absorption cell. The dimensions are indicated on the scale in Fig. 14.9.

Baker, Bedo, and Tomboulian§ have been able to use the technique in which an absorption cell is inserted between source and spectrograph in a study of the total absorption cross-section of helium and of neon in the wavelength range 200–600 Å. The geometry of their arrangement is as indicated in Fig. 14.10. They prepared the windows for the cell by cementing electroformed mesh of 100 lines/in and 80 per cent transmission to a stainless steel flange and depositing a film of plastic,

† See pp. 1154, 1156.
‡ TOUSEY, R., JOHNSON, F. S., RICHARDSON, J., and TORAU, N., *J. opt. Soc. Am.* **41** (1951) 696.
§ BAKER, D. J., BEDO, D. E., and TOMBOULIAN, D. H., *Phys. Rev.* **124** (1961) 1471.

Formvar or Zapon, on to the mesh. It was usually difficult to obtain leak-free windows with one plastic coating and two layers blocked out radiation with wavelength greater than 350 Å. However, using a cell 1·27 cm long and 1·91 cm diameter, it was possible to work with a helium pressure no greater than 5 torr. In these circumstances the leak rate from a single-coating window was tolerable. Photographic detection

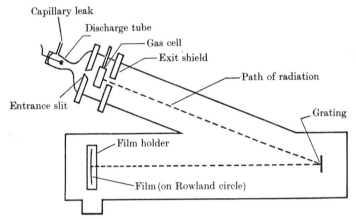

FIG. 14.10. Arrangement of apparatus used by Baker, Bedo, and Tomboulian for measurement of the total absorption cross-sections of rare gases.

was used with a normal incidence grating spectrograph except near the short wavelength end of the range for which a grazing incidence spectrometer was used.

The wavelength range now extended down to 180 Å by substituting a soft X-ray source emitting Al $L_{2,3}$ radiation for the condensed spark and an open Be–Cu photomultiplier as detector.

5.5.2. *Total absorption method for metal vapours.* Apart from the early photo-ionization observations referred to in § 5.1 all of the data available for metallic atoms have been obtained using the absorption method.

Because the hot vapour is chemically very reactive it is essential to seal off the absorption cell from the remainder of the equipment. This in itself presents difficulties because the vapour will attack the windows themselves. Braddick and Ditchburn† were able, by working with a very long absorbing path, to measure the absorption coefficient of caesium vapour at temperatures low enough for chemical changes in the windows to be of no optical significance. For no other metals is this

† BRADDICK, H. J. J. and DITCHBURN, R. W., *Proc. R. Soc.* A**150** (1935) 478.

possible as far too high temperatures are required to produce the necessary vapour pressure. Ditchburn, Tunstead, and Yates† solved the
problem by placing the metallic sources of vapour symmetrically on
either side of the centre of a long nickel tube whose ends were kept cool.
Vapour was prevented from rapid distillation to the end windows by introducing an inert filling gas into the tube.

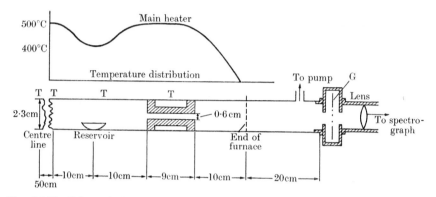

FIG. 14.11. Schematic illustration of the absorption tube used by Ditchburn and his
associates for measurement of the total absorption cross-sections of metal vapours. The
upper diagram shows a typical temperature distribution along the tube when the furnace
is on. The points at which the temperature was measured are indicated by T.

Fig. 14.11 illustrates schematically the design employed by Ditchburn
and his associates in all but the earliest experiments. The right-hand
half of the absorption tube is shown with dimensions indicated. This tube
was enclosed in a furnace up to the section indicated, giving a temperature distribution of the characteristic form shown in the upper part of
the figure. The values of the temperatures given are those employed
in studying absorption by sodium. Separate adjustment of the temperature of the metal reservoirs was provided by additional heaters.
Temperatures were measured by Chromel-Alumel thermocouples and
were controlled by hand. During an exposure they could be maintained
constant to better than 1 °C (2·7 per cent change in vapour pressure
for sodium). The filling gas used was helium at a pressure a few torr
higher than the vapour pressure of the metal.

A major problem is the determination of the effective path-length
through an absorber of this type. For this purpose we require the value
of $\int p \, dx$ integrated along the path of the light through the absorber, p
being the pressure of the vapour at a point distant x measured along the

† DITCHBURN, R. W., TUNSTEAD, J., and YATES, J. G., Proc. R. Soc. A 181 (1943)
386.

tube from some standard position. To obtain this, consider the inter-diffusion of vapour and filler gas from above a reservoir to the end of the tube.

The filler gas is supplied from a large reservoir so that when the tube is cold the pressure is P. When the metal vaporizes, some of the gas is driven out but the total pressure of gas plus metal vapour remains equal to P. Once a steady temperature is attained there is no net movement of the filler gas but the metal is transferred continually to the ends of the tube by diffusion through the gas. This is so provided the gas pressure is greater than the vapour pressure.

We consider first the usual problem of the interdiffusion of two gases. If n_1 and n_2 are the respective concentrations of the gases in a tube of uniform cross-section at a section distant x from a standard section, measured along the tube, the fluxes ϕ_1, ϕ_2 of each gas moving from left to right across the section are given by

$$\phi_1 = -D\frac{dn_1}{dx}, \qquad \phi_2 = -D\frac{dn_2}{dx} = -\phi_1, \qquad (55)$$

where D is the diffusion coefficients for mixing of the two gases.

To adapt this to the present problem, in which the vapour diffuses through the gas in which there is no net motion, it is only necessary to superpose a motion of the whole system that transfers a flux of ϕ_2 molecules of the inert gas from right to left. This will transfer at the same time $n_1\phi_2/n_2$ molecules of the vapour in the same sense, so the total flux of the vapour from left to right is

$$\phi^* = -(1+n_1/n_2)\phi_1$$
$$= -\frac{D}{\kappa T}\frac{p_1+p_2}{p_2}\frac{dp_1}{dx}. \qquad (56)$$

Here T is the absolute temperature and p_1, p_2 are the respective pressures of vapour and inert gas. Since p_1+p_2 ($=$ the constant P) and ϕ^* must be independent of x in a tube of uniform bore, we have

$$p_1 = P - p_0 e^{\alpha x}, \qquad (57)$$

where
$$\alpha = \frac{\kappa T\phi^*}{PD}, \qquad (58)$$

and p_0 is the value of p_2 at $x = 0$. At the condensing end of the tube $p_1 = 0$, while immediately above the metal reservoir p_1 is equal to the vapour pressure of the metal at the working temperature. From these two boundary conditions p_0 and α may be determined, so giving p_1 as a function of x as required.

For wavelengths longer than 2050 Å (the quartz u.v.) the source of light was a low-voltage hydrogen arc used with a Zeiss medium quartz spectrograph. The absorption tube was closed by silica windows. In the vacuum ultra-violet below 2250 Å a high-voltage hydrogen discharge tube was used as a source with a fluorite spectrograph of the type described by Cario and Schmidt-Ott.† The beam was collimated and focused by fluorite lenses.

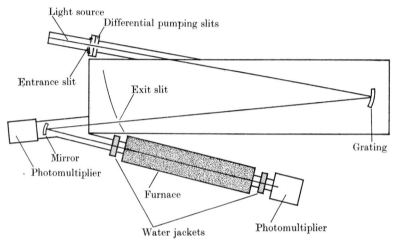

Fig. 14.12. Arrangement of apparatus used by Hudson for measurement of the total absorption cross-sections of metal vapours.

In the quartz region rotating sectors were used for calibration. These were replaced in the vacuum ultra-violet by grids.

In the experiments carried out by Ditchburn and his collaborators photographic detection has been used. Hudson‡ has more recently made measurements using essentially similar technique except that the beam intensity was measured photoelectrically. Fig. 14.12 illustrates the general arrangement typically employed in these experiments.

One difficulty is the background intensity of infra-red radiation from the furnace. To reduce the importance of this background the light from the exit slit of the monochromator was partially reflected from a concave mirror so as to be focused on a sodium salicylate (see p. 1081) screen in front of the photomultiplier at the far end of the absorption tube. This made it possible to provide radiation baffles in front of the photomultiplier with narrow openings, thereby limiting the infra-red

† Cario, G. and Schmidt-Ott, H. D., Z. Phys. 69 (1931) 719.
‡ Hudson, R. D., Phys. Rev. 135 (1964) A1212.

radiation that could be recorded. In addition it was found necessary to place a filter (Kodak Wratten No 47B) between the sodium salicylate screen and the photomultiplier that had a pass band of about 100 Å centred at the wavelength at which the fluorescence from sodium salicylate is a maximum. The intensity of the incident radiation could be monitored throughout an experiment by a photomultiplier behind the mirror through which a smaller hole was bored to let through an observable fraction of the incident light. The mirror was coated with aluminium for observations between 3700 and 1500 Å. For shorter wavelengths down to 900 Å the reflectivity dropped very rapidly as the furnace temperature was raised and the aluminium was replaced by gold the reflectivity of which was unaffected under these conditions.

An important possibility which must be allowed for is the presence of metallic molecules in the vapour, particularly for the alkali metal vapours (apart from lithium), because the atomic absorption close to the threshold is very weak. If n_a, n_m are the concentrations of atomic and molecular species respectively and Q_a, Q_m the corresponding absorption cross-sections

$$\ln(I_0/I) = (n_a Q_a + n_m Q_m)L,$$

where L is the absorption path length, I_0, I the incident and transmitted light intensities.

According to the law of mass action

$$n_m/n_a^2 = K(T),$$

where K is the equilibrium constant. We therefore have

$$(1/n_a L)\ln(I_0/I) = Q_a + n_a K Q_m. \tag{59}$$

Ditchburn and his collaborators[†] then assumed that K is effectively constant over the temperature range covered in their experiments. Then, by plotting the left-hand side as a function of n_a, a linear relation should be found, the intercept giving Q_a and the slope Q_m. However, the assumption of effectively constant K is not justified, as may be seen from Fig. 14.13, which shows K as a function of $1/T$ using data for sodium from the JANAF Thermochemical Tables.[‡] Hudson,[§] in his experiments using photoelectric detection, therefore derived Q_a and Q_m from (59) by a least squares fit of observations taken at twenty or more different temperatures. The introduction of this technique for separating atomic and molecular absorption has proved particularly important near the very low (probably zero) minimum in the absorption cross-sections for

† loc. cit., p. 1094.
‡ The Dow Chemical Company, Midland, Michigan (1962).   § loc. cit., p. 1096.

sodium and potassium near the threshold (see Figs. 14.32 and 14.33).

Apart from the need for separation of molecular and atomic absorption, the accuracy of the observations depends very much on that of the vapour pressure tables available for each metal.

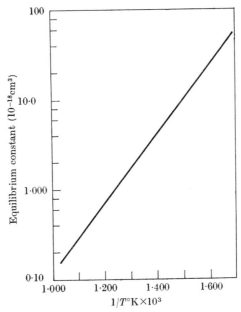

FIG. 14.13. Variation of equilibrium constant for
$Na_2 \rightleftharpoons 2Na$ with $1/T$.

5.5.3. *Direct measurement of photo-ionization.* A comprehensive series of observations of total absorption and photo-ionization cross-sections for a number of atoms and molecules, over the wavelength range 600–1800 Å, has been carried out by Cook *et al.*,† using apparatus‡ the general arrangement of which is shown in Fig. 14.14. Essentially it was of the type in which the absorption chamber is located behind the exit slit of the monochromator. No windows were used so that reliance was placed on differential pumping to isolate the source and absorbing gases from each other and from the evacuated monochromator. A pressure as low as 10^{-5} torr could be maintained in the main chamber when the source was operated with helium at 150-torr pressure.

The intensity of the radiation passing through the absorption chamber was measured either by a platinum photocathode or a photomultiplier

† See §§ 7.3, 8.3–8.10.
‡ METZGER, P. H. and COOK, G. R., *J. Quant. Spectros. Radiat. Transfer* **4** (1964) 107.

coated with sodium salicylate. Calibration of the response of these
detectors was carried out by the single ionization chamber method (see
§ 5.1) with rare gas filling. Fig. 14.15 gives the results obtained for the
photoelectric yield of platinum as a function of wavelength. The figure
also includes results obtained by other investigators[†] using the double
ionization chamber method.

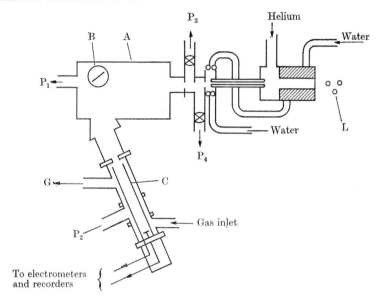

A   Vacuum ultraviolet monochromator
B   Diffraction grating on turntable
C   Absorption and photo-ionization chamber
G   To pressure gauges
L   Hg lamp in quartz tube
P   To pumping systems

FIG. 14.14. Arrangement of apparatus used by Cook *et al.* for measurement of
total absorption and photo-ionization cross-sections.

Care was taken to test for, and if necessary eliminate, various possible
sources of error. Thus it was verified that the level of scattered light was
negligible, that reflection from the platinum detector was unimportant,
that saturation was reached in the ion collection with a comparatively
low voltage, 15 V, on the collector, and that the ion current varied with
pressure as expected. It was estimated that the observed values for Q^a

† MATSUNAGA, F. M. and WATANABE, K., Private communication to Cook and
Metzger 1963; WAINFAN, N, WALKER, W. C. and WEISSLER, G. L., *J. appl. Phys,* **26**
(1953) 1318; WAINFAN, N., WALKER, W. C. and WEISSLER, G. L., *Phys. Rev.* **99** (1955)
542.

greater than 10^{-18} cm² should be correct to 15 per cent and for Q_i^{ph} to 20 per cent.

Earlier extensive observations of Q^a and Q_i^{ph} have been carried out by Weissler and by Watanabe with their associates.† The only difference in principle from the experiments described above is in the use of a double ionization chamber in place of the single chamber plus photoelectric detector.

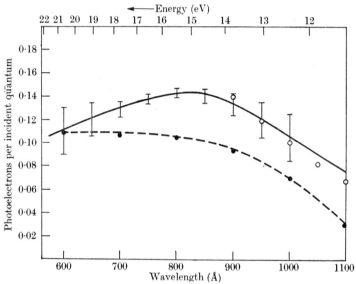

FIG. 14.15. Observed photoelectric yield of platinum. —— (with error bars) observed by Cook and Metzger. ○ observed by Matsunaga and Watanabe. —•—•— observed by Wainfan, Walker, and Weissler.

Photo-ionization cross-sections for rubidium and caesium were measured by Mohler and Boeckner‡ as long ago as 1929. They passed the radiation from a suitable source (a tungsten filament lamp, mercury arc, or hydrogen glow discharge) through a quartz monochromator so that a chosen narrow wavelength band was focused within an ionization chamber or space-charge detector containing the metal vapour at a pressure determined by the oven temperature. Absolute measurements of light intensity were made with a sensitive thermopile placed at the position of the ion detector.

5.5.4. *Experiments directed toward the determination of ionization potentials and structure near thresholds.* It is obvious that measurement of photo-ionization cross-sections as functions of wavelength offers a

† See WEISSLER, G. L., *Handb. Phys.* **21** (1956) 304 and note †, p. 1092.
‡ loc. cit., p. 1079.

method for the determination of ionization potentials that has many
advantages. In particular, whereas cross-sections for ionization by
electron impact rise gradually from the threshold, those for photo-
ionization rise much more steeply, being even of step function form in
many cases. It has indeed been pointed out in Chapter 13 (see Fig. 13.21)
that the variation of the photo-ionization cross-section near the threshold
is very similar to that of the derivative of the variation of the cross-
section for electron impact ionization with respect to electron energy.
On the other hand, the accuracy of determination of the threshold wavelength was, until recently, limited by the spacing of the spectrum lines
from the sources that were available for the relative cross-section
measurements. The resolving power in terms of photon energy at a
given wavelength is nevertheless very high as compared with most
electron sources and the absolute value of the energy is readily obtained
from wavelength calibration. Nicholson† carried out experiments to
determine what accuracy could be achieved when the work was directed
specifically towards the accurate determination of ionization potentials
and of any structure in the photo-ionization curve close to these
thresholds.

He used equipment essentially similar in principle to that of Cook
and his collaborators described above. No windows were used but the
entrance from source to monochromator and from monochromator to
ionization chamber was in each case a tunnel 1 cm in diameter, the former
being of cross-section 6×0.12 mm and the latter 6×0.25 mm. The
ionization chamber plates were of 1 cm² cross-section, 1 cm apart, and
operated with a potential difference of $7\frac{1}{2}$ V across them. The light that
passed through the chamber was monitored by a sodium-salicylate
sensitized photomultiplier. A hydrogen lamp of a type intermediate
between a simple discharge lamp and a capillary arc was used as source.
Calibration of the grating drive was carried out using different gases
in the lamp to provide standard lines. A daily check was made using
the H Lyα line. Measurement of the energy spread of the light showed
that the half width was about 4.4 Å corresponding to 0.05 eV at a mean
energy of 12 eV. Special care was taken to eliminate or allow for background effects due to scattered light and residual gas. Results were
expressed in terms of the so-called photo-ionization efficiency (PE) as a
function of quantum energy. This is the ratio of the ion current to the
intensity of transmitted light as monitored by the photomultiplier. If
the absorption in the chamber is not too great this efficiency is roughly

† NICHOLSON, A. J. C., *J. chem. Phys.* **39** (1963) 954.

proportional to the photo-ionization cross-section. Results obtained indicate an accuracy considerably better than 0·010 V in ionization potentials in most cases, and considerably better for first ionization potentials. Thus, for xenon, Nicholson finds a first ionization potential of 12·129±0·002 V to be compared with the spectroscopic value 12·1292 V. The corresponding values for the second potential are 13·426±0·007 V as compared with 13·435 V. Further discussion of Nicholson's results from xenon, oxygen, and nitric oxide will be discussed in §§ 7.3, 8.4, and 8.7 respectively.

5.5.4.1. *Use of photoelectron spectra.* One of the great difficulties in determining threshold energies precisely, particularly thresholds beyond the first, is the confusion produced by the existence of lines due to auto-ionization (see, for example, Figs. 14.26, 14.36, and 14.41). It is possible largely to avoid this confusion if the energy distribution of the photoelectrons produced by impact of photons of a fixed and definite frequency is measured. Thus if E_i is an ionization threshold energy then radiation of frequency ν will produce electrons of energy $h\nu - E_i$. The presence of these electrons will be detected by a sharp peak in the energy distribution at this energy, the sharpness depending on the energy resolution of the electron spectrometer. Any accidental coincidence of the chosen frequency with one falling within an auto-ionization line can be detected if energy distributions are measured for more than one photon frequency.

The first application of this method was made by Vilessov, Kerbatov, and Terenin† who used an apparatus of Lozier type (Chap. 12, § 7.3.1) in which the electron beam was replaced by a beam of photons issuing from a vacuum ultra-violet monochromator. Lithium fluoride windows limited the incident photon energy to less than 11·7 eV. Schoen‡ removed this limitation by avoiding the use of a window. A further step that reduced the complexity of the equipment while improving the resolution was taken by Al-Joboury and Turner§ who used an effectively monochromatic light source and a cylindrical electrode geometry for measurement of the electron energy.

In choosing this geometry account must be taken of the fact that the angular distribution of the ejected electrons is a maximum in the direction of the electric vector of the light beam. With an unpolarized beam the distribution is symmetrical about the axis of the beam and varies

† Vilessov, F. I., Kerbatov, B. L., and Terenin, A. N., *Dokl. Akad. Nauk SSSR*, **138** (1961) 1329 and **140** (1961) 797.

‡ Schoen, R. I., *J. chem. Phys.* **40** (1964) 1830.

§ Al-Joboury, M. I. and Turner, D. W., *J. chem. Soc.* (1963) 5141.

as $\sin^2\theta$, where θ is the angle the direction of motion makes with the beam. By using the Lozier method the retarding potential analysis is confined to those electrons that are produced in directions normal to the beam, whereas in a cylindrical geometry there is a certain lack of resolution due to the initial motions of the photoelectrons not all being along the electrostatic field lines. If the production of photoelectrons is confined to a small region and the analysing field system is constructed of spherical electrodes centred on this region almost all the electron trajectories will be radial so that the resolution should be improved as compared with a cylindrical system. At the same time the defects of the Lozier system as regards energy discrimination and reduced collecting power would be avoided. A spherical system of this kind was first introduced by Frost, McDowell, and Vroom.†

Al-Joboury and Turner confined their investigations to photoionization produced by the helium resonance line (584 Å). By using a high-voltage helium discharge in a Pyrex tube 10 cm long and 5 mm internal diameter they obtained a source sufficiently monochromatic for their purposes without dispersion. Their retarding potential analyser consisted of two cylindrical grids G_1 and G_2, 0·7 and 2 cm in diameter respectively, coaxial with the light beam and enclosed in the coaxial electron collector plate P 4 cm in diameter. The grids were of rhodium–platinum alloy mesh, containing 80 wires, each 0·002 in diameter, per inch. It was found necessary to gold-plate the grid wires to eliminate local field effects. Voltage calibration was carried out by comparison of the voltages at which primary peaks occurred in the electron spectra from argon, krypton, and xenon with the spectroscopic values. This involved corrections of the order of 0·05 V.

Frost *et al.* carried out some experiments with a very similar apparatus except that a microwave cavity light source, as shown in Fig. 14.16, was used. The work has been extended to cover the wavelength range 500–1000 Å by Doolittle and Schoen using a source similar to that employed by Weissler and his collaborators (see § 5.2 above), in conjunction with a Lozier type analyser.

The arrangement used by Frost *et al.* in their later experiments with a spherical grid energy analyser is illustrated in Fig. 14.16. The photon source was a microwave discharge in helium in which power from a 10-W 2450 Mc/s generator was fed to a resonant cavity through which commercial helium flowed. Almost all the radiation emitted from such a discharge below 1000 Å is in the region of the 584-Å line. The photon

† Frost, D. C., McDowell, C. A., and Vroom, D. A., *Proc. R. Soc.* A**296** (1967) 566.

stream was collimated by passage through a capillary 0·5 mm in diameter and 5 cm long, which also assisted, in conjunction with the differential pumping system, in the isolation of the source from the ionization chamber.

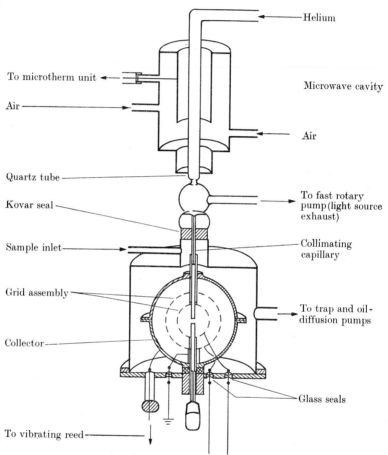

Fig. 14.16. Arrangement of apparatus used by Frost *et al.*, for retarding potential analysis of photoelectron energies.

Photo-ionization was limited to the central region of this chamber by confining the photons within two tubes of gold-plated brass except within the gap between the tubes.

The electrodes of the spherical analyser were constructed as separate hemispheres, the grids being of internal diameter 1·5 and 2 in and the outer collector electrode 3 in. The grids were pressed out of 30×30 mesh 0·005-in gauge brass and the collector turned from solid brass. Before gold plating the grids were electrolytically reduced in gauge to an optical transmission of about 75 per cent. The photon conducting tubes were 0·2-in diameter and were maintained at the same potential as the inner grid. To minimize electron reflection the collector was coated with a colloidal suspension of graphite.

Positive ions were prevented from entering the region between the outer grid and collector by a constant potential difference of 3 V between the two grids. Photoelectron-retarding potentials were applied between the outer grid and collector. The analyser was shielded from magnetic fields by enclosure within several layers of mumetal. This was found to be important for obtaining the best resolution.

The photoelectron current obtained with argon at a pressure of 10^{-4} torr was about 10^{-10} A. In all experiments it is important to check that the relative heights of the steps in the retarding potential curve are independent of the pressure. This eliminates the possibility that the negative currents collected are due to negative ions produced by attachment of the photoelectrons to the molecules of the gas under study.

A test of the performance of this type of equipment is the spectrum observed for electrons produced from argon, krypton, and xenon. In each case two peaks should be resolved corresponding to ionization leaving the ion in the $P_{\frac{1}{2}}$ and $P_{\frac{3}{2}}$ states respectively. Fig. 14.17 shows typical retarding potential curves obtained with the spherical analyser by Frost et al. The presence of the second ionization threshold is very clearly seen in each case. In Fig. 14.18 a typical differential spectrum (obtained by differentiation of the retarding potential curves with respect to retarding voltage) observed by Vroom† for argon is shown. Two peaks are clearly resolved.

Comparison between the energy difference observed between the first and second ionization thresholds and the spectroscopically determined separation between the $^{2}P_{\frac{1}{2}}$ and $^{2}P_{\frac{3}{2}}$ states of the respective ions reveals good agreement as may be seen by reference to Table 14.6.

Application of photoelectron retarding potential analysis to particular gases, using the techniques we have described will be discussed in §§ 8.4–8.11.

5.5.5. *Mass spectrometric analysis of products of photo-ionization.* Up to the present we have been dealing with the total photo-ionization cross-section that does not distinguish between the different products. To determine the cross-sections for the production of different ionic species

† VROOM, D. A., Thesis, Vancouver (1967).

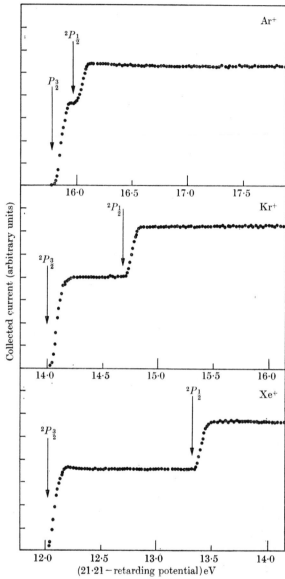

Fig. 14.17. Retarding potential curves for photoelectrons produced by 584-Å radiation in argon, krypton, and xenon, obtained using spherical electrode geometry.

it is necessary to perform a mass analysis of the ions just as in dealing with ionization by electron impact.

To carry out such an analysis is in some ways more difficult than for electron impact, because the photon fluxes obtainable are much lower

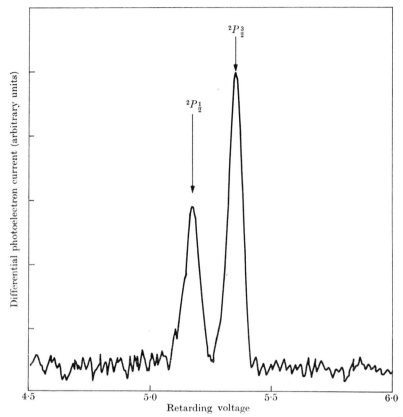

FIG. 14.18. Typical differential photoelectron spectra for argon ionized by 584-Å (21·21-eV) radiation.

TABLE 14.6

	Observed energy difference between ionization thresholds (eV)	$^2P_{\frac{1}{2}}$–$^2P_{\frac{3}{2}}$ separation determined spectroscopically (eV)
Argon	0·183±0·005	0·178
Krypton	0·676±0·006	0·67
Xenon	1·312±0·006	1·30

than the bombarding electron currents used in the type of experiments described in Chapter 3, § 2.2, Chapter 12, § 7.3.2, and Chapter 13. It is not possible to make up for this by working with higher gas-pressures because this will render serious such secondary effects as ionization by the photoelectrons.

There are, however, compensating advantages. Some of these are common to any photo-ionization method but are enhanced by the availability of mass analysis. These include high energy resolution and ease of energy determination for a photon beam and the form of the threshold law for photo-ionization. In addition, specifically on the mass analysis side, there is an advantage due to the ionizing photons being uncharged. This offers greater freedom in the design of the electrode system which draws the ions out from the ionization chamber into the mass spectrograph. This should lead to greater extraction efficiency and one markedly less dependent on the initial energy of the ions (see Chap. 12, § 7.3.2).

The first experiments aimed at a mass analysis of the products of photo-ionization were carried out in 1929 by Ditchburn and Arnot[†] to determine to what extent molecular ions K_2^+ were produced in the photo-ionization of potassium vapour. In 1932 Terenin and Popow[‡] used mass analysis to show that light of wavelength near 2000 Å dissociated thallium halides into positive thallium and negative halogen ions. No further experiments of this kind were attempted until Lossing and Tanaka[§] carried out a mass analysis of ionized fragments arising from dissociation of organic molecules by light from a krypton discharge. A number of other experiments were then initiated more or less in parallel. In some of the earlier experiments dispersion of the light was not attempted but in 1957 Hurzeler, Inghram, and Morrison[||] built an apparatus that incorporated both a monochromator and a mass spectrograph. The former was of the design introduced by Seya and Namioka[††] in which the mounting is such that the position of the entrance and exit slits and of the grating are fixed and the direction in which the diffracted beam emerges does not change with wavelength. This requires that the diffracted beam passes out in a direction making an angle 70° 15′ with the incident.

The mass spectrograph was of the Nier 60° sector type with a mass resolution of 1 in 300.

The light source, a capillary discharge in hydrogen, was isolated from the monochromator by a lithium fluoride window 1 mm thick, which limited the range of observations to wavelength greater than 1050 Å. On the other hand, no such window was included between the monochromator and the ion chamber. With entrance and exit slits from the latter of widths between 0·02 and 0·04 in the energy spread in the photon beam was about 0·05 eV at a photon energy of 10 eV.

† DITCHBURN, R. W. and ARNOT, F. L., *Proc. R. Soc.* A123 (1929) 516.
‡ TERENIN, A. and POPOW, B., *Z. Phys.* 75 (1932) 338.
§ LOSSING, F. P. and TANAKA, I., *J. chem. Phys.* 25 (1956) 1031.
|| HURZELER, H., INGHRAM, M. G., and MORRISON, J. D., ibid. 28 (1958) 76.
†† SEYA, M., *Sci. Lt, Tokyo* 2 (1952) 8; NAMIOKA, T., ibid. 3 (1954) 15.

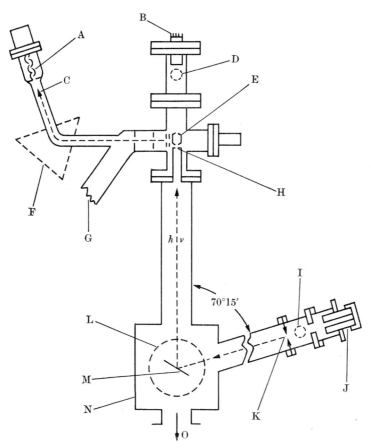

FIG. 14.19. Arrangement of apparatus used by Weissler *et al.* for mass analysis of products of photo-ionization. *A* is the electron multiplier ion detector, *B* the photomultiplier radiation monitor, *C* the ion beam, *D*, *G*, *I*, *O* connections to pumps, *E* the ion chamber, *F* a permanent magnet, *H* the monochromator exit slit, *J* the light source, *K* the monochromator entrance slit, *L* the grating turntable, *M* the grating, and *N* the Seya-type vacuum monochromator.

The volume of the light beam effective in producing ionization in the ion chamber was about 0.13 cm^3 and the gas pressures used were around 10^{-4} torr. A 16-stage electron multiplier was used to measure the final resolved ion currents. Currents greater than 10^{-20} A could be measured, the maximum currents recorded being about 10^{-15} A.

Results obtained with this apparatus are described below in § 8.12.

A little later Weissler, Samson, Ogawa, and Cook† constructed a very similar apparatus, illustrated schematically in Fig. 14.19. The only

† WEISSLER, G. L., SAMSON, J. A. R., OGAWA, M., and COOK, G. R., *J. opt. Soc. Am.* **49** (1959) 338.

important difference was the absence of any windows between source and monochromator so that experiments could be carried out down to 450 Å. Background signals in the mass spectrometer due to residual gases resulting from the absence of windows were determined by flowing argon into the ionization region and examining the mass spectrum. The energy resolution was about one-half as good as in the experiment of Hurzeler *et al.*

6. Experimental methods for measuring cross-sections for absorption of excited atoms

No direct experimental observations of cross-sections for absorption of radiation by excited atoms have been made. Indeed such measurements are only within the reach of present techniques if the atoms are in metastable states, so indirect methods must be resorted to. These depend on the observation of emission continua from high temperature sources such as shock-excited gases or high pressure arc discharges. Such continua include contributions not only from recombination of electrons to various states of ionized atoms but also from radiative attachment to neutral atoms. Accordingly we shall defer discussion of results obtained about absorption coefficients for excited atoms until Chapter 15, which is primarily concerned with radiative attachment. In Chapter 15, § 4 the techniques are discussed, while applications to oxygen, nitrogen, and chlorine are described in Chapter 15, §§ 5.2, 5.5, and 5.6.

7. Results of measurements of photo-ionization cross-sections

7.1. *Helium*

Although the calculation of the cross-section for photo-ionization of helium is relatively simple compared with that for other nonhydrogenic atoms, the accumulation of experimental data for this gas proceeded very slowly because the ionizing radiation is far in the ultra-violet—the threshold is at 504 Å. However, with the development of techniques for the study of absorption in the vacuum ultra-violet a considerable amount of data is now available. Thanks to the synchrotron radiation source (p. 1084) this has even extended to detailed investigation of some series of auto-ionization resonances, providing indeed the best opportunity for a combined theoretical and experimental study of such resonances.

The first measurement of an absorption coefficient for short-wave radiation in helium was made by Dershem and Schein† at 44·6 Å. It

---

† DERSHEM, E. and SCHEIN, M., *Phys. Rev.* **37** (1931) 1238.

was twenty-four years later before Lee and Weissler,[†] using apparatus similar to that described on p. 1091, made measurements of the cross-section from 240 to 504 Å. Further measurements have been made by Axelrod and Givens[‡] (180–300 Å). Baker, Bedo, and Tomboulian[§] covered the range 200–504 Å using the enclosed-cell technique described on pp. 1092–3. This work was later extended to 100 Å by Lowry, Tomboulian, and Ederer.[||] Meanwhile Samson and Kelly[††] also gave results down to 300 Å, using the double-ionization-chamber technique (p. 1088).

The energy resolving power in these experiments was not adequate to observe, let alone study, the auto-ionizing levels of helium just below the second ionization threshold at 65·4026 eV (189·6 Å). However, Madden and Codling[‡‡] succeeded in observing many of the bands and measured many of their properties, the method being as described in § 5.2.

We shall first ignore the possibility of auto-ionization. This introduces no difficulties at wavelengths greater than 250 Å and at smaller wavelengths refers to the 'background' cross-section between resonance lines.

Fig. 14.20 illustrates experimental values over the range 100–504 Å obtained by different observers. The agreement between different sets of results leaves a great deal to be desired. We give in Table 14.7 'experimental' values of cross-sections obtained by making a least squares fit to the observations of Tomboulian and his collaborators. These data have the justification that they not only agree well with directly calculated values but also with the oscillator-strength sum rules (see Chap. 7, § 5.2.2).

To apply these sum rules it is necessary not only to know the differential oscillator strengths but also the oscillator strengths f_{0n} for the excitation of discrete levels. Lowry et al.[||] took for the latter the calculated values of Schiff and Pekeris,[§§] which agree well with the observed values for the 1^1S–2^1P and 1^1S–3^1P transitions derived from electron impact observations (Chap. 7, § 5.6.7). Using these values, together with the differential oscillator strengths derived from direct observations, the various oscillator sums

$$S_q = \left(\sum_{n \neq 0} + \int \right) f_{0n} (E_0 - E_n)^q \qquad (60)$$

† Lee, P. and Weissler, G. L., *Phys. Rev.* **99** (1955) 540.
‡ Axelrod, N. and Givens, M. P., ibid. **115** (1959) 97.
§ Baker, D. J., Bedo, D. H., and Tomboulian, D. H., ibid. **124** (1961) 1471.
|| Lowry, J. F., Tomboulian, D. H., and Ederer, D. L., ibid. **137** (1965) A1054.
†† Samson, J. A. R. and Kelly, F. L., *G.C.A. Tech. Rept.* No. 64–3–N.
‡‡ Madden, R. P. and Codling, K., *Phys. Rev. Lett.* **10** (1963) 516.
§§ Schiff, B. and Pekeris, C. L., *Phys. Rev.* **134** (1964) A638.

were calculated for $q = -2, -1, 0,$ and 1. From the cases $q = -2$ and $q = 0$ respectively the polarizability α and number of electrons N for helium atoms may be obtained at once. From $q = -1$ the magnetic susceptibility χ may be obtained provided the matrix element

$$\sum_n \langle 0 | \mathbf{r_1} . \mathbf{r_2} | n \rangle$$

is known, $\mathbf{r_1}$ and $\mathbf{r_2}$ being the coordinates of the two electrons. This has been calculated by Vinti† as -0.073 at. u. Finally, from $q = 1$ the total

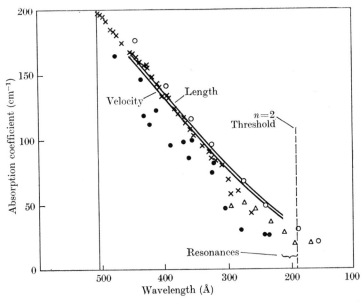

FIG. 14.20. Comparison of observed and calculated continuous absorption coefficients for helium. Observed: ○ Lowry, Tomboulian, and Ederer; × Samson and Kelly; ● Lee and Weissler; △ Axelrod and Givens. Calculated: —— Burke and McVicar (assuming length and velocity matrix elements as indicated).

energy E_t may be obtained provided $\sum_n \langle 0 | \mathbf{p_1} . \mathbf{p_2} | n \rangle$ is known, $\mathbf{p_1}$ and $\mathbf{p_2}$ being the momenta of the two electrons. This has been calculated by Kinoshita‡ as -0.6364 at. u.

In Table 14.8 values of α, χ, N, and E_t derived in this way are compared with values obtained from direct observation. In all cases there is agreement within experimental error. The contribution from discrete states to the oscillator strength sums are given in Table 14.8.

Because of its relative simplicity a number of theoretical calculations

† VINTI, J. P., *Phys. Rev.* **41** (1932) 813.   ‡ KINOSHITA, T., ibid. **105** (1957) 1490.

TABLE 14.7

Experimental and theoretical photo-ionization cross-sections for helium

Wavelength (Å)	504	443	395	356	325	275	239	181	157	93	57·2
Energy of photoelectron (at. energy units)	0·00	0·25	0·50	0·75	1·00	1·50	2·00	3·00	4·00	5·00	16·00
Cross-section in 10^{-18} cm²											
Experimental	8·37	6·56	5·25	4·30	3·56	2·50	1·84	1·12	0·770	0·220	0·040
Theoretical											
(i) C(V)	7·79	6·24		4·10		2·37	1·73	1·00	0·63	0·17	0·03
(ii) H(V)	8·38	6·69	5·36	4·35	3·57	2·48	1·79	1·02	0·64	0·16	0·03
(iii) HF(L)	7·40	6·16	5·10$_5$	4·24	3·55	2·55	1·89	1·12	0·72	0·19	0·03$_5$
(iv) HF(V)	7·16	5·89	4·85$_5$	4·03	3·38	2·43$_5$	1·81	1·07$_5$	0·69	0·18	0·03
(v) HF(A)	6·75	5·51	4·51$_5$	3·74	3·13	2·25	1·67	0·99	0·63	0·16	0·03
(vi) BM(L)		6·07	5·02	4·18	3·51	2·57	1·94				
(vii) BM(V)		5·88	4·87	4·09	3·43	2·50	1·89				

In the theoretical results (i)–(v) the ground state wave function is a six-parameter Hylleraas variational function, in (vi) and (vii) a twenty-parameter function obtained by Hart and Herzberg. The continuum wave functions used are as follows: (i) undistorted Coulomb wave, (ii) Coulomb wave distorted by static field of the He$^+$ ion, (iii)–(v) Coulomb wave distorted by static and exchange fields of He$^+$ (vi)–(vii), 1s–2s–2p close-coupling wave function. (L) (V) (A) refer to use of dipole length, velocity, and acceleration matrix elements respectively.

of absorption cross-sections for helium have been carried out, beginning in 1933 with the work of Wheeler[†] and Vinti.[‡] Both authors used the dipole length formulation (see § 2 (6)), with a variational approximation of Hylleraas type[§] for the ground state and a continuum state wave function of hydrogenic form. The next step was not taken till fifteen years later when Huang[||] carried out the first calculations using the velocity matrix formulation (see § 2 (8)) and wave functions generally

TABLE 14.8

Comparison of values of polarizability α, *magnetic susceptibility* χ, *number of electrons N, and total energy* E_t *of helium obtained from oscillator sum rules and from direct observation*

Physical quantity	From sum rule		From direct experiment
	Total	Contribution from discrete states	
α (10^{-25} cm²)	$2 \cdot 21 \pm 0 \cdot 17$	$0 \cdot 724$	$2 \cdot 07 \pm 0 \cdot 01$
χ (10^6 g mole^{-1})	$-1 \cdot 96 \pm 0 \cdot 15$		$-1 \cdot 933 \pm 0 \cdot 006$
N	$2 \cdot 05 \pm 0 \cdot 15$	$0 \cdot 431$	2
E_t (Rydbergs)	$5 \cdot 72 \pm 0 \cdot 50$	$0 \cdot 475$	$5 \cdot 76 \pm 0 \cdot 00$

similar to those of Wheeler and Vinti. Stewart and Wilkinson[††] were the first to improve the approximation for the continuum wave function. They used the 1s exchange or Hartree–Fock approximation (Chap. 8, § 5.1) together with a six-parameter Hylleraas type[§] ground-state function (see Chap. 7 (200)) and calculated cross-sections using all three matrix formulations—length, velocity, and acceleration. Their results were later extended by Stewart and Webb[‡‡] to higher energies. To test the dependence of the cross-section on the form of the free-wave function, calculations were also carried out using the hydrogenic form and also the 1s static (Hartree) approximation. Table 14.7 illustrates the results obtained with these different approximations.

Burke and McVicar[§§] took a further step in using improved wave functions. For the ground state they took the twenty-parameter functions

† WHEELER, J. A., *Phys. Rev.* **43** (1933) 258.
‡ VINTI, J. P., ibid. **44** (1933) 524.
§ HYLLERAAS, E., *Z. Phys.* **54** (1929) 347.
|| HUANG, S., *Astrophys. J.* **108** (1948) 354.
†† STEWART, A. L. and WILKINSON, W. J., *Proc. phys. Soc.* **75** (1960) 796.
‡‡ STEWART, A. L. and WEBB, T. G., ibid. **82** (1963) 532.
§§ BURKE, P. G. and McVICAR, D. D., *Proc. phys. Soc.* **86** (1965) 989.

Helium

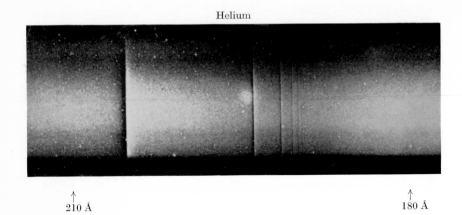

210 Å   180 Å

Fig. 14.21. Far ultra-violet absorption spectrum of helium obtained by Madden and Codling.

obtained by Hart and Herzberg† and for the free state the $1s$–$2s$–$2p$ close-coupling approximation (Chap. 8, § 5). Both length and velocity matrix elements were calculated. Although the main aim was to provide a quantitative description of the resonance effects due to auto-ionization from doubly excited levels near the second ionization threshold, the absorption cross-section was calculated from the first threshold to beyond the second.

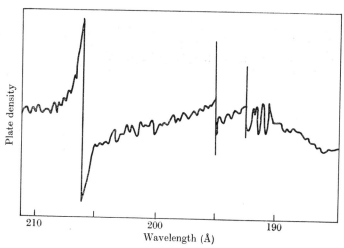

Fig. 14.22. Densitometer record of absorption of ultra-violet radiation in helium on the long wavelength side of the second ionization threshold.

It will be seen by reference to Table 14.7 that both the velocity and length results obtained with these more elaborate wave functions lie between those of Stewart and his collaborators. Towards the resonance region they begin to increase above the latter's values, probably due to a low energy tail of a resonance.

In Fig. 14.20 and Table 14.7 comparison is made with the observations of Tomboulian and his collaborators. The agreement is good except possibly at the longer wavelengths. To avoid complications due to resonance effects the comparison is not extended to wavelengths shorter than about 250 Å. It remains now to consider these effects.

Fig. 14.21 reproduces the absorption spectrum as observed by Madden and Codling‡ at wavelengths on the long-wave side of the second ionization threshold. There is definite evidence of discrete structure. This appears clearly in the densitometer record shown in Fig. 14.22. It will

† HART, J. F. and HERZBERG, G., *Phys. Rev.* **106** (1957) 79.   ‡ loc. cit., p. 1111.

be seen that the shape of the 'lines' is one of a type that is to be expected according to the theory discussed in Chapter 9, § 1.2.1 and § 4.3 above.

The structure of the auto-ionizing levels of the helium atom has already been discussed in Chapter 9, § 1. It was also shown in § 3 of that chapter that the energies and widths of the ^1S, ^1P, ^3S, and ^3P levels are given quite well if the continuum wave function is calculated in the close-coupling $1s$–$2s$–$2p$ approximation (Chap. 8, § 5). Of these it is only the $^1P^0$ levels that are involved in photo-ionization. They can be subdivided into three sets. The first two result from interaction between $2s\,np$ and $2p\,ns$ configurations giving rise to the $n^{\pm}$ levels defined by
$$n^{\pm} = 2^{-\frac{1}{2}}(2s\,np \pm 2p\,ns).$$

The third set are of the $2p\,nd$ configuration. In terms of this classification the auto-ionizing levels that give rise to the four most prominent absorption 'lines' are as given in Table 14.9. Support for this classification is given by the agreement of the observed energies with those calculated by Burke and McVicar† using the $1s$–$2s$–$2p$ close-coupling approximation. There is also good agreement with energy values for the 2^+ and 3^+ levels derived from the electron impact observations of Simpson and Mielczarek (see Chap. 1, § 6.2.3 and Chap. 9, § 3.2).

Table 14.9 includes the level widths to be expected, according to the $1s$–$2s$–$2p$ close-coupling approximation. There is a very great variation in magnitude from level to level, for reasons given in Chapter 9, § 3.1. It is also possible to obtain the line-profile index q (see (36)) for each absorption line from the calculation of the absorption cross-section using the same approximation for the continuum wave function. Two sets of calculated values are given for the different levels in Table 14.9, one set calculated using the velocity, the other the length matrix element.

The only observed line profile that has been analysed in detail is that involving the 2^+ level. It will be seen by reference to Table 14.9 that the derived values of level width Γ and line profile index q agree quite well with the calculated.

Returning to the calculated values of q, we note that, for a particular Rydberg series the values are roughly constant as expected from the theoretical considerations of § 4.3 above. The negative sign for the $\pm$ series would be expected‡ from the approximate formula (46) of Chapter 9. Thus for excitation of the $1s^2 \rightarrow 2s\,np^{\pm}$ transition in helium

---

† loc. cit., p. 1114.
‡ FANO, U. and COOPER, J. W., *Phys. Rev.* **137** (1965) A1364.

TABLE 14.9

Properties of auto-ionizing levels that give rise to resonance structure in the absorption spectrum of helium

Level designation	Resonance energy (E_r eV)			Level width (Γ eV)		Line profile index q		
	Obs. (M.C.)	Obs. (S.M.)	Calc.	Obs. (M.C.)	Calc.	Obs. (M.C.)	Calc. V	Calc. L
$2+$	$60{\cdot}135(\pm0{\cdot}015)$	$60{\cdot}1$	$60{\cdot}269$	$3{\cdot}8(\pm0{\cdot}4)\times10^{-2}$	$4{\cdot}37\times10^{-2}$	$-2{\cdot}80$	$-2{\cdot}65$	$-2{\cdot}59$
$3-$	$62{\cdot}761(\pm0{\cdot}009)$		$62{\cdot}773$		$1{\cdot}39\times10^{-4}$		$-3{\cdot}72$	$-3{\cdot}02$
$3+$	$63{\cdot}658(\pm0{\cdot}007)$	$63{\cdot}6$	$63{\cdot}691$				$-2{\cdot}51$	$-2{\cdot}44$
$4-$	$64{\cdot}144(\pm0{\cdot}015)$		$64{\cdot}134$		$5{\cdot}03\times10^{-5}$		$-3{\cdot}95$	$-3{\cdot}30$
$3d$			$64{\cdot}172$		$1{\cdot}54\times10^{-6}$		$+0{\cdot}92$	$-0{\cdot}10$
$4+$			$64{\cdot}481$		$3{\cdot}69\times10^{-3}$		$-2{\cdot}49$	$-2{\cdot}42$
$5-$			$64{\cdot}658$		$2{\cdot}30\times10^{-5}$		$-4{\cdot}04$	$-3{\cdot}42$
$4d$			$64{\cdot}676$		$7{\cdot}76\times10^{-7}$		$+1{\cdot}50$	$+0{\cdot}66$
$5+$			$64{\cdot}824$		$1{\cdot}89\times10^{-3}$		$-2{\cdot}48$	$-2{\cdot}41$

Obs. (M.C.) refers to the absorption observations of Madden and Codling, Obs. (S.M.) to the electron impact observations of Simpson and Mielczarek, and Calc. to the calculated values of Burke and McVicar. V and L distinguish calculations carried out using velocity and length matrix elements.

(46) gives, in the notation of Chapter 9, § 1.1.1,

$$q = \langle \Phi | z | \bar{\Psi}_0 \rangle / \pi \langle \psi_E | H | \phi \rangle \langle \psi_E | z | \bar{\Psi}_0 \rangle,$$

which, in terms of configuration assignments, may be written approximately[†]

$$q = \langle 2s\, np^{\pm} | z | 1s^2 \rangle / \pi \langle 1s\, Ep | H | 2s\, np^{\pm} \rangle \langle 1s\, Ep | z | 1s^2 \rangle. \qquad (61)$$

We adopt a standard normalization that all radial wave functions are positive when both electrons are near the nucleus. The factor $\langle 1s\, Ep | H | 2s\, 2p \rangle$ is then likely to be positive since it is determined mainly by the inter-electronic repulsion, which is itself positive and largest when both electrons are close to the nucleus where all the wave functions are positive. The matrix element $\langle 1s\, Ep | z | 1s^2 \rangle$, corresponding to direct photo-ionization, is also positive. On the other hand, adopting the convention that all radial wave functions are positive when both electrons are close to the nucleus the matrix element in the numerator of (61) is negative. Thus if the helium wave functions are approximated by symmetrized products of single electron wave functions then

$$\langle n^{\pm} | z | 1s^2 \rangle \simeq (2s | 1s) \langle np | z | 1s \rangle, \qquad (62)$$

where $(2s | 1s)$ denotes the overlap integral of the $2s$ and $1s$ single particle wave functions. Whereas the matrix element $\langle np | z | 1s \rangle$ is positive the overlap integral is negative. Hence $q(n^{\pm})$ is negative.

Further series of auto-ionization levels converging on ionization thresholds in which the He^+ ion is left in 3 or higher quantum states must certainly exist. Indeed, Madden and Codling[‡] have observed lines that converge on the 3- and 4-quantum thresholds. Fano and Cooper[§] have discussed the sign of the line profile index q in such cases. They estimate for the principal line of the $3s3p$ doubly excited states that q should be positive and roughly 1·7. In the same state the estimated width is 0·07 eV.

7.2. Neon

The earliest observations of absorption cross-sections in neon were those of Woernle[‖] and Dershem and Schein,[††] in both cases well in the X-ray region (see Fig. 14.23). No measurements were made in the

---

[†] This follows from the definitions in Chapter 9 (4) for ϕ, (8) for ψ_E, and (41) for Φ. In the latter the term arising from interaction with the continuum has been dropped so $\Phi = \phi$. The continuum orbital must be a p orbital as the upper state is a ^1P state.

[‡] MADDEN, R. P. and CODLING, K., quoted by FANO and COOPER, loc. cit. p. 1116.

[§] loc. cit., p. 1116.   [‖] WOERNLE, B., *Annln Phys.* 5 (1930) 475.

[††] DERSHEM, E. and SCHEIN, M., *Phys. Rev.* 37 (1931) 1238.

vacuum ultra-violet until the work of Lee and Weissler† and Ditchburn,‡ using the method described on p. 1091 in which the absorbing
gas fills the spectrograph. More recently Ederer and Tomboulian§
have measured the cross-section over the range from 80 to 600 Å,
using the isolated absorbing cell technique described on p. 1092, the
technique being generally similar to that used by Baker, Bedo, and

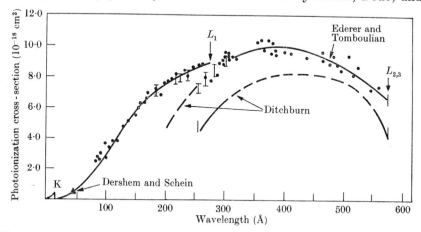

Fig. 14.23. Observed photo-ionization cross-section of neon. ● observed by Ederer
and Tomboulian using photographic detection. I observed by Ederer and Tomboulian
using a Geiger–Müller counter. ▲ observed by Dershem and Schein. — — — observed
by Ditchburn.

Tomboulian‖ for measurement of the helium cross-section (see § 7.1).
For wavelengths extending from the first threshold at 575 Å down to
225 Å, photographic detection was used in conjunction with a normal
incidence spectrograph. At shorter wavelengths two different grazing
incidence spectrographs were used, one employing photographic detection, the other using a Geiger–Müller counter.

Fig. 14.23 illustrates the results obtained by the different observers.
As for helium the results obtained using the isolated absorption cell
are higher than in the earlier work. The oscillator sum rules may be
invoked to provide a check on the accuracy of the measurements. Using
the formula (75) of Chapter 7, Ederer and Tomboulian calculated the
integral over the continuous spectrum in the sum S_0 (see (60)) from their
data extrapolated to higher frequencies by means of a power-law
representation of the cross-section. They obtained the value $10 \cdot 2 \pm 0 \cdot 4$.

† Lee, P. and Weissler, G. L., Proc. R. Soc. A220 (1953) 71.
‡ Ditchburn, R. W., Proc. phys. Soc. A75 (1960) 461.
§ Ederer, D. L. and Tomboulian, D. H., Phys. Rev. 133 (1964) A1525.
‖ loc. cit. p. 1111.

Estimates of the contributions from discrete transitions based on calculations by Cooper[†] give about 0·4 so that S_0 becomes $10·6\pm0·4$, which
is not in serious disagreement with the total number 10 of neon electrons.

In the same way, from (84) of Chapter 7, the polarizability of neon
comes out to be $0·430\pm0·020\times10^{-24}$ cm³ of which a little more than
70 per cent arises from the continuum. This is to be compared with
$0·398\times10^{-24}$ cm³ derived from observations of the refractive index.

Seaton[‡] has calculated the absorption cross-section from the first
threshold to a little beyond the second (L I) using Hartree–Fock wave
functions for the ground state. For transitions from $2p$ to d continuum
states, which give the major contributions below the L I threshold, a
Hartree–Fock wave function was also used for the continuum state to
calculate the cross-section at the threshold. It was found that this
latter function differed so little from the Coulomb form that this form
was used to extend the calculations to shorter wavelengths. Transitions
from $2p$ to s continuum states were calculated in the same way as at
the threshold and somewhat less accurately at shorter wavelengths.
Finally, for transitions from $2s$ to p continuum states, the continuum
wave functions were taken to be the same as for motion in the field of a
Na⁺ ion which were available from previous calculations.

Fig. 14.24 illustrates the comparison between the observed results
and Seaton's calculations. The agreement is not unsatisfactory, particularly when account is taken of the spread of the experimental results.

Turning now to the high-resolution spectra obtained with synchrotron
radiation, Fig. 14.25 shows a typical spectrum covering the region
250–80 Å.[§] Prominent auto-ionization lines are present, forming a
series that converges to 250 Å (48·5 eV), which is the L I limit in X-ray
level notation. This identifies the series as arising from the transitions

$$2s^2 2p^6\, {}^1S - 2s2p^6 np\, {}^1P. \tag{63}$$

Nine members of this series have been observed as well as many other
resonance anomalies.

Fano and Cooper[‖] have estimated the parameters associated with the
$n = 3$ resonance. Using rough approximations to the wave functions
involved they find that the line width is of order 0·01 eV and the line
profile index q about -2. With this value of q the contribution of the
auto-ionizing level to the cross-section is positive according to (38) so

† COOPER, J. W., *Phys. Rev.* **128** (1962) 681.
‡ SEATON, M. J., *Proc. phys. Soc.* A**67** (1954) 927.
§ MADDEN, R. P. and CODLING, K., *Phys. Rev. Lett.* **10** (1963) 516.
‖ loc. cit., p. 1116.

Neon

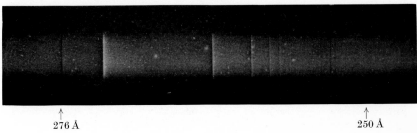

276 Å   250 Å

Fig. 14.25. Far ultra-violet absorption spectrum of neon obtained by Madden and Codling.

the resonance effects appear as enhanced absorption (see also Fig. 9.4). As we shall see, neon once again (cf. the absence of a Ramsauer–Townsend effect, Chap. 1, § 6.1 and Chap. 6, § 4.3) behaves differently from the heavier rare gases because the $2p$ wave function has no node. For argon,

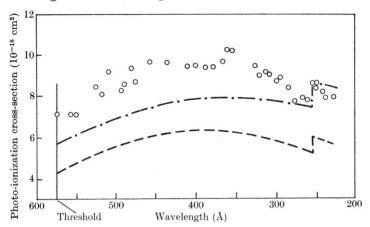

FIG. 14.24. Comparison of observed and calculated photo-ionization cross-sections of neon. ○ observed by Ederer and Tomboulian. —·—·— calculated by Seaton using dipole length matrix elements. — — — calculated by Seaton using dipole velocity matrix elements.

krypton, and xenon the outer shell p functions have one or more nodes so that in many matrix elements cancellation occurs which is absent for neon.

7.3. *Argon, krypton, and xenon*

The observations of photoelectron spectra resulting from photo-ionization of the heavier rare gas atoms by ultra-violet radiation of 584 Å wavelength have been described in § 5.5.4.1. From these observations the separation between the $^2P_{\frac{1}{2}}$ and $^2P_{\frac{3}{2}}$ states of the corresponding positive ions can be derived and agree well with those derived spectroscopically (see Table 14.6).

The classic observations of Beutler† on auto-ionization effects accompanying absorption of ultra-violet light were made in the rare gases at wavelengths between the first and second ionization thresholds. Fig. 14.26 (a) illustrates the absorption spectrum in this wavelength range for krypton observed by Metzger and Cook‡ (see p. 1098). The photo-ionization spectrum is included in the same figure, showing the close agreement in shape.

† Beutler, H., *Z. Phys.* **93** (1935) 177.
‡ Metzger, P. H. and Cook, G. R., *J. Opt. Soc. Am.* **55** (1965) 516.

Fig. 14.26 (b) shows the corresponding absorption coefficient for argon observed by the same investigators.

Close agreement is obtained between the wavelengths of the auto-ionization lines observed in these recent experiments and the much earlier

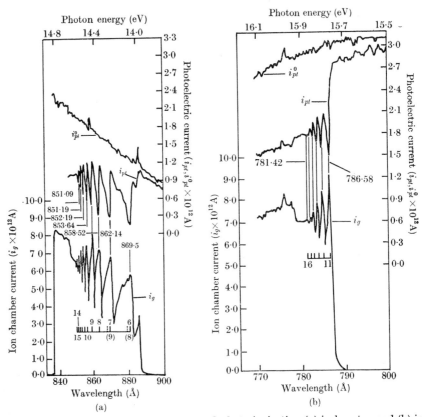

FIG. 14.26. Observations of absorption and photo-ionization (a) in krypton and (b) in argon made by Cook *et al.* i_{pt}^0 is the spectrum of the primary radiation. i_{pt} is the absorption spectrum of the transmitted radiation. i_g is the ion chamber current resulting from photo-ionization.

ones of Beutler. In argon the auto-ionization lines are identified as arising from the transition

$$(3s)^2(3p)^6\,{}^1S_0 \to (3s)^2(3p)^5md, \tag{64}$$

with m ranging from 11 to 16. For krypton the lines arise from the corresponding transition

$$(4s)^2(4p)^6\,{}^1S_0 \to (4s)^2(4p)^5md \quad (m = 6 \text{ to } 15), \tag{65}$$

and also from

$$(4s)^2(4p)^6\,{}^1S_0 \to (4s)^2(4p)^5ms \quad (m = 8 \text{ to } 9). \tag{66}$$

The xenon lines arise from the series corresponding to (65) with $m = 6$ to 17 and from members of the series corresponding to (66). Fig. 14.27 illustrates the relations of the terms for the three atoms.

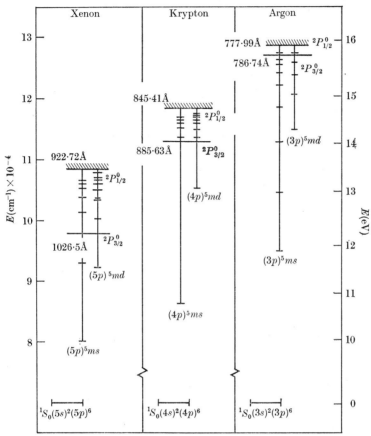

FIG. 14.27. Energy level diagrams for xenon, krypton, and argon. The auto-ionizing levels arise in the $(np)^5md$ series between the $^2P_{\frac{3}{2}}$ and $^2P_{\frac{1}{2}}$ ionization thresholds.

Table 14.10 gives some of the f-values for these auto-ionization lines, determined by Metzger and Cook.† A number of values measured for the krypton lines by Pery-Thorne and Garton‡ are also included. It will be seen that the agreement is very good between the two sets of measurements. For all these atoms the contribution to the oscillator strength sum from the auto-ionization lines is seen to be quite small.

† loc. cit., p. 1121.
‡ PERY-THORNE, A. and GARTON, W. R. S., *Proc. phys. Soc.* B**76** (1960) 833.

In addition to these auto-ionization lines close to the first threshold the heavier rare gas atoms also show strongly effects arising from auto-ionizing transitions similar to those (63) for neon. Fig. 14.28 shows spectra obtained by Madden and Codling† using synchrotron radiation. The striking feature of these spectra is that the auto-ionizing effects are reversed. Instead of introducing dark absorption lines they appear

TABLE 14.10

f-values for auto-ionization lines between the first and second ionization thresholds for argon, krypton, and xenon

Argon		Krypton			Xenon	
Wavelength (Å)	f_1	Wavelength (Å)	f_1	f_2	Wavelength (Å)	f_1
786·6	0·0039	881·0	0·0431	0·0431	995·8	0·116
784·8	0·0011	869·4	0·0231	0·0235	966·9	0·088
783·6	0·0006	862·7	0·0168	0·0143	952·1	0·042
782·7	0·0002	858·5	0·0088	0·0086	943·7	0·028
781·9	10^{-4}	855·6	0·0050	0·0051	938·5	0·014
781·4	10^{-4}	853·6	0·0024	0·0032	935·1	0·007
		852·1	0·0017	0·1021	932·6	0·004
		851·0	0·0008	0·0013	930·8	0·002
		850·1	0·0003		929·5	0·002
					928·4	0·002
					927·6	0·001

f_1 observed by Cook, Metzger, Ogawa, Becker, and Ching.
f_2 observed by Pery-Thorne and Garton.

rather as bright emission lines, corresponding to absorption windows. This is also seen very clearly in the variation of the photo-ionization cross-section of argon observed by Samson‡ and reproduced in Fig. 14.29. There is no difficulty in understanding these curious results in terms of the formula (38). When the line profile index q is much less than unity the contribution from the resonance is negative (see also Fig. 9.4), which is just as required to interpret the window effects in the spectra of Fig. 14.28. The requirement $|q| \ll 1$ is not in conflict with the fact that $q \sim -2 \cdot 0$ for the first auto-ionization line of neon for the reason already mentioned—it is only for neon that the outer p orbital has no nodes. The presence of one or more nodes in this orbital for the heavier atoms could readily account for a small value of $|q|$. Fano and Cooper§ have investigated this in more detail for argon and find that the explanation is very plausible. They also estimate the line

† loc. cit., p. 1120.   ‡ SAMSON, J. A. R., *Phys. Rev.* **132** (1963) 2122.
§ loc. cit., p. 1116.

Argon

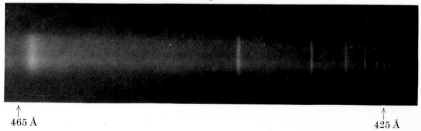

465 Å   425 Å

FIG. 14.28. Far ultra-violet absorption spectrum of argon observed by Madden and Codling.

width of the first line in argon as about 0·05 eV, much wider than for neon, which seems to be in qualitative agreement with observation.

At still shorter wavelengths, line series are observed in the krypton and xenon spectrum arising from the transitions

$$(m-1)d^{10}ms^2mp^6 \rightarrow (m-1)d^9(^2D_{\frac{3}{2},\frac{1}{2}})ms^2mp^6np, \qquad (67)$$

where $m = 4$ and 5 for krypton and xenon respectively. These lines do not possess the 'window' character of those discussed immediately

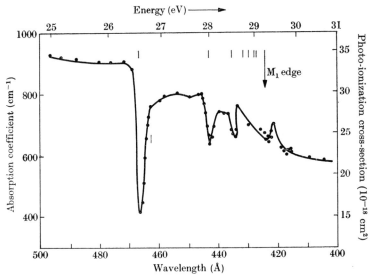

FIG. 14.29. Photo-ionization cross-section for argon between 400 and 500 Å observed by Samson.

above. They may auto-ionize through transitions involving the core only, without participation of the excited np electron. Thus

$$(m-1)d^9ms^2mp^6 \rightarrow (m-1)d^9ms^2mp^6kd, \qquad (68)$$

where kd denotes an unbound d orbital with wave number k.

It remains to consider the background absorption cross-sections. Fig. 14.30 illustrates these for wavelengths down to 600 Å observed by Cook, Metzger, and collaborators.† The earlier measurements of Lee and Weissler‡ are a little lower but the results for krypton agree closely with measurements made by Huffman, Tanaka, and Larrabee.§

7.4. Metallic atoms

7.4.1. Lithium. The absorption cross-section of lithium was first

† loc. cit., p. 1098.   ‡ LEE, P. and WEISSLER, G. L., *Phys. Rev.* **99** (1955) 540.
§ HUFFMAN, R. E., TANAKA, Y., and LARRABEE, J. C., *Bull. Am. phys. Soc.* **7** (1962) 457.

measured by Tunstead† in 1953 using the method discussed on p. 1094. His results were corrected by Marr‡ when more accurate vapour pressure tables became available. Marr‡ also repeated the measurements, obtaining values that agreed quite well with those of Tunstead at the series limit but showed a slower variation with wavelength. This

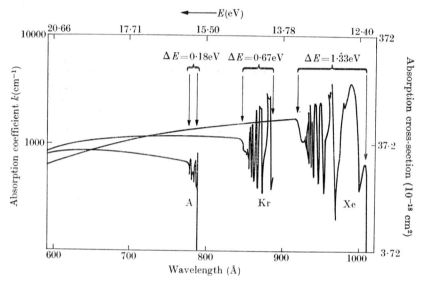

Fɪɢ. 14.30. Absorption cross-sections observed for argon, krypton, and xenon by Cook *et al.* Δ*E* denotes the energy separation of the $^2P_{\frac{1}{2}}$ and $^2P_{\frac{3}{2}}$ levels of the corresponding positive ions.

may be seen by reference to Fig. 14.31, which also includes observations made by Hudson and Carter using photomultiplier detection.§ Most of the difference between their results and those of Marr is due to the use of more recent vapour pressure data.

The comparative simplicity of the lithium atom stimulated theoretical calculation of the absorption coefficient as long ago as 1929 but it is only since 1950 that reasonably accurate theoretical results have become available. Fig. 14.31 shows a number of calculated cross-section curves obtained using different approximations as follows.

Curve (4) was obtained by Peach‖ using the general formula (31) of Burgess and Seaton,†† while curves (5) and (6) were obtained by Stewart‡‡ using the length and velocity matrix elements respectively

† Tᴜɴsᴛᴇᴀᴅ, J., *Proc. phys. Soc.* A66 (1953) 304.
‡ Mᴀʀʀ, G. V., ibid. 81 (1963) 9.
§ Hᴜᴅsᴏɴ, R. D. and Cᴀʀᴛᴇʀ, V. L., *Phys. Rev.* 137 (1965) A1648.
‖ Pᴇᴀᴄʜ, G., *Mem. R. astr. Soc.* 71 (1967) 13.   †† loc. cit., p. 1074.
‡‡ Sᴛᴇᴡᴀʀᴛ, A. L., *Proc. phys. Soc.* A67 (1954) 917.

and the following wave functions. For the ground state a separable wave function of the form

$$\psi(r_1, r_2, r_3) = (r_1 r_2 r_3)^{-1} P(1s|r_1) P(1s|r_2) P(2s|r_3) \qquad (69)$$

was taken, $\mathbf{r}_1$, $\mathbf{r}_2$, $\mathbf{r}_3$ being the coordinates of the three electrons. $P(1s|r)$ is the Hartree–Fock function for the Li$^+$ ion and $P(2s|r)$ for the series electron, as given by Fock and Petrashen.[†] The usual 1s-exchange (Hartree–Fock) approximation was used for the continuum state.

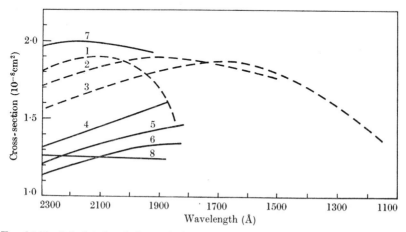

FIG. 14.31. Calculated and observed absorption cross-sections for lithium. Observed: (1) by Tunstead (as corrected by Marr), (2) by Marr, (3) by Hudson and Carter. Calculated: (4) by Peach, (5) by Stewart (dipole length), (6) by Stewart (dipole velocity), (7) by Tait (dipole length), (8) by Tait (dipole velocity).

Finally, curves (7) and (8) refer to corresponding results obtained by Tait[‡] in which he used the same continuum state function as Stewart but replaced the ground state function by a much more complicated one given by James and Coolidge,[§] which allows for correlations between the inner and outer electrons.

It will be seen that, while the use of the more elaborate bound state function has little effect on the dipole velocity results, it does lead to a greater modification of those obtained using the dipole length formulation. The experimental results fall between curves (7) and (8), which is not unsatisfactory.

7.4.2. *Sodium*. Fig. 14.32 illustrates the absorption cross-section of sodium as a function of wavelength observed by Ditchburn, Jutsum,

† FOCK, V. and PETRASHEN, M. J., *Phys. Z. SowjUn.* 8 (1935) 547.
‡ TAIT, J. H., *Atomic collision processes*, p. 586 (Amsterdam, 1964).
§ JAMES, H. M. and COOLIDGE, A. S., *Phys. Rev.* 49 (1936) 688.

and Marr† using the method described in § 5.5.2 and photographic detection, and by Hudson‡ using photomultiplier detection (p. 1096). The principal difference between these results arises from the mode of separation of atomic and molecular absorption. Hudson, using the procedure described on p. 1097, which allows for the variation of the equilibrium constant with temperature, finds a minimum cross-section that is zero within the experimental error. This is important because, according to theory, the minimum should be zero.

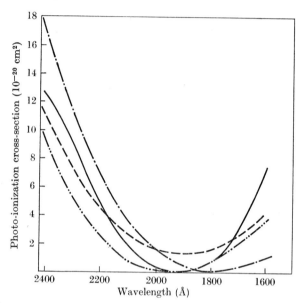

Fig. 14.32. Calculated and observed absorption cross-sections for sodium. —— observed by Hudson. — — — observed by Ditchburn, Jutsum, and Marr. —·—·— calculated by Peach from general formula (31). —··—··— calculated by Seaton.

In Fig. 14.32 are shown theoretical cross-sections calculated§ using the general formula (31) of Burgess and Seaton‖ and also by Seaton†† using the Hartree–Fock approximation for both the ground state and continuum wave functions. The general agreement with the experimental values, particularly those of Hudson which have a zero minimum, is quite good.

7.4.3. *Potassium.* Whereas the application of the general formula (31) to sodium gives quite good results, the sensitivity of the situation in

† DITCHBURN, R. W., JUTSUM, P. J., and MARR, G. V., *Proc. R. Soc.* A219 (1953) 89.
‡ HUDSON, R. D., *Phys. Rev.* 135 (1964) A1212.   § PEACH, G., loc. cit., p. 1126.
‖ loc. cit., p. 1074.   †† SEATON, M. J., *Proc. R. Soc.* A208 (1951) 418.

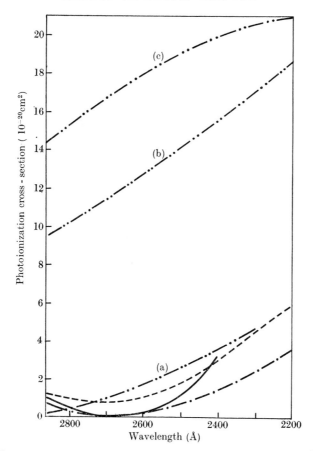

FIG. 14.33. Calculated and observed absorption cross-sections for potassium. — — — observed by Ditchburn, Tunstead, and Yates. —— observed by Hudson and Carter. — · — · — calculated by Bates. — ·· — ·· — calculated by Burgess and Seaton (a) with Hartree–Fock ground-state wave function and empirical extrapolated quantum defect chosen to give best agreement with observation ($\mu(0) = 1{\cdot}690$); (b) with Hartree–Fock ground-state wave function and extrapolated quantum defect observed from spectra ($\mu(0) = 1{\cdot}711$); (c) using the general formula (31).

potassium is so great that accurate calculation of the absorption cross-section is practically excluded.

Fig. 14.33 illustrates the cross-sections observed by Ditchburn, Tunstead, and Yates[†] using photographic detection and by Hudson and Carter[‡] using photomultiplier detection. These results agree quite well with each other and with the earlier measurements of Mohler and

† DITCHBURN, R. W., TUNSTEAD, J., and YATES, J. G., *Proc. R. Soc.* **A181** (1943) 386.
‡ HUDSON, R. D. and CARTER, V. L., *J. opt. Soc. Am.* **57** (1967) 1471.

Boeckner,† except at the threshold where the cross-section is very small. Hudson and Carter, using their improved method of separating atomic and molecular absorption, find that very close to the threshold the cross-section has, within experimental error, a zero minimum.

This smallness of the cross-section shows that strong cancellation occurs in the matrix element and hence there is extreme sensitivity to the assumed wave functions. In view of this Bates‡ carried out calculations with the dipole length formulation, using a Hartree ground-state wave function and a continuum function obtained by solving the wave equation for motion in the static Hartree field of the potassium ion together with a polarization potential of the form

$$V_p = -\tfrac{1}{2}\alpha e^2/(r^2+l^2)^2, \tag{70}$$

where α is the polarizability of the ion and l a cut-off parameter. α was regarded as a variable parameter. The closest approximation to the observed result was obtained with $\alpha = 1\cdot55\times10^{-24}$ cm³, less than half the observed value, and is shown on Fig. 14.33. It shows that the observed results fall within the theoretically likely possibilities.

Burgess and Seaton§ have also examined the theoretical possibilities. They took a Hartree–Fock ground state function and a free wave function of the same form as in the derivation of their general formula (31). However, they allowed the extrapolated quantum defect $\mu(0)$ to vary from its actual value of $1\cdot711$ so as to obtain the best agreement with observation. This occurred, as seen in Fig. 14.33, when $\mu(0) = 1\cdot690$, the agreement being very much better than with the correct value for μ. Nevertheless, the sensitivity is shown by the large change in cross-section obtained with a change of μ by a little more than 1 per cent. In this very sensitive case direct application of the general formula (31) gives very unsatisfactory results as shown in Fig. 14.33.

7.4.4. *Rubidium and caesium.* Observed results for these two atoms are shown in Fig. 14.34 (a)† and (b)†‖ respectively.

7.4.5. *Magnesium and calcium.* Fig. 14.35 illustrates the absorption cross-section of magnesium observed by Ditchburn and Marr†† using the method described on p. 1094. This is, from the theoretical point of view, a case of moderate sensitivity, so the general formula of Burgess

† Mohler, F. L. and Boeckner, C., *J. Res. natn. Bur. Stand.* **3** (1929) 303.
‡ Bates, D. R., *Proc. R. Soc.* A**188** (1947) 350.
§ Burgess, A. and Seaton, M. J., *Mon. Not. astr. Soc.* **120** (1960) 121.
‖ Braddick, H. J. J. and Ditchburn, R. W., *Proc. R. Soc.* A**150** (1935) 478.
†† Ditchburn, R. W. and Marr, G. V., *Proc. phys. Soc.* A**66** (1953) 655.

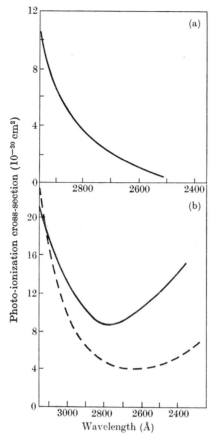

FIG. 14.34. Observed absorption cross-sections for rubidium and caesium. (a) rubidium: —— observed by Mohler and Boeckner. (b) caesium: —— observed by Braddick and Ditchburn; ———— observed by Mohler and Boeckner.

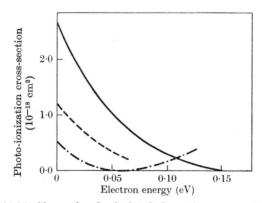

FIG. 14.35. Observed and calculated absorption cross-sections for magnesium. ———— observed by Ditchburn and Marr. —— calculated by Burgess and Seaton using general formula (31). —·—·— calculated by Burgess and Seaton using Hartree–Fock wave function for the ground state.

and Seaton† does not give very good results, as may be seen from Fig. 14.35. Replacement of the ground-state function by a Hartree–Fock function does not improve the agreement very much (see Fig. 14.35).

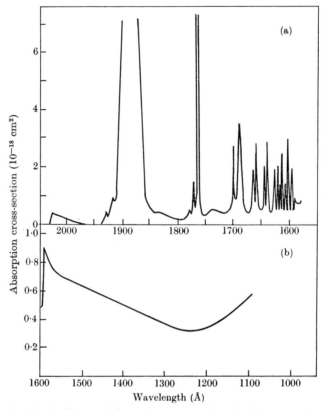

Fig. 14.36. Observed absorption cross-section of calcium as a function of wavelength. (a) for wavelengths > 1600 Å. (b) for wavelengths < 1600 Å.

The absorption cross-section of calcium exhibits strong auto-ionization lines quite close to the first threshold. This may be seen by reference to Fig. 14.36 (a), which shows the cross-section as a function of wavelength in the range from the threshold at 2050 to 1600 Å observed by Ditchburn and Hudson‡ using the method described in § 5.5.2. The lines that appear are grouped in pairs forming two series which have been identified as arising from the transitions

$$(4s)^2\,{}^1S_0 \to 3dnp\,{}^1P_1$$
$$\to 3dnp\,{}^3P_1,$$

† loc. cit., p. 1074.
‡ Ditchburn, R. W. and Hudson, R. D., *Proc. R. Soc.* **A256** (1960) 53.

the former being much stronger for the same value of n. Table 14.11 gives the respective f values for a number of these lines.

These auto-ionization line series converge to a limit at 1589 Å. No further lines were found at shorter wavelength, down to the limit of observation at 1100 Å, the observed cross-section varying as shown in Fig. 14.36 (b).

TABLE 14.11

Observed f-values of auto-ionization lines in the $(4s)^2 {}^1S_0 \rightarrow 3dnp\,{}^1P_1$ *and* ${}^3P_1$ *series*

	f	
n	Singlet series	Triplet series
5	0·024	0·0002
6	0·004	0·0001
7	0·002	0·0001
8	0·001	0·0$_4$5
9	0·0005	0·0$_4$3
10	0·00016	
11	0·0$_4$6	

Moores[†] has calculated the photo-ionization cross-section for ${}^1S$ to ${}^1P$ transitions, allowing for auto-ionization by using the extrapolated quantum-defect method for coupled channels as described in Chapter 9, § 10.5 to obtain wave functions for the continuum states. For the ground state a function of the form

$$\Psi = a_1\psi_1 + a_2\psi_2 + a_3\psi_3$$

was taken, where ψ_1, ψ_2, and ψ_3 correspond to configurations in which the outer two electrons are in $4s^2$, $4p^2$, and $3d^2$ orbitals respectively. The functions ψ_1, ψ_2, ψ_3 and the coefficients a_1, a_2, and a_3 were then determined from the Hartree–Fock calculations of Chisholm and Öpik.[‡]

Fig. 14.37 illustrates the comparison between the cross-sections calculated in this way and those observed by Ditchburn and Hudson, while in Table 14.12 comparison is made between calculated and observed wavelengths,[§] half widths,[||] and f-values[||] of the auto-ionization peaks. In view of the complexity of the atom concerned the agreement is quite good. At wavelengths near 1750 Å the observed results show two close peaks instead of a single one as calculated. This is almost

† MOORES, D. L., *Proc. phys. Soc.* **88** (1966) 843.
‡ CHISHOLM, C. D. H. and ÖPIK, U., ibid. **83** (1964) 541.
§ GARTON, W. R. S. and CODLING, K., ibid. **86** (1965) 1067.
|| DITCHBURN, R. W. and HUDSON, R. D., loc. cit. p. 1132.

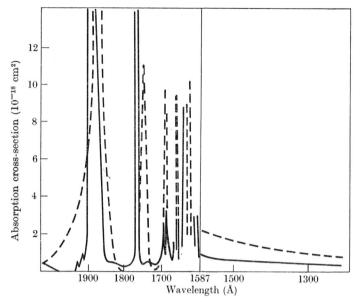

Fig. 14.37. Comparison of observed and calculated absorption cross-sections for calcium. ——— observed by Ditchburn and Hudson. — — — calculated by Moores.

TABLE 14.12

Comparison of observed and calculated wavelengths, half width, and f-values of auto-ionization peaks in calcium

Wavelength (Å)		Half-width (cm⁻¹)		f-value	
Obs.	Calc.	Obs.	Calc.	Obs.	Calc.
1882	1873	615	1174	0·024	0·029
1765		69		0·004	
	1745		473		0·0077
1740					
1689·0	1688·6	275	241	0·002	0·0035
1638·3	1657·9	160	140	0·001	0·0019
1640·1	1639·4	100	89	$0·0_3 5$	0·0012
1627·9	1627·3	80	60	$0·0_3 16$	$0·0_3 8$
1619·5	1618·9	40	43	$0·0_4 6$	$0·0_3 6$

certainly due to interaction with a foreign term converging on the $4s$ limit. Moores, in a further calculation including now three coupled channels, was able to reproduce the observed splitting of the line, but the absolute magnitudes of the cross-section at the two peaks still do not agree well with the observed. This is not surprising because the coupling with the additional channel cannot be determined very well.

There is no evidence of perturbation due to it in the region of pure bound states and so it is necessary to work from the perturbation of the positions and widths of the resonances above the first ionization limit.

7.4.6. *Other metallic atoms.* Absorption cross-sections have been measured also for strontium,† cadmium,‡ zinc,§ and thallium.‖ Some results have also been obtained for indium.††

7.5. Atomized gases

7.5.1. *Atomic hydrogen.* Beynon and Cairns‡‡ have measured the absorption cross-section of atomic hydrogen at a number of wavelengths. The general arrangement of the absorption cell they used, together with the atomizing r.f. discharge, is shown in Fig. 14.38. A mixed stream of atomic and molecular hydrogen from the discharge flowed through the cell. To define the path length of the radiation through the cell, a tube of silver foil on which atomic hydrogen recombines rapidly was mounted at each end of the cell (A_1 and A_2 in Fig. 14.38) as shown. The total pressure (of order 3×10^{-1} torr) in the cell was measured by a McLeod gauge and the degree of dissociation D by a Wrede gauge.§§ This was constructed by incorporating thin glass films, 10–15 μm thick pierced with three small holes about 50-μm diameter, in the wall of the absorption cell near the entrance and exit. Atomic hydrogen passing through the holes was recombined by passage through a silver foil. At the working pressure the atoms and molecules flowed effusively through the holes at a rate depending on their mass. This caused a pressure difference Δp to build up across the holes. The degree of dissociation, by volume, was then given by

$$D_{\rm v} = 3{\cdot}41\Delta p/p.$$

Δp was measured with a micromanometer.‖‖

Apart from a high degree of cleanliness no other precautions were taken to minimize recombination in the absorption cell.

The light source was a high-voltage spark in argon rendered monochromatic by a 1-m normal-incidence vacuum spectrograph. The absorption cell was connected to the monochromator through two pairs

---

† HUDSON, R. D. and YOUNG, P. A., *Sci. Rep. Space Tech. Lab.* 9803–6004–RU000, 1963.

‡ ROSS, K. J. and MARR, G. V., *Proc. phys. Soc.* **85** (1964) 193.

§ MARR, G. V. and AUSTIN, J. M., in course of publication.

‖ MARR, G. V., *Proc. R. Soc.* A**224** (1954) 83; MARR, G. V. and HEPPINSTALL, R., *Proc. phys. Soc.* **87** (1966) 293.

†† MARR, G. V., ibid. A**67** (1954) 196.

‡‡ BEYNON, J. D. E. and CAIRNS, R. B., ibid. **86** (1965) 1343.

§§ GREAVES, J. C. and LINNETT, J. W., *Trans. Faraday Soc.* **55** (1959) 1338.

‖‖ BEYNON, J. D. E. and CAIRNS, C. B., *J. scient. Instrum.* **41** (1964) 111.

of slits but not otherwise isolated. Radiation intensity was measured by a photomultiplier, sensitized to far ultra-violet radiation by coating its window with sodium salicylate. Measurements at a particular frequency were made in a sequence, first of the intensity with the cell evacuated, then with molecular hydrogen flowing through with the

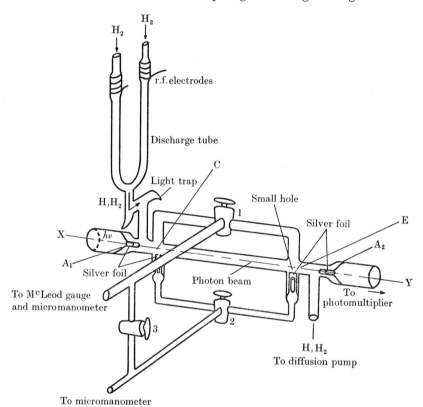

FIG. 14.38. Arrangement of the atomizing discharge and absorption cell in the experiments of Beynon and Cairns.

discharge off and at a pressure measured at the point C. The discharge was then switched on, the pressure p and degree of dissociation D measured at C and then the transmitted light intensity. p and D were then measured at E and also the transmitted light intensity once more. The discharge was then switched off and the pressure and light intensity measured again.

In analysing the data, allowance had to be made for the variation of pressure along the cell. This was done assuming a linear pressure drop.

The first measurements were made at 850·6 Å. At this wavelength the absorption cross-section of H_2 was observed to be pressure-dependent so that it was necessary to ensure that the pressure of H_2 with the discharge off was nearly equal to the partial pressure of H_2 with the discharge on. Later observations were made at 840·0 and 826·3 Å.[†] Results are given in Table 14.13.

TABLE 14.13

Absorption cross-section of atomic hydrogen

Wavelength (Å)	826·3	840·0	854·6
Observed cross-section (10^{-18} cm²)	3·8±0·8	2·9±0·5	5·1$_5$±0·2
Calculated cross-section (10^{-18} cm²)	4·84	5·06	5·30

When allowance is made for the difficulty of the experiment the comparison with the theory is not unsatisfactory. It is difficult to check the observed values of the absorption cross-section for H_2 in this wavelength range as it varies very rapidly (see Fig. 14.41).

7.5.2. *Atomic oxygen.* Special interest is attached to the determination of the absorption coefficient of atomic oxygen because of the important role it plays as a major atmospheric constituent above 100-km altitude. Experimental measurement is difficult. The results obtained by Cairns and Samson using the technique described below are not yet of high accuracy. A great deal of importance is therefore attached to accurate calculations for atomic oxygen.

The first attempt to do this was made in 1939 by Bates, Buckingham, Massey, and Unwin[‡] with the dipole length formulation, using Hartree-field wave functions for both the ground and continuum states so that electron exchange effects were ignored. This neglect meant that no discrimination was made between transitions from the ground $2s^2 2p^4\,{}^3P$ states to the three states $2s^2 2p^3\,{}^4S$, ${}^2D$, ${}^2P$ belonging to the ground configuration of the O^+ ion. They obtained a cross-section of $1·25 \times 10^{-17}$ cm² at the threshold.

Bates and Seaton[§] later improved these calculations at the first $({}^3P \to {}^4S)$ threshold by taking exchange into account, but a thoroughgoing calculation extending this work out to much higher photon energies was not carried out till some years later by Dalgarno, Henry, and

† BEYNON, J. D. E., *Proc. phys. Soc.* **89** (1966) 59.

‡ BATES, D. R., BUCKINGHAM, R. A., MASSEY, H. S. W., and UNWIN, J. J., *Proc. R. Soc.* A**170** (1939) 322.

§ BATES, D. R. and SEATON, M. J., *Mon. Not. R. astr. Soc.* **109** (1949) 698.

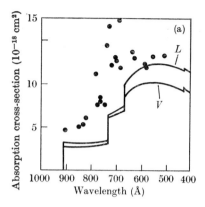

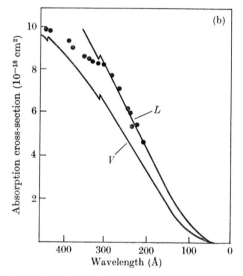

FIG. 14.39. The absorption cross-section of atomic oxygen. Calculated by Dalgarno, Henry, and Stewart: L = (dipole length), V = (dipole velocity). ● observed by Cairns and Samson.

Stewart.† They calculated both the ground state and continuum wave functions, using the Hartree–Fock method employing both the dipole length and velocity formulations. These results are shown in Fig. 14.39 (a) for wavelengths extending from the first threshold to 550 Å and in Fig. 14.39 (b) out to very small wavelengths. It is of interest to note from Fig. 14.39 (a) that at the third threshold, beyond which transitions to all terms of the ground configuration of O^+ are possible,

† DALGARNO, A., HENRY, R. J. W., and STEWART, A. L., *Planet. Space Sci.* **12** (1964) 235.

the total absorption cross-section is quite close to that calculated by Bates and Massey ignoring exchange.

Cairns and Samson† have made the first measurements of the absorption cross-section of atomic oxygen, which they produced in an electrodeless microwave discharge in oxygen at 2450 Mc/s. To increase the degree of dissociation, traces of water vapour were not removed and helium was mixed with the oxygen. The partially dissociated gas flowed through an absorption cell 30 cm long, along which a monochromatic photon beam could be passed and detected by a windowless photomultiplier.

The concentration of oxygen atoms at the centre of the absorption cell was determined by titration with nitrogen dioxide.‡ This depends on the reactions

$$O + NO_2 \rightarrow NO + O_2, \tag{71 a}$$

$$O + NO \rightarrow NO_2 + h\nu. \tag{71 b}$$

The luminescent glow due to the reaction (71 b) may readily be observed. If a gradually increasing pressure of NO_2 is injected into a stream of atomic oxygen the glow will first brighten, reach a maximum intensity, and then fall sharply to zero when the NO_2 concentration equals that of the atomic oxygen with which it mixes in the stream. Variation of the concentration of atomic oxygen along the absorption cell could be observed by adding nitric oxide to the gas stream and observing the variation along the cell of the intensity of light emitted through (71 b).

The chief difficulty of the experiment is due to the production of metastable $O_2\,^1\Delta_g$ molecules in the electrodeless discharge. The presence of these molecules was confirmed by photo-ionization measurements made by introducing ion-collecting electrodes within the absorption cell. When the discharge was on, the ionization threshold was reduced 0.99 ± 0.04 eV below that for ground state O_2. This is very close to the excitation energy of $O_2\,^1\Delta_g$.

To separate effects due to these metastable molecules use was made of the observations of Elias, Ogryzlo, and Schiff§ that while O atoms recombine readily on a mercuric oxide surface little deactivation of $O_2(^1\Delta_g)$ occurs. A glass tube coated with mercuric oxide was therefore introduced into the flow system either up- or downstream from the discharge. Photo-ionization measurements were made that showed that the concentration of $O_2(^1\Delta_g)$ molecules flowing through the absorption chamber increased by 20 ± 4 per cent when the mercuric oxide was

† Cairns, R. B. and Samson, J. A. R., *Phys. Rev.* **139** (1963) A1403.
‡ Kaufman, F., *Proc. R. Soc.* A247 (1958) 123; *J. chem. Phys.* **28** (1958) 352.
§ Elias, L., Ogryzlo, E. A., and Schiff, H. I., *Can. J. Chem.* **37** (1959) 1680.

placed downstream. This was probably due to recombination of some oxygen atoms to form metastable instead of ground state molecules. No appreciable concentration of other metastable molecules was observed in the photo-ionization measurements. It was also verified that the concentration of metastable helium atoms in the helium–oxygen mixture downstream from the discharge was not large enough to produce appreciable absorption. If I_ν, I_0 are the transmitted and incident intensities of radiation respectively, l the light path through the absorbing cell, Q_a, Q_m, Q_m^* the absorption cross-sections of O, $O_2(^3\Sigma_g^-)$, and $O_2(^1\Delta_g)$ respectively we have, when the mercuric oxide is upstream from the discharge,

$$\ln(I_0/I_\nu') = l\{n'(O_2)Q_m + n'(O_2^*)Q_m^* + n'(O)Q_a\},$$

where $n'(O_2)$, $n'(O_2^*)$, $n'(O)$ refer to the concentrations of $O_2(^3\Sigma_g^-)$ and $O_2(^1\Delta_g)$ and O respectively. Similarly, when it is placed downstream

$$\ln(I_0/I_\nu'') = l\{n''(O_2)Q_m + n''(O_2^*)Q_m^* + n''(O)Q_a\}.$$

As the inflow of molecular oxygen into the system remains constant

$$N(O_2) = n'(O_2) + n'(O_2^*) + \tfrac{1}{2}n'(O)$$
$$= n''(O_2) + n''(O_2^*) + \tfrac{1}{2}n''(O),$$

where $N(O_2)$ is the concentration of $O_2(^3\Sigma_g^-)$ molecules when the discharge is off.

Finally, from the photo-ionization measurements which indicate a 20 per cent increase in the concentration of $O_2(^1\Delta_g)$ molecules when the mercuric oxide is placed downstream, we have

$$\{n'(O_2^*) - n''(O_2^*)\}/n''(O_2^*) = -\tfrac{1}{6}.$$

Using these equations Q_a can be obtained in terms of measured quantities I_0, I_ν', I_ν'', $n'(O)$, $n''(O)$, and $N(O_2)$.

The results obtained in this way are shown in Fig. 14.39. It will be seen that they lie above the best calculated values and indicate the possible existence of a more complicated structure near the second and third absorption edges. At the highest observed frequencies the agreement with calculation carried out with the dipole length matrix element is good but may be fortuitous. It is encouraging that the discrepancies between theory and experiment are not greater.

7.5.3. *Atomic nitrogen.* Ehler and Weissler† have obtained some evidence about the absorption cross-section of atomic nitrogen by observing the absorption of ultra-violet radiation by the plasma of a Phillips ion gauge operating in nitrogen. The ion gauge was placed between the

† EHLER, A. W. and WEISSLER, G. L., *J. opt. Soc. Am.* **45** (1955) 1035.

spark discharge source and the vacuum spectrograph, the light pressing through two small holes in each cathode. By working with the electrodes at a separation of 58 cm a long absorbing path through the discharge plasma was available. The gauge operated with a magnetic field of around 2000 gauss. Electric and magnetic shielding of the gauge discharge from the source was secured by means of a copper envelope sandwiched between two mumetal cylinders enclosing the source. To prevent charged particles passing from the source to the ion gauge Alnico magnets were placed at the source exit slits.

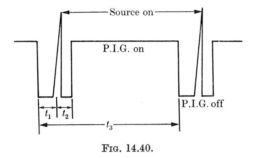

FIG. 14.40.

Two sets of observations were made. In the first, comparison was made of the absorption with and without the ion gauge operating. For the second set, the source and the gauge discharge were interrupted at regular intervals but with an adjustable phase difference so that the gauge discharge was on for around 15 ms and then turned off for several ms prior to the source being turned on. About 1 ms after the source was switched off the gauge was switched on once more. This cycle was repeated 60 times/s. Observations of the total light intensity from both the plasma and the source were observed with a photomultiplier. The shape of the variation during a typical cycle is shown in Fig. 14.40. From the sharpness of the jumps up and down in intensity when the gauge discharge was turned on and off it seems that no excited or even metastable atoms persisted 0·1 ms after the gauge discharge was off.

Similarly, by collecting ions during the off period with a suitable negative potential on one of the gauge electrodes it was found that all ionization disappeared equally quickly after turning off the gauge discharge.

Measurements were then made of the absorption at intervals 0·5, 1·1, and 4·5 ms after the gauge discharge was turned off. No difference was found between the results obtained but the absorption was somewhat less than when the gauge discharge was on.

It is reasonable in interpreting these results to assume that the difference between the absorption observed when the gauge discharge has been off for a second or more and when it is on arises from the presence of atomic nitrogen. Let a, b, c be the respective fractional concentrations of N_2, N, and ionized species respectively in the gauge when the discharge is on. The change in absorption under these conditions is equivalent to a change in the absorption cross-section/N_2 molecule from the normal value $Q(N_2)$ to Q_1 where

$$Q_1 = aQ(N_2) + bQ(N), \tag{72}$$

$Q(N)$ being the absorption cross-section for N atoms.

Within a few ms of the switching-off of the gauge discharge the ions will have recombined to produce N and N_2 so that the relative contributions from atoms and molecules will have changed to give an effective absorption cross-section Q_2 where

$$Q_2 = (a+cx)Q(N_2) + \{b+c(1-x)\}Q(N), \tag{73}$$

with $0 \leqslant x \leqslant 1$.

From (72) and (73)

$$Q(N) = \frac{-Q_1 + Q_2 - cxQ(N_2)}{c(1-x)}.$$

It is not easy to determine c and x. Ehler and Weissler give reasons why c was probably about 0·15. Earlier investigators have found that it is unlikely to be greater than 0·20. The absorption observations showed that $Q_1 - Q_2 = 5 \cdot 1 \times 10^{-18}$ cm^2 at a wavelength of 700 Å for which $Q(N_2) = 2 \cdot 1 \times 10^{-17}$ cm^2. We then have the following values of $Q(N)$ according to different assumptions about c and x.

c	x	$Q(N) \times 10^{-17}$ cm^2
0·15	0	1·7
0·15	$\frac{1}{2}$	1·3
0·20	0	1·27
0·20	$\frac{1}{2}$	0·44

The absorption cross-section for atomic nitrogen has been calculated by Bates and Seaton† taking the ground-state wave function as given by the Hartree–Fock field and the continuum wave function of Coulomb form. They found a value at the threshold of 10^{-17} cm^2, which is quite consistent with the range of possible values deduced from Ehler and Weissler's observations.

† BATES, D. R. and SEATON, M. J., *Mon. Not. R. astr. Soc.* **109** (1949) 698.

8. Experimental and theoretical studies of photo-ionization of molecules

8.1. *General theoretical considerations*

As pointed out on p. 1078 the photo-ionization cross-section of a molecule will in general be less than the total absorption cross-section because of the possibility of photodissociation processes occurring which lead only to neutral fragments. It is necessary then to obtain information specifically about the photo-ionization, expressed either as a cross-section or as a ratio to the total absorption (the photo-ionization efficiency). A further important feature of molecular absorption is the greatly increased scope for auto-ionization. It is already difficult to interpret data on appearance potentials of higher ionization processes in complex atoms because of the presence of auto-ionization lines (see, for example, Chap. 3, § 2.5.3, and § 7.3 of this chapter). The situation is much more difficult with molecules (see, for example, Fig. 14.41). With inadequate resolution very misleading results may be obtained. The problem is particularly acute when the aim is to disentangle detailed structure arising from excitation of various vibrational levels from that due to auto-ionization. For this purpose the observation of photoelectron spectra is specially important as it provides a means of observing the vibrational thresholds without confusion with auto-ionization. Many examples of the power of this technique are discussed below.

The theory of photo-ionization of diatomic molecules follows on exactly the same lines as that discussed in Chapter 12, § 4.1 for excitation by electron impact. Corresponding to (11) of Chapter 12, the cross-section for a transition from a molecular state $nvJM$ to one $n'v'J'M'$, where $nvJM$ refers to electronic (n), vibrational (v), and rotational (JM) quantum numbers, due to absorption of radiation of frequency ν, is given (see Chap. 7 (62)) by

$$Q^{n'v'J'M'}_{nvJM}$$

$$= \frac{16\pi^3 e^2 mk\nu}{3ch^2} \left| \int \chi_{nv}(R)\chi_{n'v'}(R)\mathbf{M}(R,\Theta,\Phi)\rho_{JM}(\Theta,\Phi)\rho_{J'M'}(\Theta,\Phi)\,\mathrm{d}\mathbf{R} \right|^2,$$

(74)

where

$$\mathbf{M}(R,\Theta,\Phi) = \int \psi_n(\mathbf{r}_i,\mathbf{R})(\textstyle\sum \mathbf{r}_i)\psi_{n'}^*(\mathbf{r}_i,\mathbf{R})\textstyle\prod_i \mathrm{d}\mathbf{r}_i.$$

(75)

Here m is the electron mass and k the wave number of the ejected electron, and the molecular wave function for the $nvJM$ state is written as in (8) of Chapter 12 as

$$\psi_n(\mathbf{r}_i, \mathbf{R})\chi_{nv}(R)\rho_{JM}(\Theta, \Phi),$$

$\mathbf{r}_i$ being the aggregate of coordinates of the molecular electrons relative to the nuclei, R the nuclear separation, and Θ, Φ the polar angles of the nuclear axis relative to an axis fixed in space.

Just as in Chapter 12, § 4.1 we obtain for the cross-section summed over all final rotational states

$$Q_{nvJM}^{n'v'} = \frac{16}{3}\frac{\pi^3 e^2 mk v}{ch^2} \int \left| \int \chi_{nv}(R)\chi_{n'v'}(R)\mathbf{M}(R,\Theta,\Phi)\rho_{JM}(\Theta,\Phi)\ \mathrm{d}R \right|^2 \mathrm{d}\Omega. \tag{76}$$

If $\mathbf{M}(R, \Theta, \Phi)$ varies only gradually with R, we have, as in Chapter 12 (17),

$$Q_{nvJM}^{n'v'} = \frac{16}{3}\frac{\pi^3 e^2 mk v}{ch^2} (p_{nv}^{n'v'})^2 \int |\mathbf{M}(R_0,\Theta,\Phi)\rho_{JM}(\Theta,\Phi)|^2\ \mathrm{d}\Omega, \tag{77}$$

where

$$p_{nv}^{n'v'} = \int \chi_{nv}(R)\chi_{n'v'}(R)\ \mathrm{d}R \tag{78}$$

and is the same vibrational overlap integral as that which determines the relative population of final vibrational states in excitation by electron impact. R_0 is the equilibrium value of R in the initial electronic state.

The cross-section summed over all final vibrational and rotational states is

$$Q_{nv}^{n'} = \frac{16}{3}\frac{\pi^3 e^2 mk v}{ch^2} \int\int |\ \chi_{nv}(R)\rho_{JM}(\Theta,\Phi)\mathbf{M}(R,\Theta,\Phi)|^2\ \mathrm{d}\Omega \mathrm{d}R, \tag{79}$$

provided the difference in frequency of radiation required to excite the separate vibrational states is small compared with the frequency required to produce the electronic transition according to the Franck–Condon principle.

Again, if $\mathbf{M}(R,\Theta,\Phi)$ varies only gradually with R,

$$Q_{nv}^{n'} \simeq \frac{16}{3}\frac{\pi^3 e^2 mk v}{ch^2} \int |\mathbf{M}(R_0,\Theta,\Phi)\rho_{JM}(\Theta,\Phi)|^2\ \mathrm{d}\Omega. \tag{80}$$

The same formula may be used for any value of R, giving what is often referred to as the fixed nucleus approximation—the formula is the same as that which would be obtained if the nuclei did not vibrate.

In all the cases above the integration over the orientation angles of the nuclear axis may be carried out without difficulty by noting that

$$\mathbf{r}_i = x_i \sin\Theta \cos\Phi\ \mathbf{i} + y_i \sin\Theta \sin\Phi\ \mathbf{j} + z_i \cos\Theta\ \mathbf{k},$$

where x_i, y_i, z_i are cartesian coordinates referred to the centre of mass of the molecule as origin and the nuclear axis as z-axis.

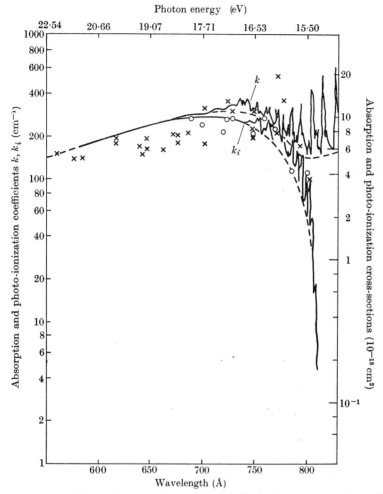

FIG. 14.41. Observed absorption and photo-ionization cross-sections of molecular hydrogen. —— Cook and Metzger, × Lee and Weissler, ○ Wainfan, Walker, and Weissler.

8.2. *Hydrogen* (H$_2$), *deuterium* (D$_2$), *and deuterium hydride* (HD)

Fig. 14.41 illustrates the absorption and photo-ionization cross-section of molecular hydrogen observed by Cook and Metzger† over the wavelength range from 830 to 600 Å. The first ionization threshold is indicated and it will be seen that some ionization is observed at longer wavelengths. This is ascribed to the presence of some molecules in excited rotational states.

For comparison, some earlier observations of the absorption cross-

† Cook, G. R. and Metzger, P. H., *J. opt. Soc. Am.* **54** (1964) 968.

section by Lee and Weissler† (see also Chap. 13, Table 13.7) and of the photo-ionization cross-section by Wainfan, Walker, and Weissler‡ are also included in Fig. 14.41. The agreement is fair.

It will be seen that there is a considerable structure in both cross-sections near the threshold. As the wavelength decreases the structure disappears and the photo-ionization efficiency becomes effectively 100 per cent.

The electron spectrometer has been applied by Al-Joboury and Turner,§ by Doolittle and Schoen,‖ and by Frost et al.,†† to seek for the

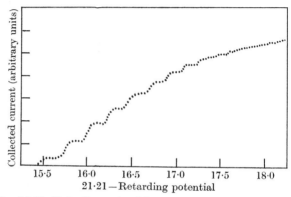

Fig. 14.42. Retarding potential curves for photoelectrons produced by photo-ionization of H_2 by the 584-Å line of helium, as observed by Frost, McDowell, and Vroom.

thresholds for excitation of the different vibrational levels of the H_2^+ $^2\Sigma_g^+$ upper state. All were successful in observing breaks in the retarding potential curve for the electrons produced by photo-ionization of H_2 by the 584-Å line of helium. Fig. 14.42 illustrates the results obtained by Frost et al.,†† using the spherical analyser described on pp. 1103–5, which gives very good energy resolution. The energy separations of the levels and the relative intensities of excitation derived from these results are given in Table 14.14. Similar results for the isotopic molecules HD and D_2 are also given in Table 14.14. Comparison is made with energy separations calculated from the theoretical potential energy curve for the $X^2\Sigma_g^+$ state of the molecular ion. In all cases the agreement is very good out to the largest values of v', the vibrational quantum number in the final state. Comparison is also made with relative transition

† LEE, P. and WEISSLER, G. L., Astrophys. J. 115 (1952) 570.
‡ WAINFAN, N., WALKER, W. C., and WEISSLER, G. L., Phys. Rev. 99 (1955) 542.
§ AL-JOBOURY, M. I. and TURNER, D. W., J. chem. Soc. (1963) 5141.
‖ DOOLITTLE, P. H. and SCHOEN, R. I., Phys. Rev. Lett. 14 (1965) 348.
†† FROST, D. C., McDOWELL, C. A., and VROOM, D. A., Proc. R. Soc. A296 (1967) 566.

TABLE 14.14

The energy separations and relative transition probabilities of different vibrational levels associated with the $X^1\Sigma_g^+ \rightarrow X^2\Sigma_g^+$ photo-ionizing transitions between H_2, D_2, and HD molecules and their respective positive ions

Final vibrational level v'		0	1	2	3	4	5	6	7	8	9	10
$H_2(X^1\Sigma_g^+) \rightarrow H_2^+(X^2\Sigma_g^+)$												
Energy separation of neighbouring levels (eV)	calc.†		0·269	0·254	0·239	0·223	0·208	0·193	0·177	0·161	0·146	
	obs.		0·27	0·25	0·24	0·23	0·21	0·20	0·19	0·17	0·16	
Relative transition probability	calc.†	0·471	0·894	1·000	0·868	0·651	0·446	0·289	0·181	0·112	0·068	
	calc.‡	0·509	0·909	1·000	0·885	0·697	0·514	0·365	0·254	0·175		
	obs.	0·48	0·90	1·00	0·91	0·74	0·58	0·44	0·28	0·15		
$HD(X^1\Sigma_g^+) \rightarrow HD^+(X^2\Sigma_g^+)$												
Energy separation of neighbouring levels (eV)	calc.†		0·235	0·223₅	0·212	0·200	0·189	0·177₅	0·166	0·154	0·143	
	obs.		0·23	0·23	0·22	0·20	0·20	0·19	0·17	0·16	0·15	
Relative transition probability	calc.†	0·362	0·792	1·000	0·962₅	0·787₅	0·578	0·398	0·261	0·167		
	obs.	0·37	0·84	1·00	0·96	0·84	0·71	0·60	0·45	0·32		
$D_2(X^1\Sigma_g^+) \rightarrow D_2^+(X^2\Sigma_g^+)$												
Energy separation of neighbouring levels (eV)	calc.†		0·194	0·186	0·178	0·171	0·163	0·155	0·148	0·140	0·132	0·125
	obs.		0·20	0·19	0·18	0·17	0·17	0·16	0·15	0·15	0·14	0·13
Relative transition probability	calc.†	0·220	0·590	0·890	1·000	0·935	0·773	0·580	0·416₅	0·283	0·186	
	calc.‡	0·238	0·604	0·888	1·000	0·962	0·837	0·682	0·531	0·402	0·298	
	obs.	0·19	0·56	0·86	1·00	0·94	0·87	0·77	0·66	0·55	0·39	

† WACKS, M. E., *J. Res. natn. Bur. stand.* **68A** (1964) 631.　　　‡ DUNN, G. H., *J. chem. Phys.* **44** (1966) 2592.

probabilities calculated from the formula (77), using vibrational overlap integrals $p_{nv}^{n'v'}$ calculated by Wacks and by Dunn as explained in Chapter 13, § 1.4.2. Agreement is good for $v' = 0$ to 4 but becomes less satisfactory for higher v'. This may be largely due to the assumption made in deriving (77) that the matrix element $M(R, \Theta, \Phi)$ varies very slowly with the frequency of the ionizing radiation, which is taken to have a constant mean value.

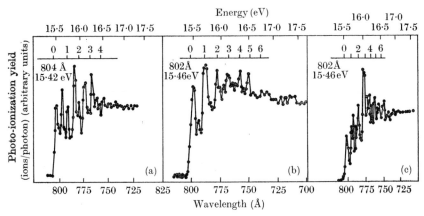

Fig. 14.43. Photo-ionization efficiency curves for (a) H_2, (b) HD, and (c) D_2 observed by Dibeler, Reese, and Krauss. The locations on the energy scale of the different vibrational levels of the $X\,^2\Sigma_g^+$ state of the ion are indicated in each case.

The fact that structure in the photo-ionization cross-section does not arise only from excitation of different vibrational levels is clearly seen in the observations of Dibeler, Reese, and Krauss.† They carried out a mass analysis of the positive ions produced from H_2, HD, and D_2 as a function of the frequency of the photo-ionizing radiation over the range 815–700 Å (15·2–17·7 eV). Fig. 14.43 shows their results for H_2^+, HD$^+$, and D_2^+. To facilitate identification of the processes occurring at the various maxima the locations of the vibrational levels of the $X\,^2\Sigma_g^+$ state of H_2^+ and of D_2^+, to which direct photo-ionization occurs, are shown in the same diagram. Although several of the peaks are clearly associated with vibrational excitation there are others that fall between vibrational thresholds and are almost certainly due to auto-ionization. Thus for H_2^+ there is a strong peak at 795 Å. Also the peak at 804 Å is relatively much higher than expected from the Franck–Condon principle for a photo-ionizing transition to the $v' = 0$ level of $H_2^+(X\,^2\Sigma_g^+)$. It is probably enhanced by an auto-ionizing peak falling at very nearly the same energy. Reference to spectroscopic data suggests that the auto-ionizing level

† DIBELER, V. H., REESE, R. M., and KRAUSS, M., *J. chem. Phys.* **42** (1965) 2045.

in this case is the $v' = 6$ level of $H_2 D^1\Pi_u$. It is to be noted that the resonance between this level and the $v' = 0$ level of the $^2\Sigma_g^+$ state of the molecular ion becomes less close in proceeding from H_2 to D_2, and reference to Fig. 14.43 shows that the peak near 800 Å becomes relatively smaller in the same sense.

The behaviour of the electron impact ionization efficiency curve for H_2 near the threshold was discussed in Chapter 13, § 1.4.2. McGowan, Fineman, Clarke, and Hanson† showed that their observed curve agreed well with that obtained by integrating the photo-ionization efficiency curve of Dibeler, Reese, and Krauss (see Fig. 13.21). In particular, the maxima appearing in the curve shown in Fig. 14.43 become bumps in the integrated curve that is reproduced in the electron-impact curve. However, Kerwin, Marmet, and Clarke‡ in a different electron impact experiment observed only vibrational excitation, while Briglia and Rapp§ observed no departures at all from linearity in the electron-impact ionization-efficiency curve. The clear observation and identification both of peaks in the photo-ionization efficiency curves, due to vibrational excitation and to auto-ionization, only deepens the mystery attending the interpretation of the electron impact data.

The first theoretical calculation of the photo-ionization cross-section of H_2 was made by Shimizu.‖ He used the approximation (77) and was primarily concerned with the estimation of the proportional yield of H^+ and H_2^+ ions when the ionizing photons had very short wavelengths (60 Å). Flannery and Öpik†† have carried out the first calculations for wavelengths ranging from the threshold to 930 Å which may be directly compared with the observations described above.

They used both approximations (76) and (77) applied to the transitions from the $X^1\Sigma_g^+$ state of H_2 to the $1s\sigma_g {}^2\Sigma_g^+$ state of H_2^+ (see Fig. 13.39) down to wavelengths of 600 Å. They obtained a third set of results by representing the vibrational wave functions $\chi_{n'v'}$ in (76) by delta functions at the classical closest distance of approach (see Chap. 12, § 3.2 and Chap. 13, § 1.4.2).

For the ground state of H_2 the electronic wavefunction used was taken to be of the form given by Weinbaum†† (see Chap. 10 (43) and (44 b)).

† McGowan, J. W., Fineman, M. R., Clarke, E. M., and Hanson, H. P., Phys. Rev. 167 (1968) 52.
‡ Kerwin, L., Marmet, P., and Clarke, P., Can. J. Phys. 39 (1961) 1240.
§ Briglia, D. D. and Rapp, D., Phys. Rev. Lett. 14 (1965) 245.
‖ Shimizu, M., J. phys. Soc. Japan 15 (1960) 1440.
†† Flannery, M. R. and Öpik, U., Proc. phys. Soc. 86 (1965) 491.
‡‡ Weinbaum, S., J. chem. Phys. 1 (1933) 593.

The continuum wave function was taken to be of the form

$$\psi = \psi_{1s\sigma}(\mathbf{r}_1)F(\mathbf{r}_2) + \psi_{1s\sigma}(\mathbf{r}_2)F(\mathbf{r}_1),$$

where $\psi_{1s\sigma}(\mathbf{r})$ denotes the wave function of the ground state of H_2^+ and $\mathbf{r}_1$, $\mathbf{r}_2$ are the coordinates of the two electrons. As in atomic calculations of dipole matrix elements, the main contribution comes from large values of the electronic coordinates. For large r and fixed R, the continuum orbital may be represented to a good approximation by the wave function for motion, with the same energy, of an electron in the field of two fixed centres, each with a charge $+\frac{1}{2}e$ at a certain separation R'. This is such that the quadrupole moment $\frac{1}{4}eR'^2$ is equal to the true quadrupole moment of H_2^+ when the nuclear separation is R. In this way the actual field of H_2^+ is reproduced, at large R, as far as terms of order r^{-6}.

The vibrational wave functions for the ground state were obtained by a perturbation method. The potential energy function in which the nuclear vibration takes place was represented by

$$U(R) = \tfrac{1}{2}k(R - R_0)^2 - g(R - R_0)^3,$$

where the cubic term is treated as small, R_0 is the equilibrium separation, and k and g are constants that depend on the rotational quantum number J. For the final state of H_2^+ this method is too crude and the vibrational wave functions were obtained by direct numerical solution of the appropriate wave equation with the exact potential energy function included.

Fig. 14.44 shows the photo-ionization cross-section calculated by three different approximate methods, assuming the molecule to be initially in the ground rotational state.

It will be seen that over the entire frequency range studied, the delta-function approximation gives results that differ little from those obtained with the more elaborate approximation (76). The approximation (77) of regarding the nuclei as fixed at the initial equilibrium separation R_0 gives good results except at frequencies below that which gives the peak cross-sections.

Comparison with the observed cross-sections reveals good agreement, considering the complexity of the calculations and the consequent need for the introduction of fairly drastic approximations. The observed cross-section is considerably larger near the threshold than the best calculated value (from (76)) but this is not surprising because the evidence discussed above (see Fig. 14.43) shows that auto-ionization, which is not allowed for in the theory, is important in this region.

8.3. H_2^+

The distribution of the potential-energy curves for this molecule has been discussed in Chapter 13, § 2.1 (see Fig. 13.39). Photon absorption by molecules in their ground $1s\sigma_g\,^2\Sigma_g^+$ electronic states will almost always lead to dissociation either into H and H^+ or $2H^+$, just as for excitation by proton impact.

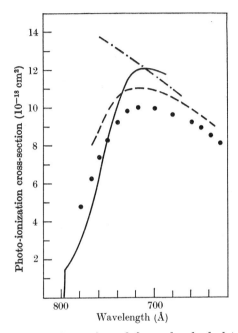

FIG. 14.44. Comparison of observed and calculated photo-ionization cross-sections of H_2. —— calculated from (76), — · — · — calculated from (77), — — — calculated from (76) using delta-function approximation, ● observed by Cook and Metzger.

The strongest of the transitions from the $1s\sigma_g$ state is to $2p\sigma$, which is an entirely repulsive state. Bates[†] has calculated the optical oscillator strengths for this transition in the fixed nucleus approximation (see (80) et seq.) as a function of the nuclear separation R. These calculations were extended to the $1s\sigma_g$–$2p\pi_u$ transition by Bates, Darling, Hawe, and Stewart.[‡] In both cases accurate two-centre wave functions were used for the electron. Some of their results are given in Table 14.15.

It was suggested that continuous absorption due to excitation of the

† BATES, D. R., *J. chem. Phys.* **19** (1951) 1122.
‡ BATES, D. R., DARLING, R. T. S., HAWE, S. C., and STEWART, A. L., *Proc. phys. Soc.* A **66** (1953) 1124.

$1s\sigma$–$2p\sigma$ transition in H_2^+ might be of astrophysical importance. Buckingham, Reid, and Spence[†] therefore carried out detailed calculations of the absorption cross-section for H_2^+ molecules as a function of frequency and gas temperature, it being supposed that the rotational and vibrational degrees of freedom were in temperature equilibrium.

TABLE 14.15

Calculated oscillator strengths for $1s\sigma_g$–$2p\sigma$ and $1s\sigma_g$–$2p\pi_u$ transitions in H_2^+ for different nuclear separations, in the fixed nucleus approximation

Nuclear separation R (a_0)	0·0	0·4	0·8	1·2	1·6	2·0	3·0	4·0	5·0
$1s\sigma$–$2p\sigma$	0·139	0·175	0·240	0·293	0·317	0·319	0·289	0·218	0·175
$1s\sigma$–$2p\pi_u$	0·28	0·32	0·37	0·41	0·44	0·46	0·48	0·46$_5$	0·43

TABLE 14.16

Mean absorption cross-section (in 10^{-18} cm^2) of H_2^+ at different gas temperatures due to excitation of the $1s\sigma$–$2p\sigma$ transition

Temp. (°K)	Photon wavelength (Å)							
	4545	3846	2500	1818	1429	1176	1000	869·6
0				0·24	2·53	7·06	5·97	2·71
1600			0·01	0·59	3·64	7·24	5·51	2·07
2500	0·052	0·102	0·64	2·24	4·38	5·08	3·45	1·12

They used the optical oscillator strengths as functions of nuclear separation calculated by Bates[‡] and as given in Table 14.15, but evaluated accurately the vibration and rotation overlap integrals for a number of initial and final states of nuclear motion sufficient for an average over the thermal distribution among the initial levels to be taken. Their results are given in Table 14.16. They agree well with the results of a semi-classical calculation by Bates.[§]

Bates, Öpik, and Poots[||] have also calculated cross-sections for photo-ionization of H_2^+ from the ground vibrational level of the $1s\sigma$ state. They again used exact two-centre wave functions but used the delta-function approximation (§ 4.1 of Chap. 12, p. 830) for evaluating the overlap integrals (78). They find that the cross-section remains very

[†] BUCKINGHAM, R. A., REID, S., and SPENCE, R., *Mon. Not. R. astr. Soc.* **112** (1952) 382.
[‡] loc. cit., p. 1151.   [§] BATES, D. R., *Mon. Not. R. astr. Soc.* **112** (1952) 40.
[||] BATES, D. R., ÖPIK, U., and POOTS, G., *Proc. phys. Soc. A* **66** (1953) 1113.

small from the threshold to a wavelength of about 1400 Å. At shorter wavelengths it increases rapidly to a maximum of about 6.7×10^{-19} cm^2 at 1200 Å, after which it decreases gradually, being about 2.8×10^{-19} cm^2 at 860 Å.

Experimental study of photodissociation and photo-ionization of H_2^+ encounters similar difficulties to the corresponding study of excitation and dissociation by electron impact, a detailed account of which has been given in Chapter 13, § 2. Experiments have been carried out by Dunn, von Busch and van Zyl† following preliminary experiments by Dunn,‡ the arrangement being similar to that for the electron impact studies except that the electron beam was replaced by a beam of photons. This was produced by a xenon arc operated at 5000 W, chopped mechanically so as to use phase-sensitive detection, and then passed through a monochromator. The wavelength range studied, from 2537 to 5790 Å, was such that photodissociation but not photo-ionization could occur. The results obtained were compared with theoretical expectation in which the initial distribution of vibrational levels was assumed to be the same as in the analysis of the electron impact data (Chap. 13, § 2). According to the theory, calculated using accurate electronic and vibrational wave functions, the dissociation cross-section has a maximum of 2.94×10^{-18} cm^2 at 2000 Å and falls monotonically at larger wavelengths. The observed cross-sections agree with the calculated within 15 per cent.

8.4. *Oxygen*

Special interest attaches to detailed study of the absorption and photo-ionization cross-sections of O_2 because of its importance in the earth's upper atmosphere.

We discuss first the behaviour between the first photo-ionization threshold (1026 Å, 12·07 eV) and about 885 Å (14 eV). A great deal of structure was observed in this region as long ago as 1935 in the course of absorption measurements by Price and Collins.§ Several further sets of observations have been made since that time, including observations of the photo-ionization. The agreement between the different results is striking, as may be seen by reference to Table 14.17, in which comparison is made between wavelengths of the principal maxima as

† DUNN, G. H., VON BUSCH, F., and VAN ZYL, B. *Abstracts of Vth International Conference on Physics of Electronic and Atomic Collisions*, Leningrad, 1967, p. 610.
‡ DUNN, G. H., *Atomic collision processes*, ed. McDOWELL, M. R. C., p. 997 (North Holland, Amsterdam, 1964).
§ PRICE, W. C. and COLLINS, G., *Phys. Rev.* **48** (1935) 714.

TABLE 14.17

Comparison of wavelengths of maxima in absorption and photo-ionization of O_2 (1000–800 Å) as measured by different observers

Observers	Measurements	Wavelengths (Å)														
Price and Collins†	Absorption	1003·7	993·0	983·1	972·6	965·4	956·1	947·9	938·9	930·5	924·5	917·3	909·9			
Matsunaga and Watanabe‡	Absorption	1004·0	993·3	983·1	927·8	965·8	955·9	947·9	939·2	930·7	924·7	917·1	909·8	901·1	891·5	885·6
Watanabe and Marmo§	Absorption	1003·8	993·4	982·5	972	964·7	954·7	947·0	938·0	931	924·6	916·0	909·9	900·6		
Nicholson‖	Photo-ionizing efficiency	1003·7	993·5	983·0	972·7	965·1	956·2	947·0	938·2	931·0	924·1	916·0	910·1	901·6	892·1	885·3
Cook and Metzger†† {	Absorption and photo-ionization	1003·71	993·04	983·09	972·57	965·44	956·11 } 957·21 }	947·87	938·88	930·58	924·47	916·34	909·17	900·5	893·8	885·4

† PRICE, W. C. and COLLINS, G., *Phys. Rev.* **48** (1935) 714.
‡ MATSUNAGA, F. M. and WATANABE, K., *Sci. Rep. No. 5 Contract No. AF.* 19 (604) 4576 (Geophys. Rev. Directorate, Hawaii, 1961).
§ WATANABE, K. and MARMO, F. F., *J. chem. Phys.* **25** (1956) 965.
‖ NICHOLSON, A. J. C., ibid. **39** (1963) 954.
†† COOK, G. R. and METZGER, P. H., *J. chem. Phys.* **41** (1964) 321.

determined by several different investigators. Since the peaks are observed at the same wavelength both for total absorption and for photo-ionization there is no doubt that they arise from an ionizing process. Before discussing the interpretation further it is to be noted

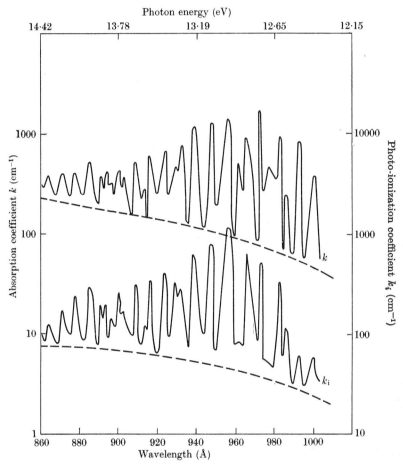

FIG. 14.45. Absorption and photo-ionization coefficients, k and k_i respectively for oxygen in the wavelength range from 1000 to 860 Å as observed by Cook and Metzger.

that, in the most recent investigations, the positions of a number of smaller subsidiary peaks have been determined. Thus Cook and Metzger† observed such maxima at 985·80, 975·32, 933·27, and 922·85 Å.

Fig. 14.45 illustrates the absorption and photo-ionization coefficients observed by Cook and Metzger over the wavelength concerned. The

† loc. cit., p. 1154.

photo-ionizing efficiency is about 60 per cent at 1025 Å and then decreases gradually to about 25 per cent at 850 Å. In comparison with results obtained by other observers there is good general agreement with the absorption spectrum observed by Watanabe and Marmo† but, particularly in the range 850–920 Å, the minima observed by Cook and Metzger are higher. This also applies by comparison with observations by Lee‡ and by Clark§ in the same range. On the other hand, Wainfan,

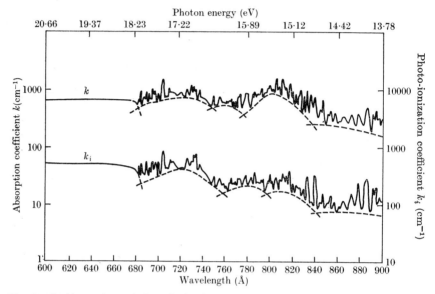

Fig. 14.46. Absorption and photo-ionization coefficients, k and k_i respectively, for oxygen in the wavelength range from 900 to 600 Å as observed by Cook and Metzger.

Walker, and Weissler‖ observe even higher minima. There is good agreement, as far as the photo-ionization coefficient is concerned, with the result of Watanabe and Marmo.†

Turning now to the absorption of photo-ionization at shorter wavelengths, Fig. 14.46 gives the results of Cook and Metzger, down to 600 Å. Below 680 Å the structure disappears and both the absorption and the photo-ionization cross-sections vary slowly from there to 600 Å, the photo-ionizing efficiency being about 80 per cent. These results agree quite well with those of Wainfan et al.‖ but are somewhat lower in absolute magnitude.

† WATANABE, K., and MARMO, F. F., J. chem. Phys. 25 (1956) 965.
‡ LEE, P., J. opt. Soc. Am. 45 (1955) 703.
§ CLARK, K. C., Phys. Rev. 87 (1952) 271.
‖ WAINFAN, N., WALKER, W. C., and WEISSLER, G. L., J. appl. Phys. 24 (1953) 1318; Phys. Rev. 99 (1955) 542.

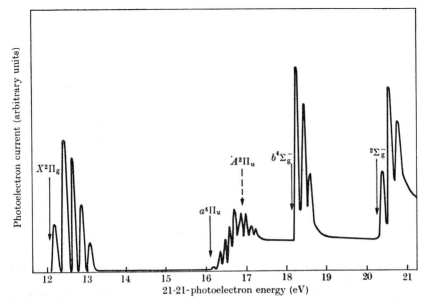

Fig. 14.47. Energy spectrum of photoelectrons produced by photo-ionization of O_2 by 584-Å radiation as observed by Frost, McDowell, and Vroom.

TABLE 14.18
Ionization potentials for O_2

Final ionic state	Ionization potentials (V)			
	Spectroscopic[†]	Photo-ionization[‡]	Photoelectron energy spectrum	
			Al-Joboury et al.[§]	Frost et al.[∥]
$X^2\Pi_g$		12·075	12·10	12·10±0·01
$a^4\Pi_u$	16·07		16·26	16·13±0·01
$A^2\Pi_u$	16·824			
$b^4\Sigma_g^-$	18·173		18·18	18·19±0·01
$^2\Sigma_g^-$	20·308		20·31	20·30±0·01

The energy spectrum of photoelectrons produced by photo-ionization of O_2 by radiation of wavelength 584 Å has been observed by Al-Joboury, May, and Turner,§ using a cylindrical analyser and by Frost et al.∥ using a spherical analyser. Fig. 14.47 reproduces a spectrum observed by the latter investigators by differentiation of the appropriate retarding

† HUFFMAN, R. E., LARRABEE, J. C., and TANAKA, Y., J. chem. Phys. 40 (1964) 356.
‡ WATANABE, K., ibid. 26 (1957) 542.
§ AL-JOBOURY, M. I., MAY, D. P., and TURNER, D. W., J. chem. Soc. (1965) 616.
∥ FROST, D. C., McDOWELL, C. A., and VROOM, D. A., Proc. R. Soc. A296 (1967) 566.

potential curve. The relation of the energy peak to the electronic state in which the O_2^+ is left after ionization (see Fig. 13.69) is indicated. Comparison of the structure shown in Fig. 14.47, which arises only from excitation of different vibrational levels, with that shown in Figs. 14.45 and 14.46 shows without doubt that most of the peaks seen in the latter must arise from auto-ionization.

Table 14.18 shows the comparison between the ionization potentials derived from photoelectron spectra and those obtained spectroscopically. The agreement is very satisfactory.

Table 14.19

Energy separations and relative transition probabilities of different vibrational levels associated with $X^3\Sigma_g^- \to X^2\Pi_g$ *photo-ionizing transitions in* O_2

Final vibrational level (v')		0	1	2	3	4
Energy separation between neighbouring levels (eV)	calc.[†]	—	0·228	0·225	0·220	0·216
	obs.	—	0·23±0·01	0·23±0·01	0·22±0·02	0·21±0·02
Relative transition probability	calc.[†]	0·603	1·000	0·673	0·236	0·045
	obs.	0·43	1·00	0·93	0·43	0·14

In Table 14.19 the energy separations of neighbouring vibrational levels associated with the ground $^2\Pi_g$ state of O_2^+ as well as the relative transition probabilities to these levels are given and compared with calculated values.

As far as vibrational spacing is concerned the agreement with the calculated values is excellent, but for the relative transition probabilities it is not very good. This may be partly due to the approximate nature of the calculation in which the potential energy curve for the O_2^+ state is represented by a Morse potential.

The relative transition probabilities to different electronic states of O_2^+ have been determined from 584 Å radiation by Frost et al.[‡] Schoen[§] and Blake and Carver,[||] using cylindrical analysers, have measured these probabilities over a wider wavelength range. Fig. 14.48 illustrates the results obtained which, while still not very precise or consistent, show the possibilities inherent in the technique.

† WACKS, M. E., *J. chem. Phys.* **41** (1964) 930.   ‡ loc. cit., p. 1157.
§ SCHOEN, R. I., *J. chem. Phys.* **40** (1964) 1830.
|| BLAKE, A. J., and CARVER, J. H., ibid. **47** (1967) 1038.

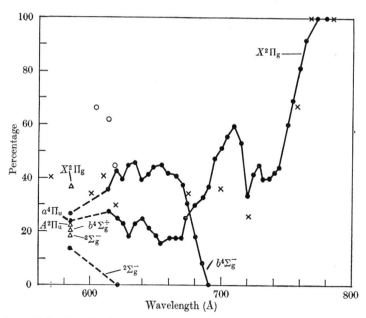

FIG. 14.48. Fractional percentage of photo-ionization of O_2 leaving O_2^+ in different electronic states as a function of wavelength. —●— observed by Blake, × observed by Schoen ($X^2\Pi_g$), ○ observed by Schoen ($b^4\Sigma_g^- + {}^2\Sigma_g^-$), △ observed by Vroom.

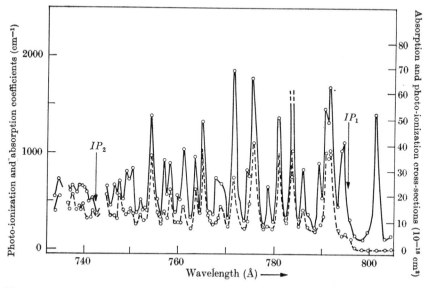

FIG. 14.49. Absorption and photo-ionization coefficients for nitrogen in the wavelength range from 735 to 805 Å as observed by Cook and Ogawa. The first ionization threshold (IP_1) is at 795·9 Å, the second (IP_2) at 742·7 Å. —— absorption coefficient, — — — photo-ionization coefficient.

8.5. *Nitrogen*

Absorption and photo-ionization of N_2 is also of special interest for upper atmospheric physics. As usual we consider first the behaviour

TABLE 14.20

Observed absorption cross-sections and photo-ionization efficiencies for N_2

Wavelength (Å)	Absorption cross-section (10^{-18} cm^2)		Photo-ionization efficiency (%)	
	Cook and Ogawa	Samson and Cairns	Cook and Ogawa	Samson and Cairns
790·2				45
790·2		22·7		
790·4	19·3		64·0	
790·5			27·0	
787·2			61	
788·4	8·00		85·5	
787·7		8·34	89	
779·8		12·8		65
779·9		12·8		
781·1	50·6		74·0	
779·7	12·3		77·0	
774·5	30·3	34·0	3·3	40
765·14		85·4		77
765·2	48·7		80·0	
764·36		13·5	80·0	69
764·5	15·3		93	
762·7	13·4		66·0	
763·6	35·4		65·0	
763·34		27·3		80
762·7	11·9		65·0	
761·2	38·7		66·0	
762·0		27·8		46
761·13		40·1		55
761·2	38·7		42·5	
760·445		19·8		57
760·229		19·8		
760·3	19·2		53·0	
759·4	14·0	11·6	86	73
758·67		23·9		75
759·4	14·0		73·0	
758·3	33·9		70·0	

of the cross-sections near the first ionization threshold (795·9 Å). Fig. 14.49 illustrates the structure in the photo-ionization cross-section down to 735 Å and its correlation with the absorption spectrum as observed by Cook and Ogawa.† Table 14.20 gives absolute values of the absorption cross-sections and photo-ionizing efficiency observed by Cook and Ogawa

† COOK, G. R., and OGAWA, M., *Can. J. Phys.* **43** (1965) 256.

and also by Samson and Cairns,† using the double ionization-chamber technique (see § 5.3). On the whole the agreement is remarkably good.

Fig. 14.50 shows the extension of the absorption and photo-ionization coefficients down to 600 Å, as observed by Cook and Metzger. In this wavelength range the threshold for ionization to the $B^2\Sigma_u^+$ state of N_2^+

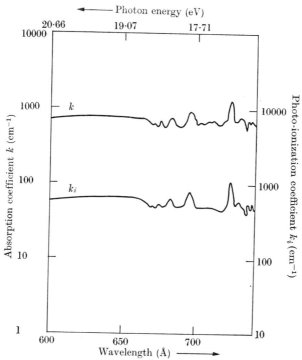

FIG. 14.50. Absorption and photo-ionization coefficients for nitrogen in the wavelength range from 600 to 750 Å as observed by Cook and Metzger.

(see Fig. 13.46) at 18·7 eV occurs but it is not distinguishable in either the absorption or photo-ionization coefficients. The same applies to the threshold for ionization to the $A^2\Pi_u$ state at 16·7 eV (742·7 Å) indicated in Fig. 14.49.

The power of the technique of photoelectron spectroscopy in the observations of threshold energies is well seen by reference to Fig. 14.51 (a), which reproduces a retarding potential curve for the photo-electric current produced by 584-Å radiation in N_2, observed by Frost et al.‡ using their spherical analyser. Not only are the three ionization

† SAMSON, J. A. R. and CAIRNS, R. B., *J. geophys. Res.* **69** (1964) 4583.
‡ loc. cit., p. 1154.

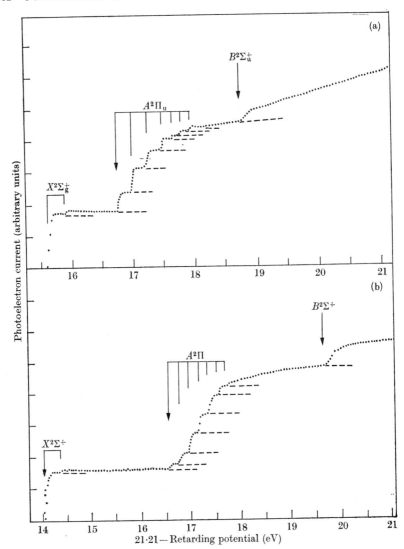

Fig. 14.51. Retarding potential curve for the photoelectron current produced by 584-Å radiation in N_2 and in CO, as observed by Frost, McDowell, and Vroom. (a) N_2, (b) CO.

thresholds observed quite clearly but also those for excitation of the different vibrational levels. It is quite clear that much of the structure seen in Figs. 14.49 and 50 arises from auto-ionization.

Table 14.21 shows the comparison between the ionization potentials derived from photoelectron spectra and those obtained spectroscopically. The agreement is very satisfactory.

Vibrational structure for the $X^1\Sigma_g^+ \to X^2\Sigma_g^+$ transition is not well developed because the equilibrium separation is nearly the same for the neutral and ionized molecular states (see Fig. 13.46). The energy separation between the 0 and 1 vibrational levels for the $X^2\Sigma_g^+$ state as derived from the photoelectron spectrum is 0·35 eV, which compares well with

TABLE 14.21

Ionization potentials for N_2

Final ionic state	Ionization potential (V)		
	Spectroscopic	Photoelectron energy spectrum	
		Al-Joboury et al.[†]	Frost et al.[‡]
$X^2\Sigma_g^+$	15·576	15·57	15·58±0·01
$A^2\Pi_u^+$	16·693	16·72	16·60±0·01
$B^2\Sigma_u^+$	18·757	18·72	18·80±0·01

the calculated 0·29 eV. Quantitative agreement is not found for the relative transition probabilities to the 1 and 0 levels, the observed[‡] being 0·054 and the calculated[§] 0·100, but both agree in predicting a low excitation of upper vibrational levels and a factor of 2 in magnitude is not serious as both observation and theory will be less accurate under these circumstances.

For the $X^1\Sigma_g^+ \to A^2\Pi_u$ transitions the vibrational structure is well developed, corresponding to a considerable difference in equilibrium separation of the neutral and ionized molecular states (see Fig. 13.46).

In Table 14.22 the energy separations of neighbouring vibrational levels associated with the $A^2\Pi_u$ states of N_2, as well as the relative transition probabilities to these levels, are given and compared with spectroscopic or calculated values.

There is good agreement as far as it goes with spectroscopic determinations of energy separations. The comparison of observed and calculated relative transition probabilities reveals a similar situation to that for H_2, HD, and D_2 (see Table 14.14). There is good agreement for lower vibrational levels but the observed probabilities are considerably

[†] AL-JOBOURY, M. I. and TURNER, D. W., *J. chem. Soc.* (1963) 5141.

[‡] FROST, D. C., McDOWELL, C. A., and VROOM, D. A., loc. cit., p. 1157.

[§] NICHOLLS, R. W., *J. Res. natn. Bur. Stand.* 65A (1961) 451; HALMANN, M. and LAULICHT, I., *J. chem. Phys.* 43 (1965) 1503.

higher than the calculated as for the higher levels. This may be due to the same reasons.

Fig. 14.52 illustrates the results obtained by different investigators†‡ for the relative transition probabilities to different electronic states of N_2^+. The agreement between these results is more satisfactory than for O_2^+ (see Fig. 14.48).

8.6. Carbon monoxide

Fig. 14.53 illustrates the absorption and photo-ionization coefficients of carbon monoxide in the wavelength range 900–700 Å as observed by Cook, Metzger, and Ogawa.§ Because of the complicated structure,

TABLE 14.22

Energy separations and relative transition probabilities of different vibrational levels associated with $X^1\Sigma_g^+ \to A^2\Pi_u$ photo-ionizing transitions in N_2

Final vibrational level (v')		0	1	2	3	4	5	6
Energy separation between neighbouring levels (eV)	spectroscopic		0·24	0·23	0·23			
	photoelectron spectra†		0·23	0·23	0·22	0·21	0·20	0·19
Relative transition probability	calc.‖	0·787₅	1·000	0·725	0·398	0·185	0·077	0·030
	photoelectron spectra†	0·85	1·00	0·71	0·46	0·21	0·14	0·08

comparison with the results obtained by other observers is difficult, but the agreement with the data of Sun and Weissler†† is not unsatisfactory. At shorter wavelengths comparison is easier as auto-ionization is less important, and in Fig. 14.54 we show the results obtained by Cook *et al.*,§ Sun and Weissler,†† and Huffman, Larrabee, and Tanaka‡‡ for the absorption coefficient.

We would expect that, as usual, carbon monoxide should behave in a rather similar way to nitrogen. A clear example of this is provided by the comparison between the retarding potential curve observed for the photoelectron current produced by 584 Å radiation in CO (Fig. 14.51 (b)) with that in N_2 (Fig. 14.51 (a)). The same features appear in both. Thus the vibrational structure is well developed for the excitation of the $A^2\Pi$ state of CO^+ but not for $X^2\Sigma^+$ and $B^2\Sigma^+$ just as in N_2.

† FROST, D. C., McDOWELL, G. A., and VROOM, D. A., loc. cit., p. 1157.
‡ SCHOEN, R. I., *J. chem. Phys.* **40** (1964) 1830; BLAKE, A. J. and CARVER, J. H. *J. chem. Phys.* **47** (1967) 1038.
§ COOK, G. R., METZGER, P. H., and OGAWA, M., *Can. J. Phys.* **43** (1965) 1706.
‖ NICHOLLS, R. W., *J. Res. natn. Bur. Stand.* **65A** (1961) 451; HALMANN, M. and LAULICHT, I., *J. chem. Phys.* **43** (1965) 1503.
†† SUN, H. and WEISSLER, G. L., *J. chem. Phys.* **23** (1955) 1625.
‡‡ HUFFMAN, R. E., LARRABEE, J. C., and TANAKA, Y., ibid. **40** (1964) 2261.

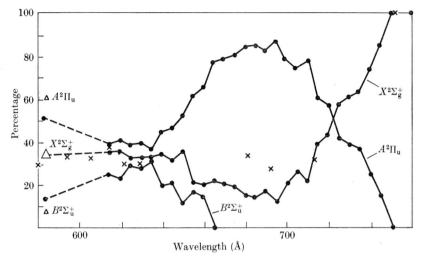

FIG. 14.52. Fractional percentage of photo-ionization of N_2 leaving N_2^+ in different electronic states, as a function of wavelength. —●— observed by Blake and Carver. ✕ observed by Schoen. △ observed by Frost, McDowell, and Vroom.

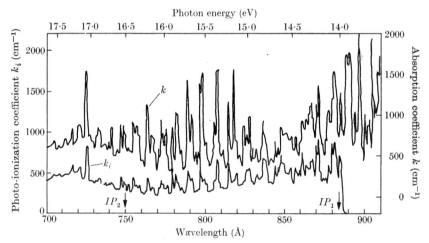

FIG. 14.53. Absorption and photo-ionization coefficients, k and k_i respectively, for carbon monoxide in the wavelength range 700–900 Å as observed by Cook et al. The first and second ionization thresholds are indicated as IP_1, IP_2 respectively.

Table 14.23 shows the comparison between the ionization potentials derived from the photoelectron spectra and those obtained spectroscopically. As for N_2 the agreement is very satisfactory, particularly with the photoelectron data of Frost et al.†

For the $X^1\Sigma^+ \rightarrow X^2\Sigma^+$ transition, in which only the 0 and 1 final vibrational levels are resolved, the energy separation (0·32 eV) is to be

† FROST, D. C., McDOWELL, C. A., and VROOM, D. A., loc. cit., p. 1157.

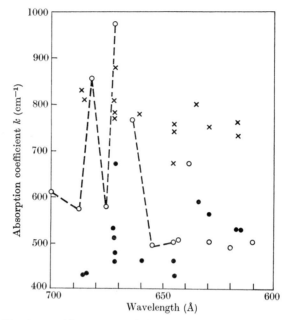

Fig. 14.54. Absorption coefficients for CO in the wavelength range 600–700 Å observed by ○ Cook *et al.*; ● Sun and Weissler; × Huffman *et al.*

Table 14.23
Ionization potentials for CO

Final atomic state	Ionization potential (V)		
	Spectroscopic	Photoelectron energy spectrum	
		Al-Joboury *et al.*†	Frost *et al.*‡
$X^2\Sigma^+$	14·013	13·89	14·01
$A^2\Pi$	16·536	16·58	16·55
$B^2\Sigma^+$	19·674	19·67	19·67

compared with 0·27 eV calculated by Wacks.§ The relative transition probability to the $v = 1$ state as compared with that to $v = 0$ is observed to be 0·039, which compares remarkably well with the calculated 0·038.

In Table 14.24 the energy separations of neighbouring vibrational levels associated with the $A^2\Pi$ state of CO as well as the relative transition

† Al-Joboury, M. I., May, D. P., and Turner, D. W., *J. chem. Soc.* (1965) 616.
‡ loc. cit., p. 1157.
§ Wacks, M. E., loc. cit., p. 1158.

TABLE 14.24

Energy separations and relative transition probabilities of different vibrational levels associated with $X^1\Sigma^+ \to A^2\Pi$ photo-ionizing transitions in CO

Final vibrational level		0	1	2	3	4	5	6
Energy separation between neighbouring levels (eV)	calc.		0·190	0·187	0·184	0·180	0·177	0·173
	obs.		0·20	0·19	0·19	0·19	0·18	0·17
Relative transition probability	calc.	0·367	0·819	1·000	0·893	0·653	0·417	0·241
	obs.	0·38	0·83	1·00	0·89	0·65	0·49	0·28

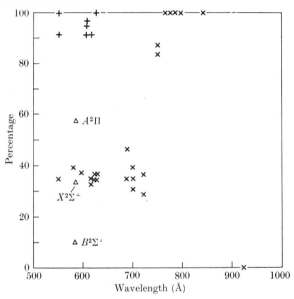

Fig. 14.55. Fractional percentage of photo-ionization of CO leaving CO⁺ in different electronic states, as a function of wavelength. △ observed by Vroom. ✕ observed by Schoen ($X^2\Sigma^+$). + observed by Schoen ($B^2\Sigma^+$).

probabilities to these levels are given and compared with calculated values.† The agreement for both these quantities is very good and for the transition probabilities persists out to much higher vibrational levels than for H_2 and N_2.

Fig. 14.55 illustrates the results obtained by different investigators‡§ for the relative transition probabilities to different electronic states of CO⁺.

† WACKS, M. E., loc. cit., p. 1158.
‡ FROST, D. C., McDOWELL, C. A., and VROOM, D. A,. loc. cit., p. 1157.
§ SCHOEN, R. I., loc. cit. p. 1164.

8.7. *Nitric oxide*

The first ionization potential of nitric oxide is as low as 9·4 eV, corresponding to a wavelength of about 1320 Å. Because of this it may be ionized by Ly α radiation of hydrogen at 1216 Å, thereby providing a method of measuring the intensity of such radiation that is of particular importance in solar physics. Furthermore, it seems likely that the main source of ionization in the earth's atmosphere at altitudes of around

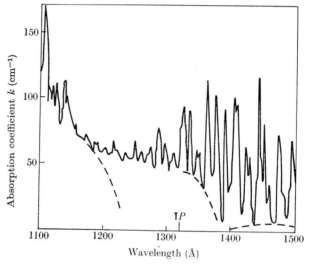

FIG. 14.56. Absorption coefficient of nitric oxide for radiation in the wavelength range 1100–1500 Å as observed by Marmo. The first ionization threshold is indicated as IP.

80 km is photo-ionization of the small proportion of nitric oxide present at these altitudes by solar Ly α-radiation, which penetrates to this depth through an absorption window.

Fig. 14.56 shows the absorption coefficient between 1100 and 1500 Å observed by Marmo,† using an apparatus in which the absorption cell with lithium fluoride windows was placed between the exit slit of a monochromator and the photomultiplier detector. The contribution to the absorption due to photo-ionization, though certainly considerable at wavelengths shorter than 1320 Å is not easy to estimate. Thus there is clear evidence of a continuum with a threshold near 1400 Å, which is presumably due to some photodissociation process. The relative contribution from this continuum and the first ionization continuum below 1300 Å leads to a total background continuum which increases slowly as the wavelength falls.

† MARMO, F. F., *J. opt. Soc. Am.* **43** (1953) 1186.

Evidence about the photo-ionization yield has been obtained by Nicholson† who measured the so called 'photo-ionization efficiency' by the method described in § 5.5.4. Provided the absorption is not too great this efficiency is proportional to the fractional photo-ionization yield. Fig. 14.57 shows the 'photo-ionization efficiency' observed by Nicholson. To normalize these results use may be made of the photo-ionization yield measurements of Watanabe‡ who found that this increases from

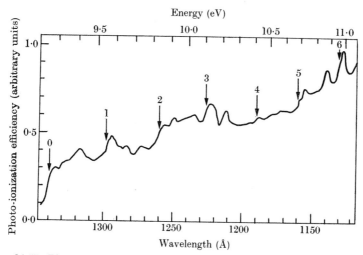

FIG. 14.57. Photo-ionization efficiency of nitric oxide for radiation in the wavelength range 1350–1120 Å as observed by Nicholson.

60 per cent at 1050 Å to nearly 90 per cent near 1160 Å. Referring to Fig. 14.56, we see that this indicates that at the wavelength of Ly α (1216 Å) the photo-ionization coefficient should be about 30 cm^{-1} corresponding to a cross-section of $1 \cdot 12 \times 10^{-18}$ cm^2.

The locations of the vibrational levels of the ground state of NO$^+$ relative to different peaks in the ionization efficiency curve are indicated.

Between 1000 and 700 Å both the total absorption and photo-ionization coefficients have been measured by Metzger, Cook, and Ogawa,§ and their results are illustrated in Fig. 14.58. They show a rich band structure overlying a number of continua. There is good agreement with less complete data obtained by Sun and Weissler.‖ The photo-ionization yield over the same wavelength range is shown in Fig. 14.59. This agrees fairly well with the results of Walker and Weissler.††

† NICHOLSON, A. J. C., *J. chem. Phys.* **39** (1963) 954.
‡ WATANABE, K., ibid. **22** (1954) 1564.
§ METZGER, P. H., COOK, G. R., and OGAWA, M., *Can. J. Phys.* **45** (1967) 203.
‖ SUN, H. and WEISSLER, G. L., *J. chem. Phys.* **23** (1955) 1372.
†† WALKER, W. C. and WEISSLER, G. L., ibid. **23** (1955) 1962.

Fig. 14.60 shows a retarding potential curve observed for the photo-electron current produced by 584 Å radiation in NO.† This shows clear evidence of changes of slope due to production of NO^+ ions in the $X^1\Sigma^+$ ground state as well as in a number of excited states identified

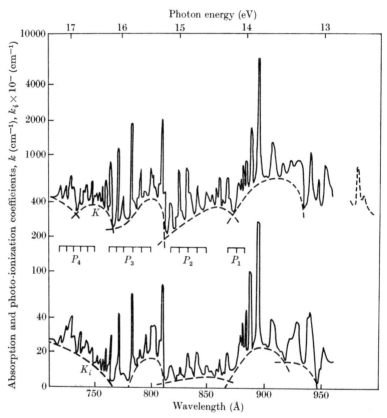

FIG. 14.58. Absorption and photo-ionization coefficients k and k_i respectively for nitric oxide in the wavelength range 700–1000 Å, as observed by Metzger *et al.*

as shown in the diagram. A well-developed vibrational structure is associated with production of ground state NO^+.

Table 14.25 shows the comparison between the ionization potentials derived from the photoelectron spectra and those observed spectro-scopically. The agreement is quite good throughout though the photo-electron threshold near 15·7 V has not been observed spectroscopically. In Table 14.26 the energy separations of neighbouring vibrational levels associated with the ground $X^1\Sigma^+$ state of NO^+, as well as the relative

† FROST, D. C., McDOWELL, C. A., and VROOM, D. A., loc. cit., p. 1157.

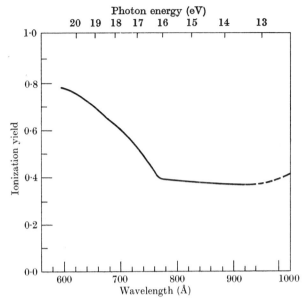

FIG. 14.59. Photo-ionization yield in nitric oxide in the wavelength range 700–1000 Å as observed by Metzger *et al.*

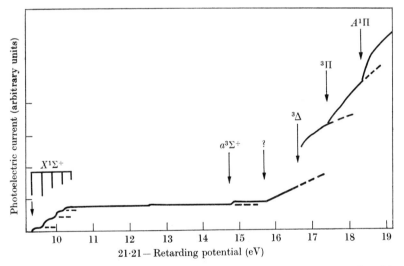

FIG. 14.60. Retarding potential curve for the photoelectron current produced by
584 Å radiation in NO, as observed by Frost, McDowell, and Vroom.

transition probabilities to these levels, are given and compared with
spectroscopic or calculated values. The agreement is not unsatisfactory.
Frost *et al.* also measured the relative transition probabilities to the

different electronic states for ionizing radiation at 584 Å. They found that these probabilities were as $0.36:0.06:1.00:0.51$ for the $^1\Sigma^+$, $a^3\Sigma^+$, $^3\Delta$, and $A^1\Pi$ states respectively.

TABLE 14.25

Ionization potentials for NO

Final ionic state	Ionization potential (V)		
	Spectroscopic	Photoelectron energy spectrum	
		Al-Joboury et al.[†]	Frost et al.[‡]
$X^1\Sigma^+$	9·26	9·23	9·32
$a^3\Sigma^+$	14·2		14·84
?		15·4	15·72
$^3\Delta$	16·55	16·55	16·62
$^3\Pi$	16·9		17·18
$A^1\Pi$	18·4	18·24	18·24

TABLE 14.26

Energy separations and relative transition probabilities of different vibrational levels associated with $X^2\Pi \to X^1\Sigma^+$ *photo-ionizing transitions in* NO

Final vibrational level		0	1	2	3	4
Energy separation between neighbouring levels (eV)	Spectroscopic Photoelectron[‡] spectra		0·291	0·286	0·283	0·278
			0·30	0·29	0·28	0·28
Relative transition probability	Calc.[§] Photoelectron[‡] spectra	0·478	1·000	0·917	0·484	0·163
		0·59	1·00	0·81	0·40	0·16

8.8. *Water vapour*

The first ionization potential is at 12·59 V, corresponding to a photo-ionization threshold wavelength of 985 Å. Fig. 14.61 shows the absorption and photo-ionization coefficients and the photo-ionization yield in the wavelength region from this threshold down to 600 Å as observed by Metzger and Cook.[||] The absence of any band structure is noteworthy. These results agree quite well with those of Wainfan *et al.*[††]

† AL-JOBOURY, M. I., MAY, D. P., and TURNER, D. W., loc. cit., p. 1166.
‡ FROST, D. C., McDOWELL, C. A., and VROOM, D. A., loc. cit., p. 1157.
§ WACKS, M. E., loc. cit., p. 1158.
‖ METZGER, P. H., and COOK, G. R., *J. chem. Phys.* **41** (1964) 642.
†† WAINFAN, N., WALKER, W. C., and WEISSLER, G. L., *Phys. Rev.* **99** (1955) 542.

but not with those of Astoin,† which show many bands below 1000 Å.

Photoelectron spectra have been observed by Al-Joboury and Turner‡ and by Blake,§ using cylindrical analysers. The former authors found, for 584 Å radiation, three peaks at 12·61, 14·23, and 18·02 eV, the first corresponding to the spectroscopically determined ionization potential referred to above. Blake found the first of these peaks at

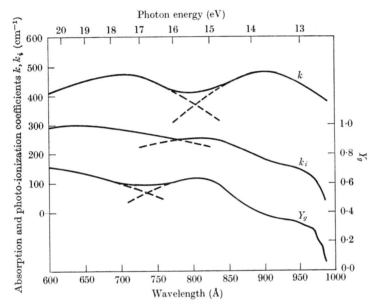

Fig. 14.61. Absorption and photo-ionization coefficients k and k_i and the photo-ionization yield (Y_g) for water vapour in the wavelength range 1000–600 Å, as observed by Metzger and Cook.

12·7 eV and the second at 14·9 eV, and two further maxima not well resolved at 18·6 and 20·7 eV. These are to be compared with the values 12·60, 14·35, and 16·34 eV observed by Frost and McDowell‖ in electron impact experiments (see also Chap. 13, § 7.2).

8.9. *Ammonia*

Fig. 14.62 shows the absorption and photo-ionization coefficients in the wavelength range 1250 to 1070 Å observed by Watanabe.†† The

† Astoin, N. *C.r. hebd. Séanc. Acad. Sci. Paris*, **242** (1956) 2327.
‡ Al-Joboury, M. I. and Turner, D. W., *J. chem. Soc.* (1964) 4434.
§ Blake, A. J., Thesis, Adelaide, 1965.
‖ Frost, D. C. and McDowell, C. A., *Can. J. Chem.* **36** (1958) 39.
†† Watanabe, K., *J. chem. Phys.* **22** (1954) 1564.

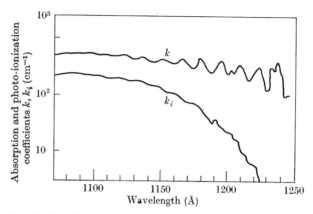

FIG. 14.62. Absorption and photo-ionization coefficients for ammonia, in the wavelength range 1250–1070 Å as observed by Watanabe.

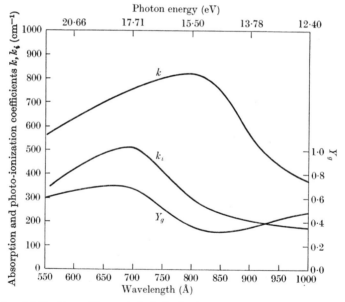

FIG. 14.63. Absorption and photo-ionization coefficients k and k_i and the photo-ionization yield (Y_g) for ammonia, in the wavelength range 1000–550 Å as observed by Metzger and Cook.

first ionization potential is at 10·16 eV (or 1220 Å) and it will be seen that the band structure apparent at the longer wavelengths becomes more diffuse towards shorter wavelengths as the ionization threshold is passed. Below 1000 Å no evidence of band structure is discernible as may be seen by reference to Fig. 14.63, which shows the absorption coefficients in the range 1000–550 Å, as well as the photo-ionization yield, observed by Metzger and Cook.†

Two ionization potentials have been observed by photoelectron spectroscopy. According to Al-Joboury and Turner‡ these occur at 10·16 and 15·02 V, while Vroom,§ using a spherical analyser, finds

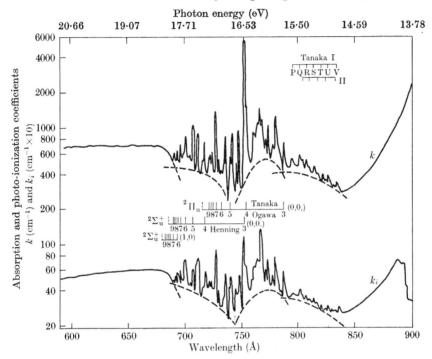

Fig. 14.64. Absorption and photo-ionization coefficients k and k_i for carbon dioxide in the wavelength range 900–600 Å, as observed by Cook et al. The positions of lines of observed Rydberg series are indicated for comparison.

them at 10·35±0·01 and 14·95±0·02 V. These are associated with two different electronic states of NH_3^+ and Vroom finds that, after ionization by 584 Å radiation, it is seven times more likely that the NH_3^+ will be in the upper than in the lower (ground state). This may be connected with a difference in the equilibrium geometrical configuration of the ground states of NH_3 and NH_3^+.

8.10. *Carbon dioxide*

Fig. 14.64 shows the absorption and photo-ionization coefficients for the wavelength range from the final ionization limit at 900 Å down to

† loc. cit. p. 1172.
‡ AL-JOBOURY, M. I. and TURNER, D. W., *J. chem. Soc.* (1964) 4434.
§ VROOM, D. A., Thesis, Vancouver, 1967.

600 Å observed by Cook *et al.*† Although from the ionization limit to about 840 Å no band structure is observed, a weak band structure is superposed on the continuum from 840 to 790 Å, while from 790 to 690 Å there is very strong band absorption. At the maximum in this region at 752 Å the photo-ionization yield is only 21·5 per cent.

From observed photoelectron spectra due to ionization by 584 Å radiation Al-Joboury, May, and Turner‡ find four ionization thresholds at 13·68, 17·23, 18·08, and 19·29 eV.

TABLE 14.27

Ionization potentials (V) for N_2O and NO_2

N_2O							
Al-Joboury and Turner§	12·82	16·37	17·67	20·10			
Vroom‖	12·90	16·41	17·74	20·10			
Spectroscopic††	12·89	16·39		20·10			
NO_2							
Al-Joboury and Turner§	10·97	12·82	13·48	14·01	14·37	16·79	18·86
Vroom‖	10·91	12·92	13·64	14·14	14·59	17·31	18·90
Spectroscopic		11·62 } ‡‡ 12·3 } §§					18·87‖‖

8.11. *Other applications of photoelectron spectroscopy*

The powerful technique of photoelectron spectroscopy has been applied to many other molecules. Al-Joboury and Turner§ have published a table of ionization potentials for forty-eight molecules including those we have discussed above. This table includes the further molecules with less than six atoms—H_2S, CS_2, SO_2, NO_2, N_2O, and CH_4.

Using the spherical analyser, Vroom‖ has studied NO_2 and N_2O, the hydrogen halides, and the diatomic homonuclear halogen molecules. For N_2O and NO_2 the ionization potentials found in these two different investigations are compared with each other and with spectroscopic data in Table 14.27. There is very good agreement between the two sets of results obtained by photoelectron spectroscopy for each molecule. For N_2O the spectroscopic data also agree very well for the three thresholds that have been observed in this way. There has been general

† COOK, G. R., METZGER, P. H., and OGAWA, M., *J. chem. Phys.* **44** (1966) 2935.

‡ AL-JOBOURY, M. I., MAY, D. P., and TURNER, D. W., *J. chem. Soc.* (1965) 6350.

§ AL-JOBOURY, M. I. and TURNER, D. W., loc. cit., p. 1173.

‖ VROOM, D. A., loc. cit., p. 1175.

†† TANAKA, Y., JURSA, A. S., and LE BLANC, F. J., *J. chem. Phys.* **32** (1960) 1205.

‡‡ NAKAYAMA, T., KITAMURA, M. Y., and WATANABE, K., ibid. **30** (1959) 1180.

§§ PRICE, W. C. and SIMPSON, D. M., *Trans. Faraday Soc.* **37** (1941) 106 .

‖‖ TANAKA, Y. and JURSA, A. S., *J. chem. Phys.* **36** (1962) 2493.

difficulty in obtaining spectroscopic data for NO_2 and the photoelectron technique has proved particularly effective in this case.

Vroom[†] has also determined the relative transition probabilities to different final ionic states in N_2O^+ and NO_2^+ corresponding to the different ionization potentials, when the ionizing radiation is of wavelength 584 Å.

The ground state of the hydrogen halide ions is a $^2\Pi_u$ state split by spin-orbit coupling into a $^2\Pi_{\frac{3}{2}}$, $^2\Pi_{\frac{1}{2}}$ doublet. Vroom[†] has been able to measure the splitting of this doublet for HCl^+, HBr^+, and HI^+ in much the same way as the corresponding splitting of the $^2P_{\frac{3}{2}}$, $^2P_{\frac{1}{2}}$ doublet in the rare gas ions (see § 5.5.4.1 and Figs. 14.17 and 14.18). Good agreement was obtained with spectroscopic determinations of the splitting, even for HCl^+ for which it is only 0·08 eV. The first excited electron configuration is a $^2\Sigma^+$ state and the corresponding ionization thresholds were determined for all four hydrogen halides. Relative probabilities for ionization by 584 Å radiation, leaving the ions in the $^2\Pi_{\frac{3}{2}}$, $^2\Pi_{\frac{1}{2}}$, and $^2\Sigma^+$ states, were also obtained.

Extension of this work by Vroom[†] to such chemically active substances as the halogens, particularly fluorine, required special methods for introducing the halogen vapour into the photo-ionization chamber. Thus fluorine was introduced from a stainless steel container through a stainless steel inlet system including an Edwards High Vacuum stainless steel needle valve.

The halogen ions X_2^+ may be found in the $^2\Pi_g$, $^2\Pi_u$, and $^2\Sigma_g^+$ states in that order of excitation. Both the $^2\Pi_g$ and $^2\Pi_u$ states are split into doublets by spin-orbit interaction. Of these the $^2\Pi_{\frac{3}{2},g} - ^2\Pi_{\frac{1}{2},g}$ splitting was measured for Br_2^+ and I_2^+ and the $^2\Pi_{\frac{3}{2},u} - 2\Pi_{\frac{1}{2},u}$ for I_2^+, the results agreeing well with values derived from electron impact observations. Relative transition probabilities (for 584 Å radiation) to the $^2\Pi_g$, $^2\Pi_u$, and $^2\Sigma^+$ states of all four ions were measured and, for Br_2^+ and I_2^+, to each member of a resolved doublet.

8.12. *Polar photodissociation of bromine and iodine*

Morrison, Hurzeler, Inghram, and Stanton[‡] have applied the technique referred to in § 5.5.5 to investigate the nature and yield of ions produced by photo-ionization of Br_2 and I_2. Fig. 14.65 shows the results obtained for both molecules. In each case the atomic ions Br^+, I^+ are produced at a lower threshold energy than the diatomic, Br_2^+, I_2^+.

[†] loc. cit., p. 1175.

[‡] MORRISON, J. D., HURZELER, H., INGHRAM, M. G., and STANTON, H. E., *J. chem. Phys.* **33** (1960) 821.

This can be ascribed to their production by polar dissociation in which a positive and a negative ion are produced,

$$Br_2 + h\nu \rightarrow Br^- + Br^+.$$

The shape of the yield curve for the atomic ions, consisting primarily of a relatively sharp peak, indicates that the polar dissociation process

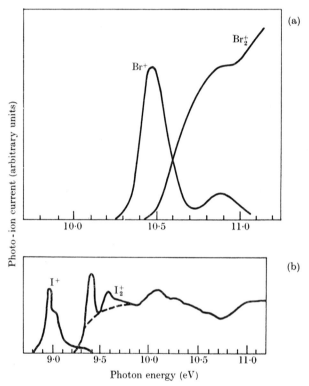

Fig. 14.65. Observed variation with photon energy of the rate of production of atomic and molecular ions of bromine and iodine by photo-ionization. (a) Bromine, (b) iodine.

proceeds through formation of an intermediate complex, which subsequently breaks up by polar auto-ionization so that we have

$$Br_2 + h\nu \rightarrow Br_2^* \rightarrow Br^- + Br^+.$$

The peak for Br^+ production occurs at 10.48 ± 0.02 eV, which represents the energy required to produce a virtual transition from the equilibrium position in the ground state to the potential energy curve in the excited state which dissociates into Br^- and Br^+. As the energy spread of the photon beam is only about 0.04 eV the tailing of the Br^+

peak on the low energy side must be ascribed to other transitions within the Franck–Condon region, which extend down to 10·28 eV at least. If the Franck–Condon region intersects the dissociation limit for $Br^+ + Br^-$ (see, for example, Fig. 12.4), then the threshold at 10·28 eV will be associated with production of Br^+ ions with negligible kinetic energy. That the Franck–Condon region does intersect the dissociation limit was established by a separate investigation in which the dissociation was produced by electron impact, the kinetic energy of the ions being measured by retarding potential analysis in a cylindrical collector surrounding the electron beam, and the nature of the ions established by means of a sector type mass analyser. The electron affinity $A(Br)$ of Br may now be derived in eV from the relation

$$A(Br) = I(Br) + D(Br_2) - 10·28,$$

where $I(Br)$ is the first ionization potential of Br (11·82 eV) and $D(Br_2)$ the dissociation energy (1·99 eV) of Br_2. This gives $A(Br) = 3·53$ eV, which agrees well with values obtained by other methods (see Chap. 15, Table 15.5).

The second smaller peak in the yield curve for Br^+ has a vertical excitation energy of 10·88±0·05 eV, 0·40 eV above that for the main reaction. This is very close to the energy difference (0·39 eV) between the $^3P_2$ and $^3P_1$ levels of Br^+.

A similar analysis carried out for I^+ gives an electron affinity

$$A(I) = 3·13 ± 0·12 \text{ eV},$$

which also agrees well with values obtained by other methods (Chap. 15, Table 15.5).

The yield curve for Br_2^+ is of the expected form for direct photoionization but that for I_2^+ shows peaks due to excitation of autoionizing levels.

9. Free–free transitions—theory of bremsstrahlung

9.1. *Introduction*

A completely free electron may not radiate or absorb energy as this would involve violation of the conservation of linear momentum. In the presence of another body, such as an atom or ion, which may take up momentum, emission or absorption of radiation may occur. Thus an electron may suffer an inelastic collision in which kinetic energy is transformed into radiant energy instead of internal motion of the atom or ion. Classically this process is one of emission of radiation due to the acceleration of the electron by the atomic field. It is often referred to as

bremsstrahlung. Absorption of light by an electron in the neighbourhood of an ion or neutral atom is usually referred to as free–free absorption although it only occurs when the electrons are not free. It is an
important mechanism of absorption in the atmospheres of stars. In
this chapter we discuss free–free transitions in the field of positive ions,
corresponding to transitions between continuum states of neutral atoms.
Transitions between such states of negative ions which may be regarded
as free–free transitions in the fields of neutral atoms will be considered
in Chapter 15.

9.2. *Theoretical formulae—non-relativistic*

In Chapter 7, § 5.2 we have summarized the formulae for the cross-
sections that specify the rates of these processes. Thus from (70) of
Chapter 7 we write $I_E^{E'} \, d\nu d\Omega$ for the cross-section for a process in which
an electron of energy E moving in the field of a positive ion of charge Ze,
radiates a photon with frequency between ν and $\nu + d\nu$ and is scattered
into the solid angle $d\Omega$ about the direction (Θ, Φ):

$$I_E^{E'} = \frac{64\pi^4 m^2 v' \nu^3 e^2}{3h^3 c^3 v} |\mathbf{r}_{EE'}|^2. \tag{81}$$

Here $E' = E - h\nu$, v and v' are the initial and final electron velocities
and

$$\mathbf{r}_{EE'} = \int \psi_E \mathbf{r} \psi_{E'} \, d\tau \tag{82}$$

where ψ_E, $\psi_{E'}$ are the initial and final electron wave functions normalized so as to have the asymptotic forms (see Chap. 7, p. 440, footnote and
Chap. 6, § 3.9)

$$\left. \begin{aligned}
\psi_E &\sim \exp i\{\mathbf{k}.\mathbf{r} + \alpha \ln(kr - \mathbf{k}.\mathbf{r})\} + r^{-1} f_E(\theta) \exp i(kr - \alpha \ln 2kr) \\
\psi_{E'} &\sim \exp[-i\{\mathbf{k'}.\mathbf{r} - \alpha' \ln(k'r + \mathbf{k'}.\mathbf{r})\}] + \\
&\qquad + r^{-1} f_{E'}(\pi - \theta') \exp i(k'r - \alpha' \ln 2k'r)
\end{aligned} \right\}, \tag{83}$$

where $\mathbf{k}$, $\mathbf{k'}$ are the wave vectors of the initial and final electron motions,

$$\cos \theta' = \cos \theta \cos \Theta + \sin \theta \sin \Theta \cos(\phi - \Phi),$$

and $\alpha = Ze^2 m / k\hbar^2$, $\alpha' = Ze^2 m / k'\hbar^2$.

If the process of bremsstrahlung is studied in terms of the intensity
of emission of radiation rather than that of inelastic electron scattering,
it is necessary to consider the differential cross-section in terms of the
angular distribution of the emitted photons. If we write

$$J_E^{E'}(\mathbf{s}; \Theta, \Phi; \vartheta, \chi) \, d\Omega d\omega d\nu \tag{84}$$

for the differential cross-section for a process in which a photon, of
frequency between ν and $\nu + d\nu$ and polarization in the direction of the
unit vector $\mathbf{s}$, is emitted in direction (ϑ, χ) within the solid angle $d\omega$

while the electron is scattered in the direction (Θ, Φ) within $d\Omega$, then

$$J_E^{E'} = \frac{16\pi^3 m^2 v' v^3 e^2}{h^3 c^3 v} |\mathbf{s} \cdot \mathbf{r}_{EE'}|^2. \tag{85}$$

Writing $\qquad \mathbf{r}_{EE'} = \mathbf{i}x_{EE'} + \mathbf{j}y_{EE'} + \mathbf{k}z_{EE'},$

where the $\mathbf{k}$-axis is taken in the direction of incidence of the electron and the $\mathbf{j}$-axis perpendicular to the plane of scattering, and summing over both polarizations for a final direction of emission of a photon of either polarization,

$$\sum_{\text{pol}} J_E^{E'} = \frac{16\pi^3 m^2 v' v^3 e^2}{h^3 c^3 v} \{2|z_{EE'}|^2 \sin^2\vartheta + (|x_{EE'}|^2 + |y_{EE'}|^2)(1+\cos^2\vartheta)\}. \tag{86}$$

In many experiments such as that concerned with the study of continuous X-ray emission the differential cross-section required is

$$\int \sum_{\text{pol}} J_E^{E'} \, d\Omega. \tag{87}$$

This can be written

$$I(\vartheta, \nu) = \{I_z \sin^2\vartheta + \tfrac{1}{2}(I_x + I_y)(1+\cos^2\vartheta)\}, \tag{88}$$

where $\qquad I_z = \dfrac{32\pi^3 \nu^3 m^2 v' e^2}{h^3 c^3 v} \int |z_{EE'}|^2 \, d\Omega. \tag{89}$

The cross-section for free–free absorption when the number of electrons present with energy between E' and $E'+dE'$ is $n(E') \, dE'$, is given, according to (71) of Chapter 7, by

$$q_E^{E'} \, dE' = \{8\pi^3 m^2 v e^2 \nu \, n(E')/3ch^3\} \int |\mathbf{r}_{EE'}|^2 \, d\omega dE'. \tag{90}$$

In the most important applications to absorption in stellar atmospheres the electrons have a Maxwellian distribution of velocities and their concentration is determined by the Saha formula at the atmospheric temperature T.

On classical grounds the bremsstrahlung would be expected to be partially polarized. If an electron suffers a head-on collision with a target atom so that all the acceleration of the electron is parallel to its direction of motion, one would expect classically that the radiation emitted at right angles to the direction of incidence would be polarized in the incident direction. On the other hand, for very distant collisions, in which the main component of the acceleration is perpendicular to the direction of motion, it would be expected that the radiation emitted at right angles would be polarized with the electric vector perpendicular to the direction of incidence. For other collisions, in which the acceleration is neither parallel nor perpendicular to the incident direction, the polarization should be incomplete.

9.3. *Calculated intensity of emission of the continuous X-ray spectrum (neglecting relativity and electron spin)*

The formulae (81), (82), (86), (88) have been applied to the problem of emission of continuous X-radiation by several authors.† In most cases the screening by the electrons of the Coulomb field of the nucleus has been neglected. Under the conditions of most of the experiments this is certainly justified for the radiation near the high-frequency limit. In this case most of the radiative scattering will occur when the electrons are inside the K-shell. With this assumption ψ_E and $\psi_{E'}$ are the appropriate wave functions for electrons with positive energy in the Coulomb field of a nucleus of atomic number Z. The quantities $x_{EE'}$, $y_{EE'}$, $z_{EE'}$ have been evaluated, using these functions, by Sommerfeld and his collaborators.

An alternative approximate method, introduced by Sauter,‡ is to use for the functions ψ_E, $\psi_{E'}$ those given by Born's approximation (see Chap. 7), which represents the motion of the electron in the Coulomb field by undistorted plane waves. This method, which is valid when $2\pi Ze^2/hv \ll 1$,§ v being the velocity of the electron after the emission of bremsstrahlung, has the advantage that it provides a simpler expression for the radiant intensity and can be extended without difficulty to allow approximately for screening by the atomic electrons. It will not give accurate results near the high-frequency limit.

The effect of the retardation term omitted in obtaining (81) has been considered by Sauter.‡ It introduces a correction of the same order of magnitude as that due to relativity and has an important influence on the directional distribution of the emitted quanta.

The evaluation of the integrals (89) is a lengthy process since it has not been found possible to obtain a closed formula for $|z_{EE'}|^2$. However, these integrations have been carried out numerically in a number of cases from Sommerfeld's data for $x_{EE'}$, $y_{EE'}$, and $z_{EE'}$ by Elwert,‖ Weinstock,†† and Kirkpatrick and Wiedmann.‡‡

It is found that the quantities $I_{x,y,z}$, which determine the total intensity emitted per unit frequency range in the X-ray spectrum, are functions of V_0/Z^2 and v/v_0 only, where $V_0 e$ is the electron energy in e.s.u., so long

---

† SOMMERFELD, A., *Annln. Phys.* **11** (1931) 257; SCHERZER, O., ibid. **13** (1932) 137; SOMMERFELD, A. and MAUE, A. W., ibid. **23** (1935) 589; ELWERT, G., ibid. **34** (1939) 178; WEINSTOCK, R., *Phys. Rev.* **61** (1942) 584; KIRKPATRICK, P. and WIEDMANN, L., ibid. **67** (1945) 321.

‡ SAUTER, F., *Annln. Phys.* **18** (1933) 486; **20** (1934) 404.

§ See MOTT, N. F. and MASSEY, H. S. W., *The theory of atomic collisions*, 3rd edn, p. 111 (Clarendon Press, Oxford, 1965).

‖ loc. cit.   †† loc. cit.   ‡‡ loc. cit.

as only the scattering by a true Coulomb field is considered and no account taken of screening. If screening is allowed for by Sauter's method, it is found that Z occurs in combinations other than the ratio V_0/Z^2, but this is only of importance in the long-wave limit of the spectrum.

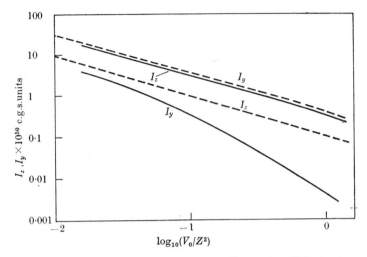

FIG. 14.66. I_y, I_z calculated as functions of $\log_{10}(V_0/Z^2)$, where V_0 is the electron accelerating potential in e.s.u. The units of I_y, I_z are ergs per steradian per unit frequency interval per bombarding electron per atom per cm² of target. —— I_z, I_y at short-wave limit; ———— I_z, I_y at long-wave limit.

Using the exact Sommerfeld formulae, Kirkpatrick and Wiedmann† calculated $I_{x,y,z}$ for a large number of values of V_0/Z^2 and ν/ν_0. Fig. 14.66 shows the values of I_z and I_y $(= I_x)$ at the extreme short-wave limit of the spectrum, $\nu/\nu_0 = 1$, as a function of V_0/Z^2. Similarly, this figure shows also the same quantities calculated at the extreme long-wave limit, $\nu/\nu_0 = 0$, using Sauter's method to allow for the effect of screening. From the exact calculations Kirkpatrick showed that, to a rough approximation, I_y and I_z for constant ν/ν_0 are inversely proportional to V_0/Z^2, this proportionality being much better in the case of I_z than of I_y. I_z is nearly constant for a large range of ν/ν_0 except near the long-wave limit. I_y decreases with increase of ν/ν_0.

Using these values of I_y, I_z, Fig. 14.67 shows the total intensity in ergs per steradian per unit frequency interval per bombarding electron per atom per cm², emitted at 90° to the direction of bombardment, as a

† loc. cit., p. 1182.

function of ν/ν_0 for a number of values of V_0/Z^2. As pointed out above, at small values of ν/ν_0, and a given value of V_0/Z^2, I_y and I_z depend upon Z. For $V_0/Z^2 = 0\cdot1359$, the lower and upper curves refer to values of Z of 8 and 25 respectively. For $V_0/Z^2 = 0\cdot00985$, the corresponding values of Z are 58 and 92.

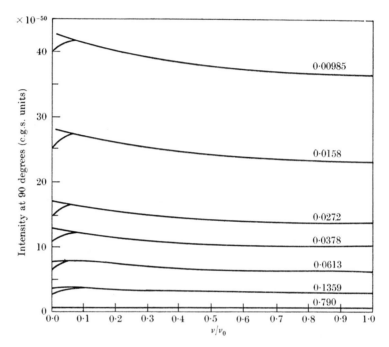

FIG. 14.67. Intensity emitted at right angles to the incident electron beam as a function of ν/ν_0. The numbers on the curves refer to different values of V_0/Z^2. The units of intensity are as in Fig. 14.66. The two parts of the curve at low frequencies refer to different values of Z. The lower and upper curves refer to values of Z of 16 and 47 respectively for $V_0/Z^2 = 0\cdot1359$, of 58 and 92 for $V_0/Z^2 = 0\cdot00985$, and to intermediate ranges of Z for the other curves.

The degree of polarization, $P(\vartheta)$, of the bremsstrahlung emitted at an angle ϑ to the electron direction is defined by the relation

$$P(\vartheta) = \frac{I_\perp - I_\parallel}{I_\perp + I_\parallel}, \tag{91}$$

where $I_\perp$ is the intensity of the component of the radiation emitted at angle ϑ with its electric vector perpendicular to the emission plane, $I_\parallel$ the intensity of the component with its electric vector in the emission plane.

The degree of polarization expected can also be obtained from Fig. 14.66 using (91). The predicted polarization is similar to that expected from classical theory. Thus I_z, the component along the direction of incidence, is much more important than I_y at the high-frequency limit where, on the classical picture, one would expect complete polarization, while I_y predominates at the low-frequency limit. According to the quantal calculations the polarization is not complete at either limit.

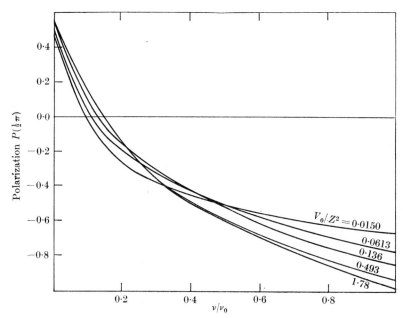

FIG. 14.68. Calculated degree of polarization $P(\tfrac{1}{2}\pi)$ of radiation emitted at 90° to the incident electron beam as a function of ν/ν_0 for a number of values of V_0/Z^2.

Calculated values of $P(\tfrac{1}{2}\pi)$ as a function of ν/ν_0 are represented in Fig. 14.68 for a wide variety of values of V_0/Z^2. At the high-frequency limit the polarization is the more complete the larger the value of V_0/Z^2. Scherzer† investigated the limiting case $V_0/Z^2 = 0$ and showed that in this case $|P(\tfrac{1}{2}\pi)|$ should be 0·6 for the entire spectrum. The calculations of Kirkpatrick and Wiedmann seem to be approaching this limiting condition as $V_0/Z^2 \to 0$.

9.4. *Theory of bremsstrahlung emission, including spin-relativity effects*

The Sommerfeld non-relativistic theory takes no account of electron spin. When this is included the expressions for bremsstrahlung cross-

† SCHERZER, O., *Annln Phys.* **13** (1932) 137.

sections become very complicated. The state of polarization of the electrons before and after scattering and of the emitted radiation, including both linear and circular polarization, have to be considered. By representing the electrons by Dirac wave functions relativistic effects are included at the same time as spin is taken into account.

The cross-section for bremsstrahlung summed over photon polarization has been treated by Bethe and Heitler,[†] using Born's approximation. This is equivalent to using the first term in an expansion in powers of $Z/137\beta$, where βc is the velocity of the electron after scattering. The expansion breaks down at the high-frequency end of the bremsstrahlung spectrum, corresponding to the scattered electron emerging with a small velocity. Fano,[‡] following a method used by Sauter[§] for calculating cross-sections for the K-shell photoelectric effect, which involves an expansion in powers of $Z/137$ and $Z/137\beta_0$, where $\beta_0 c$ is the velocity of the incident electron, was able to show that the cross-section does not tend to zero at the high-frequency limit but has a finite value, proportional to Z^3. The polarization of the bremsstrahlung has been calculated to the same approximation by Gluckstern, Hull, and Breit.[||] These calculations are valid under the condition $Ze^2/hv \ll 1$, so that they break down for large Z and small v.

Fig. 14.69 shows the polarization, $P(\vartheta)$, of the bremsstrahlung calculated for incident electron energies of 0·1 and 0·5 MeV for a range of angles without screening. Comparison with the non-relativistic calculations illustrated in Fig. 14.68 for $\vartheta = \frac{1}{2}\pi$ shows reasonable agreement at the lower energy except that the degree of polarization at long wavelengths is smaller when spin-relativity effects are included.

The situation is more complicated when the incident electron beam is polarized. Just as elastic scattering through a given angle of polarized electrons leads to an azimuthal asymmetry of the scattered electrons, so the bremsstrahlung emission would be expected to exhibit a similar asymmetry. Approximate calculations of the asymmetry have been made by Johnson and Rozics[††] and by Sobolak and Stehle.[‡‡] Since to lowest order the asymmetry is proportional to αZ it was necessary to compute the bremsstrahlung cross-section (or at least those parts of it containing the electron spin) to one higher order in αZ than that used

† BETHE, H. and HEITLER, W., *Proc. R. Soc.* A**146** (1934) 83.
‡ FANO, U., *Phys. Rev.* **116** (1959) 1156.
§ SAUTER, F., *Annln Phys.* **9** (1931) 217; **11** (1931) 454.
|| GLUCKSTERN, R. L., HULL, M. H., and BREIT, G., *Phys. Rev.* **90** (1953) 1026; GLUCKSTERN, R. L. and HULL, M. H., ibid. 1030.
†† JOHNSON, W. R. and ROZICS, J. D., ibid. **128** (1962) 192.
‡‡ SOBOLAK, E. S. and STEHLE, P., ibid. **129** (1963) 403.

to derive the Bethe–Heitler formula. The terms involving the electron spin are found to contribute to the cross-section the term

$$\mathscr{I}_E^{E'}(\omega, \nu, \boldsymbol{\zeta})\, \mathrm{d}\omega\mathrm{d}\nu = -\mathbf{n}.\boldsymbol{\zeta}\frac{r_0^2\alpha^2Z^3}{\pi}\frac{p'}{p}\frac{\mathrm{d}\nu}{\nu}\,\mathrm{d}\omega\,A(\vartheta, \nu), \qquad (92\,\mathrm{a})$$

where $\boldsymbol{\zeta}$ is a unit vector in the direction of polarization of the incident electron and $\mathbf{n} = (\mathbf{k}\times\mathbf{p'})/|(\mathbf{k}\times\mathbf{p'})|$ is a unit vector normal to the photon

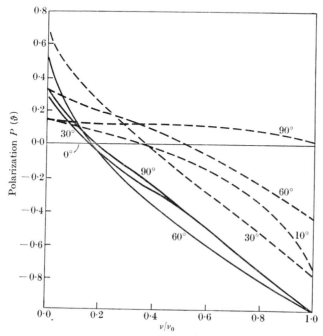

Fig. 14.69. Polarization $P(\vartheta)$ of the bremsstrahlung calculated for incident electron energies of 0·1 and 0·5 MeV for a range of angles (without screening) as a function of ν/ν_0. ——— 0·1 MeV, ———— 0·5 MeV.

production plane, $\mathbf{k}$ being the momentum of the photon and $\mathbf{p}$, $\mathbf{p'}$ those of the incident and scattered electrons. $\mathrm{d}\omega$ is the element of solid angle in the direction (ϑ, χ) of emission of the photon and r_0 $(= e^2/mc^2)$ is the classical electron radius. The Bethe–Heitler cross-section for photon emission in this direction can be written

$$I_E^{E'}(\omega, \nu)\, \mathrm{d}\omega\mathrm{d}\nu = \frac{r_0^2\alpha Z^2}{\pi}\frac{p'}{p}\frac{\mathrm{d}\nu}{\nu}\,\mathrm{d}\omega\,B(\vartheta, \nu), \qquad (92\,\mathrm{b})$$

so that the azimuthal asymmetry, $A_\mathrm{s}(\vartheta, \nu)$, can be written

$$A_\mathrm{s}(\vartheta, \nu) = -\mathbf{n}.\boldsymbol{\zeta}\alpha Z A(\vartheta, \nu)/B(\vartheta, \nu) \qquad (93)$$
$$= -\mathbf{n}.\boldsymbol{\zeta}\, s(\vartheta, \nu).$$

Fig. 14.70 shows the ratio $100A(\vartheta, \nu)/B(\vartheta, \nu)$ obtained by Johnson and Rozics for an electron velocity βc, where $\beta = 0.7$. Rozics and Johnson†
have calculated the bremsstrahlung cross-section exactly for electrons
of total energy $W = 1.25mc^2$ ($E = 128$ keV, $\beta = 0.75$), photon frequency

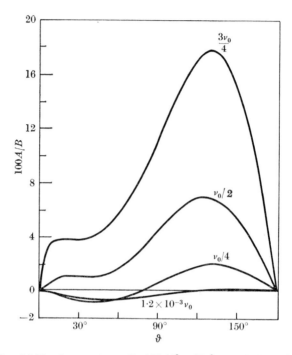

FIG. 14.70. Asymmetry ratio $100A(\vartheta, \nu)/B(\vartheta, \nu)$ calculated by
Johnson and Rozics for an electron velocity βc, where $\beta = 0.7$.
The value of ν is given in terms of ν_0 on each curve.

$\nu/\nu_0 = 0.75$ incident on gold ($Z = 79$). Fig. 14.71 shows the spin independent, $B(\vartheta)$, and spin dependent, $A(\vartheta)$, part of the bremsstrahlung
cross-section in this case as well as the ratio

$$\rho(\vartheta) = A(\vartheta)/B(\vartheta). \tag{94}$$

Comparison with Fig. 14.70 shows that Born's approximation can give
order of magnitude results only.

The polarization of the bremsstrahlung produced by polarized electrons has been discussed by several authors.‡ The polarization is

† Rozics, J. D. and Johnson, W. R., *Phys. Rev.* **135** (1964) B56.
‡ McVoy, K. W., ibid. **106** (1957) 828; McVoy, K. W. and Dyson, F. J., ibid. **106**
(1957) 1360; Claesson, A., *Ark. Fys.* **12** (1957) 569; Böbel, G., *Nuovo Cim.* **6** (1957)

expressed by specifying two components of the radiation field at right angles and their phase difference. In general the polarization is elliptical. If, however, the incident electron beam is polarized longitudinally the bremsstrahlung emitted will be circularly polarized. Studies of the helicity of electrons emitted in beta decay have increased the interest in circularly polarized bremsstrahlung and means of detecting it in recent years. McVoy† pointed out that a longitudinally polarized electron has a high probability of radiating a circularly polarized photon with the same helicity in the forward direction.

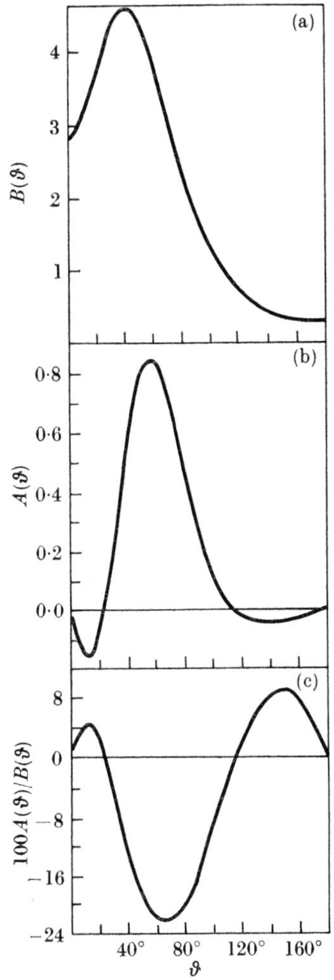

A detailed interpretation of the contributions of orbital and spin currents to the intensity and polarization of the bremsstrahlung produced has been made by Fano, McVoy, and Albers‡ in relation to a Born approximation calculation. They demonstrated the following characteristics.

(1) The contributions from spin currents vanish at low photon energies but generally predominate at high photon energies. They tend to contribute to the emission of photons in the higher energy part of the spectrum, making above average angles with the directions either of the incident or scattered electrons.

(2) The contributions from orbital currents yield bremsstrahlung plane polarized in either of the planes containing the photon direction and the incident or scattered electron direction. (Owing to

Fig. 14.71. (a) Spin-independent function $B(\vartheta)$, (b) spin-dependent function $A(\vartheta)$, (c) asymmetry ratio $100A(\vartheta, \nu)/B(\vartheta, \nu)$ for electrons with total energy $W = 1.25mc^2$ ($E = 128$ keV), as calculated by Rozics and Johnson.

nuclear recoil these planes are not the same.) The spin currents contri-

1241; FRONSDAL, C. and ÜBERALL, H., *Phys. Rev.* **111** (1958) 580; BANERJI, H., *Proc. natn. Inst. Sci. India* **26A** (1960) 502.

† McVoy, K. W., *Phys. Rev.* **106** (1957) 828.
‡ FANO, U., McVoy, K. W., and ALBERS, J. R., ibid. **116** (1959) 1159.

bute bremsstrahlung that is circularly polarized. Unpolarized electrons cannot emit circularly polarized bremsstrahlung.†

(3) For the case where the electron after scattering is moving parallel to the incident electron direction, Born's approximation predicts the vanishing of bremsstrahlung in this direction.

The most extensive numerical calculations of the polarization of bremsstrahlung from polarized electrons are those of Fronsdal and Überall‡ for incident electrons of kinetic energy 0·5 MeV and 2·5 MeV, using Born's approximation. Fig. 14.72 shows the results of their calculations for 2·5-MeV electrons. If $I_\perp$, $I_\parallel$, I_c are respectively the differential cross-sections for bremsstrahlung polarized perpendicular or parallel to the production plane or for circularly polarized bremsstrahlung, corresponding to a given angular and energy range of bremsstrahlung emission, the linear (P_l) and circular (P_c) polarization of the bremsstrahlung are defined by the relations

$$P_l = \frac{I_\perp - I_\parallel}{I_{\mathrm{BH}}}, \qquad (95\,\mathrm{a})$$

$$P_c = \frac{I_c}{I_{\mathrm{BH}}}, \qquad (95\,\mathrm{b})$$

where I_{BH} is the corresponding differential cross-section obtained using the Bethe–Heitler formula. The polarization of the incident electron beams is specified by the spin vector $\boldsymbol{\zeta}$,§ with longitudinal and transverse components ζ_l, ζ_t respectively. ψ is the angle between the direction of the transverse component and the normal to the bremsstrahlung production plane.

The cross-section I_c is further broken down into two parts, viz. $I_c = I_{ct} + I_{cl}$, corresponding to contributions to the cross-section for the emission of circularly polarized bremsstrahlung arising from the transverse and longitudinal components of the incident electron polarization, so that

$$P_{ct} = I_{ct}/I_{\mathrm{BH}}, \qquad P_{cl} = I_{cl}/I_{\mathrm{BH}}. \qquad (95\,\mathrm{c})$$

The circular polarization is seen to arise predominantly from the longitudinal component of the incident electrons, to be largest at the high energy limit of the spectrum, and not to depend greatly on the angle of emission. The linear polarization, on the other hand, is greatest at

---

† This was first pointed out by GLUCKSTERN, R. L., HULL, M. H., and BREIT, G., *Phys. Rev.* **90** (1953) 1026.

‡ FRONSDAL, C. and ÜBERALL, H., *Phys. Rev.* **111** (1958) 580.

§ This is the polarization vector in the laboratory system. In the rest system of the electron it transforms to $\boldsymbol{\zeta}^0 = \boldsymbol{\zeta} - \mathbf{p}\boldsymbol{\zeta}.\mathbf{p}/E(E+mc^2)$, E, $\mathbf{p}$ being the total energy and momentum of the incident electron, respectively.

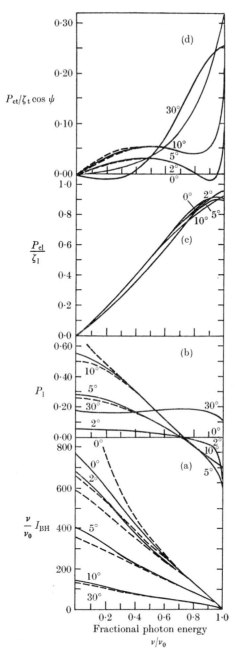

FIG. 14.72. Properties of the bremsstrahlung emitted at various angles ζ to the incident direction of electrons of initial kinetic energy $5mc^2$ (2.5 MeV) as functions of the fractional photon energy $h\nu/5mc^2$. (a) $(\nu/\nu_0)I_{BH}$, where I_{BH} is the differential cross-section per unit solid angle, per unit fractional photon energy range in units $Z^2 \times 1.844 \times 10^{-28}$ cm². (b) The linear polarization P_l. (c) The circular polarization P_{cl} for incident electrons 100 per cent longitudinally polarized. (d) The circular polarization P_{ct} for incident electrons 100 per cent transversely polarized.

the low energy end of the spectrum and is markedly dependent on emission angle.

Kresnin† has calculated the change of polarization of the incident electron beam following bremsstrahlung emission, using Born's approximation. In the non-relativistic limit the polarization of the beam does not change. For the emission of very low energy bremsstrahlung quanta he found that the spin polarization vector in the rest system would be rotated through an angle ϕ around the normal to the plane of scattering, given by

$$\tan \phi = \frac{(\gamma-1)\{(\gamma+1)+(\gamma-1)\cos\Theta\}}{\{2\gamma+(\gamma^2-1)\cos\Theta+(\gamma-1)^2\cos^2\Theta\}} \sin\Theta, \qquad (96)$$

where $\gamma = E/mc^2$ and Θ is the angle of scattering.

Much less attention has been paid to the study of bremsstrahlung emission in electron–electron scattering. Classically, and for low energy encounters, the cross-section for this process is very small, since the two-electron system has a zero dipole moment. Quantum-mechanically, different investigators have obtained somewhat conflicting results‡ and there seem to be no experimental results available.

10. The experimental study of the emission of bremsstrahlung

10.1. *Introduction*

An adequate experimental study of the bremsstrahlung process would require the measurement, for longitudinally and transversely polarized electron beams, of fixed energy, the energy and angular distributions and the state of polarization of both the scattered electrons and the photons produced. Although the techniques are available for carrying out such a comprehensive study it has not yet been done and most attention has been concentrated on particular aspects important in other applications. We deal in § 13 with bremsstrahlung emission from ionized gases where bremsstrahlung studies can contribute to knowledge of electron concentrations in high temperature plasmas. Bremsstrahlung emission and the related continuous absorption by hot gases in which the electrons concerned are accelerated by the fields of neutral atoms are discussed in Chapter 15.

Bremsstrahlung emission is of particular importance in X-ray physics

† Kresnin, A. A., *Zh. éksp. teor. Fiz.* **37** (1959) 872 (*Soviet Phys. JETP* **10** (1960) 621).

‡ Katzenstein, J., *Phys. Rev.* **78** (1950) 161; Redhead, M. L. G., *Proc. phys. Soc.* A66 (1953) 196; Fedyushin, B. K., *Zh. éksp. teor. Fiz.* **22** (1952) 140; Garibyan, G. M., ibid. **24** (1953) 617. The position is reviewed by Stabler, R. C., *Nature, Lond.* **206** (1965) 922.

in relation to the emission of continuous X-radiation. Interest in studies of bremsstrahlung polarization has been stimulated in recent years as a means of studying the state of polarization of the electrons producing it.

The continuous spectrum of X-radiation is of great practical importance and its properties have been studied over a long period by many investigators. Most of the early measurements involved the use of thick targets, however, and although they may be of considerable practical importance, they shed little light on the fundamental process involved because of the complexity of the phenomena associated with the passage of the electron beam through the target material.

Measurements of the continuous X-radiation from a thin target have been made in recent years. Problems which have been the subject of investigation in such experiments are:

(a) the measurement of distribution of intensity in the continuous X-ray spectrum excited by homogeneous electrons;

(b) the variation with electron energy of the intensity of continuous X-radiation in a definite frequency range (thin target X-ray isochromat);

(c) the dependence of X-ray continuous spectrum intensity on atomic number of the target material;

(d) the degree of polarization of the emitted X-radiation and its variation with frequency and electron energy;

(e) the variation of intensity of the emitted radiation with direction of emission relative to the direction of the incident electron beam and the dependence of this directional distribution on frequency and electron energy.

10.2. *Experimental arrangements*

10.2.1. *Measurement of the continuous X-ray spectrum intensity.* The type of arrangement used for the study of the intensity of the continuous X-radiation from thin targets is shown in Fig. 14.73. This is the apparatus used by Harworth and Kirkpatrick† for thin targets of nickel.

A very similar arrangement has been used to study characteristic radiation from thin targets in the experiments of Webster and his colleagues (see Chap. 3, § 5.1).

The target T, consisting of a film of nickel about 200 Å thick formed by evaporation on a cellulose acetate backing foil, was mounted inside a hollow aluminium cylinder C, which screened the fragile target from

† HARWORTH, K. and KIRKPATRICK, P., *Phys. Rev.* **62** (1942) 334.

the destructive effects of bursts of gas in the apparatus. Electrons from
the cathode K were focused through a hole in the anode A and fell on
the target. By means of a small bias voltage applied between the filament
and shield of the cathode, the focal spot on the target could be maintained
about 1 cm in diameter for all accelerating potentials. The position of
the cathode was adjustable by means of the sylphon bellows B.

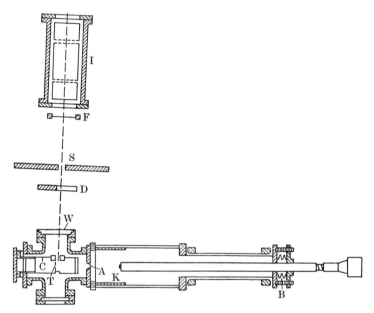

Fig. 14.73. Apparatus of Harworth and Kirkpatrick for studying the
X-ray continuous spectrum from thin targets.

The X-rays passed out through the window W and were collimated by
means of the slit S so that a beam was selected making an angle of 93·5°
with the direction of the cathode rays. The intensity of the beam was
measured by means of the ionization chamber I. The length of exposure
was defined by the clock-controlled shutter D. Before entering the argon-
filled ionization chamber the X-rays passed through filters F. First an
iron filter was used to absorb selectively the Ni K-radiation from the
target. In order to select a narrow frequency range of the continuous
spectrum of reasonable intensity, Ross's method of balanced filters was
used. The principle of this method has been described in Chapter 3,
§ 5.1. Balanced filters of Ag and Pd selected a small band of wavelengths
near 0·479 Å, while another pair of Se and As filters selected a narrow
band near 1·01 Å. After a series of measurements the effect of stray

radiation not originating at the targets was allowed for by repeating the measurements with the target removed. Under no conditions did this correction exceed 3 per cent of the thin target intensity.

Corrections were made for the finite thickness of the target using methods similar to those described in Chapter 3, § 5. The calculated path-length of electrons exceeded the target thickness by about 1 per cent at 50 kV and by about 8 per cent at 15 kV.

With this apparatus, relative measurements were made of the intensity of radiation in the bands defined by the Ross filters, for different electron energies. Absolute measurements of these intensities have been made with a similar apparatus by Smick and Kirkpatrick† for nickel and by Clark and Kelly‡ for aluminium.

A similar type of arrangement was used much earlier for similar measurements with aluminium by Kulenkampff,§ using as target an aluminium foil about 6000 Å thick. For a target of this thickness true thin target conditions are not obtained. In Kulenkampff's arrangement the angle between the emitted X-ray beam and the direction of the incident electrons could be varied in order to study the angular distribution of the emitted X-radiation.

More recently, Kulenkampff and his co-workers‖ have returned to the experimental study of the intensity of the continuous spectrum, using foils of thickness a few hundred Å or less satisfying requirements for true thin target conditions. The quantum energy of the X-radiation was measured, using a proportional counter consisting of a copper tube of wall thickness 0·3 cm and 31 cm long and of 10 cm internal diameter. The central wire of diameter 0·05 mm was surrounded at either end by a cylindrical shielding electrode at the same potential as the wire to enable the active length of the counter to be exactly defined. The counter was filled with argon gas at a pressure of 720 torr and alcohol at a vapour pressure of 20 torr. The X-radiation entered through a window in one of the end plates of the counter and passed along the counter parallel to the wire and about 2 cm from it. Pulses from the counter were analysed using a kick sorter.

Amrehn and Kulenkampff measured the continuous X-ray spectra emitted at right angles to the incident electron direction for incident electron energies of 25 keV and 34 keV incident on targets of C, Al, Ne,

† SMICK, E. and KIRKPATRICK, P., *Phys. Rev.* **60** (1941) 162.
‡ CLARK, J. C. and KELLY, H. R., ibid. **59** (1941) 220.
§ KULENKAMPFF, H., *Annln Phys.* **87** (1928) 597.
‖ AMREHN, H. and KULENKAMPFF, H., *Z. Phys.* **140** (1955) 452; AMREHN, H., ibid. **144** (1956) 529; KERSCHER, R. and KULENKAMPFF, H., ibid. **140** (1955) 632.

Ag, and Au. Kerscher and Kulenkampff studied the dependence of the X-ray spectrum from Al produced by 34-keV electrons on the emission angle. They used a special X-ray tube operated in a magnetic field so that the electrons before and after passage through the thin target move on a circular path of radius 5 cm. Their arrangement is shown in

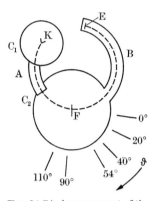

FIG. 14.74. Arrangement of the apparatus used by Kerscher and Kulenkampff to study dependence of X-ray spectrum from Al on emission angle for incident electron energy 34 keV.

Fig. 14.74. The vacuum chamber consisted of two brass cylinders C_1 and C_2, joined together by the tubes A and B. The cathode K was situated at the centre of C_1, the target foil F at the centre of C_2, while the current was measured at the collecting electrode E. The advantage of this arrangement is that it enables measurement of the emitted X-rays both at the 0° and 180° positions. A number of slits 5 mm high covered with aluminium foil 0·05 mm thick defined the angle of emission of the X-rays investigated. By placing the proportional counter behind these slits in turn it was possible to study the X-ray emission at a large range of angles between 0° and 180°.

Unfortunately, in these experiments the pulse length (1 ms) was too long at the maximum counting rates employed (80/s), so that the results were unreliable at the higher counting rates. Later work of Kulenkampff and Röss† is not open to this criticism.

Differential cross-sections for continuous X-ray emission in different directions from thin metal foils of Be, Al, and Au have been measured by Motz‡ (0·5- and 1·0-MeV electrons), by Starfelt and Koch§ (2·7- to 9·7-MeV electrons), and by Motz and Placious‖ (50-keV electrons). The bremsstrahlung spectrum was measured by means of a scintillation spectrometer consisting of a NaI crystal, photomultiplier, and a differential pulse height analyser able to detect photons with energy as low as 7 keV. With such an arrangement the energy resolution is not very good at low energies (37 per cent for 22-keV photons) but more satisfactory at higher energies. Great care was taken in estimating corrections for finite target thickness, spectrometer detection efficiency, window

† KULENKAMPFF, H. and RÖSS, D., *Z. Phys.* **161** (1961) 424.
‡ MOTZ, J. W., *Phys. Rev.* **100** (1955) 1560.
§ STARFELT, N. and KOCH, H. W., ibid. **102** (1956) 1598.
‖ MOTZ, J. W. and PLACIOUS, R. C., ibid. **109** (1958) 235.

absorption, etc., so that absolute cross-sections for bremsstrahlung emission could be obtained.

In recent years there has been some interest in the extreme long wavelength region of the continuous X-ray spectrum from solid targets. Peterson and Tomboulian[†] have studied the spectrum in the region 80 to 180 Å for electrons of energy from 0·6 to 3·0 keV incident on Mg, Al, Mn, Cu, Ge, and Ag. The targets consisted of freshly evaporated layers 400 Å thick on steel but since all the electrons stopped in the layer they constituted effective thick targets. The spectra were recorded photographically by means of a grazing incidence (4·64°) vacuum spectrograph with a concave glass grating (30 000 lines per in, 154-cm radius).

Arakawa *et al.*[‡] studied the region 2500–5500 Å by means of a vacuum ultra-violet spectrometer with an aluminized replica grating, a Glan prism for polarization measurements, and a quartz window photomultiplier detector. Electrons of energy in the range 25 to 100 keV incident on unbacked Au foils of thicknesses 340 and 530 Å

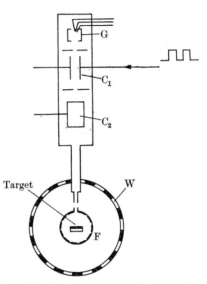

Fig. 14.75. Arrangement of apparatus used by Boersch *et al.* to study X-ray emission in different directions from silver and nickel targets.

were used to produce the X-rays. Absolute measurements of the light intensity emitted were made.

A similar wavelength range has been covered by Boersch *et al.*[§] for electrons in the energy range between 6 and 40 keV incident on silver and nickel targets freshly formed by evaporation. The targets were effectively thick targets (thickness 2–5 μm) but the experimental method employed was interesting. The arrangement is shown in Fig. 14.75. Electrons from the gun G, after acceleration through the required potential difference, passed through two sets of parallel condenser plates C_1, C_2 that could be used to deflect it in two directions at right

† PETERSON, T. J. and TOMBOULIAN, D. H., *Phys. Rev.* **125** (1962) 235.

‡ ARAKAWA, E. T., EMERSON, L. C., HAMMER, D. C., and BIRKHOFF, R. D., ibid. **131** (1963) 719.

§ BOERSCH, H., DOBBERSTEIN, P., FRITZSCHE, D., and SAUERBREY, G., *Z. Phys.* **187** (1965) 97.

angles before hitting the target. The radiation was analysed by means of a monochromator and photomultiplier. The target was surrounded by a cooled cylindrical cage, which protected it from deposition of pump oil and also served as a Faraday cage F to measure the incident electron current. Openings in the cylinder allowed the electrons to enter and the radiation to escape in the required direction. A series of quartz windows W in the outer cylinder of the vacuum system enabled the bremsstrahlung emission in different directions to be studied. The electron beam was pulsed at a frequency of 825 c/s by applying a deflecting potential of this frequency to C_1 and the photomultiplier currents were amplified by a narrow band amplifier of the same frequency locked to the potential applied to C_1. This enabled the very weak signals from the bremsstrahlung radiation to be distinguished from the background.

The measurement of the intensity variation with incident electron energy of bremsstrahlung of a fixed narrow frequency range (isochromats) is of interest. Many such measurements have been made in the low-frequency region where the interest is in the observation of structure near the threshold.†

The excitation of the nucleus ^{115}In by hard X-rays has been used by several investigators‡ as a detector of X-radiation of a definite frequency from a thin target using higher-energy incident electrons. Metastable indium decays to normal indium, with the emission of γ-rays and internal conversion electrons, with a half-life of 4·1 h. The first activation level from which it can be produced has an excitation energy of $1\cdot04\pm0\cdot02$ MeV and a width of only 4 mV. As a result it can be used to study X-rays in an extremely narrow frequency band at 0·0118 Å.

In the experimental arrangement of Miller and Waldman§ a foil of gold, thickness 0·02 mm, was placed on one side of a carbon disc thick enough to absorb electrons of about 2-MeV energy. On the other side was mounted a thin indium foil that was water-cooled. X-rays excited in the gold foil on the carbon disc activated the indium. To correct for the excitation by X-rays produced in the carbon a second experiment was carried out without the gold foil. The activity of the indium was measured by a Geiger counter. For electrons of energy above 1 MeV the gold foil acted as a true 'thin target'.

† OHLIN, P., *Ark. Mat. Astr. Fys.* A29 no. 3 (1943); A31 no. 9 (1945); BEARDEN, J. A. and SCHWARZ, G., *Phys. Rev.* **79** (1950) 674; SANDSTROM, R., *Ark. Fys.* **18** (1960) 305; CLAUS, H. and ULMER, K., *Phys. Lett.* **12** (1964) 170; ULMER, K. and VERNICKEL, H., *Z. Phys.* **153** (1958) 149.

‡ COLLINS, G. B. and WALDMAN, B., *Phys. Rev.* **59** (1941) 109; MILLER, W. C. and WALDMAN, B., ibid. **75** (1949) 425; KORSUNSKY, M. I., WALTHER, A. K., IVANOV, A. V., ZYPKIN, S. I., and GANENKO, V. E., *Fiz. Zh.* **7** (1943) 129. § loc. cit.

This method is adapted to the measurement of thin target X-ray isochromats for electrons in the energy range 1·04 to 1·42 MeV. At 1·42 MeV a second activation level of ^{115}In can be excited so that the measurements at higher energies no longer refer to a single wavelength band.

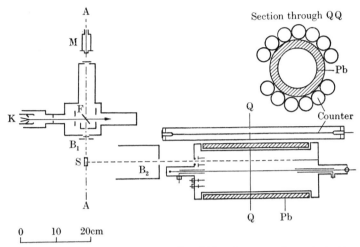

Fig. 14.76. Apparatus of Kulenkampff and Zinn for measuring polarization of radiation emitted from very thin foils of Al$_2$O$_3$ and Au by impact of electrons of energy 35 and 45 keV.

10.2.2. *Measurement of the polarization of the bremsstrahlung.* The degree of polarization of the emitted radiation has been studied by Piston[†] and Boardman[‡] using the method originally employed by Barkla[§] to demonstrate X-ray polarization. Kulenkampff and Zinn[||] have also used Barkla's method for measuring the polarization of the bremsstrahlung emitted from very thin foils of Al$_2$O$_3$ (51 μg cm^{-2}) and Au (26, 37, and 59 μg cm^{-2}), by electrons of energy 35 and 45 keV. Their apparatus is shown in Fig. 14.76. The Al$_2$O$_3$ foil was mounted on a ring, 3 cm in diameter and made of 0·2-mm diameter aluminium wire. The gold foils were formed by evaporation on a thin (1000 Å) plastic carrier foil. The foils were mounted at an angle of 45° to the direction of the electron beam from the cathode K, which struck the target in a spot of diameter 3 mm. The radiation from the plastic backing was quite negligible in comparison with that from the gold foil. X-radiation emitted at right angles to the direction of the electron beam passed

† Piston, D. S., *Phys. Rev.* **49** (1936) 275.
‡ Boardman, B. F., ibid. **60** (1941) 163.
§ Barkla, C. G., *Phil. Trans. R. Soc.* **204** (1905) 467.
|| Kulenkampff, H. and Zinn, W., *Z. Phys.* **161** (1961) 428.

through the collimating hole B_1 and fell on the beryllium scatterer S in the form of a disc 2 cm diameter and 0·5 cm thick. Scattered radiation from S collimated by the hole B_2 entered a proportional counter of the type used by Kulenkampff and Amrehn in their measurements of the bremsstrahlung spectrum (see § 10.2.1). The proportional counter was surrounded by a lead cylinder to shield out scattered radiation or soft cosmic radiation. Thirteen Geiger–Müller counters surrounding the lead cylinder were used to exclude the penetrating component of the cosmic radiation by requiring anticoincidence between these counters and the proportional counter. The X-ray intensity was monitored by means of another Geiger counter M. For radiation of the energy used in these experiments Thomson-scattering is predominant and the intensity of radiation scattered through 90° in the arrangement shown in the figure is proportional to the intensity, $I_\perp$, of the component polarized with its electric vector perpendicular to the electron beam. By rotating the proportional counter assembly through 90° about the axis AA, the intensity of the scattered radiation is proportional to the intensity $I_\parallel$ of the component polarized with its electric vector parallel to the electron beam. The degree of polarization P in this case is then given by

$$P = \frac{I_\perp - I_\parallel}{I_\perp + I_\parallel}. \quad \text{(97) (Compare with (95).)}$$

Motz and Placious† have used Compton scattering in two directions at right angles to measure the degree of polarization of bremsstrahlung produced by the impact of electrons of energy in the range 50 keV to 1 MeV on thin targets of Be, Al, and Au. The principle of their Compton polarimeter is shown in Fig. 14.77. Electrons incident on the thin target emit X-rays that are scattered from an anthracene crystal through an angle δ. A photomultiplier associated with the crystal enables the Compton electrons ejected in the scattering process to be detected and their energies measured. The scattered X-rays are detected by a sodium iodide crystal with associated photomultiplier. By rotating the detector through an angle ϕ about the line joining the target to the anthracene crystal the azimuthal distribution of the X-rays scattered through the angle δ can be determined. ($\phi = 0$ for the case where the centres of the target, anthracene crystal, and NaI crystal are coplanar.)

In order to select an energy interval out of the continuous spectrum the polarimeter was employed as a double-crystal Compton spectro-

† Motz, J. W., *Phys. Rev.* **104** (1956) 557; Motz, J. W. and Placious, R. C., ibid. **112** (1958) 1039; *Nuovo Cim.* **15** (1960) 571.

meter. Pulses from the anthracene crystal in a prescribed pulse-height interval were used to gate the coincidence circuit of the photomultiplier associated with the NaI crystal, thus ensuring that the events observed correspond to the production of Compton electrons with kinetic energy corresponding to the scattering of photons of the selected energy through the angle δ. For photons of energy k, polarized in the bremsstrahlung production plane, the ratio R of the expected scattered intensities at the

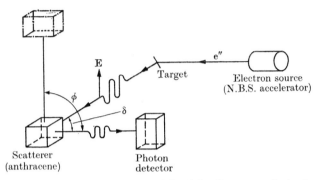

FIG. 14.77. Illustrating the principle of the Compton polarimeter used by Motz and Placious.

NaI crystal for the azimuths $\phi = \frac{1}{2}\pi$ and $\phi = 0$ was calculated using the Klein–Nishina formula.[†] If r is the measured value of this ratio, the degree of polarization P could be calculated from the expression

$$P = \left(\frac{R+1}{R-1}\right)\left(\frac{1-r}{1+r}\right). \qquad (98)$$

Kulenkampff, Leisegang, and Scheer[‡] have measured the degree of polarization of bremsstrahlung by studying the directional distribution of photoelectrons ejected by it from the K-shell of argon atoms. The bremsstrahlung produced by the impact of 34-keV electrons on a thin Al target (thickness 280 Å) at right angles to the electron beam, passed through a Wilson cloud chamber containing a mixture of hydrogen with 4 per cent argon. Photo-ionization events were selected in which the photoelectron was accompanied by a short stub at the origin, caused by an Auger electron of energy 3 keV produced by the radiationless refilling of the argon K-shell after ionization. The K-shell Auger yield for argon is 93 per cent so that most argon atoms that have undergone photo-ionization in the K-shell carry this short stub at their origin as a

[†] HEITLER, W., *The theory of radiation*, 3rd edn, p. 217 (Clarendon Press, Oxford, 1954).
[‡] KULENKAMPFF, H., LEISEGANG, S., and SCHEER, M., *Z. Phys.* **137** (1954) 435.

signature. Sommerfeld and Schur[†] have shown that according to non-relativistic theory, photoelectrons ejected from the K-shell have a $\cos^2\theta$ distribution, where θ is the angle between the direction of ejection and that of the electric vector of the radiation. This has been confirmed experimentally by Kirkpatrick.[‡] Kulenkampff *et al.* measured the photoelectron range and the angular distribution of the direction of initial motion of the photoelectrons in the selected events, projected on a plane perpendicular to the direction of the bremsstrahlung beam. If φ is the angle a photoelectron makes with the direction of maximum emission and if, for a given energy range of the bremsstrahlung as determined by the ranges of the ejected photoelectrons, the distribution can be expressed in the form

$$I(\varphi) = A + B\cos^2\varphi, \qquad (99\,\text{a})$$

the polarization is given by

$$P = B/(2A+B).\text{[§]} \qquad (99\,\text{b})$$

This method suffers from the disadvantage that a large effort is needed to obtain adequate statistics. About 1000 events were observed in the experiments of Kulenkampff *et al.* but since these were distributed over the whole range of bremsstrahlung energy the statistical accuracy was not high.

The existence of a spin-dependent part of the cross-section for Compton scattering has also been used by Goldhaber, Grodzins, and Sunyar[||] to study the circular polarization of bremsstrahlung produced by longitudinally polarized electrons. Their electrons were obtained from a radioactive source of $^{90}\text{Sr} + {}^{90}\text{Y}$ in equilibrium. High energy beta rays (maximum kinetic energy 2·24 MeV) were emitted from the decay of ^{90}Y. These gave rise to bremsstrahlung in the monel metal container in which the source was enclosed. The beta rays are expected to be partially longitudinally polarized. To demonstrate this they used an analyser, shown in Fig. 14.78, consisting of a cylindrical electromagnet that could be magnetized to saturation either parallel or antiparallel to the photon direction. If the bremsstrahlung is circularly polarized the Compton scattering cross-section in the iron, and thence the counting rate of the NaI counter placed below the magnet, will depend on the direction of polarization of the iron. If N_+, N_- are respectively the counting rates

† SOMMERFELD, A. and SCHUR, G., *Annln Phys.* **4** (1930) 409.

‡ KIRKPATRICK, P., *Phys. Rev.* **38** (1931) 1938.

§ KULENKAMPFF *et al.* actually calculated P from the ratio of numbers of events in fixed angular ranges (53°) about 0° and 180° and about 90° and 270° respectively.

|| GOLDHABER, M., GRODZINS, L., and SUNYAR, A. W., ibid. **106** (1957) 826.

with the field pointing up or down the quantity

$$\delta = \frac{N_- - N_+}{\frac{1}{2}(N_- + N_+)}, \tag{100}$$

gives a measure of the proportion of the bremsstrahlung circularly polarized.

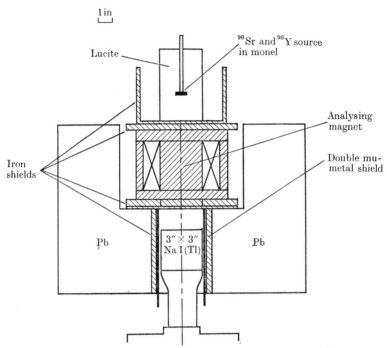

Fig. 14.78. Arrangement used by Goldhaber, Grodzins, and Sunyar for analysing the degree and sense of circular polarization of bremsstrahlung from a β^- source.

10.2.3. *Measurement of the angular distribution of electrons after radiative scattering.* Less attention has been given to the study of electrons scattered after emission of bremsstrahlung. Febel and Knop† have measured the angular distribution of 1-MeV electrons scattered in Al, Ag, and Au foils ($0\cdot1$–$2\cdot3$ mg/cm²) after emitting bremsstrahlung photons in the energy range between 20 and 100 keV. Their apparatus is shown in Fig. 14.79. Electrons from a 1-MeV van de Graaff generator were passed through a magnetic spectrometer which selected a band of electron momentum, $\Delta p/p = 2\cdot2 \times 10^{-3}$. The beam was incident on the scattering foil. Electrons scattered through a selected angle, θ, passed

† FEBEL, A. and KNOP, G., *Z. Phys.* **174** (1963) 257.

through the second magnetic analyser and those that had undergone a given energy loss fell on a plastic scintillator with associated photomultiplier. A Faraday cage set to collect electrons that passed through the foil without deviation enabled the incident current to be measured.

10.2.4. *Measurement of the asymmetry of bremsstrahlung produced by transversely polarized electrons.* The asymmetry of bremsstrahlung emitted by transversely polarized electrons of energy 300 keV incident on foils of Pb, Cd, Cu, and Zn (see § 9.4) has been investigated by

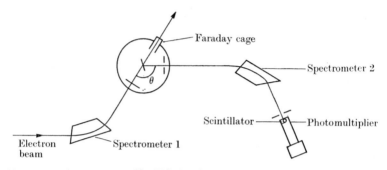

Fig. 14.79. Apparatus used by Febel and Knop for measurement of the angular distribution of 1-MeV electrons scattered in Al, Ag, and Au foils.

Güthner.† He used a radioactive ^{90}Sr+^{90}Y source of electrons, selecting mono-energetic electrons of energy 300±10 keV by deflexion through 120° in an electric field.

Fig. 14.80 illustrates the apparatus used. Electrons from the radioactive source S were focused by the lens system L on the entrance slit A of the 120° electrostatic deflector. The potential difference applied between the deflector plates deflected electrons of the desired energy (300 keV in this experiment) through the exit slit B and the slit system C. Passing out of the vacuum system through a thin window they traversed a proportional tube counter G before passing through the converter F, consisting of thin layers of Pb, Cd, and Zn evaporated on a disc of paraffin wax thick enough to reduce the electrons to rest. Bremsstrahlung photons produced in the converter were detected by means of a sodium iodide crystal and photomultiplier PM, set to receive photons emitted in a direction making an angle 22±20° with the electron beam direction. The photomultiplier was operated in coincidence with the tube counter and the photomultiplier assembly could be rotated about the electron beam direction in order to measure the azimuthal distribution of the

† Güthner, K., *Z. Phys.* **182** (1965) 278.

bremsstrahlung emission. About one-third of the bremsstrahlung photons were produced in the paraffin-wax backing of the converter foils and the background correction was obtained by measuring the bremsstrahlung due to the paraffin wax alone.

Electrons of velocity v emitted from the source are expected to be polarized longitudinally, the degree of polarization P being equal in magnitude to v/c $(= 0.77$ for 300-keV electrons). The electrostatic deflector changes the direction of motion of the electrons through $120°$

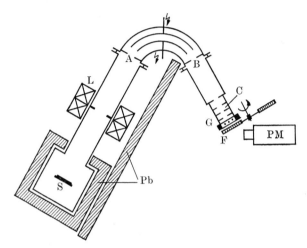

Fig. 14.80. Apparatus used by Güthner to investigate the asymmetry of bremsstrahlung produced by transversely polarized electrons.

without changing the spin direction, so that the electron beam striking the converter foil is transversely polarized. A certain amount of depolarization is expected to occur in passing through the counter tube and windows, but the degree of transverse polarization of the beam at the converter foil was expected to be approximately 0.70.

10.3. *Discussion of results of experiments on the continuous X-ray spectrum*

10.3.1. *Distribution of intensity in the X-ray spectrum.* As early as 1917 Webster† suggested that the observed shape of the continuous X-ray spectrum from a thick target could be explained on the assumption that the distribution in intensity from a thin target increased to a sharp maximum at the short-wavelength limit of the spectrum and then

† WEBSTER, D. L., *Phys. Rev.* **9** (1917) 220.

decreased with increasing λ like $1/\lambda^2$, i.e.

$$I_\lambda \, d\lambda = \begin{cases} a \, d\lambda/\lambda^2 & (\lambda > \lambda_0), \\ 0 & (\lambda < \lambda_0), \end{cases} \tag{101}$$

or, expressed in terms of a frequency distribution,

$$I_\nu \, d\nu = \begin{cases} b \, d\nu & (\nu < \nu_0), \\ 0 & (\nu > \nu_0), \end{cases} \tag{102}$$

where a and b are constants.

The exact non-relativistic calculations of Kirkpatrick and Wiedmann on the basis of Sommerfeld's theory show that the bremsstrahlung intensity decreases with increasing frequency. This decrease is most rapid for bremsstrahlung emission at small angles and is expected to be quite slow for X-rays emitted at $90°$ to the direction of the X-ray beam. Qualitatively these predictions are borne out by the experimental results. Fig. 14.81 shows a comparison with the theory of the results of Motz and Placious[†] for bremsstrahlung emitted at different angles. The absolute differential cross-section $I(\nu, \vartheta)$ per unit frequency range per unit solid angle multiplied by νZ^{-2} is shown as a function of photon energy for X-rays produced by the impact of electrons of energy 50 keV on targets of aluminium ($17 \, \mu g/cm^2$) and gold ($10 \, \mu g/cm^2$). The experimental results are compared with the non-relativistic calculations of Kirkpatrick and Wiedmann[‡] shown by the full line. The measured cross-sections show the predicted decrease with increasing frequency but the absolute magnitudes do not agree with those calculated. A simple relativistic correction was made by multiplying the cross-sections by $(1 - \beta_0 \cos \vartheta)^{-2}$, where β_0 is the velocity of the incident electrons. This factor was obtained by comparing the relativistic and non-relativistic Born-approximation bremsstrahlung calculations.[§] The corrected cross-sections are only rough estimates of the cross-sections that would be obtained using relativistic Coulomb wave functions and taking account of retardation. Nevertheless the correction does improve the agreement with the measured values. For both Al and Au the closest agreement between theory and experiment is obtained for emission angles of about 50 degrees. Similar measurements of absolute differential cross-sections for bremsstrahlung production by 500-keV electrons on thin targets of Al and Au are shown in Fig. 14.82.[||] Comparison is made

† Motz, J. W. and Placious, R. C., *Phys. Rev.* **109** (1958) 235.
‡ loc. cit., p. 1182.
§ Heitler, W., *The quantum theory of radiation*, 3rd edn, p. 248 (Clarendon Press, Oxford, 1954).
|| Motz, J. W., *Phys. Rev.* **100** (1955) 1560.

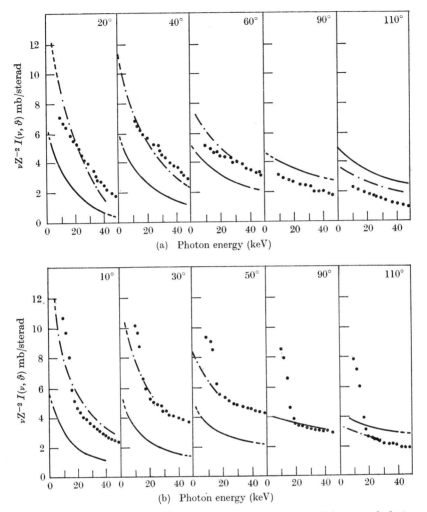

FIG. 14.81. Intensity $I(\nu, \vartheta)$ of bremsstrahlung emitted through impact of electrons of energy 50 keV from (a) Al, and (b) Au as a function of photon energy at various angles. ..... observed by Motz and Placious. —— calculated by Kirkpatrick and Wiedmann. —·—·— calculated with simple relativistic correction.

with the Born approximation calculations of Sauter.† There are large discrepancies between the calculated and measured cross-sections for large Z, especially near the high-frequency limit of the spectrum. Born's approximation would be expected to be poor under these conditions (see p. 1182).

Fig. 14.81 and 14.82 illustrate the approximate proportionality of

† SAUTER, F., *Annln Phys.* **20** (1934) 404.

the bremsstrahlung cross-sections to Z^2, a dependence that was known already in early studies of X-ray emission.†

10.3.2. *Directional distribution of the emitted radiation.* The non-relativistic theory of the continuous X-ray spectrum takes account

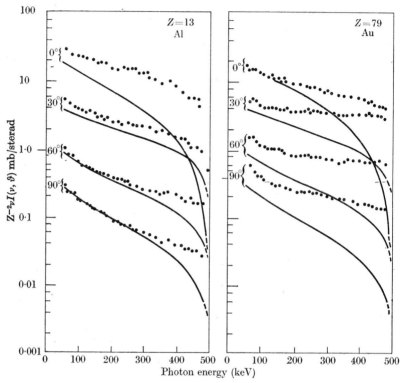

FIG. 14.82. Intensity $I(\nu, \vartheta)$ of bremsstrahlung emitted through impact of electrons of energy 500 keV from Al and Au as a function of photon energy at various angles. ..... observed by Motz and Placious. ——— calculated by Sauter using Born's approximation.

only of dipole transitions and predicts an angular distribution of intensity of emitted radiation per unit solid angle at an angle ϑ of the form

$$I(\vartheta) = (I_z - I_y)\sin^2\vartheta + 2I_y, \qquad (103)$$

giving a maximum at right angles to the incident radiation. When account is taken of retardation effects (including terms of the first order in $\beta = v/c$, where v is the electron velocity), Sommerfeld‡ showed that a maximum intensity of radiation was to be expected in the forward

† BEATTY, R. T., *Proc. R. Soc.* A89 (1913) 314.
‡ SOMMERFELD, A., *Annln Phys.* 11 (1931) 257.

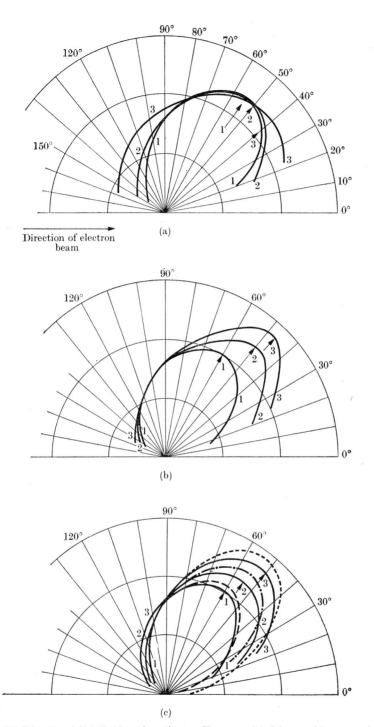

Direction of electron
beam

(a)

(b)

(c)

Fig. 14.83. Directional distribution of continuous X-rays emitted from a thin magnesium target. (a) Electron energy 31 keV : (1) $\lambda = 0{\cdot}43$ Å ; (2) $\lambda = 0{\cdot}65$ Å ; (3) $\lambda = 1{\cdot}60$ Å ; (b) $\lambda = 0{\cdot}65$ Å, electron energy : (1) 20 keV, (2) 31 keV, (3) 40 keV. (c) Distribution at high-frequency limit : (1) 20 keV, (2) 31 keV, (3) 40 keV. Broken curves show the angular distributions calculated from Scherzer's theory : — — — 20 keV ; —·—·— 31 keV ; - - - - 40 keV. The arrows show the directions of maximum intensity.

direction, at an angle ϑ_1 with the direction of incidence, where

$$\cos \vartheta_1 = 2\beta/(1-I_y/I_z).\qquad(104)$$

A directional distribution of this type was found by Kulenkampff[†] and Duane.[‡] A comprehensive study of the directional distribution of the continuous radiation from a target of magnesium less than 1000 Å thick has been made by Böhm.[§] In these measurements the Ross-filter method was employed to isolate a small region of the spectrum. Fig. 14.83 shows some typical polar curves obtained by Böhm. Fig. 14.83 (a) shows the intensity per unit solid angle of the radiation emitted in different directions for an electron energy of 31 keV and mean wavelengths of 0·43, 0·65, and 1·60 Å. For these three cases the direction of maximum intensity tends to shift towards larger angles as the wavelength decreases.

Fig. 14.83 (c) shows the intensity distribution at the high-frequency limit for electron energies of 20, 31, and 40 keV. The broken curves in this figure show the theoretical directional distribution deduced by Scherzer[‖] allowing for retardation and relativistic effects. The agreement is seen to be very good in this case. It is clear that the curves of Fig. 14.83 (c) pass through zero for angles of 0° and 180°. This does not appear to be the case for wavelengths away from the high-frequency limit.

The remarkable agreement between the experiments and the theory at the high-frequency limit is shown in Fig. 14.84, where the angle of maximum intensity at the high-energy limit is compared with Sommerfeld's theory. Fig. 14.84 includes also Kulenkampff's[†] measurements using an Al foil, 6000 Å thick. At long wavelengths, however, the calculated direction of maximum intensity comes out smaller than that observed. Similar studies by Motz and Placious[††] for thin Al and Au targets bombarded by 50-keV electrons show good agreement with the Sommerfeld–Kirkpatrick–Wiedmann cross-sections, including the simple relativistic correction in the case of gold, but the agreement is less good for aluminium.

10.3.3. *Measurement of X-ray isochromats.* The shape of the variation of intensity with electron energy for a definite frequency band is given fairly well by theory. Thus Fig. 14.85 shows the curves observed for nickel by Harworth and Kirkpatrick[‡‡] for two frequency bands

  [†] KULENKAMPFF, H., *Annln Phys.* **87** (1928) 597.
  [‡] DUANE, W., *Proc. natn. Acad. Sci. U.S.A.* **13** (1927) 662; **14** (1928) 450.
  [§] BÖHM, K., *Annln Phys.* **33** (1938) 315.
  [‖] SCHERZER, O., ibid. **13** (1932) 137.
  [††] MOTZ, J. W. and PLACIOUS, R. C., *Phys. Rev.* **109** (1958) 235.
  [‡‡] HARWORTH, K. and KIRKPATRICK, P., ibid. **62** (1942) 334.

centred about wavelengths of 0·497 and 1·010 Å respectively. These curves show that, when the tube potential is increased through the value V_{min}, which is just sufficient to excite the frequency band in question, the intensity rises sharply to a maximum and then decreases, approximately proportional to V^{-1} for further increase of the tube potential V.

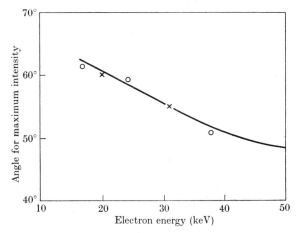

FIG. 14.84. Illustrating the variation with electron energy of the angle between the direction of the electron beam and that of maximum emission of radiation at the high-frequency limit. ○ Observed values for Al (6000 Å thick) (Kulenkampff). ✕ Observed values for Mg (1000 Å thick) (Böhm). —— theoretical curve.

The broken lines of Fig. 14.85 show the calculated curves for thin nickel targets. The curve for radiation centred about a wavelength of 0·497 Å is arbitrarily adjusted to fit the experimental curve at one point. Then without any further adjustment, the theoretical curve for 1·010 Å is as shown. The shape of both curves shows good agreement between theory and experiment, but the relative intensities for the two frequencies are not in very good agreement with theory.

Ohlin[†] showed that when the variation of intensity of a bremsstrahlung isochromat with electron energy was studied near the threshold, using high electron energy resolution, a characteristic fine structure was observed. In general, a sharp peak is observed within a few volts of the threshold, followed by further structure in the region some tens of volts above. Fig. 14.86, taken from a paper by Sandström,[‡] shows the variation with electron energy near the threshold of isochromats produced on tungsten. Curves are shown for isochromats with threshold at 1·25 kV

† OHLIN, P., *Ark. Mat. Astr. Fys.* A29 (1943) no. 3; A31 (1945) no. 9.
‡ SANDSTRÖM, R., *Ark. Fiz.* 18 (1960) 305.

(Ulmer and Vernickel†), 3·7 kV (Sandström‡), and 8 and 19 kV (Bearden and Schwarz§). The curves are plotted so that the first maximum comes at zero volts for each of the curves, which are displaced vertically relative to each other. The rise from threshold to the first maximum is sharpest for the isochromats of lowest energy. The so-called 'Ohlin structure' has been studied, using thick targets, and has been

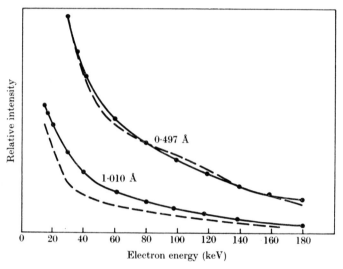

Fig. 14.85. Intensity of X-ray emission at a given frequency as a function of electron energy, for nickel. The minimum energies required to excite the two bands of radiation shown are 12·2 keV and 24·9 keV respectively. ——— observed (Harworth and Kirkpatrick). ——— calculated.

accounted for in terms of characteristic electron energy losses in the target so that, after traversing some distance in the target, electrons of the initial energy Ve are present together with others of energy $(V-v_1)e$, $(V-v_2)e,...$, where $v_1 e$, $v_2 e,...$ are characteristic energy losses in the metal. The phenomenon does not appear to be well understood and its detailed discussion is outside the scope of this book. It would be useful, however, to investigate how structure depends on the target thickness for these targets. The effect introduces a fine structure in the upper energy region of the continuous X-ray spectrum, produced by electrons of a given energy, and complicates the determination of the constant h/e by studying this upper energy limit.

† ULMER, K. and VERNICKEL, H., Z. Phys. 153 (1958) 149.
‡ SANDSTRÖM, R., loc. cit. p. 1211.
§ BEARDEN, J. A. and SCHWARZ, G., Phys. Rev. 79 (1950) 674.

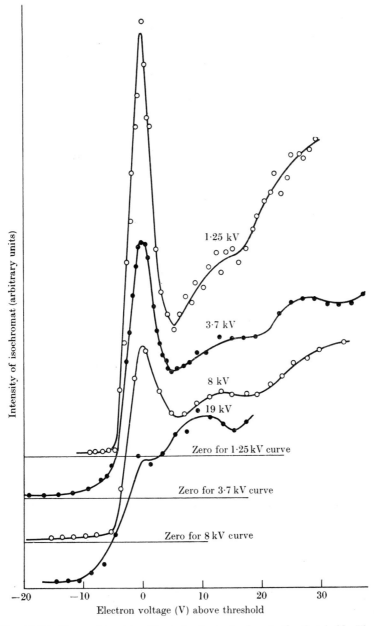

Fig. 14.86. Shape of tungsten X-ray isochromats close to the threshold. The isochromats with threshold at 1·25 kV were observed by Ulmer and Vernickel, those at 8 and 19 kV by Bearden and Schwarz, and at 3·7 kV by Sandström. The electron voltage scale is adjusted so the first maximum occurs at 0 for all isochromats.

10.3.4. *The low energy end of the continuous X-ray spectrum.* Relative intensity measurements of soft continuous X-radiation (80–180 Å) produced on targets of Mg, Al, Mn, Cu, Ge, and Ag by electrons of energy 600 eV to 3 keV have been carried out by Peterson and Tomboulian.† Fig. 14.87 shows the results obtained for a Mn target bombarded with 630-eV electrons, the soft X-rays being measured at an angle of 95 degrees to the incident electron direction. The agreement with the Sommerfeld–Kirkpatrick–Wiedmann calculations shown by the solid curve is remarkably good, particularly since the target, although only 400 Å thick, has

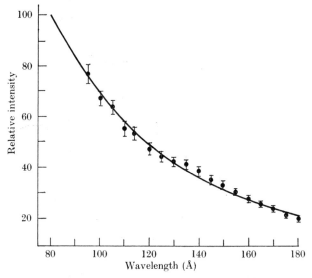

Fig. 14.87. Intensity of soft X-rays emitted from a Mn target at 95° to the direction of incidence of electrons of energy 630 eV. —— calculated. ⬤ observed (Peterson and Tomboulian).

to be considered as effectively a thick target for electrons of this energy. The good agreement can be understood since the detailed shape is very insensitive to the electron energy.

In studying bremsstrahlung emission in the extreme long-wavelength region, care has to be taken to separate the bremsstrahlung from other sources of soft radiation. For example, Ferrell‡ predicted that plasma oscillations induced by charged particles passing through thin metal foils should decay by the emission of monochromatic photons at the plasma frequency. Again, Ginzburg and Frank§ predicted the produc-

† PETERSON, T. J. and TOMBOULIAN, D. H., *Phys. Rev.* **125** (1962) 235.
‡ FERRELL, R. A., ibid. **111** (1958) 1214.
§ GINZBURG, W. and FRANK, I., *Zh. éksp. teor. Fiz. SSSR* **16** (1946) 15.

tion of transition radiation in passing from one medium to another. Both these sources of soft radiation must be separated out from the soft bremsstrahlung spectrum.

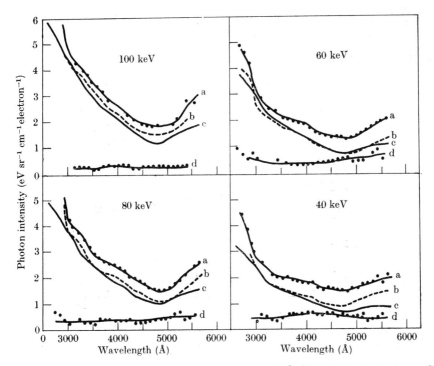

FIG. 14.88. Spectral distribution of radiation from 340-Å gold foil due to electrons of different incident energies : a = observed intensity in the parallel plane $(I_{||})$; d = experimental intensity in the perpendicular plane $(I_{\perp})$; b = curve a—curve d $(I_{||} - I)$; c = transition radiation (theory).

In their measurements of optical bremsstrahlung from gold foils 340 and 530 Å thick bombarded by electrons of energy 25–100 keV, Arakawa et al.† separated out transition radiation by making use of the fact that it should be polarized in the plane of production. Fig. 14.88 shows the spectral distribution obtained, using electrons of a number of energies, for components of the optical radiation polarized (a) in the parallel plane, (d) in the perpendicular plane. Assuming the bremsstrahlung is unpolarized, the subtraction of (d) from (a) should give the part of the radiation to be ascribed to transition radiation. This is given by the dotted curve (b). The predicted distribution of the transition

† ARAKAWA, E. T., EMERSON, L. C., HAMMER, D. C., and BIRKHOFF, R. D., *Phys. Rev.* **131** (1963) 719.

radiation, according to the theory of Ginzburg and Frank, is given by curve (c). The agreement (in absolute intensity) is very satisfactory.

Boersch et al.† have sought to separate out the transition radiation from bremsstrahlung, making use of the different dependence of the intensities of the two kinds of radiation on the electron energy. While the intensity of transition radiation increases proportional to electron energy E, the bremsstrahlung intensity, except close to the threshold,

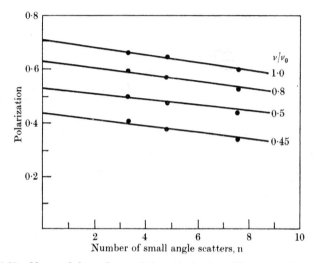

FIG. 14.89. Observed dependence of the polarization of bremsstrahlung on the (gold) target thickness t expressed in terms of the number of small angle scatters.

goes more like $1/E$, so that analysis of the energy dependence enables the two to be separated out. Boersch et al. also observed a third radiation from silver targets, with peak intensity near 3500 Å and half-width 200–300 Å, emitted when the electrons grazed the target at an angle of 0·5° to the target surface. The intensity of this radiation was 100 times greater than that of transition radiation at normal incidence. The wavelength and incident-angle dependence, as well as the strong dependence on surface contamination, suggested the interpretation of this radiation as due to plasma excitation in the silver target, as proposed by Ferrell.

10.3.5. *Polarization of bremsstrahlung.* The early experiments on bremsstrahlung polarization gave conflicting results. This seems to have been caused by the use of targets of finite thickness. Owing to the scattering of the electrons in the target material much of the radiation coming from some depth in the target is produced by electrons not moving

† BOERSCH, H., DOBBERSTEIN, P., FRITZSCHE, D., and SAUERBREY, G., Z. Phys. **187** (1965) 97.

parallel to the direction of incidence. Fig. 14.89, obtained by Kulen-kampff and Zinn,† illustrates the effect for gold foils of different thick-nesses. It was comparatively straightforward in this case to obtain the true polarization by extrapolating back to zero target thickness.

Fig. 14.90 shows the results obtained by Kulenkampff and Zinn,† using Al_2O_3 and Au foils, for 35–45 keV electrons, and also the results of

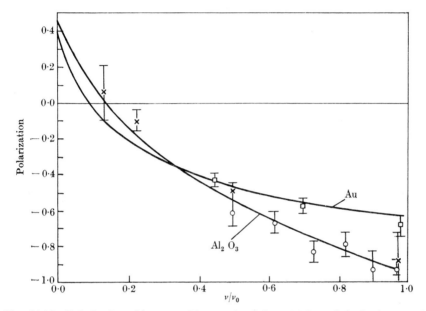

FIG. 14.90. Polarization of bremsstrahlung emitted from Al_2O_3 and Au by impact of electrons with energy in the range 34–45 keV. ⌀ observed by Kulenkampff, Leisegang, and Scheer for Al and 34 keV electrons. ✕ observed by Kulenkampff and Zinn for Al_2O_3 and electrons with energy in the range 35–45 keV. ▢ observed by Kulenkampff and Zinn for Au and electrons with energy in the range 35–45 keV. The curves are calculated by the Sommerfeld theory.

Kulenkampff, Leisegang, and Scheer‡ for Al by a different method (see p. 1201) using 34-keV electrons. The comparison with the predic-tions of the Sommerfeld theory§ is seen to be quite good.

A systematic comparison of the observed polarization with the calcula-tions of Gluckstern and Hull,‖ based on Born's approximation, over a wide range of initial electron energies and emission angles has been made by Motz and Placious†† for Be, Al, and Au.

† KULENKAMPFF, H. and ZINN, W., *Z. Phys.* **161** (1961) 428.
‡ KULENKAMPFF, H., LEISEGANG, S., and SCHEER, M., ibid. **137** (1954) 435.
§ KULENKAMPFF, H., SCHEER, M., and ZEITLER, E., ibid. **157** (1959) 275.
‖ GLUCKSTERN, R. L. and HULL, M. H., *Phys. Rev.* **90** (1953) 1030.
†† MOTZ, J. W. and PLACIOUS, R. C., *Nuovo Cim.* **15** (1960) 571.

Fig. 14.91 shows the variation of polarization with emission angle for X-rays of frequency $\nu = 0.9\nu_0$ for incident electron energies of 50 keV (Fig. 14.91 (a)) and 500 keV (Fig. 14.91 (b)). The Gluckstern–Hull theoretical curve, which takes no account of screening, fits the experimental results poorly at the lower energy, in contrast with the good

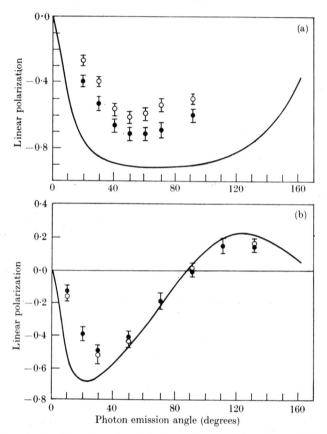

FIG. 14.91. Variation of polarization with angle for X-rays of frequency $\nu = 0.9\nu_0$ emitted due to impact of electrons (a) of 50 keV, (b) of 500 keV, incident energy. ——— calculated by Gluckstern and Hull. ⌀ observed $Z = 79$ Motz and Placious. ◉ $Z = 4$ Motz and Placious.

results obtained with the Sommerfeld theory (see Fig. 14.90). At the higher energy, however, the Gluckstern–Hull curve gives a good representation of the results for beryllium, but less good for gold, as would be expected for calculations based on Born's approximation. For low energy bremsstrahlung ($\nu = 0.1\nu_0$) Fig. 14.92 shows the results obtained from aluminium and gold foils for an incident electron energy of 500 keV, compared with the Gluckstern–Hull calculations both without and with screening corrections.

Kulenkampff, Scheer, and Zeitler† have shown that the Sommerfeld theory suggests that the polarization should be maximum in a direction ϑ^*, given by $\cos \vartheta^* = \beta_0$, where $\beta_0 c$ is the incident electron velocity. This seems to agree with the low energy results reasonably well.

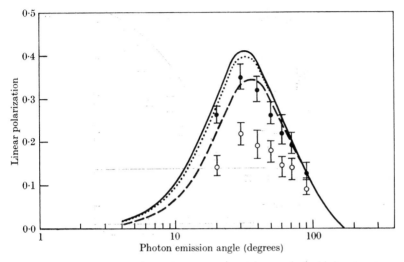

FIG. 14.92. Angular dependence of the bremsstrahlung linear polarization for an incident electron kinetic energy of 500 keV.

● observed for an Al target (0·6 mg/cm²)   ⎱ and a photon energy range
○ observed for an Au target (0·43 mg/cm²) ⎰ between 45 and 55 keV.

—— calculated by Gluckstern and Hull for a photon energy of 50 keV. • • • • calculated allowing for screening, for Al. — — — calculated allowing for screening, for Au.

In the work described above, unpolarized electrons were used. Little systematic work seems to have been carried out for bremsstrahlung production by electrons transversely polarized, but measurements are available establishing the circular polarization of bremsstrahlung from longitudinally polarized electrons. Fig. 14.93 shows the results obtained in the experiment of Goldhaber, Grodzins, and Sunyar‡ for the difference in transmission through iron, magnetized parallel and antiparallel to their direction of motion, of bremsstrahlung quanta produced by the passage of electrons from an $^{90}Y + ^{90}Sr$ source, expected to be longitudinally polarized. The full line shows the expected value for 100

† KULENKAMPFF, H., SCHEER, M., and ZEITLER, E., Z. Phys. 157 (1959) 275.
‡ GOLDHABER, M., GRODZINS, L., and SUNYAR, A. W., Phys. Rev. 106 (1957) 826.

per cent circularly polarized photons of the quantity δ (p. 1203). The results indicate that the bremsstrahlung is circularly polarized with spin antiparallel to the direction of motion, but that the degree of polarization is less than 100 per cent. This experiment establishes, not only the

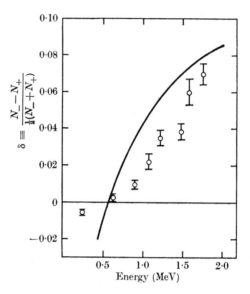

Fig. 14.93. Comparison of observed values for δ (see p. 1203), and the computed magnet response curve as a function of electron energy for 100 per cent circularly polarized photons with spin antiparallel to the direction of propagation.

longitudinal polarization of fast beta rays, but also that such longitudinally polarized electrons emit circularly polarized photons.

10.3.6. *Asymmetry of bremsstrahlung from transversely polarized electrons.* As pointed out above (§ 9.4) the bremsstrahlung produced by transversely polarized electrons should show azimuthal asymmetry exactly analogous to the asymmetry observed in the elastic scattering of such electrons (see Chap. 5, § 6). Writing for the differential cross-section $I(\vartheta, \varphi, \nu)\,d\Omega d\nu$ for emission in direction ϑ, azimuthal angle φ, relative to direction of polarization, frequency range $d\nu$,

$$I(\vartheta, \varphi, \nu) = I_0(\vartheta, \varphi, \nu)(1 - PS\sin\varphi), \tag{105}$$

where P is the degree of polarization of the electron, Güthner observed the expected asymmetry of bremsstrahlung production on foils of Pb, Cu, Zn, Cd, with $\vartheta = 45°$. His results were consistent with an asymmetry proportional to Z and of magnitude:

$$S = -(0.133 \pm 0.020)Z \quad \text{for electrons of energy 60–240 keV,}$$

$$S = -(0.102 \pm 0.029)Z \quad \text{for electrons of energy 15–60 keV.}$$

10.3.7. *The angular dependence of the radiative energy loss of electrons.*
So far we have discussed exclusively the radiation produced in the bremsstrahlung process. Much less attention has been paid to the study of the scattering of the electrons themselves in the process. Febel and Knop[†] studied the radiative scattering of 1-MeV electrons in Al, Ag, and Au foils. After allowing for effects of multiple Coulomb scattering Fig. 14.94 shows the ratio, χ, of $I(\theta, \epsilon)\,d\theta\,\Delta\epsilon$, the differential cross-section for scattering through an angle θ into solid angle $d\omega$ with energy loss between ϵ and $\epsilon + \Delta\epsilon$ to the differential elastic scattering cross-section for scattering into the same angular range. The three curves correspond to electron scattering through $30°$, $60°$, and $90°$. The results are compared with the calculations of Racah[‡] who calculated the distribution of energy losses using Born's approximation. The agreement appears satisfactory.

Although considerable progress has been made in the study of bremsstrahlung processes in recent years, the number of variable parameters to be investigated is so large that many aspects remain for future investigation.

11. Free–free absorption

11.1. *Absorption in the field of bare positive ions*

The free–free absorption cross-section (Chap. 7, § 5.2.1 (71)) per electron per unit energy range in the field of an unscreened point charge $+Ze$ may be written in the form[§]

$$q_E^{E'} = \frac{2^7 \pi^3 \alpha a_0^5}{3\sqrt{3}\, Z^5} \frac{g(\sigma, \epsilon)}{\sigma^3 \epsilon^{\frac{1}{2}}}, \tag{106}$$

where α is the fine structure constant, $Z^2\epsilon$ and $Z^2\epsilon'$ are the initial and final electron energies in rydbergs, and $\sigma = \epsilon' - \epsilon$. Thus

$$Z^2\sigma = h\nu/h\nu_0, \tag{107}$$

where $h\nu_0$ is the ionization energy of hydrogen.

† Febel, A. and Knop, G., *Z. Phys.* **174** (1963) 257.
‡ Racah, G., *Nuovo Cim.* **11** (1934) 477.
§ Menzel, D. H. and Pekeris, C. L., *Mon. Not. R. astr. Soc.* **96** (1935) 77.

$g(\sigma, \epsilon)$, the Gaunt factor, is a complicated function of ϵ and ϵ' which, under many circumstances, is not greatly different from unity (cf. (16)), in which case (106) reduces to the formula given by classical theory.

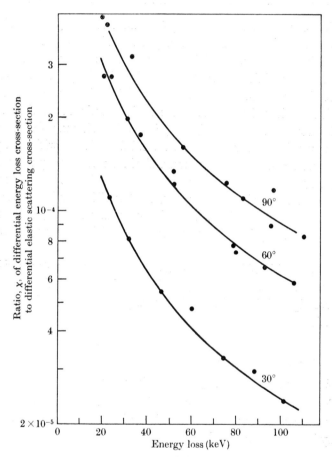

Fig. 14.94. Ratio χ of $I(\theta, E)\,\mathrm{d}\theta\Delta E$ to the differential elastic cross-section. ● observed by Febel and Knop. —— calculated by Racah.

More than twenty approximate formulae for the non-relativistic Gaunt factor, valid for specified ranges of electron energy or of bremsstrahlung wavelength, have been given in the literature. These have been summarized in a review article by Brussaard and van de Hulst.†

The free–free absorption cross-section per positive ion for a medium containing n_e electrons per cm³, with energy distribution function

† Brussaard, P. J. and van de Hulst, H. C., Rev. mod. Phys. 34 (1962) 507.

$f(\epsilon)\,d\epsilon$, is given by

$$n_e \int q_\epsilon^\epsilon f(\epsilon)\,d\epsilon. \tag{108}$$

In the important case in which $f(\epsilon)$ is Maxwellian about a temperature T we write the cross-section per positive ion in the form

$$n_e \mathcal{Q}(\sigma, T). \tag{109}$$

$\mathcal{Q}$ has been calculated by Karzas and Latter[†] and by Peach,[‡] using the

TABLE 14.28

			$Z^4\sigma^3\mathcal{Q}$			
$Z = 1$						
$\sigma = 0.05$	0.10	0.15	0.20	0.25	0.30	
T (°K)						
4000	1.827	1.810	1.808	1.809	1.810	1.812
5000	1.654	1.632	1.627	1.626	1.626	1.626
6000	1.527	1.501	1.494	1.491	1.490	1.490
7000	1.430	1.400	1.391	1.387	1.385	1.385
8000	1.352	1.320	1.309	1.304	1.301	1.300
9000	1.289	1.253	1.241	1.235	1.232	1.230
10 000	1.235	1.197	1.184	1.177	1.173	1.171
11 000	1.189	1.149	1.135	1.127	1.123	1.120
12 000	1.149	1.108	1.092	1.084	1.079	1.076
13 000	1.115	1.071	1.055	1.046	1.041	1.037
$Z = 2$						
	0.05	0.10	0.15	0.20	0.25	0.30
T (°K)						
4000	1.756	1.766	1.774	1.781	1.787	1.791
5000	1.576	1.583	1.590	1.595	1.600	1.604
6000	1.444	1.448	1.455	1.458	1.462	1.465
7000	1.341	1.344	1.348	1.352	1.355	1.358
8000	1.259	1.259	1.263	1.266	1.269	1.271
9000	1.191	1.190	1.192	1.195	1.198	1.200
10 000	1.133	1.131	1.133	1.135	1.137	1.139
11 000	1.084	1.081	1.082	1.084	1.086	1.087
12 000	1.041	1.037	1.038	1.039	1.041	1.042
13 000	1.004	0.998_2	0.998_3	1.000	1.001	1.002

full formula (106). Table 14.28 gives the results obtained by Peach for $Z = 1$ and $Z = 2$ and a number of values of σ and T.

Similar calculations have been carried out for high temperature plasmas by Sodha[§] and by Janes and Koritz.[||]

† KARZAS W. J. and LATTER, R., *Astrophys. J. Suppl. Sci.* **55** (1961) 167.
‡ PEACH, G., *Mem. R. astr. Soc.* **71** (1967) 1.
§ SODHA, M. S., *Can. J. Phys.* **37** (1959) 1380.
|| JANES, G. S. and KORITZ, N., *J. appl. Phys.* **31** (1960) 525.

11.2. *Absorption in the fields of partially ionized atoms*

It is possible to calculate $\mathcal{Q}(\sigma, T)$ for partially ionized atoms to a good approximation by using the same technique as that of Burgess and Seaton† for bound-free absorption described in § 4.2. This has been done by Peach‡ who finds, as the equivalent of (31),

$$q_{\epsilon}^{\epsilon'} = \frac{2^4 \pi^2 \alpha}{3} \frac{a_0^5}{Z^5 \sigma^3} \sum_{l'=l\pm 1} C_{l'} \{g(\epsilon, l; \epsilon', l')\}^2,$$

where

$$g(\epsilon, l; \epsilon', l') = G(\epsilon, l; \epsilon', l') \cos[\pi\{\mu_{l'}(\epsilon') - \mu_l(\epsilon) + \chi(\epsilon, l; \epsilon', l')\}]. \quad (110)$$

Thus $\mu_l(\epsilon)$ and $\mu_{l'}(\epsilon')$ are quantum defects extrapolated to the continuum as explained in § 4.2, p. 1075 and in Chapter 6, § 3.10. The functions G and χ have been tabulated by Peach for $\sigma = 0.05\,(0.05)\,0.30$, $\epsilon = 0.00\,(0.05)\,0.40$, and $l, l' = 0, 1; 1, 0; 1, 2; 2, 1; 2, 3$ and $3, 2$. With numerical values substituted for the atomic constants

$$q_{\epsilon}^{\epsilon'} = 1.59 \times 10^{-42} Z^{-5} \sigma^{-3} \epsilon^{-\frac{3}{2}} \sum C_{l'} \{g(\epsilon, l; \epsilon', l')\}^2 \text{ cm}^5. \quad (111)$$

To examine the importance of the presence of bound electrons Peach writes for a partially ionized atom of charge $+Ze$

$$\mathcal{Q}^Z(\sigma, T) = \mathcal{Q}_{\mathrm{H}}^Z(\sigma, T)\{1 + D(\sigma, T)\}, \quad (112)$$

where $\mathcal{Q}^Z(\sigma, T)$ is the value for a bare nucleus of the same charge Z and gives table of $D(\sigma, T)$ for $\sigma = 0.05\,(0.05)\,0.30$ and $T = 4000\,(1000)$ 13 000 °K for absorption in the fields of He^+, Li^+, C^+, N^+, O^+, Na^+, Mg^+, Mg^{++}, Al^+, Si^+, Cl^+, K^+, and Cr^+. Fig. 14.95 shows D as a function of σ for various values of T for O^+ and Ca^{++}.

12. Total absorption in an ionized atmosphere

In an atmosphere at a temperature T containing a single atomic species the concentrations N_0, N_+, N_e of neutral atoms, singly ionized atoms, and electrons will be related by Saha's equation

$$\frac{N^+ N_e}{N_0} = \frac{2\mathcal{Z}_+}{\mathcal{Z}_0}\left(\frac{2\pi m \kappa T}{h^2}\right)^{\frac{3}{2}} e^{-I/\kappa T}, \quad (113)$$

where $\mathcal{Z}$ and $\mathcal{Z}_0$ are the partition functions for the positive ion and neutral atom respectively and I is the ionization energy.

The population of different excited states will follow a Boltzmann distribution so that, if N_n is the concentration of atoms in the nth excited state,

$$\frac{N_n}{N_0} = \frac{w_n}{w_0} e^{-(E_n - E_0)/\kappa T}, \quad (114)$$

† loc. cit., p. 1074.   ‡ PEACH, G., *Mon. Not. R. astr. Soc.* **130** (1965) 361.

where E_n, E_0 are the energies of the nth excited state and of the ground state respectively and w_n, w_0 the corresponding statistical weights.

Absorption of radiation by a partially ionized atmosphere will arise both from bound–free and from free–free transitions. The effective absorption cross-section per neutral atom due to the former source can

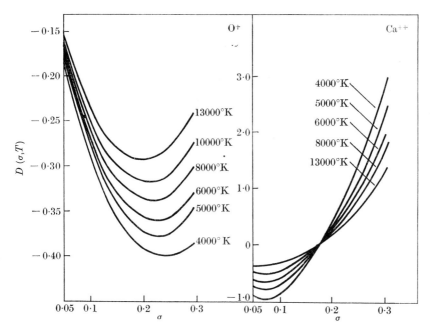

FIG. 14.95. D as a function of σ for various values of T, for O^+ and Ca^{++}, as calculated by Peach.

be calculated, using the method described in § 4.2, to calculate the absorption cross-sections for each excited atom and (114) to determine the relative concentration of these excited atoms. The free–free contribution can be obtained using the Saha equation (113) and the analysis of § 11.2.

Peach[†] has carried out such calculations for atmospheres of C, N, O. Mg, Mg$^+$, Si, Cl, and Ca$^+$ atoms and ions for six different temperatures over a wide range of wavelengths. Experimental data with which to compare these calculations are available for O,[‡] N,[§] and Cl.[||] They have been derived from observation of the emission spectra of high-pressure arc discharges by the technique described in Chapter 15, § 4, allowance

† PEACH, G., Mem. R. astr. Soc. 71 (1967) 29.
‡ BOLDT, G., Z. Phys. 145 (1959) 330.   § Ibid. 319.
|| HENNING, H., ibid. 169 (1962) 467.

having been made for the contribution due to radiative attachment. Fig. 14.96 (a), (b), and (c) shows the comparison between Peach's results for these three atoms and those obtained experimentally. The results are presented in the form of curves for each temperature, representing the effective absorption cross-section per atom as a function of wavelength. Discontinuities appear when, as the wavelength decreases, new states begin to contribute to photo-ionization. The nature of the states involved is indicated in Fig. 14.96 at each discontinuity.

The agreement between theory and experiment is quite good for O and N, but less satisfactory for Cl. The failure for this atom may be partly due to inadequacy of the theory for a relatively complex atom and partly to special experimental difficulties.

Measurements have also been made for argon[†] but no theoretical values are available for comparison.

13. Collective effects in bremsstrahlung emission from a plasma

The emission of bremsstrahlung from a plasma is modified by collective effects in which the electron can be considered to interact with the plasma as a whole. For the low-frequency end of the bremsstrahlung spectrum with $h\nu \leqslant 0.1\kappa T$, Theimer[‡] has derived for the total bremsstrahlung energy emitted per unit volume per unit time in the frequency range $d\nu$

$$I(\nu)\,d\nu = \frac{4e^6}{3c^3\pi m^2}\left(\frac{2m}{\pi\kappa T}\right)^{\frac{1}{2}} n_e\, n_i \ln(p_{max}/p_{min})\,d\nu, \qquad (115)$$

where n_e, n_i are respectively the electron and ion concentration, m the electron mass, T the temperature, and p_{max} and p_{min} maximum and minimum impact parameters. $p_{min} = \frac{3}{2}h/m\bar{v}$, where $\bar{v}$ is the Maxwellian r.m.s. velocity, is approximately equal to the mean de Broglie wavelength of the electrons.

The parameter p_{max}, which takes account of collective aspects of bremsstrahlung emission, depends on three characteristic lengths—the wavelength λ, the Debye shielding length $l \{= (\kappa T/4\pi e^2 n_e)^{\frac{1}{2}}\}$, and the electron mean free path l_e—being given by

$$p_{max} = \frac{(v/c)\lambda}{\{(1+v\lambda/\pi c l_e)^2 + 4v^2\lambda^2/\pi^2 c^2 l^2\}^{\frac{1}{2}}}. \qquad (116)$$

The spectral distribution of the radiation that leaves the plasma is greatly modified by absorption and re-emission processes. Below the

† SCHLÜTER, D., Z. Astrophys. 56 (1962) 43.

‡ THEIMER, O., Ann. Phys. 22 (1963) 102; see also OSTER, L., Physics Fluids 7 (1964) 263.

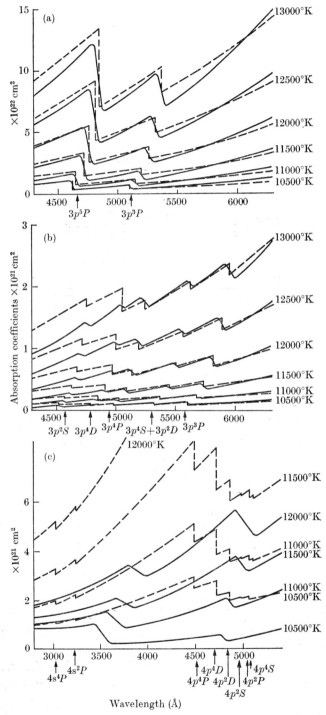

FIG. 14.96. Comparison of theoretical and observed absorption co-efficient of ionized atmospheres of (a) O, (b) N, (c) Cl, at different temperatures as indicated. ——— observed, ——— theoretical.

plasma frequency $\nu_p = (e^2 n_e/\pi m)^{\frac{1}{2}}$, the attenuation length is of the order of one wavelength and the plasma is nearly opaque. For $\nu \geqslant \nu_p$, on the other hand, the plasma is transparent.

Measurements of the intensity of continuous radiation in the infra-red have been used to estimate the electron density n_e of a high-density plasma. Fig. 14.97 illustrates the arrangement used by Dushin,

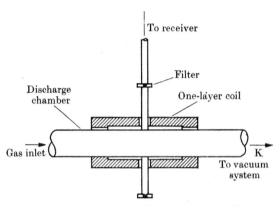

FIG. 14.97. Arrangement used in the experiments of Dushin *et al.*

Kononenko, Pavlichenko, and Nikolskii.[†] An alternating magnetic field of frequency 116 kc/s, provided by a single layer coil surrounding the discharge tube, produced an electrodeless discharge in hydrogen. The length of the fused quartz discharge tube was 80 cm and its diameter was 10 cm and the hydrogen flowed through it at a constant rate. A silicon filter, transparent in the wavelength region 1·5 to 15 μm, was placed in a side tube between the discharge tube and the liquid nitrogen cooled germanium–gold alloy photo-resistor used as detector. The signal from the detector was displayed on an oscilloscope after passage through a pre-amplifier. The detector and measuring arrangement was calibrated using black-body radiation. The electron density, n_e ($= n_i$) was estimated using the expression (115).

The plasma density was estimated by two independent means, viz. by measuring the Stark broadening of the H_β line and the transmission of microwave signals of wavelength 8 mm and 2 mm respectively. Fig. 14.98 shows the consistency of the density measurements by the three methods.

Bekefi, Hirshfield, and Brown[‡] have pointed out that absorption and

† DUSHIN, L. A., KONONENKO, V. I., PAVLICHENKO, O. S., and NIKOLSKII, I. K., *Optik. Spektrosk.* **19** (1965) 674 (*Optics Spectrosc. Wash.*) **19** (1965) 378).

‡ BEKEFI, G., HIRSHFIELD, J. L., and BROWN, S. C., *Physics Fluids* **4** (1961) 173.

re-emission processes of bremsstrahlung in plasmas could lead to population inversion and negative radiation temperature in the plasma, as a result of which the bremsstrahlung is greatly enhanced by stimulated emission. They derived the expression for the radiation temperature, T_r,

$$\kappa T_r = -m \int_0^\infty R(v)f(v)v^5 \, dv \bigg/ \int_0^\infty R(v)\frac{\partial f(v)}{\partial v} v^4 \, dv, \tag{117}$$

where $\quad f = A \exp\{-b(v/\bar{v})^d\} \qquad (A, b, d \text{ positive constants}) \tag{118}$

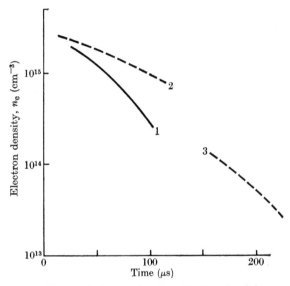

FIG. 14.98. Change of electron density with time in a plasma. The results obtained from bremsstrahlung emission measurements (curve 1) are compared with those obtained from Stark broadening of H_β (curve 2) and transmission of microwave signals of wavelength 2 mm (curve 3).

is the velocity distribution function of the electrons and

$$R(v) = \frac{4\pi^2\nu^2 Q_d(v)}{4\pi^2\nu^2 + \{n_e v Q_d(v)\}^2}, \tag{119}$$

where $Q_d(v)$ is the momentum transfer cross-section.

Integrating the denominator of (117) by parts gives

$$\kappa T_r = \frac{m \int_0^\infty R(v)f(v)v^5 \, dv}{f(0)\{R(v)v^4\}_{v=0} + \int_0^\infty f(v)(\partial/\partial v)\{R(v)v^4\} \, dv} \tag{120}$$

and T_r can be negative only if $(\partial/\partial v)\{R(v)v^4\} < 0$ over some finite region of electron velocity. In a weakly ionized gas (n_e small) $R(v) = Q_d(v)$ and $v^4 Q_d(v)$ is unlikely to have a negative slope. For very low frequencies, however, such that $n_e v Q_d(v) \gg 2\pi\nu$, $v^4 R(v)$ is proportional to $v^2/Q_d(v)$ and for gases exhibiting a pronounced Ramsauer–Townsend effect the slope of this could well be negative.

A very large radiation enhancement attributed to this effect has been observed by Tanaka, Honzawa, and Takayama† in Xe, Kr, and A plasmas but not in plasmas of Ne and He, for which there is no marked Ramsauer–Townsend effect. The peak power in the anomalous emission was more than 10 dB above the time-average value and was observed in the very low frequency range 40–70 Mc/s, well below the collision frequency (180 Mc/s). The anomalous emission was very sensitive to the discharge current and pressure.

A similar enhancement was observed in the cyclotron radiation emitted in the presence of a magnetic field.

14. Two-quantum photo-ionization

Until the invention of the laser the possibility of observing processes that could only occur in collisions with two quanta did not arise. However, the light intensity already attainable by lasers is sufficiently great to observe these processes under suitable circumstances, and this has already been done for the 2-quantum photodetachment from I^- ions by ruby laser light (Chap. 15, § 8). Some evidence has also been obtained of 2-quantum photo-ionization of caesium. We shall now discuss the theoretical expectations concerning the probability of 2-photon photo-ionization by quanta with energy below the threshold for single ionization—the conditions most favourable for experimental study. This will be followed by a brief description of the caesium experiments, but a detailed account of 2-quantum photodetachment will be referred to Chapter 15, § 8.

Consider a 2-quantum photo-ionization process in which E_0 is the energy of the initial atomic state and $h\nu$ the photon energy, so that the wave number of the ejected electron is given by

$$k'^2 = (2m/\hbar^2)(2h\nu + E_0). \tag{121}$$

The 2-quantum process proceeds through virtual transitions to intermediate atomic states of energy E_j. The amplitude for a virtual transition from the initial state to the intermediate state will be proportional to

† Tanaka, S., Honzawa, T., and Takayama, K., J. phys. Soc. Japan 20 (1965) 2099.

$\langle \mathbf{e} . \mathbf{r} \rangle_{j0}$, the dipole matrix element between the two states in the direction of polarization $\mathbf{e}$ of the incident photons. Similarly the amplitude for the virtual transition from the intermediate to the final state will be proportional to $\langle \mathbf{e} . \mathbf{r} \rangle_{fj}$. In fact, according to second-order perturbation theory, the cross-section per unit light intensity is given by

$$Q^{\mathrm{i}(2)} = \frac{\alpha a_0^2}{4\pi} \frac{J}{J_0} \int \left| \sum_j \frac{\langle \mathbf{e} . \mathbf{r} \rangle_{fj} \langle \mathbf{e} . \mathbf{r} \rangle_{j0}}{E_0 - E_j + h\nu} \right|^2 h\nu k' \, d\omega. \qquad (122)$$

In the matrix elements the wave functions for the initial and intermediate states are normalized to unit density and, for the final state, to represent a Coulomb distorted plane wave of unit amplitude with the appropriate spherical wave (see (83)). The angular integration is over the direction of motion of the ejected electron. The matrix elements, the energies, and the wave number k' are all expressed in atomic units. α is the fine structure constant $e^2/\hbar c = 137 \cdot 12$. J/J_0 is the ratio of the light intensity J to the atomic unit of intensity J_0. This is such that the field strength in the radiation field is e^2/a_0^2 ($5 \cdot 14 \times 10^9$ V/cm) giving

$$J_0 = 7 \cdot 02 \times 10^{16} \ \mathrm{W/cm^2}.$$

This gives a rough indication of the radiant intensities that would need to be attained in order that the cross-section should approach that for a 1-quantum process. However, one or more of the denominators in (122) will in general be small so that intensities considerably less than J_0 may be sufficient.

It is often convenient to express the rate of a 2-quantum process in terms of a transition probability per ion given by

$$W^{\mathrm{i}(2)} = \delta F^2, \qquad (123)$$

where F is the flux of photons. In terms of $Q^{\mathrm{i}(2)}$ given by (122)

$$\delta = \frac{Q^{\mathrm{i}(2)}}{J} \frac{J}{F}$$

$$= \frac{Q^{\mathrm{i}(2)}}{J} 1 \cdot 6 \times 10^{-19} V \ \mathrm{cm^4/s}$$

$$= \frac{Q^{\mathrm{i}(2)}}{J} 1 \cdot 97_5 \times 10^{-15} \lambda^{-1} \ \mathrm{cm^4/s}, \qquad (124)$$

where V is the photon energy in eV, λ the wavelength in Å, and J is in W cm^{-2}.

Zernik† has applied the formula (122) to calculate $Q^{\mathrm{i}(2)}/J$ for 2-quantum ionization of metastable hydrogen atoms. Fig. 14.99 shows the results obtained for the wavelength range from 4000 to 7300 Å. The resonant peaks occur when $h\nu = E_j - E_0$.

† ZERNIK, W., *Phys. Rev.* **135** (1964) A51.

Thus the peaks at 6500 and 4850 Å arise from vertical transitions to the $3p$ and $4p$ states respectively. The order of magnitude of the quantity δ, the rate of transition per ion for unit photon flux, is also to be noted from Fig. 14.100. Finally, for ruby laser light at 6943 Å, $Q^{i(2)}$ is the same as that for single-photon ionization of the $3p$ state when the light intensity is 6×10^{11} W/cm².

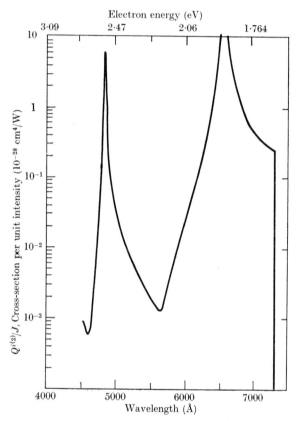

Fig. 14.99. Calculated cross-sections $Q^{i(2)}/J$ per unit intensity for 2-quantum photo-ionization of metastable H atoms.

In a later paper† these calculations were extended to ionization of a hydrogen atom in a state of any principal quantum number n.

With a view to experimental study of the process Hall‡ has calculated the 2-quantum photo-ionization cross-section for caesium. His results, using the $6P$ to $12P$ levels as intermediate states, are illustrated

† ZERNIK, W. and KLOPFENSTEIN, R. W., J. Math. Phys. 6 (1965) 262.
‡ HALL, J. L., J. quant. Electron. 2 (1966) 361.

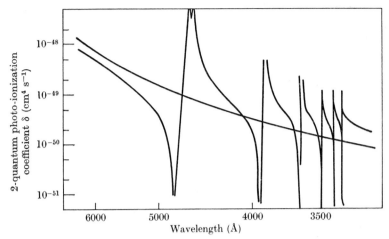

FIG. 14.100. Calculated 2-quantum photo-ionization coefficients δ for Cs atoms. $(\delta = 1 \cdot 97_5 \times 10^{-15} \lambda^{-1} (Q^{i(2)}/J)\ \mathrm{cm}^4/\mathrm{s}).$

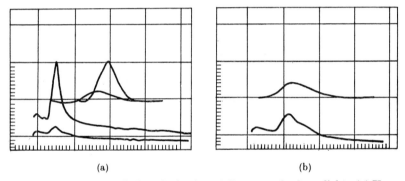

FIG. 14.101. Records of photo-ionization of Cs vapour by laser light. (a) Upper traces due to two successive 5311 Å laser pulses (99 kW/div and 20 ns/div). Lower traces are the corresponding ionization current signals showing non-linear variation with laser power ($1 \cdot 3 \times 10^{-8}$ A/div and 5 μs/div). (b) Upper trace due to 3470-Å laser pulses (20 kW/div and 20 μs/div). Lower trace is the corresponding ionization current (10^{-6} A/div and 2 μs/div).

in Fig. 14.100. As with ionization of $H(2s)$, resonance peaks occur corresponding to each of these intermediate states.

Observation of this process through irradiation of caesium vapour by laser light is difficult because of complications due to:

(a) stray single-quantum photoelectric emission from scattered light,

(b) evaporation of ionized material from the windows,

(c) transient photoconductivity of the caesium-coated glass envelope,

(d) presence of molecular caesium,

(e) non-reproducibility of mode structure of the laser.

Nevertheless, Hall was able to obtain ionization signals that could be positively identified as having arisen from 2-quantum processes. In the experiment, the laser beam produces ions in the vapour that are collected in a shielded system so that no induced current is recorded until the ions reach the shield gauze. As the cross dimensions of the pulsed laser beam are small compared to the separation of the collector plates, time of flight mass analysis of the ions could be carried out. Fig. 14.101 (a) reproduces records taken with 5300-Å photons arising from the second harmonic of a Q-switched neodymium laser. Two successive laser pulses are shown as the upper traces. The ionization signal currents are displayed as the lower traces and it is clear that the signal strength varies faster than linearly with laser power, indicating that the major source is a 2-quantum process.

Fig. 14.101 (b) shows records obtained using the second harmonic of a Q-switched ruby laser, at 3470 Å. Again the upper trace is the laser pulse and the lower the ionization signal. While the main peak in the latter varies quadratically with laser power the second smaller peak shows a linear variation and is probably due to single-quantum ionization of Cs_2. If this is so the dissociation energy of Cs_2^+ must be at least 0·77 eV.

Much more definite measurements of transition probabilities have been made for 2-quantum photodetachment from I^-. These are discussed in Chapter 15, § 8.

COLLISIONS INVOLVING ELECTRONS AND PHOTONS—PHOTODETACHMENT AND RADIATIVE ATTACHMENT

1. Introduction—theoretical behaviour

IN the preceding chapter we discussed radiation processes involving electrons in so-called free states of motion in the fields of positive ions. These included photo-ionization of neutral and ionized atoms and molecules and the inverse process of radiative recombination of electrons to positive ions as well as bremsstrahlung and free–free absorption in the presence of positive ions. We now consider similar processes in which the 'free' electron is moving in the field of a neutral atom or molecule. These include photodetachment of electrons from negative ions and the inverse process of radiative attachment. Free–free absorption and bremsstrahlung by electrons in the presence of neutral atoms are also considered.

While the same theoretical formulae apply to these processes as for those discussed in Chapter 14, the fact that the 'free' electron no longer moves in a long-range field due to the net unscreened charge on the positive ion but one of much shorter range, introduces important differences of detail.

For bound-free transitions that are responsible for photodetachment and radiative recombination the relevant cross-sections are determined by the dipole matrix element

$$\mathbf{r}_{iE} = \int \psi_i^* \, \mathbf{r} \psi_E \, d\tau, \tag{1}$$

where ψ_i is the wave function for the bound state of the negative ion and ψ_E that for the motion of an electron of energy E in the field of the neutral atom or molecule. A variational approximation must be used for the wave function ψ_i. The problem of determining ψ_E is exactly the same as that of determining the wave function describing the collisions of electrons of energy E with the neutral atom or molecule. Much of Chapter 8 has been concerned with just this latter problem.

At high energies ψ_E may be represented by a plane wave but under conditions in which Born's approximation is not valid more elaborate functions must be used, as described in Chapter 8.

Because approximate wave functions must be used in (1) it is often useful to work in terms not of $\mathbf{r}_{iE}$, but of the equivalent velocity or acceleration matrix elements as given by (8) of Chapter 14. Examples of the effective use of these alternative forms are given in § 5. Good use may also be made of the sum rules given in Chapter 7, § 5.2.2 (see also Chap. 14, § 7.1), which take on a specially simple form for negative ions because of the absence of bound excited states.

The threshold behaviour of photodetachment and attachment cross-sections is different from that for photo-ionization, a consequence of the fact that ψ_E has the asymptotic form of a plane wave and not that of a wave modified even at large distances by a long-range Coulomb field.

According to Chapter 14 (5) the photodetachment cross-section

$$Q^{\mathrm{d}} = \frac{32\pi^4 m^2 e^2 v v}{3ch^3} \, |\mathbf{r}_{iE}|^2, \tag{2}$$

where v is the velocity of the ejected electron and ν the photon frequency. The variation with the frequency close to the threshold is therefore determined by that of

$$v\nu|\mathbf{r}_{iE}|^2. \tag{3}$$

To examine this it is convenient to expand $\psi_E(r, \theta)$ in the form

$$\psi_E = r^{-1} \sum i^l(2l+1)G_l(r)P_l(\cos\theta). \tag{4}$$

If the outer electron in the negative ion is in an s-orbital the only term in (4) that contributes to $\mathbf{r}_{iE}$ is G_1. At low electron energies this will give a contribution proportional to k, the electron wave number, if the effective interaction in which the electron moves is that of a short-range field. In that case we would have from (2), remembering $v = k\hbar/m$,

$$Q^{\mathrm{d}} \propto \nu k^3$$

$$= \nu(\nu-\nu_0)^{\frac{3}{2}} \tag{5a}$$

$$= (A/\lambda)\{(\lambda_0-\lambda)/\lambda\lambda_0\}^{\frac{3}{2}}, \quad (s-p), \tag{5b}$$

where ν_0, λ_0 are the threshold frequency and wavelength respectively.

If, on the other hand, the outer electron in the negative ion occupies a p-orbital, then terms in (4) with $l = 0$ and $l = 2$ contribute and of these the first gives the slowest variation of Q^{d} with electron wave number, as k.† We then have for this case

$$Q^{\mathrm{d}} \propto \nu(\nu-\nu_0)^{\frac{1}{2}} \tag{6a}$$

$$= (B/\lambda)\{(\lambda_0-\lambda)/\lambda\lambda_0\}^{\frac{1}{2}}, \quad (p-s). \tag{6b}$$

† G_0 tends to a constant for small k.

Both (5) and (6) have been derived on the assumption that the effective atomic interaction is short range. As pointed out in Chapter 6, § 3.5 this is not correct, because dipole polarization of the atom by the 'free' electron gives rise to an interaction falling off as r^{-4} for large r. It is unlikely that this will modify the behaviour, for small k, of $G_0(r)$, so that (6) should be valid, but that (5) will not be affected is by no means so sure. Thus the effective range expansion of $k^{2l+1}\cot\eta_l$, where η_l is the usual phase shift (Chap. 6, § 3.5), is not valid for $l > 0$ when the polarization is allowed for. However, Geltman† has produced reasons that suggest that (5) is in fact correct, though the range of frequency above the threshold for which it gives a good representation of the behaviour may be quite small.

TABLE 15.1

Values of index appearing in threshold law for molecular photo-detachment

$|\lambda_0^*|$			Heteronuclear molecules	Homonuclear molecules	
				g symmetry	u symmetry
0			1	3	1
$|\lambda_0^*| \geqslant 1$	$|\lambda_0^*|$	even	$2|\lambda_0^*|-1$	$2|\lambda_0^*|-1$	$2|\lambda_0^*|+1$
	$|\lambda_0^*|$	odd	$2|\lambda_0^*|-1$	$2|\lambda_0^*|+1$	$2|\lambda_0^*|-1$

In both cases (5) and (6) the cross-section vanishes at the threshold, in contrast to the finite value of the photo-ionization cross-section under the same circumstances (see Chap. 14, § 2.1). Also the threshold behaviour depends on the orbital angular momentum state of the outer electron in its bound state, the rise from zero being more rapid when it is a p- than when it is an s-electron. In fact Q^d has an infinite slope at the threshold for the former case.

Geltman† has examined the threshold behaviour to be expected when the negative ion is diatomic. The result depends on the quantum number λ_0^*,‡ which specifies the angular momentum about the molecular axis of the detached electron in its initial bound state. It also depends on the molecular symmetry. Expressing the threshold behaviour in the form

$$Q^d \propto \nu(\nu-\nu_0)^m, \tag{7}$$

the values of m found by Geltman are as given in Table 15.1.

As there is usually only one bound state of a negative ion, the radiative attachment cross-section Q^a is obtained directly from that

† GELTMAN, S., *Phys. Rev.* **112** (1958) 176.
‡ We denote this quantum number by λ_0^* to distinguish it from the threshold wavelength λ_0.

for photodetachment from the ground state† by use of the detailed balance relation (see (7) of Chap. 14)

$$Q^{a} = \frac{2h^{2}\nu^{2}}{m^{2}v^{2}c^{2}} \frac{g_{-}}{g_{0}g_{e}} Q^{d}, \tag{8}$$

where g_{-}, g_{0}, g_{e} are the respective statistical weights of the ground states of the negative ion, neutral atom, and electron respectively. Thus we have for the threshold behaviour of Q^{a}, using (5), (6), and (8),

$$Q^{a} \propto k(k^{2}+k_{0}^{2})^{3} \qquad \text{for capture into an } s\text{-orbital}, \tag{9a}$$

$$\propto k^{-1}(k^{2}+k_{0}^{2})^{3} \quad \text{for capture into a } p\text{-orbital}, \tag{9b}$$

k being the initial wave number of the incident electron.

Once again there is a clear difference between the cases of capture into s- and p-orbitals.

In general, the photodetachment cross-section well above the threshold will be of the order 10^{-18}–10^{-19} cm^2 and that for radiative attachment of the order 10^{-21}–10^{-22} cm^2.

We now discuss the measurement and calculation of these cross-sections. Experimental techniques for the study of photodetachment and radiative attachment will first be described. The results obtained will then be discussed for separate atomic and molecular ions in relation to the theory. This will be followed with an account of the theory of free–free transitions with particular reference to the case of hydrogen and its applications to solar physics. Finally, 2-quantum photodetachment, which has been definitely observed for I$^-$ using a ruby laser beam, will be discussed.

2. The crossed-beam method for measuring photodetachment cross-sections

The use of crossed-beam techniques for measuring collision cross-sections between species, neither of which can be concentrated in bulk, has been discussed in earlier chapters—elastic and inelastic collisions of electrons with atomic hydrogen, oxygen, and nitrogen in Chapters 1 and 3, with metastable helium in Chapter 4, with ionized helium and neon in Chapter 3, and with molecular hydrogen ions in Chapter 13. In all cases use was made of phase-sensitive frequency-modulated detection to discriminate between the wanted signal and background noise. In fact the first use of this technique was made in the study of photodetachment, the crossed beams being a light beam and one of

---

† This is only true provided the photon frequency is below the threshold value for a detachment process that leaves the atom in an excited state.

H$^-$ ions. These pioneering experiments were carried out by Branscomb and Fite[†] and later extended to study many other photodetachment processes. Apart from the difficulties inherent in the use of the modulated crossed-beam method there are many additional problems introduced by the fact that one of the colliding beams is a stream of photons. Results are required as a function of frequency and of high accuracy, at least in relative magnitude. No light source of continuously variable, effectively monochromatic radiation in much of the wavelength range of interest is available and considerable pains had to be taken to obtain accurate results nevertheless. In any case, unusually powerful light sources were required to make the experiments practicable, particularly at first. At a later stage the alternating current electrometer amplifier used to detect the detached electron beam was replaced by an electron multiplier. This made it possible to work with weaker ion beams, thereby extending the scope of the experiments to a wider range of ionic species, or with less intense light sources so that effectively monochromatic radiation could be used. We shall describe first the technique used in the earlier experiments in which different wavelength bands were isolated by filters and the detachment cross-section as a function of frequency derived from analysis of data obtained using the radiation transmitted by a selection of these filters. With this technique satisfactory results have been obtained for H$^-$ and O$^-$ in particular. Improved techniques, using nearly monochromatic radiation, have been applied to I$^-$ and to the molecular ions OH$^-$ and OD$^-$ and will be described in §§ 5.6 and 6 respectively.

The chance that a single negative ion of velocity v passing through the photon beam should suffer photodetachment is given by

$$P = v^{-1} \int\int Q^{\mathrm{d}}(\lambda) n_\nu(\lambda, x) \, \mathrm{d}\lambda \mathrm{d}x, \tag{10}$$

where $n_\nu(\lambda, x) \, \mathrm{d}\lambda \mathrm{d}x$ is the flux density of photons, at a point along the path of the ion at a distance between x and $x+\mathrm{d}x$ from some reference position, in the wavelength range λ to $\lambda+\mathrm{d}\lambda$, and $Q^{\mathrm{d}}(\lambda)$ is the photodetachment cross-section for radiation of wavelength λ. Taking 300 eV as a typical ion energy and Q^{d} as of the order 10^{-18} cm^2, we find that P is about $2 \cdot 5 \times 10^{-7}$ per cm of ion path per W/cm^2 photon flux density. At the working background pressure ($\sim 10^{-6}$ torr) the chance of detachment by collision with impurities is $\sim 10^{-5}$ per cm of path. The optical system that was devised illuminated about 2 cm of ion path with about $\frac{1}{4}$ W/cm^2. The need for the modulation and phase-sensitive detector is obvious.

† Branscomb, L. M. and Fite, W. L., *Phys. Rev.* **93** (1954) 651A.

Two sources of noise limit the performance of the detecting system—thermal noise in the preamplifier and noise in the ion source. As a compromise between reduction of the latter, which was smaller at lower frequencies, and maintaining a high input impedance at the collector, the photon beam was chopped at a frequency of 450 c/s. For optimum signal-to-noise ratio the two noise sources should be of comparable intensity. This requires ion beam currents at least as great as 10^{-8} A.

For many ions such as H^-, O^-, OH^-, it is not difficult to obtain beams of this intensity but for others, such as C^-, it is difficult to obtain more than 10^{-10} A. To work with such ions the background pressure was reduced to 10^{-7} torr, so reducing the signal-to-noise ratio by about ten. A further advantage was gained by replacing the electron collector and preamplifier by an electron multiplier, as mentioned earlier.

In typical experiments[†] ions were produced in a glow discharge at pressures between 25 and 75×10^{-6} torr and emitted through a hole in the anode, 0·02-in diameter and 0·1 in long, with energy between 300 and 500 eV and energy spread of 25–50 eV. Referring to Fig. 15.1, the issuing ions were accelerated by about 3000 V at electrode 1, focused into a parallel beam by the decelerating lens formed by electrodes 1 and 2, and accelerated in a parallel beam by lenses formed by electrodes 3 to 8 to enter a 90° sector mass-analyser. To avoid serious defocusing when carrying out a mass-scan electrostatically, a two-dimensional lens was inserted between electrode 8 and the sector. This consisted of two parts, one tied to electrode 8 and the other to the sector, so a potential difference could be maintained between it and electrode 8. After passing through the sector the beam entered an axially symmetric lens system consisting of electrodes 10–14, which refocused it into the reaction chamber with a kinetic energy of a few hundred eV. The whole system was designed neither for high resolution nor for accurate determination of relative abundances of ions but for high transmission.

Electrons produced by photodetachment in the reaction chamber were separated from the ions and trapped by a magnetic field of 40 gauss and an electric field of 15 V/cm, both perpendicular to the ion and photon beams. In later experiments they were then accelerated to an energy of 500 eV and focused on the cathode of a ten-stage Ag–Mg electron multiplier.

To obtain relative measurements of the photodetachment cross-section at different wavelengths, the procedure was now as follows. Allowing for variation of flux density across the ion beam, the effective

† SMITH, S. J. and BRANSCOMB, L. M., *Rev. scient. Instrum.* **31** (1960) 733.

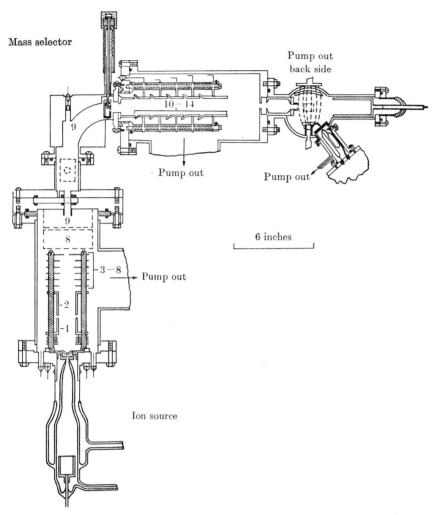

Mass selector

Pump out
back side

Pump out

Pump out

Pump out

6 inches

Pump out

Ion source

FIG. 15.1. Arrangement of the ion source in crossed-beam photodetachment experiments.

detachment probability per ion when a narrow-band pass filter m is introduced in the photon beam as indicated in Fig. 15.2 is given by

$$P_{\text{eff,m}} = \frac{CW}{v} \int Q^{\text{d}}(\lambda)\phi(\lambda)T_{\text{m}}(\lambda)\lambda \, \text{d}\lambda, \tag{11}$$

where W is the unfiltered beam power and $\phi(\lambda)$ is the spectral distribution of the unfiltered radiation at the ion beam. C is a geometrical factor and $T_{\text{m}}(\lambda)$ defines the band-pass distribution of a particular filter introduced as indicated in Fig. 15.2. If $T_{\text{m}}(\lambda)$ is chosen so that Q^{d} varies

U

linearly over the range of λ involved then we find

$$vP_{\text{eff,m}} = CWQ(\lambda_{\text{m},2})\lambda_{\text{m},1} \int \phi(\lambda)T_{\text{m}}(\lambda) \, \mathrm{d}\lambda, \qquad (12)$$

where

$$\lambda_{\text{m},s} = \int \phi(\lambda)T_{\text{m}}(\lambda)\lambda^s \, \mathrm{d}\lambda \Big/ \int \phi(\lambda)T_{\text{m}}(\lambda)\lambda^{s-1} \, \mathrm{d}\lambda. \qquad (13)$$

Hence

$$Q(\lambda_{\text{m},2}) = vP_{\text{eff,m}} \Big/ CW\lambda_{\text{m},1} \int \phi(\lambda)T_{\text{m}}(\lambda) \, \mathrm{d}\lambda$$

$$= vj_{\text{e,m}}/j_{\text{i}} \, C\lambda_{\text{m},1} \, W_{\text{m}}, \qquad (14)$$

where $j_{\text{e,m}}$ is the current of detached electrons, j_{i} the ion current, and W_{m} the total power in the photon beam filtered through the band-pass

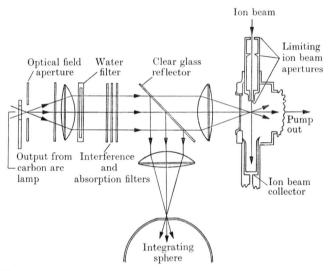

FIG. 15.2. Arrangement of the optical system in crossed-beam photo-detachment experiments.

filter m. Relative measurements of Q thus depend on relative measurements of $vj_{\text{e,m}}/j_{\text{i}} W_{\text{m}}$. Careful checks were made to establish that the factor C does not depend appreciably on wavelength while time dependent variations were averaged out.

Twenty-six band-pass filter combinations, including coloured glass absorption filters, sharp cut-off interference reflection filters, and water cells, were used to cover the wavelength range from 0·4 to 1·7 μm.

Fig. 15.2 illustrates the arrangement of the optical system that focused the light from a carbon arc (replaced by a xenon lamp in later experiments) on the ion beam in the reaction chamber. Between the two lenses the light beam was parallel for about 8 in, which made it possible to insert multilayer interference filters, glass absorption filters, and water cells in the path of the beam. This was necessary in order to determine the dependence of the photodetachment cross-section on frequency.

The total power in the light beam transmitted through any particular combination of filters was measured in the following way. A sheet of clear glass 1 mm thick was placed diagonally in the light beam as shown in Fig. 15.2 to reflect off about 8 per cent of the radiation into a lens identical with the first, which focuses it through a 1-in hole into a spherical cavity of 4-in outside diameter. If r is the radius of the sphere and R the reflecting power of the inner surface, then the brightness I of the interior of the sphere, which is indirectly illuminated, is given by

$$I = \frac{S}{4\pi r^2} \frac{1}{1-R},\tag{15}$$

where S is the total radiation power.

The inner surface of the sphere was coated with a white ceramic for which R was equal to 0·85 within a few per cent over the relevant range of wavelength.

Under typical circumstances I was about 4×10^{-4} W/cm^2 and was measured by a bolometer, of receiving area 0·07 cm^2 and a sensitivity of about 0·4 V(r.m.s.)/W at 450 c/s, inserted through a $\frac{1}{2}$-in hole in the sphere.

Because the bolometer signal is proportional to $1/(1-R)$ it is sensitive to even a small dependence of R on wavelength. There are also other smaller sources of optical selectivity in the system so that it was necessary to calibrate it by reference to measurements with a non-selective detector. This was done by measuring, as a function of wavelength, the ratio W'/W of the bolometer reading to the intensity, measured calorimetrically, of the primary beam at the position of the ion beam. The calorimeter consisted of a copper disc on a blackened face of which the beam was incident normally. Such measurements were carried out using each of twenty-six narrow band-pass filters.

Denoting by $W'(\lambda_m)$ the bolometer reading when the radiation is passed through a filter for which the mean transmitted wavelength is λ_m ($\lambda_{m,1}$ in (14)), the ratio

$$T'(\lambda_m) = W'(\lambda_m)/W$$

was measured for different filters, and hence λ_m. The filter for which $\lambda_m = 5280$ Å was used as a control and $T'(\lambda_m)$ was measured relative to $T'(5280$ Å$)$. To eliminate slow-drift effects the different filters were each compared six times in a cyclic manner with the control. Fig. 15.3

illustrates results obtained for $T'(\lambda_m)/T'(5280 \text{ Å})$ as a function of λ_m for twenty-four filters. These were compared with the spectral variation of $1/(1-R)$ (see (15)) determined from the measured reflectivity of the integrating sphere and with the measured spectral response of the bolometer to parallel light. From time to time checks were made that the calibrated ratios remained unchanged.

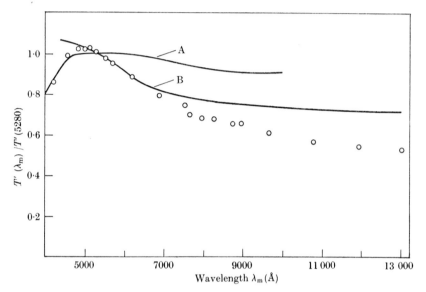

FIG. 15.3. Illustrating the calibration of the bolometer monitoring system for different band-pass filters. ○, measured values for different filters. Curve A, measured spectral variation of $1/(1-R)$ for the integrating sphere. Curve B, measured spectral variation of bolometer response to parallel light.

A modified procedure was necessary in studying the behaviour of the cross-section near the threshold. This will be described below (§ 5.2) in discussing the observations on O⁻.

To make absolute measurements, only a limited number of which are required, the carbon arc was replaced by a tungsten lamp, very much more stable in operation. With this lamp the diameter of the ion beam was irradiated uniformly so the detachment probability P given by (10) was the same for all ions. We then have for the ratio j_e/j_i of detached electron to ion current

$$j_e/j_i = v^{-1} \int Q^d(\lambda) \left\{ \int n_\nu(\lambda, x) \, dx \right\} d\lambda, \tag{16}$$

with $\qquad \int n_\nu(\lambda, x) \, dx = \phi(\lambda) T(\lambda)(\lambda/hc) \int w(x) \, dx, \tag{17}$

$w(x)$ being the incident power density at the ion trajectory, $T(\lambda)$ the transmission of the lamp envelope and other optical elements, and $\phi(\lambda)$ the spectral distribution at the source normalized so that

$$\int \phi(\lambda)T(\lambda)\, \mathrm{d}\lambda = 1.$$

$\int w(x)\,\mathrm{d}x$ was measured with an absolute radiometer, $\phi(\lambda)$ obtained from measurement of the colour temperature of the lamp and the emissivity of tungsten, and $T(\lambda)$ measured with a spectrophotometer. From observed values of $j_\mathrm{e}/j_\mathrm{i}$ and v, $Q^\mathrm{d}(\lambda)$ could then be determined in principle by inversion of the integral equation (16). In practice the easiest method is to substitute theoretical values for $Q^\mathrm{d}(\lambda)$ and check the consistency.

3. Shock-wave method for observing emission and absorption spectra of negative ions

It is possible by means of shock-wave heating (see Chap. 17, § 2.2) to produce an atmosphere of negative halogen atomic ions with concentrations of order $10^{15}/\mathrm{cm}^3$ over a region with lateral dimensions of order 10 cm. As photodetachment cross-sections are of the order $10^{-17}\,\mathrm{cm}^2$ absorption of light by the negative ions should be observable photographically.

The equilibrium constant K for the reaction

$$\mathrm{CsI} \rightleftharpoons \mathrm{Cs^+} + \mathrm{I^-} \tag{18}$$

in the gas phase may be calculated from the known rotational, vibrational, and electronic energy levels in the usual way. It is then found that
$$K = \frac{n(\mathrm{Cs^+})n(\mathrm{I^-})}{n(\mathrm{CsI})} = \begin{cases} 5 \times 10^{17} & \text{at 3000 °K,} \\ 2{\cdot}6 \times 10^{19} & \text{at 4000 °K,} \end{cases}$$

$n(X)$ being the concentration of the species X. If we take $n(\mathrm{Cs^+}) = n(\mathrm{I^-})$ then, for $n(\mathrm{CsI}) = 10^{17}/\mathrm{cm}^3$, $n(\mathrm{I^-})$ is between 10^{17} and $10^{18}/\mathrm{cm}^3$ over the temperature range concerned. This will be modified by the reaction

$$\mathrm{Cs^+} + \mathrm{I^-} \rightarrow \mathrm{Cs} + \mathrm{I}, \tag{19}$$

for, ignoring the presence of free electrons or $\mathrm{I_2}$ molecules so that $n(\mathrm{Cs}) = n(\mathrm{I})$ as well as $n(\mathrm{Cs^+}) = n(\mathrm{I^-})$, we find, from the equilibrium constant for (19), that $n(\mathrm{I})/n(\mathrm{I^-})$ varies from about 13 at 3000 °K to 9 at 4000 °K. From similar considerations applied to the reactions

$$\mathrm{I} + \mathrm{e} \rightleftharpoons \mathrm{I^-}, \tag{20 a}$$

$$\mathrm{I} + \mathrm{I} \rightleftharpoons \mathrm{I_2}, \tag{20 b}$$

it seems that, while $n(I_2)$ is negligible at the temperatures concerned, $n(e)$ might be important at 4000 °K or higher. Allowing for these factors, values of $n(I^-)$ of the required order should be attainable.

These considerations were sufficiently promising for an experimental programme† to be undertaken in which the high temperatures required were obtained by shock-wave heating. Not only have good absorption spectra been observed photographically for the halogen negative ions but it has also proved possible to observe emission due to electron capture.

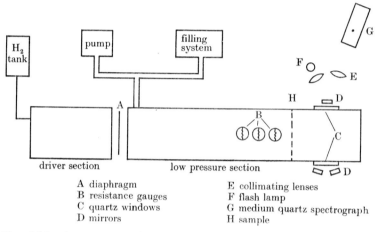

A diaphragm   E collimating lenses
B resistance gauges   F flash lamp
C quartz windows   G medium quartz spectrograph
D mirrors   H sample

Fig. 15.4. Arrangement of apparatus for observation of the absorption coefficient of the vapour released by shock heating of alkali halide salts.

Fig. 15.4 illustrates schematically the apparatus used. The shock tube was of steel with interior cross-section $8 \cdot 5 \times 8 \cdot 5$ cm². The driver section was 5 ft long and could be filled with hydrogen to pressures between 200 and 400 lb/in². This section was isolated from the low-pressure section, which was initially filled with argon at pressures between 15 and 25 torr and was in five parts, each 3 ft long. One part carried three resistance gauges B, made of platinum painted in thin strips on glass, and another carried fused silica windows C on the top and opposite sides of the tube. Three front-surfaced mirrors D, forming a multiple-reflection system, were carried by a small aluminium beam mounted on the tube. The salt specimen H ($0 \cdot 1$–$1 \cdot 0$ g) was deposited on cellulose tissue or thin perforated aluminium foil clamped in the tube 90 cm up-stream from the observation windows.

† Berry, R. S., Reimann, C. W., and Spokes, G. N., J. chem. Phys. 35 (1961) 2237; Berry, R. S. and Reimann, C. W., ibid. 38 (1963) 1540.

Passage of the shock past each resistance gauge was recorded by a voltage change, displayed as a pulse on an oscilloscope triggered by the first arriving signal. The oscilloscope itself supplied a gate signal to a variable delay circuit from which the flash-lamp F was triggered. By adjustment of the multiple reflection system to give four traverses of

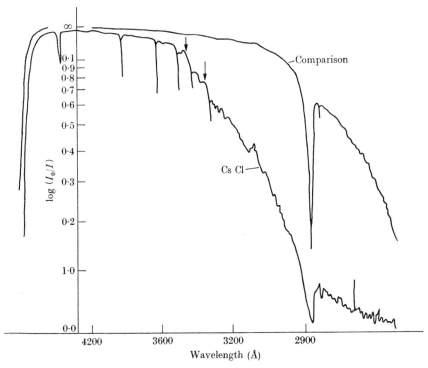

Fig. 15.5. Microdensitometer tracing of the absorption spectrum of the vapour from CsCl evaporated by shock heating. A comparison spectrum taken without any CsCl present is also shown. The arrows indicate photodetachment thresholds for Cl⁻.

the tube the optical path length was about 34 cm. After these terminals the light was dispersed by a Bausch and Lomb medium quartz spectrograph and recorded photographically. Comparison spectra were taken both of the flash-lamp alone and when the supporting material was used without the alkali halide salt.

Typical absorption spectra obtained are shown in Figs. 15.5 and 15.6. Many lines appear due to absorption by neutral alkali-metal atoms. These include not only the principal series but also forbidden lines and lines arising from transitions between excited states. Other atomic lines are also observed and, when cellulose tissue is used as backing, bands of OH and C_2 appear. The strong continuous absorption is obvious, exhibiting a very rapid rise of extinction as the frequency increases beyond a threshold value. These continua are not at all similar to those that arise from atomic ionization, normal molecular dissociation, or alkali halide dissociation. The form of the continua is the same for a given halogen (cf. Fig. 15.6) independent of the nature of the alkali metal. Moreover, the extinction increases with driving pressure and

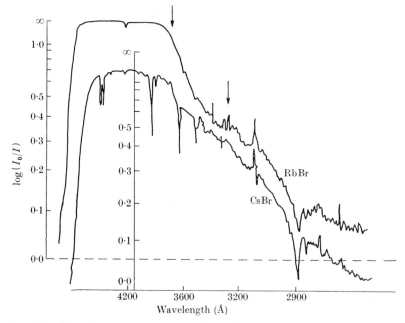

Fig. 15.6. Microdensitometer tracings of the absorption spectrum of the vapour from RbBr and CsBr, evaporated by shock heating. The arrows indicate photo-detachment thresholds for Br^-.

hence shock temperature, and shows a dependence on size of salt sample that correlates well with that of the intensity and width of the alkali metal lines.

All of this evidence implies an atomic source, coming from the salts. In any case neutral or ionized halogen molecules are ruled out, as are also neutral halogen atoms, because their absorption falls in the ultra-violet (the concentration of excited atoms which will absorb in the visible is far too small). The same considerations apply *a fortiori* to halogen positive ions.

This all points strongly to negative halogen atomic ions as the source of absorption. Since in the halogen atoms photodetachment involves the active electron in a p–s transition, the detachment cross-section increases as $(\nu - \nu_0)^{\frac{3}{2}}$ from the threshold (see (6)), ν_0 being the threshold frequency. The infinite slope of the cross-section frequency curve at the threshold is in agreement with the characteristic feature of the continua—the rapid increase of extinction with frequency above the threshold.

Photodetachment absorption spectra have been observed for all four halogen ions, although it was necessary for F^- to use very strong shocks to produce a sufficient concentration. Electron affinities have been derived in all cases and, in addition, for F^-, Cl^-, and Br^- it has also been possible to determine the fine-structure splitting in the ground state of the neutral atom. These results will be discussed in § 5.6.

The temperatures produced by strong shocks, with 320–400 lb/in² driving pressure, are high enough to generate considerable concentrations of free electrons. Berry and David† have been able to observe in their shock-tube apparatus emission continua due to radiative capture of electrons by neutral Cl, Br, and I atoms. For this purpose the only changes in the arrangement indicated in Fig. 15.4 were the elimination of the flash-light and the use of the mirrors of the multiple-path cell to trap light and direct it to the spectrograph. In addition the spectra were taken with a high-speed spectrograph which gave a time resolution of 20 μs or so. From these spectra, examples of which are shown in Fig 15.7, it is possible to determine the threshold wavelengths and hence again the electron affinities. The results will be discussed in § 5.6 below.

Successful observations of emission continua due to capture of electrons by neutral atoms in a shock-excited gas were first made a little earlier by Weber for hydrogen.‡ As in the experiments described above he used hydrogen as driving gas, but a mixture of 80 per cent krypton and 20 per cent hydrogen as the low-pressure gas. Working between a driving pressure of 100 atm and a low pressure of 28 torr and using a closed low-pressure section so that the pressure and temperature would be enhanced by a reflected shock wave, it was possible to obtain temperatures of 6000 °K or more. These are high enough to generate a considerable concentration of free electrons from the krypton but are not so high as to reduce the population of H^- to negligible proportions.

It is possible, from measurement of the velocity of either the primary or the reflected shock wave and knowledge of the initial conditions of

† BERRY, R. S. and DAVID, C. W., *Atomic collision processes*, p. 543 (North Holland, Amsterdam, 1964).

‡ WEBER, O., *Z. Phys.* **152** (1958) 281.

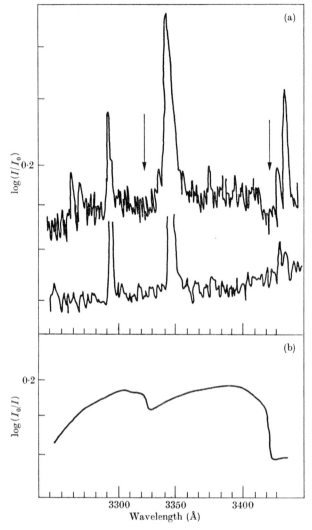

Fig. 15.7. (a) Microdensitometer tracings of the emission spectrum
from shock-heated RbCl (upper curve) and RbI (lower curve). The
onset wavelengths for two continua in the former case are indicated
by the arrows. (b) Relative intensity distribution in the continua
present in the upper tracing of (a), allowing for plate calibration.

temperature and pressure, to determine the temperature and pressure
in the respective shock (see Chap. 17, § 2.2). Given this, the Saha
equation may be used to calculate the concentrations n_e, $n(H)$, $n(Kr)$,
$n(H^+)$, $n(Kr^+)$, and $n(H^-)$ (see (21), (22), and (23)).

The shock wave velocities were measured optically, photographically with a film rotating at 132 rev/s viewing the shock space through a side window, and photoelectrically by observing the passage of the light pulse past two windows 50 cm apart along the shock tube.

As a typical example, working between 100 atm and 28 torr the velocity of the reflected shock was found to be 10^4 cm/s, giving a temperature of 8600 °K in the shock front and a pressure of $2 \cdot 95 \times 10^7$ dyn/cm^2. This gives the particle concentrations/cm^3:

$$n_e = 4 \cdot 8 \times 10^{16}, \quad n(\mathrm{H}) = 8 \cdot 3 \times 10^{18}, \quad n(\mathrm{H^+}) = 3 \cdot 9 \times 10^{15},$$
$$n(\mathrm{H^-}) = 1 \cdot 45 \times 10^{14}, \quad n(\mathrm{H_2}) = 1 \cdot 66 \times 10^{19}, \quad n(\mathrm{Kr^+}) = 4 \cdot 4 \times 10^{16}.$$

Under these conditions the intensity to be expected from the H$^-$ capture continuum at wavelengths between 4400 and 5500 Å is about thirty times greater than from any other source—principally the continua due to electron capture by H$^+$ and Kr$^+$. At longer wavelengths an appreciable contribution is made by free–free (bremsstrahlung) transitions.

Fig. 15.8 (a) shows the calculated intensities as functions of wavelength. The Balmer discontinuity (N) is very prominent in the H$^+$ capture continuum (labelled H) but when all continuum emission is added it is not very apparent. The observed variation of total intensity with wavelength is shown in Fig. 15.8 (b) and is very similar to that predicted, particularly in that the discontinuity, smoothed out by line broadening, represents a rise of only about 25 per cent as compared with an expected factor of 10 if the H capture continuum (labelled H$^-$) were absent.

4. Observations of negative ion emission continua from high-pressure arc discharges

Historically, the first direct observations† of the H$^-$ emission continuum were made in 1951 in Lochte-Holtgreven's laboratory at Kiel. In one case‡ the source was a spark discharge in hydrogen in a Geissler tube at pressures ranging from 20 to 60 atm, and in another a spin-stabilized arc discharge through hydrogen at a pressure between 0·4 and 1·0 atm.§ Further experiments used as sources a high-pressure hydrogen arc‖ (10–140 atm) and an arc in water vapour†† at a pressure between 25 and 150 atm.

In later experiments, results of which are described in §§ 5.2, 5.5, and 5.6, a wall-stabilized arc was used. Thus, in experiments in oxygen,

† LOCHTE-HOLTGREVEN, W., *Naturwissenschaften*, **38** (1951) 258.
‡ FUCHS, R., *Z. Phys.* **130** (1951) 69.
§ LOCHTE-HOLTGREVEN, W. and NISSEN, W., *Z. Phys.* **133** (1952) 124.
‖ NISSEN, W., ibid. **139** (1954) 638.   †† PETERS, T., ibid. **135** (1953) 573.

the walls consisted of centrally bored copper discs, isolated from each other. These were water-cooled to obtain the maximum possible duration for the arc. Oxygen was admitted tangentially through a hole of diameter 7 mm and length 5 cm. A stream of compressed air was admitted obliquely into the central holes to ensure removal of hot gas

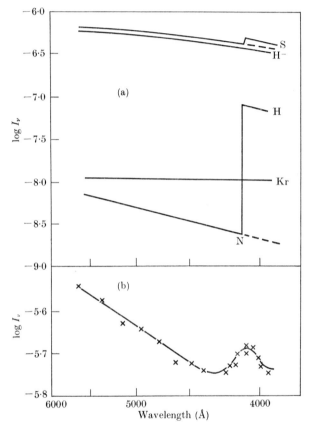

Fig. 15.8. (a) Calculated intensity I_ν in quanta/cm³/s per unit frequency range, of emission spectra from a shock-heated mixture of hydrogen and krypton. The contributions from H, Kr, and H⁻ continua are respectively indicated as well as the total, S. N denotes the Balmer limit. (b) Observed intensity I_ν.

layers, which would otherwise cause the emitting region to extend beyond the arc column. The arc current was 80 A at the operating pressure of 1 atm.

Optical observations were made 'end-on' through one of the pierced electrodes, so that an image of the arc column with about fivefold magnification was produced at the slit of the spectrograph, which had a dispersion of about 80 Å/mm at 5000 Å. As a radiation standard an image of the anode hole spot of a carbon arc could be produced by the same optical system.

The major problem in all this work is to disentangle the different continua. We describe briefly the technique that has been found effective and applied to arc discharges in oxygen† and nitrogen‡ by Boldt and chlorine by Henning.§ In all cases the arc was water-cooled and wall-stabilized and operated at a pressure of about 1 atm. The arc current was between 50 and 100 A.

In an arc of this kind thermodynamical equilibrium applies at a certain temperature T, which is so high that the concentration of molecular species is negligible. The concentrations of electrons (n_e), positive ions (n_+), neutral atoms (n_0), and negative ions (n_-) are given in terms of the Saha relations:

$$\frac{n_+ n_e}{n_0} = S_0(T), \qquad \frac{n_0 n_e}{n_-} = S_-(T), \tag{21}$$

where

$$S_0 = \frac{2(2\pi m\kappa T)^{\frac{3}{2}}}{h^3} \frac{Z_+(T)}{Z_0(T)} e^{-\chi_0/\kappa T}, \tag{22}$$

$$S_- = \frac{2(2\pi m\kappa T)^{\frac{3}{2}}}{h^3} \frac{Z_0(T)}{Z_-(T)} e^{-\chi_-/\kappa T}. \tag{23}$$

Here Z_0, Z_+, Z_- are the partition functions for the neutral atom, positive ion, and negative ion at temperature T. χ_0 is the ionization energy and χ_- the electron affinity of the atom. Given the temperature T and the pressure p, the concentrations of all four species can be determined from the equations (21)–(23), the gas law, and the plasma condition

$$n_+ = n_e + n_-. \tag{24}$$

The temperature is determined by measuring the intensity, of a chosen multiplet, emitted per unit solid angle from a length l of the arc light. In terms of the transition probability A_{nm} for the multiplet, the intensity is given by

$$I_{nm} = \frac{1}{4\pi} \frac{hc}{\lambda} A_{nm} n_m(T) l, \tag{25}$$

where n_m is the concentration of atoms in the mth state. Also

$$n_m = \frac{n_0(T)}{Z_0(T)} g_m e^{-E_m/\kappa T}, \tag{26}$$

where g_m and E_m are the respective statistical weight and energy of the mth state. Hence, if A_{nm} is known, T can be determined from I_{nm}.

† BOLDT, G., Z. Phys. **154** (1959) 319.   ‡ Ibid. **154** (1959) 330.
§ HENNING, H., ibid. **169** (1962) 467.

Emission continua arise from three sources—free–free transitions of electrons in the field of the positive ions, the recombination continuum due to capture by positive ions, and the electron affinity continuum due to capture by neutral atoms. By avoiding measurement at too long wave lengths (beyond 6500 Å) the contribution from free–free transitions in the field of the neutral atoms is kept low enough to be ignored.

Under the thermal equilibrium conditions we can express the intensities of the radiation in the respective continua, at temperature T, as

$$\left.\begin{array}{l} I_\lambda^{ff}(T) = \mathscr{Q}_\lambda^{ff}(T)n_e(T)n_+(T) \\ I_\lambda^{fb}(T) = Q_\lambda^{fb}(T)n_0(T) \\ I_\lambda^{fa}(T) = Q_\lambda^{fa}n_-(T) \end{array}\right\} B_\lambda(T)(1-e^{-hc/\lambda\kappa T})l, \qquad (27)$$

where
$$B_\lambda(T) = (8\pi hc/\lambda^5)(e^{hc/\lambda\kappa T}-1)^{-1}. \qquad (28)$$

$\mathscr{Q}_\lambda^{ff}(T)$ is the absorption cross-section per electron per positive ion defined in Chapter 14, § 11.1. $Q(T)_\lambda^{fb}$ is a mean absorption cross-section for photo-ionization defined by

$$n_0(T)Q_\lambda^{fb}(T) = \sum_s n_s(T)Q_\lambda^{(s)}, \qquad (29)$$

where $n_s(T)$ is the concentration and $Q_\lambda^{(s)}$ the photo-ionization cross-section of an atom in the sth excited state. Finally, Q_λ^{fa} is the cross-section of the negative ion for photodetachment. Because there are no stable excited states of the negative ion Q_λ^{fa} is independent of T.

The sum $I_\lambda(T)$ of all these intensities is directly measured and $I_\lambda^{ff}(T)$ is directly estimated using $\mathscr{Q}_\lambda^b(T)$ as given from the general formula of Peach (Chap. 14, § 11.1). I_λ^{ff} may then be subtracted to give

$$\{B_\lambda(1-e^{-hc/\lambda\kappa T})l\}^{-1}\{I_\lambda^{fb}(T)+I_\lambda^{fa}(T)\} = Q_\lambda^{fb}(T)n_0(T)+Q_\lambda^{fa}n_-(T) \quad (30)$$
$$= Q_\lambda(T)n_0(T), \quad \text{say.}$$

Since $Q_\lambda^{(s)}$ rises from zero as λ decreases beyond a threshold value $\lambda_0^{(s}$ given by
$$hc/\lambda_0^{(s)} = E_s,$$

the form of $Q_\lambda^{fb}(T)n_0(T)$ will be as shown in Fig. 14.96. On the other hand, Q_λ^{fa} will vary smoothly with λ, having the general form shown in Fig. 15.12. It is not difficult, given such a curve, to sort out the contributions from the separate recombination continua, leaving a background that represents the electron affinity continuum.

From this Q_λ^{fa} may be derived. An important check is that, when this process is carried out at different temperatures, the same values of Q_λ^{fa} should be obtained.

Results obtained by this method for oxygen, nitrogen, and chlorine are discussed in §§ 5.2, 5.5, and 5.6 respectively.

5. Results for atomic negative ions

5.1. H⁻

The relative photodetachment cross-section for H⁻ was first measured by Branscomb and Fite[†] in 1954. Absolute values were obtained a little later by Branscomb and Smith,[‡] who measured the cross-section for radiation of wavelength greater than 4000 Å by the method described in § 2. Taken together with relative measurements this gave $Q^d = 3 \cdot 28 \pm 0 \cdot 3 \times 10^{-17}$ cm² at the standard wavelength, 5280 Å. The accuracy of relative measurements was improved by Smith and Burch[§] who measured the cross-section from 4000 to 13 000 Å relative to that at 5280 Å with probable errors of only 2 per cent. Experimental results are shown in the curves of Fig. 15.9, which shows the comparison with various theoretical calculations.

The electron affinity capture continuum of H⁻ has been observed in the spectra of strong shock waves (see § 3), and of arcs and sparks (§ 4). In all cases the observed intensities agree well with those estimated from the crossed-beam photodetachment observations. Cross-sections for radiative attachment for various electron energies are given on p. 1260 after the discussion of the theory of photodetachment and the comparison between theory and experiment. It will be found that the position is now very satisfactory and that the cross-sections are known quite accurately.

A remarkable amount of effort has been devoted to the calculation of cross-sections for absorption of radiation by H⁻ ions, stimulated by the great importance of this particular process in the solar atmosphere.

It was first pointed out by Wildt[‖] in 1939 that, due to capture of electrons released from metallic atoms of low ionization potential, a significant concentration of H⁻ might occur in the solar atmosphere, large enough to be the main source of opacity in the infra-red region of the solar spectrum.

Rough calculations[††] of the relevant absorption cross-section had already been made at that time using crude wave functions, but were sufficient to indicate the importance of Wildt's suggestion. In 1940 Bates and Massey[‡‡] recalculated the cross-section, using for the bound-state wave function a 3-parameter variational function due to

† BRANSCOMB, L. M. and FITE, W. L., *Phys. Rev.* **93** (1954) 651A.
‡ BRANSCOMB, L. M. and SMITH, S. J., ibid. **98** (1955) 1028.
§ SMITH, S. J. and BURCH, D. S., *Phys. Rev.* **116** (1959) 1125.
‖ WILDT, R., *Astrophys. J.* **89** (1939) 295.
†† JEN, C. K., *Phys. Rev.* **43** (1933) 540; MASSEY, H. S. W. and SMITH, R. A., *Proc. R. Soc.* **A155** (1936) 472.   ‡‡ MASSEY, H. S. W. and BATES, D. R., ibid. **91** (1940) 202.

Hylleraas[†] and taking a plane wave for the free state. The latter approximation, while simple, is really quite good, as the transition concerned is from a bound state to a free p-state. As discussed in Chapter 8, § 2.7, the p-state phase shifts for electron scattering by hydrogen atoms are always small. Bates and Massey showed in fact that no serious error could arise from the free-state approximation. Nevertheless the calculated cross-sections, while of about the correct magnitude for the solar application, showed a maximum at a wavelength near 4000 Å, a few thousand Å shorter than indicated from the solar emission.

In 1942 Williamson[‡] repeated the calculations using a 6-parameter bound-state function of Hylleraas type and thereby shifted the calculated wavelength of maximum absorption to about 5000 Å, a distinct but not adequate improvement.

The use of an 11-parameter function by Henrich[§] further improved the situation. However, Chandrasekhar[||] pointed out that large contributions come to the dipole-length matrix element from values of $r > 10a_0$, whereas in the expression for the total energy < 5 per cent comes from such large values. It is therefore quite possible to be misled about the accuracy of a variational wave function for dipole-length calculations. In an effort to improve the position Chandrasekhar introduced the equivalent dipole velocity and acceleration formulae in which there is a greater emphasis on contributions from smaller values of r. On the other hand, with these formulae depending on small r it was no longer so certain that plane waves could be used for the free state.

In 1958 Chandrasekhar and Elbert[††] carried out calculations using the 20-parameter bound-state wave function of Hart and Herzberg.[‡‡] They considered that the value of the velocity operator calculated with this wave function and either plane waves or waves distorted by the Hartree atomic field (without allowance for exchange) was not likely to be substantially improved upon by any subsequent improvements in bound or free wave functions.

At that time the observed photodetachment cross-section was not accurate enough to detect any failing of these theoretical results. However, in 1959, refinements introduced by Smith and Burch, discussed above, led to measurements of the cross-section, relative to that for a

† Hylleraas, E. A., *Z. Phys.* **60** (1930) 624.
‡ Williamson, R. E., *Astrophys. J.* **96** (1942) 438.
§ Henrich, L. R., ibid. **99** (1944) 59 and 318.
|| Chandrasekhar, S., ibid. **102** (1945) 223.
†† Chandrasekhar, S. and Elbert, D. D., ibid. **128** (1958) 633.
‡‡ Hart, J. F. and Herzberg, G., *Phys. Rev.* **106** (1957) 79.

wavelength of 5280 Å, which were accurate to 3 per cent. These measurements revealed discrepancies with theory that stimulated a further batch of even more elaborate calculations concentrating particularly on the use of more accurate free-state wave functions.

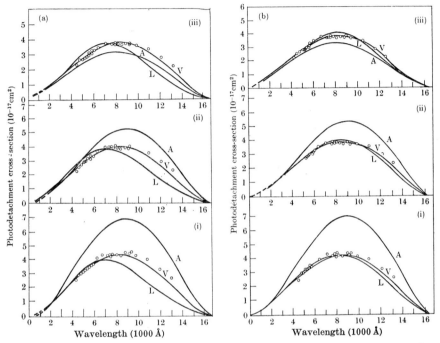

FIG. 15.9.   Observed and calculated photodetachment cross-sections for H⁻. Curves L, V, A, calculated in terms of dipole length, velocity, and acceleration matrix elements respectively. ○, observed by Smith and Burch, normalized to agree with the velocity curve at 5280 Å. (a) calculated using the 20-parameter bound-state wave function of Hart and Herzberg

   (i) using plane waves   ⎱ for the
   (ii) using 1s approximation   ⎰ continuum
   (iii) using 1s–2s–2p close-coupling approximation   ⎰ wave functions.

(b) calculated using the 70-parameter bound-state wave function: (i), (ii), (iii), continuum wave functions as in (a).

Fig. 15.9 (a) and (b) show the results of two series of calculations. In Fig. 15.9 (a) the bound-state function is the 20-parameter Hylleraas type function obtained by Hart and Herzberg,† while in Fig. 15.9 (b) the results all refer to a 70-parameter bound-state function of the form

$$\psi(\mathbf{r}_1, \mathbf{r}_2) = e^{-\alpha(r_1+r_2)} \sum_{lmn} \chi_{lmn}(r_1^m r_2^n + r_1^n r_2^m) r_{12}^l, \qquad (31)$$

† loc. cit., p. 1256.

with $\alpha = 0 \cdot 7$ at.u., used by Schwartz.† Results are then given in each series using

(i)   plane waves;‡
(ii) the $1s$ approximation§ (see Chap. 8, § 2.2.3) in which direct and exchange distortion of the wave by the atom in its $1s$ state is included;
(iii) the $1s$–$2s$–$2p$ close-coupling approximation§ (see Chap. 8, § 2.2.3).

In all cases the results obtained with the dipole length, velocity, and acceleration operators are shown separately, as well as the experimental results of Smith and Burch normalized in each case to the calculated results from the velocity formula at a wavelength of 5280 Å. As a further check we have the absolute value $3 \cdot 28 \pm 10$ per cent $\times 10^{-17}$ cm² of the cross-section at 5280 Å observed by Branscomb and Smith.

The first striking conclusion is the superiority of the velocity formula, which gives good results with the 20-parameter bound-state functions and plane waves, both as regards the shape and absolute magnitude of the cross-section. It is also clear that, as the approximate wave functions are improved, the calculated results with the different operators approach each other, though very gradually. The inclusion of coupling with $3s$–$3p$ and $3d$ states in the free-wave function makes little further improvement in the situation shown in Fig. 15.9 (b) (iii).

It is to be expected that the acceleration operator results—depending on the behaviour at small r—should be sensitive to the form of the free wave function. On the other hand, the length operator is insensitive to the free wave function but depends much more strongly on the form assumed for the bound-state function. Both features are closely shown in Fig. 15.9.

Summarizing the results, it seems that the shape of the photodetachment cross-section as a function of frequency, as given by the experimental results, agrees so well with the velocity operator theory, which in turn seems to be reliable throughout, that they must surely be close to the true values. The absolute values given by the most sophisticated calculations are probably accurate to a few per cent.

An interesting check on the cross-section is given by the sum rules

---

† Quoted by GELTMAN, S., *Astrophys. J.* **136** (1962) 935.

‡ With Hart and Herzberg functions—length and velocity matrix elements, CHANDRA-SEKHAR, S., *Astrophys. J.* **128** (1958) 114; with Schwartz function, GELTMAN, S., loc. cit.

§ With Hart and Herzberg functions—length and velocity matrix elements, JOHN, T. L., *Mon. Not. R. astr. Soc.* **121** (1960) 41; all other cases, DOUGHTY, N. A., FRASER, P. A. and McEACHARN, R. P., ibid. **132** (1966) 255.

that must be satisfied.† The first one is simply the form taken by the oscillator strength sum in a case in which no discrete excited states exist. We simply have then, if $df/d\nu$ is the differential oscillator strength for photodetachment, that

$$\int_{\nu_0}^{\infty} \frac{df}{d\nu}\, d\nu = 2, \tag{32}$$

ν_0 being the threshold frequency. In terms of the photodetachment cross-section this becomes (see Chap. 7 (82))

$$\frac{mc}{\pi e^2} \int_{\nu_0}^{\infty} Q^{\mathrm{d}}(\nu)\, d\nu = 2. \tag{33}$$

It must be remembered, however, that $Q^{\mathrm{d}}(\nu)$ in this formula includes processes that leave the atom in an excited state or even ionized. Hence if we substitute for Q^{d} the observed cross-section, extrapolated out to high frequencies, we must expect that the $=$ sign should be replaced by $>$. This imposes limits on the possible magnitude of $Q^{\mathrm{d}}(\nu)$.

Further sum rules may be derived. They are of the form (see Chap. 7, § 5.2.2)

$$\left(\frac{e^2}{ha_0}\right)^s \frac{mc}{\pi e^2} \int_{\nu_0}^{\infty} Q^{\mathrm{d}}(\nu)\nu^{-s}\, d\nu = A_s, \tag{34}$$

in which the A_s are determined by the ground-state wave function for H⁻ only. The simplest case (32) corresponds to $s = 0$. Other cases, for small negative and positive s are:

$$A_1 = \frac{2}{3a_0^2}\langle(\mathbf{r}_1+\mathbf{r}_2)^2\rangle_{00}, \qquad A_2 = \alpha/a_0^3,$$

$$A_{-1} = \tfrac{2}{3}\{(E_T/E_{\mathrm{H}})-a_0^2\langle\mathrm{grad}_1.\mathrm{grad}_2\rangle_{00}\}, \qquad A_{-2} = \frac{4\pi a_0^3}{3}\langle\delta(\mathbf{r}_1)+\delta(\mathbf{r}_2)\rangle, \tag{35}$$

where α is the polarizability and E_T the total energy of H⁻. E_{H} is the energy $e^2/2a_0$ of the ground state of H.

A_{-2} and A_{-1} have been calculated by Pekeris‡ using his elaborate 70-parameter wave function for the H⁻ ground state, while A_1, A_2, and A_3 have been calculated by Schwartz§ using the 70-parameter function (31). Table 15.2 compares these values with those derived from the velocity operator method with good wave functions.

† See GELTMAN, S., *Astrophys. J.* **136** (1962) 935.
‡ PEKERIS, C. L., *Phys. Rev.* **112** (1958) 1649.
§ SCHWARTZ, C., ibid. **123** (1961) 1700.

It will be seen that, in proceeding towards negative values of s, which throws increasing weight on the high frequencies, the sum evaluated using the cross-section calculated only for a process that leaves the atom in its ground state, becomes progressively much smaller than the true sum. This indicates considerable contributions from transitions

TABLE 15.2

Values of A_s for H$^-$ photodetachment

s	Calculated from ground state wave functions	Calculated by integration of calculated Q^d from ground state
3	4000	3670†
2	212	200†
1	15·1	14·1†
0	2	1·76‡
−1	1·01	0·61‡
−2	2·76	1·18‡

that leave the atom in an excited state. Comparison with corresponding results on photo-ionization of neutral helium (see Chap. 14, Table 14.8) indicates that for H$^-$ the contribution from transitions to excited states is relatively much more important.

TABLE 15.3

Cross-sections for attachment of electrons to normal hydrogen atoms

Electron energy (Rydbergs)	0·01	0·03	0·05	0·1	0·2	0·4	0·6	0·8
Electron energy (eV)	0·135	0·406	0·677	1·354	2·708	5·416	8·124	10·832
Attachment cross-sections (10^{-23} cm^2)	0·456	0·570	0·583	0·547	0·455	0·315	0·300	0·270

The extent to which the observed cross-section enables one to account for the solar continuous emission in the visible and near infra-red depends on allowance for the contribution from free–free transitions. Discussion of this aspect will therefore be deferred till § 7.

Table 15.3 gives the cross-section for radiative attachment of electrons to hydrogen atoms for a number of electron energies. These have been derived from the best theoretical results for the photodetachment cross-section using the relation (8).

† DOUGHTY, N. A., loc. cit., p. 1258.
‡ DALGARNO, A. and EWART, R. W., *Proc. phys. Soc.* **80** (1962) 616; GELTMAN, S., oc. cit.

5.2. O⁻

Because of its possible importance in the ionosphere, the photodetach-ment of electrons from O^- ions was one of the earliest investigated by the crossed-beam method, the first observations being made by Branscomb and Smith† in 1955. At that time the accepted value for the electron affinity of atomic oxygen was 2·2 eV, based largely on electron impact studies (see Chap. 13, § 4.4.1) but apparently confirmed by Metlay and Kimball‡ using the equilibrium constant method. Branscomb and Smith obtained a substantially lower electron affinity of 1·45 eV, which caused much surprise and required very strong evidence for its accep-tance. As a result, very careful measurements were made of the threshold behaviour, which confirmed this result and led to higher precision.§ The photodetachment value is now fully accepted. The errors in the impact observations and the equilibrium constant method are reason-ably well understood. A full historical account is given in Chapter 13, § 4.4.1.

The technique employed for studying the behaviour of the cross-section near the threshold was necessarily different from the combined-filter method used for higher frequencies and described in § 2. This is because in the case of O^- the active electron makes a transition from a p- to an s-state, so the cross-section Q^d has an infinite slope at the threshold (see § 1 (6)).

The problem is further complicated by the fine structure of the initial O^- and final O levels. A level diagram illustrating the possible transitions is shown in Fig. 15.10. The electron affinity is strictly the energy difference between the ground $^2P_{\frac{3}{2}}$ state of O^- and the ground $^3P_2$ state of O, but the lowest energy involved in transitions between sub-states, the threshold energy for detachment, arises from the $^2P_{\frac{1}{2}} \rightarrow {}^3P_2$ states. By employing a polynomial interpolation from the sequence of isoelectronic systems Mg^{+++}, Na^{++}, Ne^+, and F the energy difference between the $^2P_{\frac{1}{2}}$ and $^2P_{\frac{3}{2}}$ states is estimated to be about 0·026 eV (230 cm⁻¹), which corresponds to a wavelength difference of about 160 Å near the threshold.

When allowance is made for these different transitions with appro-priate threshold and statistical weights as shown in Fig. 15.10, the total cross-section, apart from a small toe at the threshold and nearby,

† BRANSCOMB, L. M. and SMITH, S. J., *Phys. Rev.* **98** (1955) 1028 and 1127.
‡ METLAY, M. and KIMBALL, G. E., *J. chem. Phys.* **16** (1948) 774.
§ BRANSCOMB, L. M., BURCH, D. S., SMITH, S. J., and GELTMAN, S., *Phys. Rev.* **111** (1958) 504.

follows the same threshold law (see (6)) as that given for a single transition. As a good approximation

$$Q^d(\lambda) = \begin{cases} (\gamma B/\lambda)\{(\lambda_1-\lambda)/\lambda_1\lambda\}^{\frac{1}{2}} & (\lambda_0 < \lambda < \lambda_1), \\ (\gamma/\lambda)\{(\lambda_0-\lambda)/\lambda_0\lambda\}^{\frac{3}{2}}+(\gamma A_1/\lambda)\{(\lambda_0-\lambda)/\lambda_0\lambda\}^{\frac{3}{2}} & (\lambda < \lambda_0), \end{cases} \quad (36)$$

where A_1 and B are adjustable parameters, γ is a constant, and λ_0 and λ_1 are the threshold wavelengths for photodetachment from the $^2P_{\frac{1}{2}}$ and $^2P_{\frac{3}{2}}$ states respectively.

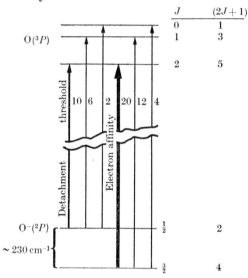

FIG. 15.10. Energy level diagram showing the fine structure of the ground states of O and O⁻.

To obtain the threshold behaviour the following procedure was adopted. Turning to the analysis of § 2 we can no longer assume (see (12)) that Q^d varies linearly over the range of λ covered by a particular band-pass filter. Instead of the band-pass filters a set of seven filters were made with transmittances $T_m(\lambda)$ of the form shown in Fig. 15.11. In all cases the overlap with wavelengths shorter than the threshold at 8500 Å is not greater than 1500 Å. Over the narrow wavelength range between 7000 and 8500 Å the spectral distribution of the carbon arc is nearly constant so (11) becomes, for the detachment probability with a particular filter combination,

$$P_{\text{eff},m} = \frac{CW}{v} \bar{\phi} \int Q^d(\lambda)T_m(\lambda)\lambda \, d\lambda, \quad (37)$$

where $\bar{\phi}$ is the mean value of ϕ in the wavelength interval 7500–8500 Å.

The threshold shape is then found by solving the seven integral equations (37) for each filter, using the approximation (36) for $Q^d(\lambda)$. A very good fit to the observed data can be found in this way, providing in itself considerable justification of the method.

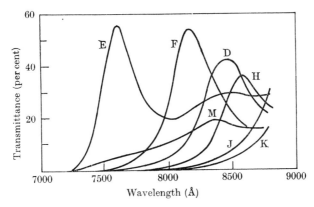

FIG. 15.11. Transmittance of filters used in measurements on photodetachment from O^-.

By this procedure the threshold wavelength was found to be

$$\lambda_0 = 0\cdot846\pm0\cdot003 \ \mu\text{m},$$

giving for the electron affinity of O

$$A(O) = 1\cdot465\pm0\cdot005 \ \text{eV}. \tag{38}$$

An absolute cross-section was obtained by the method described on pp. 1240–2 and at λ 5280 Å is $6\cdot3\times10^{-18}$ cm². A check on this was made by comparing the photodetachment probabilities of O^- and D^- with a narrow band-pass filter centred at 5375 Å. The ratio $P(O^-)/P(D^-)$ was found to be $0\cdot20\pm0\cdot03$, giving $Q^d(D^-) = (3\cdot18\pm0\cdot48)\times10^{-17}$ cm², which is consistent with the data for H^- discussed above.

Relative cross-sections above the threshold were measured by Smith† to an accuracy of 2 per cent, up to 5200 Å. By using a scanning monochromator the relative measurements were extended to the ultra-violet as far as 4250 Å. The final curve showing the variation of cross-section with wavelength is illustrated in Fig. 15.12.

A second threshold is apparent at 3600 Å, corresponding to a photon

† SMITH, S. J., Proc. 4th Int. Conf. Ioniz. Phenom. Gases, Uppsala, 1959, p. 219 (North Holland, Amsterdam, 1959).

energy of 3·43 eV. This can be interpreted as the onset for

$$O^- + h\nu \to O(^1D) + e. \tag{39}$$

Since $O(^1D)$ lies 1·96 eV above the ground ^3P state the threshold for (39) should occur at $1·47 + 1·96 = 3·43$ eV just as observed. The good numerical agreement provides additional support for the electron affinity (38).

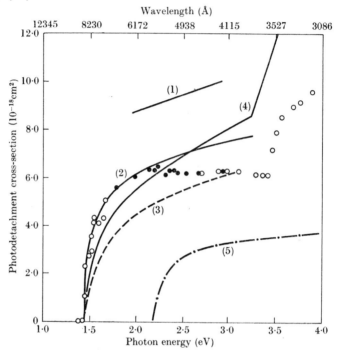

FIG. 15.12. Photodetachment cross-sections for O^-; ● ○ observed by Branscomb, Smith, and Tisone; (1) derived from analysis of arc emission spectra by Boldt; (2) calculated by Cooper and Martin; (3) calculated by Klein and Brueckner; (4) calculated by Robinson and Geltman; (5) calculated by Bates and Massey (assuming the electron affinity = 2·2 eV).

Boldt† has disentangled the O^- electron affinity continuum from the continua emitted in the visible from an arc in oxygen operated near atmospheric pressure. The detachment cross-section that he derives from his data is shown in Fig. 15.12. As the accuracy claimed for this determination is not greater than 30 per cent the results are consistent with those obtained by the crossed-beam method. It is of interest to note that, with the arc temperature 11 000 °K, the concentrations of neutral O atoms and O^- ions were $6·00 \times 10^{17}$ and $1·07 \times 10^{13}/cm^3$

† BOLDT, G., Z. Phys. 154 (1959) 319.

respectively and the relative intensities of the electron affinity, re-combination, and free–free (in the field of O$^+$) continua were as 61 : 28 : 11—the continuum due to electron capture by the neutral atoms is much the strongest. The intensity variation of the recombination continuum O$^+$ derived by Boldt from his experiments is compared with the theoretical variation, calculated by Peach, in Fig. 14.96 (a) and the agreement is not unsatisfactory.

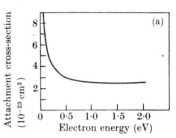

Fig. 15.13 shows the radiative attachment cross-section Q^a and rate coefficient vQ^a, where v is the electron velocity, as a function of electron energy derived from the observed photodetachment cross-section Q^d, through the relation (8), taking $g_- = 6, g_0 = 9$, no allowance being made for the energy separation of the fine structure states.

The first detailed calculations of Q^d for atomic oxygen using elaborate wave functions were carried out by Bates and Massey† with the dipole length formula. They used Hartree–Fock wave functions for the ground state of O$^-$, and for the continuum state a function allowing for direct and exchange interaction with the Hartree–Fock field

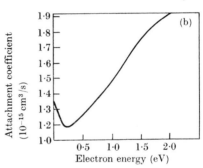

FIG. 15.13. Cross-sections and rate coefficients for radiative attachment of electrons to normal O atoms. (a) Attachment cross-section. (b) Attachment rate coefficient.

of O(2^3P) and also for polarization of the atom by the empirical method discussed in Chapter 8, § 9.2. They treated the polarizability α as an adjustable parameter for reasons explained in that section. If we take their results for $\alpha = 5.7a_0^3$, which is probably close to the true polarizability, the derived photodetachment cross-section is as shown in Fig. 15.12 and is rather too small. Klein and Brueckner‡ obtained better agreement with a more empirical calculation. Instead of Hartree–Fock wave functions for the ground state of O$^-$ they ignored exchange effects completely and treated the problem on a one-electron basis. The electron was taken to move in the Hartree–Fock field of the atom

† BATES, D. R. and MASSEY, H. S. W., *Phil. Trans. R. Soc.* A**239** (1943) 269.
‡ KLEIN, M. M. and BRUECKNER, K. A., *Phys. Rev.* **111** (1958) 1115.

together with a polarization term, the polarization being adjusted to give a binding energy equal to the observed electron affinity. Both bound- and free-state wave functions were derived from this same interaction. These calculations were repeated and extended by Cooper and Martin† and by Robinson and Geltman,‡ whose results are also shown in Fig. 15.12. The agreement with the observations is not unsatisfactory but in view of the empirical nature of the method there is clearly room for improved calculations based on theoretical rather than empirical wave functions.

The value of the integral (see (33))

$$\frac{mc}{\pi e^2} \int_{\nu_0}^{\infty} Q^{\mathrm{d}}(\nu)\,\mathrm{d}\nu \qquad (40)$$

for this case has been investigated by Branscomb and Smith.§ If the range of integration is from the threshold to a wavelength of 4000 Å the integral is only 0·02. This is very small in view of the fact that the integral over the whole range should equal 9, the number of electrons in O^-. Smith‖ extended the cross-section out to photon energies beyond 130 eV approximately, using Klein–Brueckner wave functions for the ground state and plane waves for the free state. The value of (40) is then increased to between 4 and 5, which is at least half of the full value when allowance is made for excitation and ionization of the residual O atoms in the detachment process.

5.3. C^-

The chief difficulty in measuring the photodetachment cross-section for C^- is that of obtaining a sufficiently intense source of the ions. These were generated in a hot cathode discharge run in carbon monoxide at a pressure of about 5×10^{-2} torr. After mass analysis a current usually of about 5×10^{-10} A, occasionally as large as 3×10^{-9} A, was obtained. This involved working with currents about two orders of magnitude smaller than for H^- and O^-.

One power of ten was made up by reducing the background pressure to 10^{-7} torr. This was achieved by constructing the reaction chamber and its components entirely of refracting metals and using metal gaskets as vacuum seals so that the whole could be baked out at 300 °C for 12 h.

† Cooper, J. W. and Martin, J. B., *Phys. Rev.* **126** (1962) 1482.
‡ Robinson, E. J. and Geltman, S., ibid. **153** (1967) 4.
§ Branscomb, L. M. and Smith, S. J., loc. cit., p. 1261.
‖ Smith, S. J., loc. cit., p. 1263.

The second improvement was to replace the electron collector and preamplifier by an electron multiplier.

In other respects the technique used by Seman and Branscomb[†] was essentially the same as in the other crossed-beam experiments described above. At wavelengths well beyond the threshold there is no difficulty. The general shape of the cross-section is essentially the same as for O^-. Up to photon energies of 3·0 eV no second threshold was found. This

TABLE 15.4

Values of the electron affinity of carbon

Method	$A(C)$ (eV)
Photodetachment	1·25±0·03[†]
Electron impact on CO ⎱	1·11±0·05[‡]
See Chap. 13, § 5.2 ⎰	1·33±0·18[§]
Mass spectrometer analysis of C^- sublimed from a carbon filament	1·2[‖]
Isoelectronic extrapolation	⎧ 1·24[††] ⎩ 1·21[‡‡]

confirms that the ground state of C^- is ^4S as for the isoelectronic N atom. If this is so the only possible transition is to C ^3P. On the other hand, if the ground state were ^2D two transitions of approximately equal strength would be possible, namely $^2D \rightarrow {}^3P$ and $^2D \rightarrow {}^1D$. These would have thresholds 1·24 eV apart and the second should certainly have been observed if it existed.

Determination of the threshold energy was complicated by the fact that some photodetachment continued to be observed down to very long wavelengths, at least as far as 2 μm. The weak signals at these wavelengths nevertheless continued to be proportional to ion current and photon flux and were quite reproducible for each filter combination. No corresponding signal was obtained below the threshold when O^- ions were substituted for C^-. It seems probable that they arose from the presence of C^- ions in a metastable state, possibly ^2D, with a very low binding energy relative to C^3P.

Allowing for the presence of these weak signals extending well below the main threshold, it was found that the threshold wavelength for

[†] SEMAN, M. L. and BRANSCOMB, L. M., *Phys. Rev.* 125 (1962) 1602.
[‡] LAGERGREN, C. R., Ph.D. Thesis, University of Minnesota (1955).
[§] FINEMAN, M. A. and PETROCELLI, A. W., *Bull. Am. phys. Soc.* 3 (1958) 258.
[‖] HONIG, R. E., *J. chem. Phys.* 22 (1954) 126.
[††] EDLÈN, B., ibid. 33 (1960) 98.
[‡‡] JOHNSON, H. R. and ROHRLICH, F., ibid. 30 (1959) 1608.

detachment from $C^-(^4S)$ is 0.99 ± 0.02 μm, giving for the electron affinity $A(C)$, 1.25 ± 0.03 eV. This is compared with values obtained by other methods in Table 15.4. Apart from the electron impact values, obtained as described in Chapter 13, § 5.2 and which are subject to considerable uncertainty, the agreement among the other values is very good.

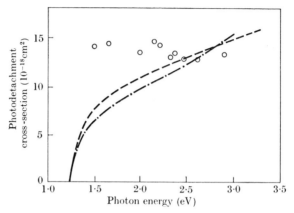

FIG. 15.14. Photodetachment cross-section for C^-. ○ observed by Seman and Branscomb. —————— calculated by Cooper and Martin. —·—·—·— calculated by Robinson and Geltman.

The observed photodetachment cross-section is shown in Fig. 15.14. On the same diagram are also shown two calculated cross-sections, by Cooper and Martin† and by Robinson and Geltman,‡ using the same semi-empirical procedure as for O^-.

5.4. S^-

Preliminary observations§ of photodetachment from S^- by the crossed-beam method have been sufficient to determine the threshold wavelength. This gives the electron affinity $A(S)$ as 2.07 ± 0.07 eV.

5.5. *Metastable* N^- (?)

Boldt‖ has applied the high-pressure arc discharge technique to observe the continua produced in nitrogen. He finds that there is present, in addition to the free–free and recombination continua, a further continuum with characteristics very similar to the electron-affinity spectrum in oxygen. In fact at 10 000 °K the intensity in this

† loc. cit., p. 1266.   ‡ loc. cit., p. 1266.
§ BRANSCOMB, L. M. and SMITH, S. J., *J. chem. Phys.* **25** (1956) 598.
‖ BOLDT, G., *Z. Phys.* **154** (1959) 330.

continuum at 5300 Å is slightly more intense than the recombination continuum and each more than three times as intense as the free–free continuum.

The results could be accounted for by the presence of a negative ion with a detachment energy about 1 eV. There seems no possibility that a stable N^- ion exists with this electron affinity. On the other hand, Bates and Moiseiwitsch[†] have pointed out that the $2p^4\,^1D$ state of N^- is likely to be metastable. This state may be produced when an electron is captured by N atoms in the metastable $2p^3\,^2D$ state, releasing about 1 eV energy. The total energy is, however, greater than that of the $2p^3\,^4S$ ground state of the neutral atom, which lies 2·36 eV below the ^2D state. Eventually the ^1D state of N^- will decay by auto-ionization

$$N^-(^1D) \to N(^4S)+e,$$

but the rate will be very slow because the reaction necessarily involves a change in total spin. It seems probable then that Boldt observed the continuum arising from capture of electrons by N atoms in the ^2D metastable state. The intensity variation of the recombination continuum N^+ derived from these experiments is compared with the theoretical variation, calculated by Peach, in Fig. 14.96 (b) and the agreement is not unsatisfactory.

5.6. *The halogen negative ions* F^-, Cl^-, Br^-, I^-

Absorption of radiation by all of these negative ions has been observed by Berry and his collaborators,[‡] using the shock-wave method described in § 3. The Cl^- electron affinity continuum has been observed by Henning,[§] using the high-pressure arc method described in § 4, while photodetachment from I^- has been investigated by the crossed-beam method (§ 2).

Starting first with I^-, Steiner, Seman, and Branscomb[||] studied photodetachment from an I^- beam, the source of the ions being a hot-cathode discharge through a mixture of iodine vapour and ammonia.

They first carried out observations using a 2000 W d.c. xenon arc discharge lamp as photon source in place of the carbon arc lamp, but later substituted a 2500 W lamp that made it possible to use a monochromator. In this way the detachment current was measured over the wavelength region from 4000 Å (3·1 eV, close to the threshold) to 3000 Å with a photon beam of half-width 33 Å. Even higher resolution (about

† BATES, D. R. and MOISEIWITSCH, B. L., *Proc. phys. Soc.* A68 (1955) 540.
‡ loc. cit., p. 1246.   § loc. cit., p. 1253.
|| STEINER, B., SEMAN, M. L., and BRANSCOMB, L. M., *J. chem. Phys.* 37 (1962) 1200.

18 Å) was used in scanning between 4100 and 3900 Å. The intensity
of the resolved light beam was monitored by splitting into two fractions
before entering the interaction chamber. The non-interacting fraction
was reflected from a diffuse reflecting white vitreolite surface and a
fraction of this detected with a masked phototube. As a check the output from the tube was compared frequently with that from a calorimeter
situated at the main image of the monochromator. No attempt was

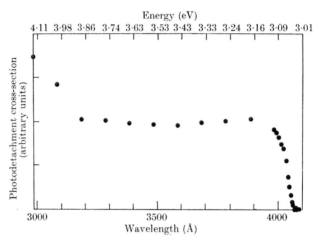

FIG. 15.15. Variation with wavelength of the photodetachment
cross-section for I⁻ as observed by Steiner, Seman, and Branscomb.

made to determine the absolute value of the detachment cross-section
Q^d. Fig. 15.15 shows the observed variation of Q^d with frequency.
There is clear evidence of a second threshold but it could not be located
with precision.

Special attention was paid to accurate determination of the first
threshold energy. To do this allowance must be made for the finite
resolving power of the monochromator, which may be expressed in
terms of a slit function $S(\mu)$ so that the measured cross-section Q^d_m at
a nominal wavelength λ is given in terms of a true cross-section Q^d at a
precise wavelength by

$$Q^d_m(\lambda) = \int\limits_{-\infty}^{\infty} Q^d(\lambda+\mu)S(\mu)\,d\mu.$$

It was verified that in practice the folding-in of $S(\mu)$ has little effect
except within 10 Å of the threshold. In this region, knowing $S(\mu)$, Q^d
was unfolded by a trial-and-error procedure. The unfolded curve giving

$Q^{\mathrm{d}}(\lambda)$ was found to be represented very well over the rising portion by the form (see § 1 (6)) $Q^{\mathrm{d}} = a(k+bk^2),$

where k is the wave number of the detached electron. This gives the threshold as 4053 ± 3 Å ($3{\cdot}059\pm0{\cdot}002$ eV) in which an additional uncertainty of 3 Å in the monochromator setting has been included.

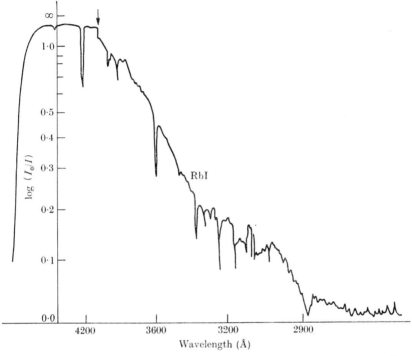

Fig. 15.16. Microdensitometer tracing of the absorption spectrum of the vapour from RbI, evaporated by shock heating. The photodetachment threshold is indicated by the arrow.

This is to be compared with 4031 ± 8 Å derived from the earlier experiment using filters and fitting the data to a step-function behaviour at threshold. In these latter experiments the absolute cross-section between the threshold and 3600 Å was measured by comparison with the known cross-section for photodetachment of H^- between 3600 and 4900 Å. The magnitude, assuming the step-function form, was $2{\cdot}1\pm0{\cdot}5\times10^{-17}$ cm^2.

Fig. 15.16 illustrates the absorption spectrum obtained for RbI by Berry, Reimann, and Spokes† using the shock-wave method. From such observations the threshold energy was determined as $3{\cdot}063$ eV,

† BERRY, R. S., REIMANN, C. W., and SPOKES, G. N., *J. chem. Phys.* **37** (1962) 2278.

which agrees quite closely with the photodetachment results. Both are substantially lower than those obtained by other methods, as may be seen by reference to Table 15.5. This seems to be a general tendency of the shock wave results both for absorption and emission (see below), even though it is less marked in the other cases.

TABLE 15.5

Electron affinities of halogen atoms (in eV)

Method	F	Cl	Br	I
Surface ionization[†]	$3 \cdot 62 \pm 0 \cdot 09$	$3 \cdot 76 \pm 0 \cdot 007$	$3 \cdot 51 \pm 0 \cdot 007$	$3 \cdot 17 \pm 0 \cdot 05$
Surface ionization[‡] assuming $A(\mathrm{Br}) = 3 \cdot 50$ eV	$3 \cdot 47$	$3 \cdot 71$		$3 \cdot 23$
Surface attachment[§]			$3 \cdot 49 \pm 0 \cdot 02$	
Lattice energies[‖]	$3 \cdot 48$	$3 \cdot 69$	$3 \cdot 45$	$3 \cdot 14$
Photo-ionization of halogen molecule X_2[††] (see Chap. 14, § 8.12)			$3 \cdot 53 \pm 0 \cdot 12$	$3 \cdot 13 \pm 0 \cdot 12$
Shock wave absorption	$3 \cdot 448$[‡‡]	$3 \cdot 613$[§§]	$3 \cdot 363$[§§]	$3 \cdot 063$[§§]
Shock wave emission[‖‖]		$3 \cdot 610$	$3 \cdot 366$	$3 \cdot 063$
Crossed beam photodetachment[†††]				$3 \cdot 059 \pm 0 \cdot 02$
Empirical extrapolation	$3 \cdot 50$[‡‡‡] $3 \cdot 47$[§§§] $3 \cdot 34$[‖‖‖]			

For the other three halogen ions a second threshold was observed in the absorption cross-section observed by the shock-wave method. Figs. 15.5 and 15.6 show this feature in the cross-section–wavelength curves for Cl⁻ and Br⁻. This second threshold (see also Fig. 15.15) arises from detachment leaving the atom in the ^2P state. The observed separations agree closely with the known fine-structure splittings as shown in Table 15.6. The electron affinities derived from the wavelength of the first threshold are given in Table 15.5. They certainly confirm

† BAILEY, T. L., *J. chem. Phys.* **28** (1958) 792.

‡ BAKULINA, I. N. and IONOV, N. I., *Dokl. Akad. Nauk SSSR* **105** (1955) 680.

§ DOTY, P. M. and MAYER, J. E., *J. chem. Phys.* **12** (1944) 323.

‖ CUBICCIOTTI, D., ibid. **34** (1961) 2189.

†† MORRISON, J. D., HURZELER, H., INGHRAM, M. G., and STANTON, H. E., ibid. **33** (1960) 821.

‡‡ BERRY, R. S. and REIMANN, C. W., ibid. **38** (1963) 1540.

§§ BERRY, R. S., REIMANN, C. W., and SPOKES, G. N., ibid. **37** (1962) 2278.

‖‖ BERRY, R. S. and DAVID, C. W., loc. cit., p. 1249.

††† STEINER, B., SEMAN, M. L., and BRANSCOMB, L. M., ibid. **37** (1962) 1200.

‡‡‡ EDLÈN, B., ibid. **33** (1960) 98.

§§§ EDIE, J. W. and ROHRLICH, F., ibid. **36** (1962) 623.

‖‖‖ SCHERR, C. W., SILVERMAN, J. N., and MATSEN, F. A., *Phys. Rev.* **127** (1962) 830.

that chlorine has the highest electron affinity of any atom but there is still a greater spread in the values obtained by different methods than would be expected from the estimated probable errors in each.

Support for the values derived from the shock-excited absorption spectra is provided from the electron affinity of Cl, Br, and I obtained by Berry and David† from the thresholds for emission of the electron

TABLE 15.6

Comparison of threshold separations in absorption spectra of halogen ions and fine-structure splittings of corresponding halogen atoms

	Threshold separation (eV)	Fine-structure splittings (eV)
F	0·052 ± 0·006	0
Cl	0·108 ± 0·007	0·1092
Br	0·457 ± 0·007	0·4568

affinity continua by shock excitation as described on p. 1249 (see Fig. 15.7 for typical spectra). The results agree closely with those obtained from the absorption spectra as will be seen from Table 15.5. It was also possible to observe the fine structure separation of the ground term of Cl and of Br from the wavelength at which a second continuum begins (see Fig. 15.7). These agreed well with the results from shock-excited absorption spectra as well as with results obtained by other methods.

The electron affinity continuum of Cl has been observed by Henning‡ in the emission from a high-pressure arc in chlorine, using the method described in § 4. The photodetachment cross-section derived is shown in Fig. 15.17. It is considered to be correct to ±50 per cent. In this case, at 11 000 °K, the affinity continuum contained 63 per cent of the total continuum emission at a wavelength of 4200 Å. The agreement of the intensity variation in the recombination continuum with the calculated variation, shown in Fig. 14.96 (c), is considerably less satisfactory than for O and N.

Photodetachment cross-sections for F⁻ and Cl⁻ have been calculated by Cooper and Martin§ and by Robinson and Geltman,‖ using the same semi-empirical method as for O⁻ and C⁻. The calculated cross-section rises less rapidly from the threshold but the order of magnitude and the relative magnitude of the absolute values for F⁻ and Cl⁻ is not inconsistent with the shock-wave observations. For I⁻, Robinson and

---

† BERRY, R. S. and DAVID, C. W., loc. cit., p. 1249.   ‡ loc. cit., p. 1253.
§ loc. cit., p. 1266.   ‖ loc. cit., p. 1266.

Geltman obtain agreement within 25 per cent with the observed values of Steiner *et al.*†

6. Results for negative molecular ions

Many additional complications are introduced when the crossed-beam method is applied to negative molecular ions. These arise from the existence of vibrational structure in the energy level system of both the negative ion and the corresponding neutral molecule. The transitions that occur in photodetachment will follow the Franck–Condon principle.

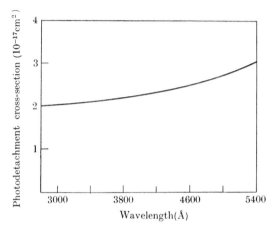

Fig. 15.17. Photodetachment cross-section for Cl^- derived
by Henning.

If the nuclear separations of the neutral molecule and the negative ion are nearly the same, the strongest transitions will not involve change of vibrational quantum number—the vertical detachment energy will be nearly equal to the electron affinity and there will be little additional structure near the threshold due to transitions to different vibrational states of the neutral molecule (see Chap. 13, § 1.3.1 for discussion of a similar effect in relation to impact of electrons with molecules).

Even if this condition is satisfied there is the uncertainty introduced by the ion source. It is unwise to assume that the molecular ions produced are all in the ground vibrational state. Vibrational excitation will usually be disposed of through radiation before the ions reach the reaction region if the molecules are heteronuclear, but the absence of a dipole moment in a homonuclear molecule reduces the vibrational transition probability so much that any vibrational excitation would

† loc. cit., p. 1269.

persist. The only way to obtain some idea as to whether such effects are important is to examine whether the observed photodetachment cross-section depends on the nature of the source. If it does not there is a reasonable chance that a variety of vibrational states are not involved.

6.1. OH$^-$ *and* OD$^-$

Because of these considerations the first negative molecular ions investigated were the heteronuclear ions OH$^-$ and OD$^-$. For these cases theoretical evidence suggested that the ion and neutral molecule have nearly the same equilibrium nuclear separations. The results obtained justified the hope that photodetachment from these ions would follow much the same lines as for atomic ions.

The first experiments were carried out by Smith and Branscomb,† using the filter method to isolate different wavelength bands. No evidence of any vibrational structure was found near the threshold that occurred at 7000 Å (1·78 eV), but the wavelength resolution was not great enough for the nature of this threshold energy to be interpreted in terms of vertical detachment energy or electron affinity. Some years later Branscomb‡ repeated the work with a hot-cathode arc for the ion source so that a monochromator could be used to provide much more highly resolved light (cf. the experiments with I$^-$ discussed in § 5.6). With this arrangement the resolution was about 50 Å which, near 7000 Å, corresponds to about 0·015 eV, much less than the vibrational quanta (0·4 eV) but not small enough to distinguish separate rotational levels.

Fig. 15.18 shows the photodetachment cross-section observed for OH$^-$ and OD$^-$ normalized by comparison with H$^-$ (see § 5.1). Earlier results obtained by Smith and Branscomb† are also shown and seem to agree well at the wavelengths concerned, which are considerably shorter than the threshold. There is clearly no evidence of vibrational structure near the threshold. Moreover, at wavelengths longer than the threshold, there is no evidence of any photodetachment process with an intensity greater than 10^{-3} of that at 6000 Å. Denoting a transition from a vibrational level v' of OH$^-$ or OD$^-$ to one v'' of OH or OD respectively, by (v'', v') it seems clear that the threshold corresponds to the (0, 0) transition—this is the only one that allows no absorption at longer wavelengths. This provides strong support for the assumption that the

† SMITH, S. J. and BRANSCOMB, L. M., *Phys. Rev.* **99** (1955) 1657A.
‡ BRANSCOMB, L. M., ibid. **148** (1966) 11.

OH⁻ and OD⁻ ions used in the experiments were actually in the ground vibrational levels.

The (1, 0) threshold should occur near 5830 Å for OD⁻ and 5500 Å for OH⁻, but no statistically significant evidence could be found of its presence. If $I(v'', v')$ is the intensity of the (v'', v') transition then

$$I(1, 0)/I(0, 0) \leqslant 0 \cdot 02. \tag{41}$$

By using the harmonic oscillator approximation for the vibrational wave functions it is found that (41) implies that ΔR_0, the difference in equilibrium nuclear separation between the ion and the corresponding neutral molecule, is $\leqslant 0 \cdot 0020$ Å for OH and $0 \cdot 0017$ Å for OD.

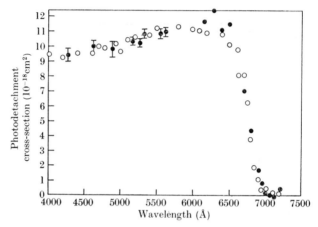

Fig. 15.18. Photodetachment cross-section for OH⁻ and OD⁻. ● observed for OH⁻ by Branscomb using a monochromator and 100 Å resolution. ○ observed for OD⁻ by Branscomb. ⬤ observed by Smith and Branscomb for OH⁻ using band-pass filters.

Analysis of the behaviour near threshold is complicated by the fact that, apart from the need to take account of rotational excitation since the effective source temperature is of the order of 1000 °K or higher, the final state produced by photodetachment is split into two substates ²Π₃ and ²Π₁ through interaction between electron spin and molecular rotation. Branscomb carried out an analysis, assuming a step function behaviour allowing for the level splitting and for rotational excitation corresponding to a temperature of 1000 °K. No allowance was made for rotational quantization. When the slit function that determines the resolution of the monochromator was folded in, very good agreement was obtained with the experimental data if the electron affinity, the energy involved in the (0, 0) transition, was taken to be $1 \cdot 83 \pm 0 \cdot 04$ eV, the same as for OH and OD within 0·01 eV. The difference in electron affinity of OH and OD is given by

$$\Delta(A) = \tfrac{1}{2}h[\{v(\mathrm{OH})-v(\mathrm{OD})\}-\{v(\mathrm{OH^-})-v(\mathrm{OD^-})\}],$$

when allowance is made for the zero point energy in each case, $v(\mathrm{OH})$, etc., being the fundamental frequencies. If the potential energy curves of the ground electronic states are the same for OH and OD and also for OH$^-$ and OD$^-$ then, since $v \propto (k/\mu)^{\frac{1}{2}}$ where k is the force constant and μ is the reduced mass,

$$v(\mathrm{OH^-})-v(\mathrm{OD^-}) = 0{\cdot}272v(\mathrm{OH^-}),$$

$$v(\mathrm{OH})-v(\mathrm{OD}) = 0{\cdot}272v(\mathrm{OH}),$$

so that $$\Delta(A) = 0{\cdot}136h\{v(\mathrm{OH})-v(\mathrm{OH^-})\}.$$

Since $$\Delta(A) = 0\pm0{\cdot}01 \text{ eV},$$

$$h\{v(\mathrm{OH})-v(\mathrm{OH^-})\} \leqslant 0{\cdot}07 \text{ eV (560 cm}^{-1}).$$

It is clear that, not only is the equilibrium nuclear separation R_0 very closely the same for the negative ion and neutral molecule, but the same applies to the fundamental vibrational frequency v. This behaviour is in agreement with the theoretical results of Cade[†] according to which $R_0(\mathrm{OH^-})/R_0(\mathrm{OH}) = 0{\cdot}992$ and $v(\mathrm{OH^-})/v(\mathrm{OH}) = 1{\cdot}009$.

6.2. SH$^-$

A related ion SH$^-$ has also been investigated[‡] using a mass analyser of higher resolution than in earlier experiments to select the ions from the wide variety produced in a hot-cathode arc discharge source containing a mixture of sulphur hexafluoride and ammonia. Photodetachment was observed over the wavelength range from a threshold at 5350 ± 100 Å ($2{\cdot}32\pm0{\cdot}05$ eV) to 3600 Å (3·4 eV). No other threshold was observed in this range. The rise from the threshold is the sharpest observed for any negative ion by the crossed-beam method. Thus the cross-section rises within 0·06 eV from the threshold to a flat maximum. Once again the evidence suggests little difference in nuclear separation between neutral molecules and negative ions.

6.3. O$_2^-$

The important case of O$_2^-$ has also been investigated but with inconclusive results. There was, in any case, no prospect of resolving vibrational structure because the frequency resolution of the filters was

† CADE, P. E., *Proc. phys. Soc.* **91** (1967) 842.
‡ STEINER, B., *Proc. 4th Int. Conf. Phys. electron. atom. collisions*, p. 108 (Quebec, 1965).

roughly equal to the vibrational frequencies. It was found by Burch, Smith, and Branscomb† that, with a beam of O_2^- ions from a source, the photodetachment probability, while decreasing steadily as the wavelength increased from 4000 to 25 000 Å, still did not exhibit a threshold. Their observations are illustrated in Fig. 15.19.

A major difficulty in the interpretation of these results is that there is no definite knowledge of the distribution among the vibrational levels of the O_2^- ions that are intersected by the photon beam. Indeed, there

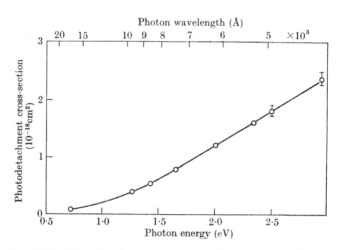

FIG. 15.19. Photodetachment cross-section for O_2^- ions as observed by Burch, Smith, and Branscomb.

may also exist metastable electronically-excited species that may contribute to the signal. Addition of D_2O vapour to the ion source, a glow discharge in O_2 at a pressure of $2 \cdot 5 \times 10^{-2}$ torr, had no effect on the results even though D_2O is known to be very effective in producing vibrational deactivation of O_2 (Chap. 17, § 3). This supports the belief that the vibrational distribution of the ions issuing from the source is determined by the production mechanism. Thus the ions approaching the anode of the discharge must have been produced close to the anode. Ions formed much earlier would suffer detachment before traversing the full length of the discharge. Even assuming this, the attachment mechanism (see Chap. 13, § 4.4) is so complicated that it is difficult to predict the vibrational distribution in the ions of the beam.

It is to be expected (see Chap. 12, p. 807) that the electron detached

† BURCH, D. S., SMITH, S. J., and BRANSCOMB, L. M., *Phys. Rev.* **112** (1958) 171; **114** (1959) 1652.

will be from a π_g orbital. The threshold law for such a case (§ 1, Table 15.1) is

$$Q^d = (E-E_0)^{\frac{3}{2}} E(A_0 + A_1 E^{\frac{1}{2}} + A_2 E + ...),\qquad (42)$$

where E is the photon energy and E_0 the threshold energy. An attempt was made to fit the data with the form (42) taking $A_n = 0$, $n > 2$, and also $A_1 = 0$. A good fit was obtained with

$$E_0 = 0.15 \pm 0.05 \text{ eV}.$$

This value differs considerably from the electron affinity $A(O_2)$ determined in other ways (see Chap. 13, §§ 4.4.3 and 10). The detachment experiments of Phelps and Pack[†] (see Chap. 19, § 4.2) give

$$A(O_2) = 0.46 \pm 0.02 \text{ eV}$$

and the technique is such that, if anything, this is a lower limit.

Less weight can be given to the electron impact determination by Curran[‡] (see Chap. 13, § 10) but this also leads to the relatively high value $A(O_2) = 0.58$ eV. Thermochemical evidence[§] gives values between 0.15 and 1.0 eV. It is possible that the relationship of the potential energy curves of O_2^- and O_2 is such that the vertical detachment energy is substantially less than the electron affinity.

6.4. NO_2^-

This ion is ubiquitous in laboratory experiments involving N_2 and O_2, and may well play an important part in the ion-electron equilibrium in the D-region of the ionosphere and in polar-cap absorption phenomena. Branscomb[‖] was unable to detect photodetachment from a beam of these ions using light of wavelength 5200 Å, showing that the vertical detachment energy is almost certainly greater than 2.5 eV.

7. Free–free transitions of electrons in the fields of neutral atoms

The theory of emission and absorption of radiation by electrons undergoing transitions between two states in the continuum has been outlined in Chapter 14, § 9. Although the results of that section refer specifically to electrons moving in the fields of positive ions, such formulae in Chapter 14 as (81) for the rate of emission and (90) for that of absorption apply without modification, except that the initial and final continuum wave functions do not have the asymptotic form of

---

[†] PHELPS, A. V. and PACK, J. L., *Phys. Rev. Lett.* **6** (1961) 111.
[‡] CURRAN, R. K., *J. chem. Phys.* **35** (1961) 1849.
[§] For references see p. 1016.
[‖] BRANSCOMB, L. M., *Threshold of space*, p. 101 (Pergamon Press, New York, 1957).

Coulomb-distorted waves but of plane waves with the accompanying spherical waves. Thus the matrix element that determines the transition rate between states of energies E, E' is

$$\mathbf{r}_{E,E'} = \int \psi_E \, \mathbf{r} \psi_{E'} \, d\tau, \qquad (43)$$

where the initial and final electron wave functions ψ_E, $\psi_{E'}$ are normalized to have the asymptotic forms

$$\psi_E \sim e^{i\mathbf{k}.\mathbf{r}} + r^{-1} f_E(\theta) e^{ikr},$$
$$\psi_{E'} \sim e^{-i\mathbf{k'}.\mathbf{r'}} + r^{-1} f_{E'}(\pi - \theta') e^{ik'r} \qquad (44)$$

in place of (83) of Chapter 14, the notation being otherwise the same.

Once more it is to be remembered that the matrix element (43) may be written in alternative forms in terms of electron velocity and acceleration operators (see Chap. 7, § 5.2.3). Whereas the length and velocity operators depend only on the electron coordinates, the acceleration operator for a system of n electrons is given by

$$\mathbf{a} = -\frac{1}{m} \sum_i \mathrm{grad}_i \, V, \qquad (45)$$

where V is the potential energy of the electrons. Since

$$V = -e^2 \sum_{i,j} (r_i^{-1} - r_{ij}^{-1}), \qquad (46)$$

we have
$$\mathbf{a} = -(e^2/m) \sum_i \mathbf{r}_i / r_i^3. \qquad (47)$$

This we refer to as the general acceleration operator. In some calculations the many-electron nature of the problem is avoided by supposing the active electron to move in an effective central potential such as that V_{H} of the Hartree field. This is then used to give an effective acceleration operator

$$\mathbf{a}_{\mathrm{H}} = -\frac{1}{m} \frac{dV_{\mathrm{H}}}{dr} \hat{\mathbf{r}}. \qquad (48)$$

7.1. Hydrogen

In § 5.1 it was pointed out that absorption by H$^-$ ions plays a vital role in determining the quality of the solar emission spectrum in the wavelength range from 4000 to 12 000 Å. It was found, however, that a further source of absorption must become dominant at wavelengths above 12 000 Å. This has been identified as due to free–free transitions in the fields of hydrogen atoms. Because of this, almost as much attention has been devoted to the theoretical study of this process under conditions of mean electron energy appropriate to solar atmospheric conditions as to the bound–free absorption process discussed in § 5.1. The problem is similar, except that the matrix elements are calculated with respect to two free wave functions.

The earliest calculations of the intensities of radiation emitted by electrons due to acceleration by the fields of neutral atoms were made by Nedelsky.† He reduced the calculation to a one-electron problem in which the electron moves in the screened field $V(r)$ of the neutral atom, treated as static, and worked with the acceleration form (48). Nedelsky used for $V(r)$ the empirical 'optical' potential derived by Allis and Morse‡ (see Chap. 6, § 4.5). Menzel and Pekeris,§ a few years later, applied Nedelsky's method to estimate the importance of free–free transitions in the fields of hydrogen atoms for the absorption of light in stellar atmospheres. Although the method could only give rough indications of the magnitudes to be expected, the results were sufficient to show that free–free transitions could play an important part. This provided sufficient stimulus for a great deal of theoretical work directed at accurate evaluation of the required absorption coefficients.

The first calculations using realistic rather than schematic atomic fields was made by Wheeler and Wildt.‖ They replaced the Allis and Morse interaction by the static field of a hydrogen atom but represented the free wave functions by plane waves. The resulting absorption coefficients (see Table 15.7) were too small to explain the observed solar absorption in the infra-red but Chandrasekhar and Breen†† repeated the calculations, replacing the plane waves by waves distorted by the static atomic field. This changed the situation very considerably, as the absorption coefficient came out nearly ten times larger than required (see Table 15.7).

Up till that time all the calculations had been carried out using the acceleration formula. It was clear in any case that by far the most important transitions as far as the solar atmosphere is concerned arise from s–p and p–s transitions of the free electrons. Ohmura and Ohmura‡‡ evaluated the contributions from these transitions by using for the first time the length operator, replacing the free s-wave by its asymptotic form and retaining the undistorted form for the p-waves. These simple

† NEDELSKY, L., *Phys. Rev.* **42** (1932) 641.
‡ ALLIS, W. P. and MORSE, P. M., *Z. Phys.* **70** (1931) 567.
§ MENZEL, D. H. and PEKERIS, C. L., *Mon. Not. R. astr. Soc.* **96** (1935) 77.
‖ WHEELER, J. A. and WILDT, R., *Astrophys. J.* **95** (1942) 281.
†† CHANDRASEKHAR, S. and BREEN, F. H., ibid. **103** (1946) 41.
‡‡ OHMURA, T. and OHMURA, H., *Astrophys. J.* **131** (1960) 8; *Phys. Rev.* **121** (1961) 513.

assumptions would not have been possible using the acceleration operator, but with the length operator the major contributions come from large distances where the free wave functions have taken up their asymptotic forms. The p-phase shifts are small, so the undistorted form assumed for them should not lead to serious error. In this way the absorption coefficient was expressed in terms of the phase shift for elastic scattering of the s-waves. It was easily possible to generalize these calculations to allow for exchange effects by taking the appropriate weighted mean of the squared matrix-elements for the cases in which the total spin quantum number is 1 and 0, the separate matrix elements being determined by the corresponding phase shifts $\eta_0^{\pm}$ (see Chap. 8, § 2.6). The resulting absorption coefficients were somewhat smaller than those obtained by Chandrasekhar and Breen (see Table 15.7) and fulfilled the solar requirements within the accuracy of the solar theory.

Nevertheless, considerable further improvements have been made. Calculations have been carried out using all three operators with free wave functions of comparable elaboration to those used in the bound-free calculations described in § 5.1. As there is a large number of variables we shall illustrate the relative effectiveness of these calculations in terms of a number of selected cases.

Because of the applications in mind it is convenient to express the absorption coefficient, as a function of electron temperature T, in terms of the cross-section per hydrogen atom per unit electron pressure averaged over the Maxwellian velocity distribution of the electrons (see Chap. 14, § 9).

This is convenient for adding the contributions from bound–free and free–free absorption. The photodetachment cross-section gives the absorption per H^- ion but, in thermodynamic equilibrium at temperature T, if $n(H)$, $n(H^-)$, n_e are the respective concentrations of H, H^-, and electrons,

$$n(H^-)/n(H) = p_e\,\phi(T),$$

where p_e (the electron pressure) $= n_e\,\kappa T$, and

$$\log_{10}\phi(T) = -0{\cdot}12 + \frac{5040}{T}\,A - 2{\cdot}5\,\log_{10}T,$$

where A is the electron affinity of hydrogen measured in eV. Hence the bound–free absorption coefficient per hydrogen atom per unit electron pressure is simply $\phi(T)$ times the photodetachment cross-section per H^- ion.

In Table 15.7 we give the free–free absorption coefficient per hydrogen atom per unit electron pressure calculated for a number of values of

TABLE 15.7

Free-free absorption coefficient per hydrogen atom per unit electron pressure in units 10^{-26} cm⁴ dyn⁻¹, calculated, after allowance for stimulated emission, for a wavelength of 9112·67 Å

Θ = 5040/T	Form assumed for free waves	sp+ps Operator used			pd+dp Operator used		
		Length	Velocity	Acc. (g)	Acc. (m)	Length	Velocity
0·5	Plane	0·000	0·000	2·71	0·036†	0·000	0·000
	1s (static only)	0·350	0·351	6·33	0·351‡	0·0$_3$18	0·0$_3$18
	1s } exchange	0·272	0·258	0·534		0·0$_2$597	0·0$_2$576
	1s–2s } included	0·287	0·302	0·343		0·0$_2$964	0·0$_2$911
	1s–2s–2p	0·280	0·292	0·416		0·0$_2$981	0·0$_2$953
	Asymptotic (1s–2s–2p)	0·225§					
0·8	Geltman	0·514‖				0·0$_2$79‖	
1·0	Plane	0·000	0·000	7·95	0·056†	0·000	0·000
	1s (static only)	0·975	0·979	26·40	0·979‡	0·0$_3$10	0·0$_4$9
	1s } exchange	0·570	0·529	1·76		0·0$_2$409	0·0$_2$383
	1s–2s } included	0·605	0·659	1·07		0·0$_2$762	0·0$_2$674
	1s–2s–2p	0·571	0·611	1·19		0·0$_2$736	0·0$_2$646
	Asymptotic (1s–2s–2p)	0·430§					
	Geltman						
2·0	Plane	0·000	0·000	19·0	0·084†	0·000	0·000
	1s (static only)	2·48	2·50	87·5	2·49‡	0·47	0·45
	1s } exchange	1·12	1·03	4·56		0·0$_2$245	0·0$_2$221
	1s–2s } included	1·19	1·33	2·69		0·0$_2$567	0·0$_2$467
	1s–2s–2p	1·08	1·18	2·64		0·0$_2$455	0·0$_2$333
	Asymptotic (1s–2s–2p)	0·76§					
	Geltman	1·36‖					

All values given except those indicated as follows, were calculated by DOUGHTY, N. A. and FRASER, P. A., *Mon. not. R. astr. Soc.* **132** (1966) 267.

† WHEELER, J. A. and WILDT, R., loc. cit., p. 1281.   § OHMURA, T. and OHMURA, H., loc. cit., p. 1281.
‡ CHANDRASEKHAR, S. and BREEN, F. H., loc. cit., p. 1281.   ‖ GELTMAN, S., *Astrophys. J.* **141** (1965) 376.

$\Theta = 5040/T$, around the temperatures of importance for solar applications but for a single wavelength 9112·67 Å corresponding to an electron energy change of 1·354 eV. The contributions from $sp+ps$ and $pd+dp$ transitions are shown separately.

In Table 15.7 results are given obtained with length, velocity, and acceleration operators. It will be noted that distinction is made between acc. (g) and acc. (m) operators. The former is the full two-electron operator (47) whereas acc. (m) refers to the operator taken when the problem is treated as that of the motion of a single electron in the mean static atomic field (see (48)) ignoring electron exchange.

The unreliability of the acceleration operator is manifest. The use of the full acceleration operator leads to gross errors in the $1s$ approximation. When exchange is allowed for it is no longer possible to substitute a simple alternative form and the general operator must be used. Even with the most elaborate wave functions it gives results that differ considerably from the much more nearly equal results from the length and velocity operators.

The results shown in Table 15.7 indicate that, once exchange is included, there is relatively little improvement to be gained by allowing for atom distortion and polarization through inclusion of close-coupling $1s-2s$ and $1s-2s-2p$ approximations. Doughty and Fraser† therefore completed the detailed calculations necessary for evaluation of the relevant absorption coefficient for the solar atmosphere using the $1s-2s$ (exchange) approximation for the free waves. Fig. 15 20 compares the results obtained in this way for $T = 6300\ °$K with those obtained in other approximations.

Fig. 15.21 shows the total, bound–free+free–free, absorption coefficients calculated for a temperature of 6300 °K. The bound–free results have been obtained using the 70-parameter Schwartz bound-state function and $1s-2s-2p-3s-3p-3d$ close-coupling exchange approximation for the free waves. In the same Fig. 15.21 are the values derived from analysis of the physics of the solar atmosphere.‡ The agreement is quite satisfactory, being within the uncertainty of the latter analysis. To indicate the magnitude of the improvements achieved since 1946 the total absorption coefficients obtained by addition of the bound–free values given by Chandrasekhar§ to the free–free values given by Chandrasekhar and Breen‖ are also included.

† loc. cit., p. 1283.
‡ CHALONGE, D. and KOURGANOFF, V., *Annls Astrophys.* **9** (1946) 69.
§ CHANDRASEKHAR, S., loc. cit., p. 1256.   ‖ loc. cit., p. 1281.

7.2. *Free–free transitions in the fields of complex atoms*

We have already referred (p. 1281) to the calculations of Nedelsky,[†] who evaluated cross-sections for free–free transitions of electrons in the fields of neutral atoms, using the acceleration formula (45) with the potential $V(r)$ taken to be the empirical form (110) of Chapter 6 used so successfully by Morse and Allis[‡] for interpreting the data on elastic

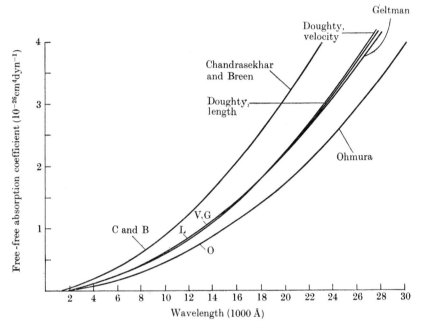

FIG. 15.20. The free–free absorption coefficient per hydrogen atom per unit electron pressure at a temperature of 6300 °K, calculated by different approximations as indicated. *References*: CHANDRASEKHAR, S. and BREEN, F. H., *Astrophys. J.* **104** (1946) 430; DOUGHTY, N. A., and GELTMAN, S., loc. cit. p. 1283, OHMURA T., *Astrophys. J.* **140** (1964) 282.

scattering of slow electrons by atoms. Nedelsky noted that the cross-section for free–free transitions could be related to that for elastic scattering under certain circumstances. This was carried further by Ohmura and Ohmura[§] for transitions in the field of hydrogen atoms, as mentioned on p. 1281. They used the dipole-length matrix element. This is determined largely by the wave functions at large distances so that when these are continuum functions they may be represented by their asymptotic forms. Suppose, for example, that the transition is one

---

† NEDELSKY, L., *Phys. Rev.* **42** (1932) 641.   ‡ loc. cit., p. 1281.
§ OHMURA, T. and OHMURA, H., *Astrophys. J.* **131** (1960) 8.

from an initial s state of energy E_0 and wave number k_0 to a final p-state of energy E_1 and wave number k_1. The matrix element

$$\mathbf{r}_{E_0 E_1} = \int \psi_0 \, \mathbf{r} \psi_1 \, d\mathbf{r},$$

where

$$\psi_0 \sim (k_0 r)^{-1} \sin\{k_0 r + \eta_0(k_0)\},$$

$$\psi_1 \sim 3i\{j_1(k_1 r)\cos \eta_1(k_1) - j_{-1}(k_1 r)\sin \eta_1(k_1)\}\cos \Theta,$$

$\eta_0(k_0)$ and $\eta_1(k_1)$ being the phase shifts for motion of electrons with the respective energy and angular momentum in the atomic field. $\cos \Theta$

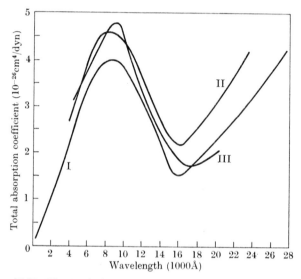

FIG. 15.21. The total absorption coefficient per hydrogen atom per unit electron pressure at a temperature of 6300 °K. I, calculated with approximations as indicated in text. II, calculated from bound–free values of Chandrasekhar and the free–free values of Chandresekhar and Breen. III, derived from analysis of absorption in the solar atmosphere.

References: CHANDRASEKHAR, S., *Astrophys. J.* **102** (1945) 395; CHANDRA-SEKHAR, S. and BREEN, F. H., ibid. **104** (1946) 430.

is given in terms of the polar angles (θ, ϕ) of the direction of motion of the scattered electrons relative to the incident direction by

$$\cos \Theta = \cos \theta \cos \theta' + \sin \theta \sin \theta' \cos(\phi - \phi').$$

If k_1 is sufficiently small we can take $\eta_1 \simeq 0$ so that we need to evaluate

$$\mathbf{r}_{E_0 E_1} = 3ik_0^{-1} \int \sin\{k_0 r + \eta_0(k_0)\}(\mathbf{r}/r)j_1(k_1 r)\cos \Theta \, d\mathbf{r}.$$

The components $x_{E_0 E_1}, y_{E_0 E_1}, z_{E_0 E_1}$ of $\mathbf{r}_{E_0 E_1}$ are now given by

$$(x_{E_0 E_1}, y_{E_0 E_1}, z_{E_0 E_1}) = (\sin \theta \cos \phi, \sin \theta \sin \phi, \cos \theta)L,$$

where $\qquad L = (4\pi i/k_0) \int\limits_0^\infty \sin\{k_0 r + \eta_0(k_0)\} r^2 j_1(k_1 r)\, \mathrm{d}r.$

The spherical Bessel function $j_1(k_1 r)$ is given by

$$j_1(k_1 r) = \frac{\cos k_1 r}{k_1 r} - \frac{\sin k_1 r}{k_1^2 r^2},$$

so that, as it stands, the integral is not convergent. However, it may be defined as the limit

$$L = \lim_{\mu \to 0} (4\pi i/k_0 k_1^2) \int\limits_0^\infty \mathrm{e}^{-\mu r} \sin\{k_0 r + \eta_0(k_0)\}(k_1 r \cos k_1 r - \sin k_1 r)\, \mathrm{d}r$$

$$= 4\pi i k_1 \sin \eta_0 / k_0 (k_0^2 - k_1^2)^2.$$

It follows that $|\mathbf{r}_{E_0 E_1}|^2$ is proportional to $k_0^{-2} \sin^2 \eta_0(k)$ and hence, since η_0 is the only significant phase shift, to $Q_{\mathrm{el}}(k_0)$, the elastic scattering cross-section. On substitution in the formula (81) of Chapter 14 for the differential cross-section per unit frequency range of emitted radiation we obtain

$$q_\nu\, \mathrm{d}\nu = 4\pi \int I_{E_0}^{E_1}\, \mathrm{d}\nu \mathrm{d}\omega = \frac{2}{3}\frac{e^2}{mc^2}\frac{\hbar}{mc}\frac{k_1^3}{k_0 \nu} Q_{\mathrm{el}}(k_0)\, \mathrm{d}\nu. \qquad (49)$$

This analysis has been generalized by Kas'yanov and Starostin[†] who find that (49) remains valid under many circumstances, provided one single partial wave predominates in the determination of the elastic scattering and that the elastic cross-section is replaced by the momentum-transfer cross-section Q_d (see Chap. 2, § 1.1).

The emission of neutral atom bremsstrahlung from a mercury-arc discharge was inferred by Bohdansky,[‡] who observed that the intensity of continuous radiation emitted increased more than linearly with pressure p. In an isothermal plasma, for a given arc current the electron and ion densities are proportional to $p^{\frac{1}{2}}$ so that the electron-ion bremsstrahlung intensity would be expected to increase as p, the electron-neutral atom intensity as $p^{\frac{3}{2}}$.

Direct emission of electron-neutral atom bremsstrahlung in the infra-red has been observed from air shock-heated to an equilibrium temperature of 6000–9000 °K by Wentink, Planet, Hammerling, and Kivel[§] and from air and nitrogen by Taylor.[||]

† KAS'YANOV, V. and STAROSTIN, A., Zh. éksp. teor. Fiz. **48** (1965) 295; Soviet Phys. JETP **21** (1965) 193.
‡ BOHDANSKY, J., Z. Phys. **149** (1957) 383.
§ WENTINK, T., PLANET, W., HAMMERLING, P., and KIVEL, B., J. appl. Phys. **29** (1958) 742.
|| TAYLOR, R. L., J. chem. Phys. **39** (1963) 2354.

In the experiment of Taylor (see Fig. 15.22) the metal shock tube was 14 ft long and of internal diameter 1·5 in. The shock wave was reflected at the end containing a window, which was made of sapphire for transmission of wavelengths less than 4·0 μm and of a special commercial material Irtram-1 for greater wavelengths. Light was observed behind the reflected shock. The shock light was focused on to the entrance slit of a monochromator containing a CaF_2 prism by means of

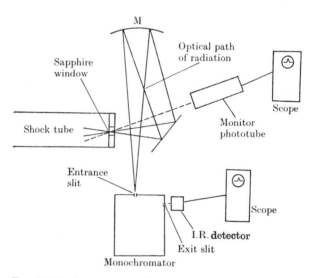

FIG. 15.22. Arrangement of apparatus used by Taylor to observe infra-red bremsstrahlung from shock-heated gases.

a concave mirror M. To avoid radiation from impurity gases near the shock tube walls, the radiation was observed along the tube axis and the solid angle chosen to ensure that radiation from the wall region did not enter the monochromator slit. The radiation was detected at the exit slit by a gold-doped germanium infra-red detector sensitive in the region 2–8 μm. The associated electronics limited the response time to less than 1·0 μs. A phototube sensitive to infra-red radiation in conjunction with a filter that passed radiation in the 0·65–1·0 μm wavelength band was used as a monitor. Standard sources of black-body radiation were used to calibrate the detector. The detector, monochromator, and shock tube optics were calibrated as a unit, thus including all corrections for reflections in the optics or for atmospheric absorption. Corrections were made for the transmission of the shock tube and windows over the spectral region employed.

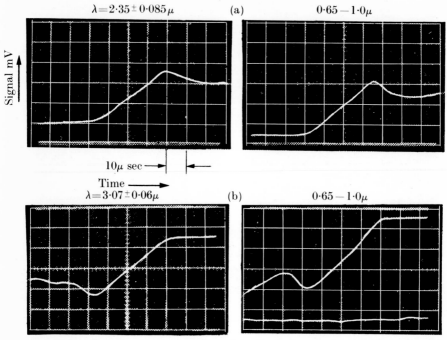

FIG. 15.23. Typical oscillograms obtained in Taylor's experiments. The left-hand signals are those received by the detector, the right-hand by the phototube which acted as a monitor.

Fig. 15.23 shows typical oscillograms for the detector and the photo-tube. The oscilloscopes were triggered about 25 μs before the incident shock reached the end of the tube. The linear rise in signal as the reflected wave recedes from the end of the tube is due to increase in length of the column of emitting gas. The phototube observed primary radiation for the $N_2(1+)$ band, which is highly temperature sensitive so that the linearity of its signal indicates uniform gas temperature behind the shock wave.

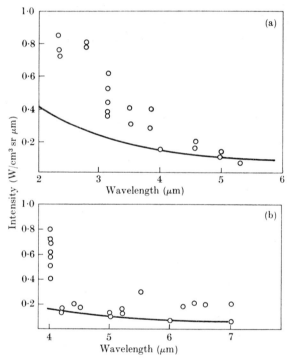

FIG. 15.24. Spectral distribution of the continuous radiation emitted from (a) nitrogen, and (b) air, at 8000 °K and densities respectively 0·8 and 0·9 times atmospheric. ○ observed by Taylor. —— calculated by Kas'yanov and Starostin.

From oscillograms such as those shown in the figure the initial slope of the best straight line is estimated. From the measured shock-wave velocity the absolute intensity per unit volume of radiating gas can then be calculated knowing the apparatus calibration, and thence the temperature of the shock-heated gas estimated.

Fig. 15.24 shows the observed spectral distribution of the continuum radiation from nitrogen and from air heated to 8000 °K, the gas densities being respectively 0·8 and 0·9 times atmospheric. The experimental intensities are compared with the predictions of the formula (49) as modified by Kas'yanov and Starostin. The electron density due to thermal ionization was assumed to be determined by the Saha equation. Observed elastic cross-sections for O (Chap. 1, § 6.1), N (Chap. 1, § 6.1), and N_2 (Chap. 10, § 3.1) were used to give Q^d. In view of the wide spread in the experimental results the agreement is not unsatisfactory. Furthermore, the contribution from bremsstrahlung in the fields of ionized atoms and molecules that may be estimated from the methods described in Chapter 14, § 9.2 is nearly two magnitudes too small to account for the observed emission at 8000 °K at a wavelength of 3 μm. Another possible source, emission of the $N_2(1+)$ band, is equally unimportant so it appears probable that radiation emitted in the fields of neutral atoms and molecules was actually observed.

8. Two-quantum photodetachment

In Chapter 14, § 14 we discussed the probability of photo-ionization of an atom by the effectively simultaneous impact of two photons. Such processes are now becoming capable of experimental study through the availability of laser sources of very high intensity. It is equally possible to consider photodetachment arising in the same way from collisions with two photons. It was pointed out by Hasted† that photodetachment in some ways presents more favourable experimental opportunities than photo-ionization because most negative ions possess no bound excited states. Ejection of an electron could therefore not take place in two successive events involving first a real excitation by one quantum and then ionization by a second quantum of the same frequency arriving at an interval within the lifetime of the excited state.

A successful experiment has been carried out on these lines by Hall, Robinson, and Branscomb‡ in which the transition rate for 2-quantum photodetachment from I⁻ by ruby laser light with a quantum energy of 1·785 eV was measured.

The general arrangement of the apparatus is shown in Fig. 15.25. It is again of the crossed-beam type with the photons now supplied from a 20 MW Q-switched ruby laser focused through an area of 10^{-2} cm². The source of the I⁻ ions was the same as that used in the single-quantum photodetachment experiments discussed in § 5.6. By means of an

---

† HASTED, J. B., Private communication.

‡ HALL, J. L., ROBINSON, E. J., and BRANSCOMB, L. M., *Phys. Rev. Lett.* **14** (1965) 1013.

'einzel' lens the I⁻ beam was focused on to an area of about 9 mm² in the region of interaction with the laser beam.

To monitor the power of the laser beam and its time-dependence, the beam was directed out of the vacuum system by a silvered mirror on to a MgO screen and thence a biplanar photodiode driving a travelling-wave oscilloscope. Absolute calibration was carried out by comparison with a liquid-cell calorimeter.

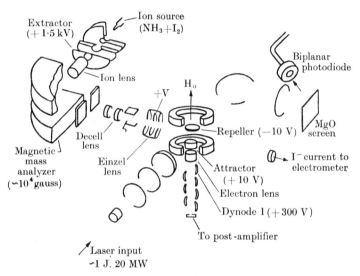

Fig. 15.25. Arrangement of apparatus used by Hall, Robinson, and Branscomb for investigating 2-quantum photodetachment from I⁻.

To examine the geometry of the crossed beams in the overlapping region a movable phosphor screen could be swung into position at an angle of 45° to both beams.

Detached electrons were extracted at right angles to both beams, along the direction of a uniform magnetic field, and focused on the first dynode of an electron multiplier. Spurious signals due to electrons emitted from the silvered mirror through the action of the laser beam were largely eliminated by biasing the interaction region to be a few volts negative.

The electron multiplier and the collection system were calibrated in the following way. The pulse-height distribution and total signals due to a single electron incident on the first dynode were first measured. These electrons were obtained through collision detachment from the ion beam by background gas at a pressure less than 10^{-7} torr. Next the electron collection efficiency was measured by substituting a weak

H⁻ beam for the I⁻ beam. Photodetachment from H⁻ ions can take place through single impact with ruby laser photons so that, with the high intensity of the laser beam, all ions passing through the beam suffered photodetachment. The total current of detached electrons was then given by the reduction of the ion current. As a final check the single quantum photodetachment cross-section of H⁻ was measured, using an attenuated laser beam, and found to agree to within 5 per cent with the known value (§ 5.1).

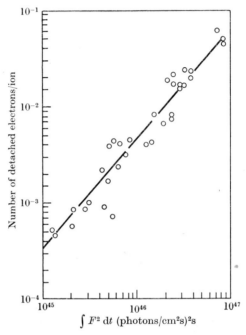

Fig. 15.27. Observed relation between photodetachment probability and $\int F^2 \, \mathrm{d}t$, where F is the laser flux.

To eliminate errors due to transit-time effects in the electron multiplier the data were taken by integrating the output from the post-amplifier of the multiplier. Both the fast output and the integral were displayed on a double-beam oscilloscope.

As the detached electron signal may have arisen from single quantum processes due to ultra-violet second-harmonic photons generated in the optical components of the system, tests were made by introducing a red absorbing filter, transparent to the ultra-violet, in the laser beam after its passage through the lenses. This reduced the signal to zero. It was also verified in the same way, by inserting a duplicate window in the

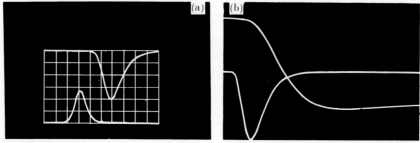

Fig. 15.26. Typical records obtained in experiments (see Fig. 15.25) on 2-quantum photodetachment from I^-. (a) Lower trace, laser output; Upper trace, 2-quantum electron signal (30 electrons/div); time base 49 ns/div. (b) Lower trace, fast 2-quantum electron signal (100 ns/div); Upper trace, integral electron signal (30 electrons/div). 2 μs/div.

beam, that no ultra-violet photons were contributed from the vacuum window.

Fig. 15.26 shows some typical records, while in Fig. 15.27 the total number of detached electrons per illuminated ion is plotted as a function of the time integral of the square of the laser flux. In a logarithmic plot the function should be a straight line of unit slope. In fact the slope is a little higher (1·12) possibly due to lack of uniformity in the ion beam.

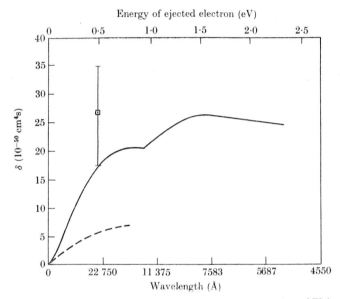

FIG. 15.28. Calculated variation of δ with frequency, where δF^2 is the rate of 2-quantum photodetachment from I^- by a laser flux F. —— with distorted waves — — — with plane waves. ⊡ observed value with range of uncertainty (the coherence factor of $\frac{2}{3}$ is not included).

From this plot we find for δ, as defined in Chapter 14 (124),

$$\delta = 350 \pm 140 \times 10^{-51} \text{ cm}^4 \text{ s}.$$

Some uncertainty arises as to the allowance to be made for laser coherence effects. For an ideal laser, producing a sinusoidal electromagnetic field, the integral over the square of the flux should be multiplied by the ratio of the mean value of $\cos^4 \omega t$ to the square of the mean value of $\cos^2 \omega t$. This introduces a factor of $\frac{2}{3}$ but a further correction is necessary if the laser is in multinode operation. A factor of 1·3 was introduced to allow for this, giving finally a corrected value

$$\delta = 180 \times 10^{-51} \text{ cm}^4 \text{ s}.$$

Geltman[†] has calculated δ using the formula (122) of Chapter 14, the only important difference being the fact that all the intermediate states are now of the continuum. In his first calculations Geltman assumed that all of these states can be represented by undistorted plane waves. This gives the particularly simple result that δF^2 (see Chap. 14 (124)) is proportional to the known single-quantum photodetachment cross-section of I^- at twice the frequency. He found

$$\delta = 56 \times 10^{-51}\ \text{cm}^4\ \text{s}$$

considerably smaller than the observed value. In a further calculation Robinson and Geltman[‡] allowed for distortion of the plane waves representing the intermediate states by the atomic field in the same way as in his calculations for the single quantum process. Summation over the intermediate states was carried out by a method due to Dalgarno and Lewis.[§] This gives the results shown in Fig. 15.28 for δ. Comparison with the observed value, allowing for the considerable uncertainty in the latter due to coherence corrections, etc., now shows quite good agreement.

[†] GELTMAN, S., *Phys Lett.* **4** (1963) 168; **19** (1965) 616.
[‡] ROBINSON, E. J. and GELTMAN, S., loc. cit. p. 1266.
[§] DALGARNO, A. and LEWIS, J. T., *Proc. R. Soc.* A**233** (1956) 70.

AUTHOR INDEX

SUBJECT INDEX

Absorption:
of radiation:
  by Ar, 1121; Ca, 1138; Cd, 1135; Co, 1164; CO_2, 1176; Cs, 1130; D_2, 1145; H, 1073; H_2, 1145; H_2^+ 1151; HD, 1145; He, 1110–13; H_2O, 1172; In, 1135; K, 1129; Kr, 1118; Li, 1125; Mg, 1130; N, 1140; N_2, 1160; Na, 1127; Ne, 1118; NH_3, 1173; NO, 1168; O, 1137; O_2, 1153; Rb, 1130; Xe, 1121.
  by atoms, cross-sections for, 1130.
  by shock-heated alkali halides, 1246–9.
  in an ionized atmosphere, 1224–6.
  in fields of bare positive ions, 1221–3.
    H atoms, 1280–4.
    neutral atoms, 1279–90.
    partially ionized atoms, 1224.
  in solar atmosphere, 1204.
of radio waves, and Luxemburg effect, 753.
of X-rays, fine structure in, 692; by $AsCl_3$, 695; Br_2, 696; $GeCl_4$, 695.

Ag:
continuous X-ray emission from, 1196, 1214, 1216.
radiative scattering of electrons by, 1203, 1221.

Air:
continuous radiation emitted from, at 8000 °K, 1289.
electrons in, characteristic energies of, 1021.
  collision frequencies of, 795, 1020.
  drift velocities of, 1020.

Al:
continuous X-ray emission from, 1195–7, 1206, 1210–11, 1214.
polarization of, 1200, 1217, 1219.
radiative scattering of electrons by, 1203, 1226.

Alkali halides, polar dissociation of, by shock waves, 1245.

Al_2O_3, continuous X-ray emission from, polarization of, 1199, 1217.

Angular distribution:
of elastically scattered electrons:
  of high energy:
    by CCl_4, 669; C_6H_6; F, 681; H_2, H_2^+, 685–90; U, 681; U_2, 684; UF_6, 681; W, 681.
    by molecules, theory of, 665–71, 678–89.

of medium energy:
  by Br_2, 692; CBr_4, CCl_4, 691–2; CH_4, CH_2, C_2H_4, 692; N_2, 690.
of low energy:
  by CH_4, 704–13; N_2, 710, 719–20.
of inelastically scattered electrons:
  after radiative scattering, 1203.
  after vibrational excitation of N_2, 740.
  by molecules, diffraction effects in, 833–7.
of products of molecular dissociation by electrons, 830–3.
of protons, from impact dissociation of H_2, 921–9.

Anharmonicity of vibration, 670.

Antibonding orbitals, 806.
in N_2 and O_2, 808.

Ar:
plasmas in, infra-red emission from, 1230.
photoelectron spectrum of, 1106.
photo-ionization cross-section of, 1121.
fine structure in, 1121.

Ar–H_2 mixtures, analysis of swarm experiments in, 775.

$AsCl_3$:
ionization of, 1056.
X-ray absorption by, 695.

Attachment, dissociative:
of electrons to molecules:
  beam experiments on, 1002.
    effect of temperature on, in O_2, 1002.
  cross-sections for, 824.
    in CO, 1028; D_2, 936; H_2, 929, 932, 935; HBr, 1062–4; HCl, 1062–4; HD, 932; HF, 1062–4; HI, 1062–4; H_2O, 1037; H_2O_2, 1042; I_2, 1057; NO, 1033; O_2, 999.
  isotope effect in, 823.
radiative, of electrons:
  threshold laws for, 1238.
  to H, 1260.

Attachment coefficient, of electrons in gases, 862:
in Ar–H_2O mixtures, 1040; CO_2, 1048; H_2O, 1019; O_2, 1006.
measurement of, by:
  diffusion methods, 870.
  electron filter method, 863.
  microwave probing of afterglows, 875.
  observations of currents in pre-breakdown discharges, 873.
  pulse methods, 864.

Au:
  continuous X-ray emission from, 1196, 1198, 1206, 1210, 1215.
  polarization of, 1199, 1200, 1217–19.
  radiative scattering of electrons by, 1203, 1221.
Autodetachment, 823.
  and dissociative attachment in H_2, D_2, and HD, 935–7.
Auto-ionizing states:
  and ionization cross-section of H_2, 914.
  and resonance effects in photo-ionization, 1076.

Balanced filter monochromator for X-radiation, 1194.
BCl_3, negative ion formation in, 1056.
Be:
  continuous X-ray emission from, 1196.
  polarization of, 1200, 1217, 1218.
Bonding orbitals, 806.
  in N_2 and O_2, 808.
Born's approximation and:
  bremsstrahlung, 1182, 1186, 1188–90, 1192, 1207, 1217.
  diffraction by molecules:
    of elastically scattered electrons, 668.
    of inelastically scattered electrons, 834.
  excitation:
    of a diatomic molecule, 828.
    of B and C states of H_2, 901.
    of $B^3\Sigma_u^-$ state of O_2, 991.
    of rotation, 747.
      for dipolar molecules, 748.
        validity of, 749.
      for quadrupolar molecules, 749.
    of vibration, 729.
  scattering of electrons by molecules, 729.
Br, electron affinity of, 1162, 1179, 1272.
Br_2:
  diffraction of inelastically scattered electrons by, 835.
  dissociation energy of, 1179.
  photoelectron spectrum of, 1177.
  polar photodissociation of, 1177.
  scattering, elastic, of electrons by, 692.
    angular distribution of, 703.
    X-ray absorption by, fine structure in, 697.
Br_2^+, doublet, splitting of, 1177.
Br^-:
  formation of:
    by polar photodissociation of Br_2, 1177.

  by shock-wave heating of alkali bromides, 1248.
  photodetachment from, 1248.
Bremsstrahlung:
  asymmetry of, 1204, 1220.
  circular polarization of 1202, 1219.
  directional distribution of, 1208.
  emission of:
    from a plasma, collective effects in, 1226, 1251.
    from an assembly of H atoms, 1251.
  polarization of, 1181, 1184, 1186–92, 1199–203, 1216–23.
  production of, by polarized electrons, 1188.
  theory of:
    non-relativistic, 1180–5.
    relativistic, 1185–92.

C:
  continuous X-ray emission from, 1195.
  electron affinity of, 1026, 1067.
  total absorption by ionized atmosphere of, 1223.
C^-:
  photodetachment from, 1240, 1266–8.
  production of, from electron impact dissociation of CO, 1028.
C^+, production of, from electron impact dissociation of CO, 1028.
Ca, absorption cross-section of, for radiation, 1132.
Ca^+, total absorption by ionized atmosphere of, 1223.
CBr_4, elastic scattering of electrons by, 691.
CCl_4:
  elastic scattering of electrons by, 669, 691, 703, 838.
  inelastic scattering of electrons by, 1055.
  ionization of, 1055.
  negative ion formation in, 1056.
  total cross-section of, for electrons, 701.
CCl_2F_2:
  ionization of, 1055.
  negative ion formation in, 1056.
CCl_3F, ionization of, 1055.
CF_3Br, negative ion formation in, 1056.
CF_3I, negative ion formation in, 1056.
CH_4:
  elastic scattering of electrons by, 692, 702, 713.
  ionization potential of, 1176.
  total cross-section of, for electrons, 700.
C_2H_4:
  elastic scattering of electrons by, 692, 703.
  total cross-section of, for electrons, 700.

PRINTED IN GREAT BRITAIN
AT THE UNIVERSITY PRESS, OXFORD
BY VIVIAN RIDLER
PRINTER TO THE UNIVERSITY